SPACE SHUTTLE MISSIONS
OF THE 80'S
PART I

AMERICAN ASTRONAUTICAL SOCIETY

Publications Office, Post Office Box 28130, San Diego, California 92128

Science and Technology

Vol. 1 Manned Space Reliability Symposium, 1965 (112 pages)—$15.00
Vol. 2 Towards Deeper Space Penetration, 1965 (182 pages)—$15.00
Vol. 3 Orbital Hodograph Analysis, S. P. Altman, 1965 (150 pages)—$15.00
Vol. 4 Scientific Experiments for Manned Orbital Flight, 1965 (372 pages)—$25.00
Vol. 5 Physiological and Performance Determinants in Manned Space Systems, 1965 (220 pages)—$15.00
Vol. 6 Space Electronics Symposium, 1965 (404 pages)—$25.00
Vol. 7 Theodore Von Karman Memorial Seminar, 1966 (140 pages)—$15.00
Vol. 8 Impact of Space Exploration on Society, 1966 (382 pages)—$25.00
Vol. 9 Recent Developments in Space Flight Mechanics, 1966 (280 pages)—$20.00
Vol. 10 Space Age in Fiscal Year 2001, 1967 (458 pages)—$30.00
Vol. 11 Space Flight Mechanics Symposium, 1967 (618 pages)—$35.00; Microfiche Suppl.—$6.00 extra
Vol. 12 The Management of Aerospace Programs, 1967 (392 pages)—$20.00
Vol. 13 The Physics of the Moon, 1967 (260 pages)—$20.00
Vol. 14 Interpretation of Lunar Probe Data, 1967 (270 pages)—$20.00
Vol. 15 Future Space Program and Impact on Range and Network Development, 1967 (583 pages)—$30.00
Vol. 16 The Voyage to the Planets, 1968 (184 pages)—$15.00
Vol. 17 Use of Space Systems for Planetary Geology and Geophysics, 1968 (623 pages)—$35.00; Microfiche Suppl.—$4.00 extra
Vol. 18 Technology and Social Progress, 1969 (170 pages)—$15.00
Vol. 19 Exobiology—The Search for Extraterrestrial Life, 1969 (184 pages)—$15.00
Vol. 20 Bioengineering and Cabin Ecology, 1969 (162 pages)—$10.00
Vol. 21 Reducing the Cost of Space Transportation, 1969 (264 pages)—$20.00
Vol. 22 Planning Challenges of the 70's in the Public Domain, 1970 (504 pages)—$30.00; Microfiche Suppl.—$12.00 extra
Vol. 23 Space Technology and Earth Problems, 1970 (418 pages)—$25.00; Microfiche Suppl.—$12.00 extra
Vol. 24 Aerospace Research and Development, 1970 (500 pages)—$30.00
Vol. 25 Geological Problems in Lunar and Planetary Research, 1971 (750 pages)—$40.00
Vol. 26 Technology Utilization Ideas for the 70's and Beyond, 1971 (312 pages)—$25.00
Vol. 27 International Cooperation in Space Operations and Exploration, 1971 (194 pages)—$16.00
Vol. 28 Astronomy from a Space Platform, 1972 (416 pages)—$30.00
Vol. 29 Space Technology Transfer to Community and Industry, 1972 (196 pages [Microfiche only])—$10.00
Vol. 30 Space Shuttle Payloads, 1973 (532 pages)—$35.00
Vol. 31 The Second Fifteen Years in Space, 1973 (212 pages)—$20.00
Vol. 32 Health Care Systems, 1974 (264 pages)—$20.00
Vol. 33 Orbital International Laboratory, Third & Fourth Oil Symposia, 1974 (322 pages)—$25.00
Vol. 34 Management and Design of Long-Life Systems, 1974 (198 pages)—$16.00
Vol. 35 Energy Delta, Supply vs. Demand, 1975 (604 pages)—Paperback—$40.00; Microfiche—$25.00
Vol. 36 Skylab and Pioneer Report, 1975 (160 pages)—$16.00
Vol. 37 Space Rescue & Safety, 1974 (294 pages)—$25.00
Vol. 38 Skylab Science Experiments, 1975 (274 pages)—$25.00
Vol. 39 Environmental Control and Agri-Technology, 1976 (346 pages)—$30.00
Vol. 40 Future Space Activities, 1976 (182 pages)—$20.00 (incl. numerical & author index for all Goddard Memorial Symposia)
Vol. 41 Space Rescue & Safety 1975, (230 pages)—$20.00

Advances in the Astronautical Sciences

Vol. 1-5 AAS Proceedings 1957-60. Available in Microfiche. $20.00 per volume.
Vol. 6   Sixth Annual Meeting Proceedings, 1961 (968 pages)—$35.00
Vol. 7-8 Third West Coast & Seventh Annual Meeting Proceedings, 1961-63. Available in Microfiche. $20.00 per volume.
Vol. 9   Fourth Western Meeting Proceedings, 1963 (910 pages)—$35.00
Vol. 10   Manned Lunar Flight, 1963 (310 pages)—$15.00
Vol. 11   Eighth Annual Meeting Proceedings, 1963 (808 pages)—$35.00
Vol. 12   Scientific Satellites, 1963 (262 pages)—$15.00
Vol. 13   Ninth Annual Meeting: Interplanetary Missions, 1963 (690 pages)—$30.00
Vol. 14   Physical and Biological Phenomena in a Weightless State, 1963 (382 pages)—$20.00
Vol. 15   Exploration of Mars, 1963 (634 pages)—$35.00
Vol. 16   Space Rendezvous, Rescue, and Recovery, 1963; Part I (1028 pages)—$35.00; Part II (380 pages)—$20.00
Vol. 17   Bioastronautics—Fundamental and Practical Problems, 1964 (128 pages)—$15.00
Vol. 18   Lunar Flight Programs, 1964 (630 pages)—$35.00
Vol. 19   Unmanned Exploration of the Solar System, 1965 (1000 pages)—$45.00
Vol. 20   Post Apollo Space Exploration, 1966; Part I (572 pages)—$35.00; Part II (648 pages)—$35.00
Vol. 21   Practical Space Applications, 1967 (508 pages)—$30.00
Vol. 22   The Search for Extraterrestrial Life, 1967 (388 pages)—$25.00; Microfiche Suppl.—$3.00 extra
Vol. 23   Commercial Utilization of Space, 1968 (512 pages, plus 24 microfiches)—$35.00
Vol. 24   Exploration of Space, 1968 (363 pages)—$20.00
Vol. 25   Advanced Space Experiments, 1969 (530 pages)—$35.00
Vol. 26   Planning Challenges of the 70's in Space, 1970 (470 pages)—$30.00; Microfiche Suppl.—$10.00 extra
Vol. 27   Space Stations, 1970 (606 pages)—$35.00
Vol. 28   Space Shuttles and Interplanetary Missions, 1970 (488 pages)—$30.00
Vol. 29   The Outer Solar System, 1971; Part I (618 pages)—$35.00; Part II (740 pages)—$35.00
Vol. 30   International Congress of Space Benefits, 1974 (528 pages)—$35.00
Vol. 31   The Skylab Results, 1975 (1174 pages, Microfiche only)—$45.00
Vol. 32   Space Shuttle Missions of the 80's, 1976 (approx. 1000 pages)—$70.00; Microfiche Suppl.—$60.00
Vol. 33   AAS/AIAA Astrodynamics Conference 1975, (approx. 500 pages)—$35.00; 1976 Microfiche Suppl.—$35.00

Special Volumes

1. Weightlessness—Physical Phenomena and Biological Effects, 1961 (182 pages)—$15.00
2. Lunar Exploration and Spacecraft Systems, 1962 (214 pages)—$15.00

DISTRIBUTED BY **UNIVELT, INC.**, P.O. Box 28130, San Diego, California 92128

AAS PRESIDENT
    Philip H. Bolger   U.S. Department of Transportation

AAS VICE PRESIDENT - PUBLICATIONS
    Dr. Charles Sheffield   Earth Satellite Corporation

SERIES EDITOR
    Dr. Horace Jacobs   Lockheed-California Company

EDITORS
    William J. Bursnall
    Dr. George W. Morgenthaler   Martin Marietta Corporation
    Dr. G. E. Simonson

ART STAFF
    J. C. Speas   Lockheed-California Company

Thanks are due Marie Heidbreder of the Martin-Marietta Corporation and Diana Law of the Lockheed-California Company for final preparation of the manuscript for the printer.

Front cover and title sheet, courtesy of Lockheed-California Company.

LARGE SPACE TELESCOPE (LST)

AN AMERICAN *Astronautical* SOCIETY PUBLICATION

Space Shuttle Missions of the 80's

**Volume 32
Part 1**

ADVANCES IN THE ASTRONAUTICAL SCIENCES

Edited by
W. J. Bursnall
G. W. Morgenthaler
G. E. Simonson

Proceedings of American Astronautical Society
Twenty First Annual Meeting

Co-Sponsored by Rocky Mountain Section of
the American Institute of Aeronautics and
Astronautics, the Denver Section of the
Institute of Electrical and Electronic
Engineers, and the Rocky Mountain Chapter
of Operations Research Society of America

August 26-28, 1976   Denver, Colorado

Distributed by UNIVELT, INC., P.O. Box 28130, San Diego, CA 92128

Copyright 1977

by

AMERICAN ASTRONAUTICAL SOCIETY

AAS Publications Office
P. O. Box 28130
San Diego, California 92128

Affiliated with the American Association for the Advancement of Science
Member of the International Astronautical Federation

Library of Congress Card No. 57-43769

PART I   ISBN 87703-078-2
PART II   ISBN 87703-087-1

Printed and Bound in the U.S.A.

FOREWORD

THEME

The Space Shuttle Transportation System is here. How can we maximize its use for the benefit of mankind?

INTRODUCTION

After several years of intense technical, planning, and fabrication efforts, the era of the Space Shuttle Transportation System is dawning. The aerospace industry and the potential users of this Space Transportation System must now concentrate on Space Shuttle Payloads and the host of experiments and special instruments that will enable society to gain a return on the imaginative investment in this great project.

Over 750 scientists, engineers, political leaders and users, both U.S. and foreign, met at the Brown Palace Hotel in Denver for the 21st Annual Meeting of the AAS to discuss ideas for Shuttle Payload Applications and to plan for future payoff missions. What are the tasks that should be accomplished on these missions? Which instruments will be needed? Who will take the lead in science missions? In Earth resources missions? In Communications experiments? In international collaborative efforts? Even though the U.S., and indeed the major industrial nations, were in a serious economic recession, and the preceding few years had seen budgetary cutbacks in the Nation's space program as a result of loss of public favor for space missions that were purely "spectaculars," a new spirit seemed to prevail at the Denver meeting. The Shuttle would be available and new missions at greater efficiency would be possible. All that was needed was imagination, decisiveness, and dedication. The attitude in Denver was positive - all systems were go!

JOINT NATIONAL MEETING

In keeping with the premise that the Space Shuttle should afford easier access to the use of space for a wide range of users, the meeting was co-sponsored with the local sections of three other major professional societies: The Rocky Mountain Section of the American Institute of Aeronautics and Astronautics; the Denver Section of the Institute of Electrical and Electronic Engineers; and the Rocky Mountain Chapter of TIMS/ORSA (The Institute of Management Sciences/Operations Research Society of America).

MEETING STRUCTURE

After the Call to Order by the General Program Chairman, Dr. George W. Morgenthaler, the attendees were welcomed by the Honorable Richard D. Lamm, Governor of Colorado. Governor Lamm challenged the scientists and engineers who had planned the Space Shuttle and its sophisticated subsystems to apply their skills equally to developing systems to help solve mankind's problems here on Earth.

He was followed by Dr. John Naugle of NASA, who characterized the Shuttle and its Payloads. The opening session also included presentations by leaders of the European consortia who are perfecting the Spacelab for Shuttle Missions, and papers by NASA officials responsible for Shuttle Payload Development and Shuttle Payload Applications.

During the next two days plenary sessions in the Main Ballroom of the Brown Palace Hotel detailed the major payloads already identified for the early Space Shuttle Missions, namely, the Large Space Telescope (LST) and the Atmospheric, Magnetospheric, and Plasmas in Space payload (AMPS). Another topic reviewed was the plan for a Space Tug. The Space Tug would be carried into orbit by the Space Shuttle, along with key payloads; the Tug would then boost these payloads from a lower parking orbit to geosynchronous or other higher energy orbits as needed.

Concurrently with the plenary sessions, technical specialist groups responsible for key Shuttle and Payload subsystems (such as avionics, communications, propulsion, life sciences, payload integration, optics, precision pointing, mechanical systems) met in smaller working groups to hear papers and to discuss problems and progress in these areas. Many other specialist sessions were held on science instrument development, detectors, and experiment data management.

Because of the importance of energy management to man's future and the applications of space technology to this critical area, special sessions were held to focus attention on how the Shuttle System and payloads may help. Great interest was shown in the possibility of an orbital solar power generator with a microwave down-link to Earth.

Two formal luncheons were part of the Symposium. The luncheon on August 26 featured Congressman Don Fuqua of Florida. He reported on the July hearings of the House Sub-Committee on Space Science and Applications. The August 28 luncheon featured Mr. William Dean, Vice President of Rockwell International, the prime contractor for the Shuttle Orbiter, who spoke on "Shuttle Missions - ROI." He was followed by a distinguished interdisciplinary panel which further discussed the nature of the "return on the Shuttle investment" the nation could receive as the actual flights occurred. The Panel included Dr. Richard Goody (Chairman), a scientist; Congresswoman Patricia Schroeder, Dr. Gerard O'Neill, scientist; Capt. Wally Schirra, astronaut; Dr. Robert Anderson, economist; and Mr. Press Layton, engineer.

That evening the traditional AAS Honors Night Banquet included the presentation of the Fellow Award to the new AAS Fellows and the awarding of the annual AAS honors. Senator Frank E. Moss of Utah, Chairman of the Senate Space Committee, spoke on "The Exploitation of Space for the Benefit of Mankind."

INTERNATIONAL ATTENDANCE

The meeting was well attended by international scientists, engineers, and industrialists active in Spacelab and other European Shuttle payload projects. They made many contributions to the planning of future international space missions.

THE SECOND NATIONAL SPACE ART SHOW

The Second National Space Art Show was a strong attraction enjoyed by the public as well as attendees at the Symposium. Mr. Doug Filter was Director. This juried show featured interpretations of space flight, and man's new space environment as perceived by leading artists.

ASTRONAUTICAL EXHIBITS

An extensive Exhibit of space systems featured entries from leading industrial firms and from government agencies. They depicted Shuttle payloads as well as next-generation space instruments for use in exploiting space for mankind.

MEETING COMMITTEE

General Program Chairman	Dr. G. W. Morgenthaler
Deputy	W. L. Kershaw
LST Program Chairman	Dr. A. Hoag
Assistant	J. C. Spencer
Tug Program Chairman	J. H. Disher
Assistant	T. J. Goyette
AMPS Program Chairman	Dr. C. A. Lundquist

Assistant	S. R. Schrock
Shuttle Payloads Chairman	M. E. Turner
AIAA - Rocky Mountain Section	Col. J. P. Wittry
ORSA/TIMS-Rocky Mountain Chapter	A. Anglund
IEEE - Rocky Mountain Section	W. B. Collins
Administration Committee Chairman	A. C. Sellke
Arrangements	S. H. Buzzard
Program & Printing	G. L. Wenner
Mailing	G. E. Simonson
Registration	D. E. Schilling
Publicity	S. Butler, D. L. Yakobson
Preprints	W. J. Bursnall
Finance	W. D. Wiley
Exhibits	H. Bass
Editors	H. E. Nylander, J. H. Nelson, W. P. Pratt
Family Program	K. Bursnall, J. E. Yakobson
Space Art Show	D. C. Filter
Audio Visual	J. Jaumotte
Protocol	S. P. DeJaeger

The enthusiasm and interest shown by the participants was infectious. The perspective gained by the formal presentations and the informal technical and applications exchanges will contribute greatly to sound planning for tne future utilization of the Space Shuttle and its payloads for the benefit of mankind.

ACKNOWLEDGMENTS

The cooperation and support of many people is required in the preparation of a book of this scope. The editors would particularly like to thank the authors, who submitted their manuscripts for publication, and Judy Gerlach, Marie Heidbreder, and Kathy Reynolds, who contributed many hours in the transcription of tapes and preparation of manuscripts.

                                                    The editors

PREFACE

At this historic meeting which featured Space Shuttle Missions of the 80's - the major space engineering program of the United States for the coming decade- some one hundred and fifty technical papers were presented. Publication of all of these papers in full would have resulted in a bulky set of books consisting of about three thousand pages. The editors therefore decided to produce in this hard-copy proceedings some thirty papers in full and all remaining papers in the form of four or five-page summaries with significant illustrations. Nearly all papers that appear in these volumes as summaries are published in full in a microfiche supplement designated as Volume 25, AAS Microfiche Series. Both hardcopy and microfiche proceedings may be ordered from Univelt Inc., P.O. Box 28130, San Diego, California 92128.

H. Jacobs
Series Editor

CONTENTS

PART I

Page No.

I. INTRODUCTION ... 1

    Welcome
        Honorable Richard D. Lamm ... 3

    Keynote Address -- The Space Transportation System and Its Payloads (AAS75-121)
        John Naugle ... 7

    Department of Defense and Shuttle (AAS75-122)
        Brigadier General Henry B. Stelling, Jr. ... 21

    Spacelab Programme - Status Review (AAS75-124)
        Bernard Deloffre ... 33

    Spacelab Status/Systems Capability (AAS75-123)
        Hans E. W. Hoffman ... 55

    Spacelab Payloads (AAS75-125)
        G. W. Sharp ... 80

    Shuttle Payload Development (AAS75-126)
        James Murphy ... 117

    Applications Payloads (AAS75-127)
        Charles W. Mathews ... 137

    Congressional Views on Long Range Space Plans
        Honorable Don Fuqua ... 159

    The Exploitation of Space for the Benefit of Mankind
        Honorable Frank E. Moss (AAS75-136) ... 165

    Shuttle Missions - Return on Investment (ROI) Panel Discussion
        William E. Dean, Chairman ... 173

II. LARGE SPACE TELESCOPE (LST) PROGRAM ... 197

    Potential for Advancement of Space Astronomy (AAS75-128)
        Arthur D. Code ... 199

Contents (cont'd)   Page No.

    Concepts of LST Operation (AAS75-129)
        C. R. O'Dell   219

    Large Space Telescope Program Status (AAS75-130)
        James A. Downey III   223

    The LST Scientific Instruments (AAS75-131)
        George M. Levin   235

LST Mirror Development

    Mirror Substrate Material and Manufacturing for the Large
    Space Telescope (AAS75-175)
        William C. Lewis   245

    Fabrication and Test of 1.8-Meter-Diameter, High Quality
    Ule Mirror (AAS75-176)
        Richard J. Wollensak and Clarence A. Rose   249

    Design and Testing with a Reflective Null System (AAS75-177)
        L. Montagnino and A. Offner   261

    Test Results on Homogeneity of Expansion for a 1.8-M Ule
    Lightweight Mirror (AAS75-178)
        G. Friedman and G. Gasser   266

LST Telescope Performance

    Science Performance Considerations for the Design of LST
        Damon D. Ostrander and James C. Tuttle (AAS75-202)   271

    Optical Performance Control (AAS75-203)
        Terence A. Facey   275

    Impact of Focal Plane Dynamics on Image Quality (AAS75-204)
        William J. Pragluski, Peter W. Abbott, Jack F. Eastman   279

    Stray Light from Out of Field Sources (AAS75-205)
        Robert J. Noll   284

    Design of Highly Stable Optical Support Structure (AAS75-206)
        Michael H. Krim   288

LST Instrument and Detector Development

    Large Format SEC Orthicon Integrating Television Sensor
    for LST (AAS75-207)
        John L. Lowrance   293

Contents (cont'd)   Page No.

    Thermo/Structural Design Considerations to Achieve the Large
    Space Telescope Line-of-Sight Requirements (AAS75-190)
        Domenick J. Tenerelli   301

    Design of Low-Thermal-Distortion LST Metering Structure
        John R. Lager (AAS75-191)   305

    3 Axis Simulation of the LST Pointing Control Subsystem –
    A Multi-Discipline Activity (AAS75-192)
        W. W. Emsley, T. D. Fehr, D. C. Fosth, and D. L. Knobbs   310

LST Data Management

    LST Data Management and Mission Operations Concept (AAS75-193)
        R. Walker, F. Hudson, and L. Murphy   314

    Data Management for Large Space Telescope (AAS75-194)
        G. R. Hope, Jr., and T. J. Rasser   318

    A Cost Effective Data Management Subsystem for the LST (AAS75-195)
        John A. Dougherty, Thomas D. Patterson, and Albert E. Cole   328

    System Consideration, Design Approach and Test of a Low Gain
    Spherical Coverage Antenna for Large Space Vehicles (AAS75-197)
        Manuel R. Moreno, Thomas D. Patterson, and Richard E.
        Ferguson   333

LST Mission Analysis and Operations

    System Application of the Fault Tolerant Memory (AAS75-296)
        L. J. Murphy   338

    A Scientific Operations Plan for the Large Space Telescope
        Donald K. West (AAS75-198)   342

    LST Operations, A Typical Day (AAS75-199)
        William J. Pragluski and Robert H. Brown   347

    Large Space Telescope Mission Analysis (AAS75-200)
        Frank M. Friedlaender   353

    Automation of the LST (AAS75-201)
        William W. Warnock and C. William Case   357

LST Maintainability and Operations

    Large Space Telescope External Interfaces (AAS75-179)
        Richard E. Collart   361

    LST Refurbishment and Support (AAS75-180)
        John Henschke   365

Contents (cont'd)   Page No.

   The Intensified Charge Coupled Device as a Photon Counting
   Imager (AAS75-208)
      Jack T. Williams   371

   The Infrared Capabilities of the Large Space Telescope
      D. E. Kleinmann (AAS75-209)   374

   Development of an Infrared Spectroradiometer (AAS75-210)
      W. H. Alff and J. G. Thunen   379

   Faint Object Spectrograph (AAS75-211)
      William P. Devereux   382

   High Speed Area Photometer Conceptual Design and Integration
      William Bloomquist and Fred Steputis (AAS75-212)   385

   High Resolution Spectrograph (AAS75-213)
      Keith Peacock   389

   The European Space Agency Study of Photon Counting Imaging
   for LST (AAS75-294)
      R. J. Laurance   393

LST Precision Pointing and Control Systems

   An Analytical and Experimental Evaluation of Actuator
   Vibration on LST Image Distortion (AAS75-184)
      A. D. Houston, L. W. Hodge, Jr., and T. J. Kertesz   405

   Development of a Large-Inertia Fine-Pointing and Dimensional
   Stability Simulator (AAS75-185)
      R. L. Gates, D. H. Wine, R. W. Seiferth, and N. A. Osborne   409

   Evaluation of Communication Antenna Drive System Design
   Requirements to Allow TDRS Tracking During LST Fine Pointing
      A. J. Besonis and C. J. Chang   (AAS75-186)   413

   A Small Instrument Pointing System for Shuttle Sortie
   Missions (AAS75-187)
      Carl W. Hendrikson and Ewald E. Schmidt   418

   A Fine-Pointing Facility for Spacelab Experiments in the
   1980's - The Instrument Pointing Subsystem (AAS75-276)
      H. Heusmann and J. Collin Jones   422

   LST Interferometric Fine Guidance Sensor (AAS75-188)
      A. B. Wissinger and R. H. Carricato   426

   Prismatic Grating Star Tracker (AAS75-189)
      Allen H. Greenleaf   430

Contents (cont'd)   Page No.

    Simulation of the On-Orbit Maintenance Cycle for LST
        J. A. Donnelly (AAS75-181)   435

    LST Power System Long Life Design Techniques (AAS75-182)
        Owen B. Smith, Richard L. Donovan, and James L. Oberg   439

    Large Space Telescope - Orbital Crew EV Maintenance
    Operations (AAS75-183)
        H. T. Fisher   443

III. ATMOSPHERES, MAGNETOSPHERES, AND PLASMAS IN SPACE (AMPS)
     PROGRAM   447

    Challenges in Space Physics (AAS75-132)
        Billy M. McCormac   449

    Challenges in the Atmospheric Sciences (AAS75-133)
        Robert E. Dickinson   463

    AMPS Science Objectives and Philosophy (AAS75-134)
        E. R. Schmerling   471

    Amps Program Status (AAS75-135)
        Rein Ise   476

AMPS Science

    Some Experiments with Energetic Particle Injectors (AAS75-215)
        W. Bernstein and John R. Winckler   487

    AMPS Experiments Involving Gas Releases (AAS75-217)
        Lewis M. Linson   501

    LIDAR Systems for AMPS (AAS75-218)
        Richard D. Hake, Jr.   506

AMPS Engineering Challenges

    Observation of Spacecraft Generated Electrostatic Fields
    In the Vicinity of the ATS 6 Satellite (AAS75-220)
        Elden C. Whipple, Jr.   510

    Measurement of Static Electric Fields at the Surface of
    Satellites (AAS75-222)
        J. E. Nanevicz   519

    Preliminary Shuttle Payload Contamination Assessment (AAS75-228)
        E. B. Ress, R. O. Rantanen, and L. E. Bareiss   539

    AMPS Data Management Concepts (AAS75-229)
        P. N. Metzelaar   543

Contents (cont'd)   Page No.

    Experiment Integration of AMPS (AAS75-230)
        Robert Witholder 549

    Ultraviolet Remote Sensing of Atmospheric Ozone from
    Payloads Using Shuttle Capabilities (AAS75-231)
        A. J. Krueger and D. F. Heath 561

    Design of High Voltage Insulation Systems for Aerospace
    Equipment (AAS75-232)
        W. G. Dunbar 567

PART II

IV.   SPACE TUG AND INTERIM UPPER STAGE (IUS) PROGRAMS   571

    Space Tug Requirements and Planning (AAS75-137)
        J. Wild and M. D. Kitchens 573

    Space Tug Baseline Description and Status (AAS75-138)
        William Teir and Alfred G. Orillion 591

    Alternate Tug/IUS Approaches (AAS75-140)
        A. O. Tischler 611

    Orbital Transportation in the 1980's and Beyond (AAS75-141)
        Hubert P. Davis 623

Space Tug Operations

    An Integrated Approach to Tug/IUS Mission Operations (AAS75-152)
        Billy S. King and Roger A. Chamberlain 645

    Tug Ground Operations - An Approach to Early Identification
    of Operational Impacts (AAS75-153)
        John L. Best 651

    Space Tug Mission Operations (AAS75-154)
        Kenneth C. Nuss 656

    Spaceflight Tracking and Data Network Support in the Space
    Tug Era (AAS75-214)
        L. R. Stelter and Robert D. Godfrey 661

Space Tug and Spacecraft Economics

    The Space Tug Economic Analysis Study - What We Learned (AAS75-147)
        Charles V. Hopkins 685

    A Survey of the Economics of Materials Processing in Space
        B. P. Miller (AAS75-149) 689

Contents (cont'd)   Page No.

   The Economics of Satellite Maintenance (AAS75-150)
   Wilfred L. DeRocher, Jr., and Richard G. Sosnay   711

Interim Upper Stage (IUS)

   Burner II Interim Upper Stage (AAS75-170)
   Henry Kudish   715

   Transtage Interim Upper Stage (AAS75-171)
   Peter B. Teets   719

   Centaur as an Initial Upper Stage for the Space Shuttle
   D. J. Jones and D. A. Heald (AAS75-172)   724

   Delta as an Interim Upper Stage (IUS) (AAS75-173)
   R. P. Dawson and J. F. Meyers   729

   Spinning Solid Perigee Stage (AAS75-174)
   H. A. Rosen, C. R. Jones, and L. M. Bronstein   733

   Agena Interim Upper Stage (AAS75-292)
   J. H. Guill   737

Space Tug Engines

   RL10 Derivatives for IUS/Tug (AAS75-155)
   J. P. B. Cuffe   759

   Advanced Space Engine Component Technology Status (AAS75-156)
   A. T. Zachary   763

   Aerospike Development Status (AAS75-157)
   J. Campbell and H. G. Diem   768

   Solid Rocket Technology Advancements for Space Tug and IUS
   Applications (AAS75-159)
   W. Ascher, R. L. Bailey, J. W. Behm, and W. Gin   772

Space Tug Propulsion Systems

   Requirements and Considerations in Selecting Space Tug
   Propulsion Systems (AAS75-160)
   Christopher J. Cohan   775

   Space Tug Propulsion Systems - Storable Versus Cryogenic
   W. E. Pipes (AAS75-161)   779

   Mixed-Mode Propulsion Systems for Full Capability Space Tugs
   R. Salkeld and R. Beichel (AAS75-162)   785

Contents (cont'd)   Page No.

    A Candidate Mission Using the Shuttle and Solar Electric
    Propulsion (AAS75-163)
        John H. Duxbury and Robert C. Finke   789

    Status Report on Nuclear Electric Propulsion Systems (NEP)
        J. W. Stearns, Jr. (AAS75-164)   799

Space Tug Mechanical and Interfaces

    Tug and Payload-to-Orbiter Interface Requirements (AAS75-165)
        Edward H. Bock   804

    Tug Payload Interfaces (AAS75-166)
        Fritz Runge   808

    Tug Rendezvous and Docking with a Spacecraft - A Remote,
    Manned Approach (AAS75-167)
        Michael J. Hurley   812

    Space Tug Thermal Control (AAS75-168)
        T. L. Ward   817

    NDT for Space Tug Thin Gage Materials (AAS75-169)
        Ward D. Rummel   822

Space Tug Avionics and Communications

    Tug Avionics System Overview (AAS75-142)
        Maurice T. Raaberg and James I. Newcomb   828

    Lightweight Fuel Cell Powerplant for Tug (AAS75-143)
        Lawrence M. Handley   833

    Space Tug Laser Gyro IMU   (AAS75-144)
        Robert F. Morrison   839

    Interferometric Landmark Tracker Applied to Precise Space
    Tug and Payload Navigation System (AAS75-145)
        D. H. Aldrich and W. F. Hubbarth   857

    LSI Computer for 1980's Space Mission (AAS75-146)
        W. A. Clapp, J. E. Saultz, and Dr. J. B. White   861

V.   SPACELAB AND AUTOMATED PAYLOADS PROGRAMS   865

Spacelab Operations

    Spacelab Task Allocation - A Preliminary Determination of
    Onboard Versus Ground Operational Priorities (AAS75-236)
        W. J. Harris and R. E. Holmen   867

Contents (cont'd)

   NASA/ESA CV-990 Airborne Simulation of Spacelab (AAS75-237)
      D. Mulholland, C. Neel, J. DeWaard, R. Lovelett,
      L. Weaver, and R. Parker ... 870

   Spacelab Payload Accommodation (AAS75-238)
      Donald M. Waltz ... 872

   Concept Verification Test: Evaluation of Spacelab/Payload
   Operation Concepts (AAS75-239)
      R. O. McBrayer and H. H. Watters ... 879

   Analysis of Extended-Duration Sortie Missions (AAS75-295)
      Robert C. Ring and Wilton C. Lide ... 905

Spacelab Payload Planning

   Space Transportation System Payloads Data and Analysis (AAS75-244)
      J. D. Peterson and H. G. Craft, Jr. ... 910

   Programmatic Aspects of German Shuttle/Spacelab Utilization
      Gottfried Greger (AAS75-245) ... 915

   Future Payload Technology Requirements (AAS75-246)
      Howard M. Ikerd and Larry R. Alton ... 929

   Computer-Aided Scheduling of Spacelab Ground Operations
      J. K. Willoughby (AAS75-247) ... 934

Spacelab Integration

   The Integration of Commercial Payloads into Spacelab (AAS75-240)
      H. L. Bloom and K. R. Taylor ... 939

   Low Cost Integration Techniques for Spacelab Payloads (AAS75-241)
      C. A. Braunwarth and T. C. Aepli ... 943

   Spacelab Payload and Program Planning in Germany (AAS75-243)
      Horst Schreiber ... 947

   Spacelab Resources (AAS75-297)
      Hans M. Kappler ... 957

Spacelab Science Payloads

   Astronomy Spacelab Payloads (AAS75-260)
      Richard Ott and Gary Wengrow ... 962

Contents (cont'd)   Page No.

    Atmospheric X-Ray Emission Experiment for Shuttle (AAS75-261)
        R. A. Goldberg, K. L. Hallam, and J. G. Emming   968

    Spacelab Ultraviolet-Optical Telescope Facility (AAS75-262)
        Murk Bottema   973

    Adaptation of an Existing Cosmic Ray Ionization Spectrometer Experiment to Spacelab (AAS75-263)
        U. R. Alvarado, J. F. Ormes, and C. V. Stahle   977

    A Large Cooled Infrared Telescope Facility for Spacelab (-234)
        Stephen G. McCarthy, Lou S. Young, and Fred C. Witteborn   981

Spacelab Life Sciences and Applications

    Application of Space Shuttle to Fundamental and Applied Microbiological Research (AAS75-255)
        Jerry V. Mayeux   985

    Life Sciences Manned Payloads for Shuttle/Spacelab (AAS75-256)
        Dennis B. Heppner, Goerge L. Drake, and Chester B. May   989

    Shuttle Bioresearch Laboratory Breadboard Simulations
        S. T. Taketa (AAS75-257)   999

    A Life Sciences Spacelab Mission Simulation (AAS75-258)
        John A. Mason, F. Story Musgrave, and Dennis R. Morrison   1000

    Space Shuttle Trace Gas Analyzer (AAS75-259)
        Wallace Dencker   1003

Spacelab Technology Payloads

    An Orbiting Molecular Shield Vaccum Facility: A Materials Laboratory in Space (AAS75-248)
        James W. Youngblood, R. A. Outlaw, Leonard T. Melfi, Jr., and John R. McIlhaney   1007

    Advanced Extravehicular Mobility Unit Technology Experiments
        Gary Wengrow (AAS75-249)   1012

    The Advanced Technology Laboratory (AAS75-250)
        C. Llewellyn and R. Milliken   1018

    Shuttle Entry Technology Payloads (AAS75-251)
        Paul M. Siemers III   1023

    Landmark Tracking Technology (AAS75-278)
        J. D. Welch, W. E. Sivertson, and R. G. Wilson   1045

Contents (cont'd) Page No.

Communications Payloads

Shuttle Communication Experiments (AAS75-252)
John J. Woodruff and Donald R. Peters 1049

Large Deployable Antenna Shuttle Experiment (AAS75-253)
R. E. Freeland, J. G. Smith, J. C. Springett, and K. E. Woo 1053

Availability of a Communications Satellite, Requirement and Feasibility (AAS75-254)
Garry D. Gordon 1057

Multidiscipline Payloads

Mission Considerations for Multidiscipline Applications Payloads (AAS75-268)
John M. Macdonald 1061

The Long Duration Exposure Facility - A Shuttle Transported Low-Cost Technology Experiment Carrier (AAS75-269)
John D. Dibattista 1073

Payload Planning for the First Spacelab Mission - A European View (AAS75-270)
A. V. Breitenstein and D. Davidts 1077

Spacelab Utilization in Different Fields of Science and Applications (AAS75-271)
Jacques Collet 1089

Early Space Station User Accommodations (AAS75-291)
Donald R. Saxton and Harry L. Wobers 1103

Automated Payloads

Automation of Space Processing Applications Shuttle Payloads
Walter E. Crosmer, Oakley T. Neau, and James Poe (AAS75-264) 1107

Shuttle Benefits for Automated Retrievable Cryogenically Cooled Payloads (AAS75-265)
John O. Simpson, Thomas M. Spencer, and William H. Follett, Jr. 1111

Retrieval of the HEAO-C Spacecraft with Space Shuttle
David H. Mitchell (AAS75-267) 1133

The Multimission Modular Spacecraft for the 80's (AAS75-235)
Robert O. Bartlett and Frank J. Cepollina 1137

Contents (cont'd)   Page No.

Automated Planetary Spacecraft

Shuttle/IUS Performance for Planetary Missions (AAS75-273)
  M. J. Cork, J. M. Driver, and J. L. Wright   1171

Possibilities for Reducing High-Energy Performance
Requirements (AAS75-279)
  G. R. Hollenbeck   1175

Impact of Space Transportation System on Planetary Spacecraft
and Missions Design (AAS75-274)
  Phillip M. Barnett   1180

Using the Shuttle for Future Advanced Planetary Missions
  L. D. Friedman and W. Scofield (AAS75-275)   1185

Panel Discussion

Can We Use the STS to Improve the Planetary Program?
  Moderator - Daniel Herman, Panel Members - John Niehoff,
  Jack Wild, Robert Parks, and Phil Culbertson   1197

VI. ENERGY IN THE SHUTTLE ERA   1221

Energy Research Overview/Alternatives for Energy Development
  Thomas J. Vogenthaler (AAS75-280)   1223

The Satellite Solar Power Station - A Focus for Future Space
Shuttle Missions (AAS75-281)
  Peter E. Glaser   1239

Nuclear Power in the Shuttle Era (AAS75-283)
  S. R. Ross   1263

Availability and Variability of Solar Energy in the Rocky
Mountain Region (AAS75-285)
  Richard C. Burriss   1279

Modeling the Western Coal Industry (AAS75-286)
  William Ganter, Claude McMillan, and Fred Glover   1283

Minimum Cost Solar Thermal Electric Power Systems: A Dynamic
Programming Based Approach (AAS75-287)
  William S. Duff   1288

100 MWe Solar Power Plant Design Configuration and Performance
  F. A. Blake (AAS75-288)   1293

Contents (cont'd)   Page No.

Minimization of Overtime Shift Hours with an L-P Model
    Gerald L. Kaes   (AAS75-289)   1301

Deposited Today - Consumable Tomorrow Sewage and Solid Waste
Hydrogenation (AAS75-290)
    Clyde W. LaGrone   1305

INTRODUCTION

WELCOME

Honorable Richard D. Lamm*

It is a pleasure for me to welcome this impressive gathering of scientists and members of the space industries and NASA here for the 21st American Astronautical Society meeting. Once again Denver is the location selected for an important gathering of all the key people involved in space exploration.

Colorado is an appropriate site for your gathering because it gives you the opportunity to see this world at its best. We proudly call attention to our beautiful state and to its cities as well as to our industries which have played such an important role in present and past space explorations. You must agree that Denver is a key city because this is the fifth major space symposium held here.

In 1961 this association gathered here to talk of a Manned Lunar Flight. In 1963 when this group met in Denver they may have thought a trip to Mars was a reality, but to some of us that sounded like Buck Rogers fiction until last week's Viking flight made it a reality. In 1965 you met here again to discuss Unmanned Explorations of the Solar System, such as the Viking flight and some of the explorations of Venus and Jupiter. In 1969 a joint meeting of your association and the Operations Research Society met to plan the challenges of the 70's in space and here you are, again, in Denver to project where we will go in space in the next decade.

---

*Governor of State of Colorado

I have often commented on our finite planet, space ship Earth, as a way of calling attention to the limits we know exist for this planet.

But we know of no such limitations on space. We simply do not know what continued explorations in space will provide as the solutions to the problems of our finite planet. But we do know that space technology and research go hand-in-hand with helping this world and helping us understand our solar system.

While the research into the unknown of our solar system has it own value, we cannot lose sight of the more **concrete** application of the space technology as it applies to life here today.

As scientists and as an industry you can help us shorten the time lag between the research and the development of related technology and the spin-off application of those techniques to every day life. We already see some very concrete examples in the space age research into solar energy and such applications will be important to this nation. We know that Martin Marietta, one of this area's major industries, is working on the development of a large power system using the sun and you will be hearing of this work from Floyd Blake at this meeting. We have just learned that one of our area schools will soon be heated by solar energy as a result of the technology developed through the space industry. We know that as a result of a fantastic system, a system of planning and thinking for space, much is being developed to help us here on Earth, now. Colorado is in the midst of the space industry with such industries as Ball Brothers, Beech and Honeywell, as well as Martin Marietta and we benefit doubly through their contribution to our economy and through the opportunity for application.

We know that in addition to energy research, we will benefit from similar space spin-offs in the areas of insulation, electrical conversion, the ability to survey the Earth's resources, communications, and even an understanding of this world's pollution problems, to name just a few areas.

Years from now we will look back and wonder how man ever managed his life on this limited planet without the tools provided by the space program. Then it will be hard to imagine a world without spacecraft just as now it is impossible to imagine a world without airplanes, television or telephones.

To quote Dr. Wernher von Braun, "The dominating problems of humanity during the next fifty years that will require such aid from space will arise from the collision between man's happy-go-lucky joy ride of unrestrained and unlimited growth -- both in numbers of people and their material expectations -- and the grim limitations of resources the Earth can provide and the wastes it can absorb."

AAS 75-121

THE SPACE TRANSPORTATION SYSTEM AND ITS PAYLOADS

John Naugle*

It is indeed a distinct pleasure for me to join with you today to discuss the Space Transportation System and the payloads which will use it as we approach the third decade of the space program. It seems like a remarkably short time ago that I attended my first general meeting covering the subject of launch vehicles and payloads. The topics at that meeting were not, however, Shuttles, Tugs, Spacelabs, Large Space Telescopes and Sortie Missions. They were, instead, considerations of basic orbital mechanics, questions about the ability of man to survive in space, considerations of the feasibility of converting a ballistic missile to a space launch vehicle, and discussions about the possibilities of getting a 150 pound spacecraft into low earth orbit. We've matured somewhat since those days. The world has witnessed a progression of space events in the intervening years ranging from the sometime successful, but frequently unsuccessful launching of those first low altitude satellites to manned roving vehicles on the lunar surface and complex spacecraft which are probing the secrets of Jupiter on their way out of the solar system. As we have moved forward in our use of space, the United States has launched 840 satellites belonging to this nation in addition to many belonging to the other free nations of the world.

Political leaders and industrialists have, as a result, become increasingly aware of the potential impact of the space program in the opportunities which are provided for constructive joint international ventures, in the advancement of fundamental scientific knowledge, in the practical applications provided by the unique viewing opportunity provided by space, by the stimulous to technology which has accompanied the space program, and finally, by the contribution which accomplishments in space have made to the feeling of national pride.

---

*Dr. Naugle is Associate Administrator, National Aeronautics and Space Administration.

The value of the knowledge gained from the exploration of the solar system and the moon is well recognized by the world scientists. Benefits derived from communication satellites, the improvement in weather forecasting, and the aids to sea-going traffic provided by navigation satellites are now accepted as commonplace. Benefits in health care, public safety, education and new products attributable to the direct use of space and to the application of technology derived from the space program are being provided every day. We are moving from the era of learning how to live and work in space to a new plateau of routine operational use of space. This transition is following a pattern similar to that which we experienced in the use of the airplane in the United States in the first half of the century. Before the first World War, the airplane was very much an experimental novelty. Then came the fascinating barnstorming days of the airplane pilots after World War I in the 1920's. In the 1930's we saw the inauguration of the use of the airplane to fly the mails and carry commercial passengers on scheduled flights. Today commerical and military transportation are built around the airplane. Similarly, many of us see the effect of the Space Transportation System on space operations as a parallel to the impact of the venerable DC-3 transport on the airline industry. The DC-3 allowed the beginning airlines to move from irregular and frequently undependable service into a truly operational business and an integral part of everyday life. Many of us see the introduction of the STS as the key to making space flight an integral part of our every day life.

My personal view of the STS, and its three major elements, about which you will be hearing a great detail today, is that it provides us with both a great opportunity and a great challenge. The opportunity is, of course, the capability, versatility and flexibility which it will afford and the effect which those characteristics can have on the cost of the use of space. The challenge which it provides is the challenge to those of us in government, universities, and the industry who plan, approve, design, test, and operate the payload programs which will use the STS to create programs which will exploit that capability.

If we don't have the foresight to see the way that space can be used effectively; or have the engineering ingenuity to design spacecraft which fully utilize the unique characteristics of intact abort, on orbit checkout, orbit-to-earth return, and the presence of man which the STS will provide; or have the skill to demonstrate the true value of space to the public and the Congress -- then we will have missed the opportunities afforded by the Space Transportation System. Each of us here today has a role in meeting that challenge. I would hope that when you leave you will better understand your role in meeting that challenge.

As you discuss the STS and its payloads you must bear in mind that more than ever before we are being challenged by the cost of space activity. During the early years of space when we were learning how to move into an entirely new medium with entirely new engineering and scientific concepts, the costs were necessarily high. The difficulty of initially penetrating beyond the earth's atmosphere and then the rapid extension of this capability to lunar distances and beyond, dictated approaches that emphasized the highest performance, lightest weight, and the utmost in reliability that could be attained from the state-of-the-art available. Now we find that the ability to take advantage of the potential of space is increasingly limited by cost. In the late 1960's NASA and the DOD squarely faced the need for operating more effectively in space by developing and focusing their efforts on reducing the cost of space operations by moving to reusable launching systems. This reusable system, now under development, is, as all of you are aware, called the Space Transportation System.

The Space Shuttle, Spacelab and a Propulsive Upper Stage form a family of vehicles which can be integrated in various combinations to accommodate the needs of the payload community.

The Space Shuttle is the key to the expansion of space activities. It combines low-cost operations with great versatility and it will provide the basis for routine operations in space.

The two fundamental characteristics of the Space Shuttle, its reusability and down payload capability, are the keys to expanding space operations. The ability to return 32,000 pounds from orbit is a unique capability which, I believe, is yet to be fully exploited by those of us engaged in spacecraft design and mission planning. I would hope that this conference would bring to light new ideas which will capitalize on this new operational mode.

We believe that most Space Shuttle missions, and certainly all delivery missions, will use a nominal crew of three or four people for seven days. But increasingly, we find that for the material processing, some of the life sciences and engineering kinds of missions, we will need to extend that capability to 30 days within the first few years of operation. Similarly I believe that we will begin to take advantage of the capability of the Shuttle to support a crew of seven with four crew members involved in payload operations.

Mission stations on board the Shuttle will accommodate the commander, pilot, mission specialists and, as I have indicated, up to four payload specialists. We are already receiving inquiries from scientists, both men and women, concerning the possibilities of their flying as payload specialists, what training they will require, and how long prior to actual flight it will take to confirm their assignment as payload specialists for a particular mission.

The Space Shuttle itself can carry scientists and their equipment in low earth orbit up to about 500 miles. However, there are many missions which require higher orbits, particularly those requiring synchronous altitudes. To achieve these higher orbits, we will ultimately use another element in the Space Transportation System -- the Propulsive Upper Stage. The Propulsive Upper Stage will be unmanned, a reusable stage that will fit in the Shuttle cargo bay and will provide the capability to place payloads in orbits beyond the capability of the Space Shuttle. Current planning calls for the development of this new vehicle -- the Tug -- to be developed in the mid 1980's to deliver payloads to synchronous orbit and retrieve spacecraft for reuse from synchronous orbit. Pending introduction of this new vehicle, an Interim Upper Stage will be developed by the DOD to be available to all users of the STS. NASA is working very closely with the Department of Defense as decisions are being made relative to the development of this stage to effectively support civil as well as military payload requirements. The minimum baseline requirement is a capability to deliver approximately 3,400 lbs to synchronous orbit with no retrieval capability.

Spacelab will, perhaps, provide the most unique capability of the STS. Ten European countries are working together within the European Space Agency to design, develop and manufacture Spacelab and they will fund all research and development costs. The United States will purchase one or more complete Spacelabs from the European consortium. In terms of responsiveness to a variety of requirements, we see the Spacelab as a modular, versatile system, providing reduced lead time for each user from the time of initial planning for the payload through payload development, system integration and flight operations, until he has flight data back in his laboratory for analysis. All of our planning is being done with this as one of the major objectives.

You will recall that three years ago we had a workshop similar to this one -- the Shuttle Sortie Workshop held at Goddard. At that time the Shuttle contractor had just been selected and we were discussing the possibility that a European consortium would build "sortie laboratories." A lot has happened since then -- as evidenced by the events of recent months.

- o A ground test version of the Shuttle main engine is undergoing firings at the National Space Technology Laboratories (NSTL) in Hancock County, Mississippi.
- o The first set of orbiter wings, the vertical tail, and fuselage midbody have been delivered to Palmdale, California, site of the final assembly.
- o The ESA/ERNO team is hard at work developing Spacelab -- having just completed the System Requirements Review (SRR).
- o The Canadian Government has just agreed to develop (at their expense) the Shuttle Remote Manipulator System.
- o The Department of Defense, in conjunction with NASA, will soon make a decision on the IUS configuration.

What this activity means is that we're a short 4 years away from the first Shuttle orbital flight. In the current STS traffic model we project the first Shuttle operational flight in 1980. During the twelve year period from 1980 to 1991 we forecast 572 Shuttle flights. Starting first with flights to qualify the IUS and the Spacelab -- we will eventually reach a flight capability of 60 missions per year.

As we look to the future in order to understand how the STS will be used, it is constructive to review briefly what we have accomplished in the Space program and to use that history as a starting point for our projection into the future. We began our space program with small automated satellites. The first one weighed only 31 pounds. These were equipped with scientific experiments which made early measurements of space characteristics, such as the Van Allen Belt. As our launch capability, and our program ambitions increased, we moved outward from low earth orbit to synchronous orbit, the moon and the near planets. The Lunar orbiter, launched in 1966, which mapped the moon, was one of the several automated precursors to the Apollo lunar landing. In the same time period we sent spacecraft to Venus and Mars. In applications, our program began with early weather and communication satellites. During the early 1960's, we also began our manned space flight program. The Mercury program was followed by the highly successful Gemini program where we gained the additional experience in manned flight so necessary for Apollo. Based on the previous manned and automated programs, the Apollo Lunar Landing in July 1969 was acknowledged as a high water mark in space flight and the successive lunar landings and exploration of the moon yielded a wealth of scientific and engineering data.

The Skylab program with its three mission periods extending to 84 days proved to be one of our most productive endeavors, both from the standpoint of manned missions and from the range of experiment data which it produced. Skylab was a showcase of future possibilities; clearly showing man's ability to carry out a variety of experimental activities in space ranging from solar astronomy through materials processing to earth resources surveys. In the weightless and high vacuum environment of Skylab, crew members have grown crystals of semi-conducting materials, of great importance to the electronic industry, which exhibit a size, structural perfection and purity unattainable on the Earth's surface. In the absence of gravity, there is no thermal convection, no sedimentation, no layering of fluids due to differences in specific weight, and no need for containers such as crucibles in melting and forming. Thus, we have created new mixtures of formerly immiscible materials and metal alloys which have been unobtainable in a gravity field. Skylab carried an experiment in electropheretic separation which indicated that the absence of gravity enhances the separation of similar organic cells and other biological materials. Although there is still much to be done to extend this encouraging experimental result to the point that it can be considered to be an operational technique, it does imply the potential of important future applications in the production of large amounts of vitally needed vaccines. The potential for commercial use of these new materials and processes is enormous.

In our review of space accomplishments, there are several unmanned applications programs we can particularly note. The ATS-6 Direct Broadcast Satellite launched last year was experimentally used to carry educational programs to junior high schools in the Rocky Mountain States. The specially prepared programs originated from a ground station here in Denver. The orbit of the ATS was recently changed and it is now repositioned in synchronous orbit at $30°E$ Longitude in order to provide direct broadcast coverage for 5,000 remote villages in India that have very limited communications links with the outside world. The programming will provide crop information, health and family planning information as well as educational activities for the children.

The LANDSAT Satellite is another of the application satellite programs which is receiving wide attention. This satellite uses multi-spectral scanners to perform a variety of observations which can be interpreted to give information on crop conditions, water pollution and natural resources.

In January of this year, we launched our second Earth Resources Satellite named LANDSAT 2. Both satellites continue to return data and are demonstrating the potential of economic returns which we believe will be measured in the hundreds of million of dollars per year.

Our new weather satellite, the Synchronous Meteorological Satellite gives us a remarkable picture of the entire western hemisphere and is another step toward the long term program to improve the accuracy of weather forecasting.

In the realm of space exploration, our spacecraft have visited Venus, Mercury and Jupiter during the past year. We have just launched one

Viking spacecraft following a number of problems at KSC which delayed the launch and forced us to launch the backup spacecraft first to be followed by what was originally the primary system. Viking, of course, will land on Mars in 1976, and among other things, search for life on the planet. Viking is the most advanced planetary spacecraft we have ever built and it will provide a new step forward in the exploration of the planets. Much of the design, testing and building of the Viking spacecraft was accomplished at the Martin Marietta facilities here in Denver. Martin Marietta also is responsible for the Titan III used as the launch vehicle. Two Mariners will be sent to Jupiter and Saturn in 1977 and two Pioneer spacecraft will explore Venus in 1978. One of the Pioneer Venus spacecraft will send four probes into the Venusian atmosphere. The other will study the Venus atmosphere from orbit.

By the 1980's, the practical and scientific use of space will be worldwide. It is reasonable to project that more nations will be participants in the space program and that those nations which are now space-faring will participate in new areas. The STS will encourage this broader involvement in space for at least a couple of reasons. One is the utility of the Space Transportation System: the STS gives us easy access into and out of space. The payloads, the scientific experiments and satellites, and the sharpened expertise of the hundred of non-astronaut experts who will make the Shuttle journeys into earth orbit could be the crux of America's technological advancement in the next 15 years.

The second reason is the economy and flexibility of the use of the STS: both from the standpoint of cost-per-flight, and, more importantly, from the opportunities which it will provide for reduction of payload costs.

Users will include commercial and private interests, including industrial corporations, research institutes and universities, and users from foreign nations. Potential governmental users are primarily those agencies involved in use of the land: the Bureau of Mines, the Bureau of Land Management, the Bureau of Reclamation, the Fish and Wildlife Service, the Food and Nutrition Service, the Soil Conservation Service, the Agricultural Research Service, the Environmental Protection Agency, the Urban Affairs Department. Other agencies which are also involved in the use of data obtained from spacecraft include the several science oriented offices within the government, HEW, DOT and, as you will hear in considerable detail, the DOD.

Beyond the 1980's our crystal ball obviously becomes somewhat fuzzy. However we can speculate about the directions in which the space program might move and what it might embrace in the nineties and at the turn of the century.

The great problems facing human society -- the growing demand for food and resources, new challenges to our environment, and the threat of unforeseen disaster, whether caused by nature or man -- demand great efforts for their solution. The basic material needs of mankind must be met, and if the space program is to play a vital role in the nation's future, it must respond to these needs.

Of particular future significance is the use of data from remote sensors in space to aid in predicting crop production and water availability, monitoring and predicting climate trends and severe storms, and monitoring the environment. Such activities promise substantial benefits to human welfare. Therefore, I foresee major increases in resources allocated to remote sensing from space.

We have a unique opportunity, I believe, to contribute to an understanding of the climate and the changes induced in it by both natural and human activities. Climatic changes have major effects not only on the weather, but also on the availability of food and water, the demand for energy and the quality of the environment. For this understanding, a major research program on the Earth's climate is warranted, within which satellite observations will play an important role.

In responding to the critical energy problem facing our society, data from space sensors have the potential of predicting water availability and climate trends and the generation of new geological maps. Beyond that, it may be technically possible and economically feasible to collect solar energy in space and transmit it to the earth, or to launch nuclear wastes from Earth based generators on trajectories carrying them outside the solar system.

The lack of gravitational effects for long periods of time is unique to space flight. The use of this characteristic will be of significant value in materials science and the understanding of fundamental physical, chemical and biological processes. Furthermore, as I mentioned earlier there is the possibility that materials processing in space could provide major economic returns. Therefore, an active research program is needed both to examine the scientific use of the orbital environment and to explore and evaluate the potential commercial opportunities of space processing.

The intellectual needs of man share importance with material needs. The Space Science program has responded to these needs during its first 15 years. A wealth of important information has been produced leading to a better understanding of our solar system and cosmos, and a deeper recognition of man's place in the Universe. The Space Science program in the 1980's and beyond must continue this purpose.

As we look to the future in Space Science, we expect to focus on specific objectives selected on the basis of their expected contribution to answer furndamental questions. It's risky to single out specific scientific questions as more important than others, but I believe that some of the more challenging include questions related to the beginning of the Universe, the nature of quasars and black holes, the evolution of planets and of their atmospheres (obviously with particular emphasis on the problems of the earth's climate) the nature of gravity, and the search for extraterrestrial life forms ranging from primitive biology to intelligence.

The desire to explore new frontiers, to occupy and to develop them, is a fundamental human characteristic. Man has entered space, and he is there to stay. Human beings will explore the solar system, perhaps not before the turn of the century, but explore we will. A portion of our research and development programs should be directed in a way which will permit that exploration to proceed when the proper time arises. When human needs demand the full development of the space frontier, we must be able to respond.

There will be an enormous increase in the amount of data gathered by space instruments, not only in programs responding to material needs, but in scientific investigations as well. This wealth of data must be translated into useful information. New automated and analytical techniques must be developed for the extraction of information, and new predictive models generated to provide information in the formats necessary for proper management and decision making.

A large number of potentially beneficial space activities requires other significant technological advances, not only for basic feasibility, but also to achieve a level of cost effectiveness necessary to justify their operation. A program of technology development is therefore envisioned, not based on single project requirements, but rather aimed at satisfying broad classes of mission needs.

It should be apparent that one of our goals must be the maintenance and expansion of space technology. It is not by chance that the most underprivileged, poverty-stricken nations are the very ones whose science and technology are the most primitive. Senator Mike Monroney once expressed the need very well; he said: "Starving our technology mortgages the future of our society."

Technology is indeed significant for our economy, for what does the U.S. have to export? We don't have many raw materials, and many nations can make things cheaper than we do. The answer is that advanced technology is really the major unique product we have to export.

A recent study by Chase Econometric Associates of the "Economic Impact of NASA R&D Spending" is noteworthy -- it presents the first application of a complex well-tested macroeconomic model to the analysis of the impact of advanced technology research and development expenditures on major economic indicators. The study shows that R&D expenditures bring very substantial returns in the form of increased gross national product by increasing productivity in the economy.

The results of these studies further validate the theory that advanced technology R&D is a principal component of economic growth. They substantiate the fact that NASA's programs, in addition to providing benefits from direct applications of aeronautical and space technologies and the advancement of scientific knowledge, make a significant contribution to the overall technological progress so important to national economic well being.

Beyond these benefits there is a broader compelling reason to ensure our leadership in the strongest technological base yet known to mankind. This is best illustrated in an example:

"Early in the 15th century, the Chinese had become a formidable maritime power, far ahead of the rest of the world in sciences and technology of the sea. Their ships ranged from Madagascar and the Arabian Gulf to the East Indies and Korea. Unlike Europeans of the time, the Chinese had learned to sail against the wind and in direct route out of sight of land. Their ships were huge, almost four times as large as the Spanish galleons of three centuries later, and were capable of carrying 2,000 passengers each on months-long voyages. They were in the forefront of a new technology.

"Abruptly, the Chinese government closed the huge shipyards in Canton and other coastal cities and used the money earmarked for maritime development for other purposes 'closer to home.' Young people's interest in the sea waned for lack of inspiration and guidance. Thus came a major turning point at a critical time in Chinese history. For within a century the Portuguese and other Europeans with inferior knowledge of the sea come to the Far East by ship, eventually to humiliate the celestial empire. The Chinese still pay the penalty for abandoning their interest in naval research and education."*

America should never have to pay a similar penalty. The STS program addresses the development of advanced technology for application to the solutions of national and international problems -- such as, controlling environmental pollution, advancing medical science, and extending education to the backward areas of the world. The nature of these problems is multidiscipline in nature and it will require multidiscipline technological breakthroughs across a broad spectrum of technology to solve them. These breakthroughs will then become the "tools" that make possible

---

*Extracted from "Space Among Us" by Charles P. Boyle, 1974.

a whole generation of unrelated breakthroughs, some of which may be in fields not directly connected to the original problem.

The far term possibilities which I have mentioned are not yet part of NASA's program or that of the federal government. I feel, however, that they are representative of the type of program which can evolve. As you participate in these discussions during the next three days and listen to the DOD involvement and planning, to the European participation in the program and to the other speakers who will discuss both the STS and its payloads; I ask each of you to keep in mind my earlier comments on the opportunity and the challenge. It is those of you who are here, and your colleagues, who will determine whether or not that challenge will be met. Each of you has a role -- I urge you to accept it.

AAS 75-122

DEPARTMENT OF DEFENSE AND THE SHUTTLE

B/Gen. Henry B. Stelling*

The Air Force, as DOD's executive agent for the Space Transportation System, has been assigned four primary tasks: (1) assure that the evolving Shuttle design will meet DOD needs, (2) develop a plan to transition our critical defense satellites from their current launch vehicles to the Shuttle, (3) develop the Interim Upper Stage, and (4) acquire and operate a Shuttle launch and landing facility at Vandenberg Air Force Base.

DOD plans for utilization of the Shuttle may be best described as a phased approach. The first phase is primarily devoted to minimizing the cost of transitioning, coupled with strong support of the NASA development effort. In the near term, the Shuttle is viewed primarily as a reusable replacement for expendable boosters. After the initial transition period, the DOD expects to exploit all the new capabilities which a manned system affords and is actively investigating the unique features of the Shuttle to revolutionize current thinking on spacecraft design.

INTRODUCTION

It gives me great pleasure to appear before you to discuss the Department of Defense role in the development of the NASA Space Shuttle and our plans for participating in the operation of the space transportation system.

The DOD was an active participant on the President's Space Task Group which recommended that the Shuttle be developed. In 1972, when the president announced his decision to authorize the development of the Shuttle, he described the potential benefits of a manned and reusable launch system which would replace all but the smallest and largest expendable launch vehicles. In short, he envisioned a national system to make space even more attractive as an operating medium to all users.

The DOD has had a continuing interest in investigating the utility of man in space. However the expense of systems designed for this purpose alone has limited our opportunities to gain space experience. The opportunity to participate with NASA in acquiring a space transportation system is an opportunity to satisfy unique objectives of both agencies and at the same time provide a national capability which can be utilized by military, civil, and international users. It is an ambitious project, with a very large price tag, and one which may not be easily evaluated in traditional cost-benefit terms. Unlike NASA, who does have a separate space mission, the DOD looks at space as a

---

*Directorate of Space Headquarters, United States Air Force

fourth operating medium whose principal use is to aid the deterrance of all levels of warfare. Accordingly, our emphasis is on the support missions of communications, navigation, meteorology, and surveillance.

Let me just say a few words on how we see the future of space systems within the DOD. The major role of military space systems in the next decade -- The Shuttle Decade -- will be to support national decision makers and terrestrial military forces. There are four basic reasons for using space systems for various military support functions: First, uniqueness -- some functions essentially can be only done from space, such as near real time warning of a ballistic missile attack; second, economics -- some functions are more cheaply done from space, such as long haul communications; third, functional effectiveness -- some functions are more effectively done from space, like meteorology, and fourth, force effectiveness enhancement -- some space functions can greatly enhance the effectiveness of terrestrial forces. The first three have been dominant until now. However, in the late 1970's and thereafter, the forth will become increasingly important.

The United States and Soviet use of and dependence on space is growing; in a decade, space systems will play a larger role in deterring warfare, will support virtually all military forces and could strongly influence the outcome of conflicts. It's within this context that the Shuttle will arrive.

Many of you may be thinking "How does this apply to Manned Systems when so much can be accomplished with automated satellites." I think a good answer to that question was supplied by Mr. Donald Hearth, NASA Deputy Director of the Goddard Space Flight Center, during his testimony before the House Space Science and Applications Subcommittee on the NASA Study - "Outlook for Space." When asked about the role man will play in space during the 1980 - 2000 time period Mr. Hearth stated, "....as we looked at the future of space, particularly at those more creative programs directed toward major exploitation of the opportunities which space provides, we inevitably found man to be an integral part of the system." Mr. Hearth added that...."the desire to explore new frontiers, to occupy and to develop them, is a fundamental human characteristic...man is in space to stay."

Before I leave this area, I feel some mention should be made of cost. Space operations have always been costly. For example, between 1974 and 1975 the number of dollars necessary to fund the procurement of a Titan IIIC launch vehicle increased by approximately $10 million, an increase of 40%. By providing routine, highly reliable transportation to and from space, along with the resulting operational flexibility and cost reductions offered by the Shuttle, space will be even more attractive as an operating medium to all potential users, including the DOD. Let me underline **highly reliable**. Today when trouble is experienced with the launch vehicle, we lose the expensive payload. I recall some years back the anguish I experienced as a program director when expensive satellites were lost because of booster failures. The Space Shuttle, of course, is being designed to eliminate such payload losses. If a failure is experienced during the Shuttle ascent phase, the Shuttle will return to the launch site with the payload intact.

The DOD strongly supports the NASA Space Shuttle development program and plans maximum utilization of its capability. Briefly stated, the Air Force, as DOD's executive agent for the Space Transportation System, has been assigned four primary tasks, which I'll now address. First, from the beginning, the Air Force has been working closely with NASA in their development effort to assure that the evolving Shuttle design will meet DOD needs. We have established DOD requirements for the Shuttle and are participating in various NASA Design Reviews and Planning activities with the objective of minimizing the cost of changing over from current expendable launch vehicles to the Shuttle. It is also a great opportunity for the two agencies with extensive space activities to share experiences and know-how. This sharing during the development phase holds the promise of reducing acquisition costs as well as the life-cycle costs of operating the Shuttle system. Coooperation and support from NASA in this effort have been excellent. Second, we are developing a plan to transition our critical National Defense Satellites from their current launch vehicles to the Shuttle in the most efficient and cost effective manner possible and are planning to use the Shuttle at Kennedy Space Center as soon as it becomes operational. Third, while our main role in the NASA Shuttle Development Program is one of support, the Air Force is actively involved in a development program of its own - The Interim Upper Stage. As Dr. Naugle explained, NASA is planning, as part of the Space Transportation System to develop a full capability upper stage called the Space Tug which will provide for the transfer, retrieval, or on-orbit servicing of payloads in high energy orbits. The Tug, however, will not be available prior to 1984. Since virtually all of the DOD payloads that will be launched from Kennedy Space Center require higher orbits than that achievable by the Shuttle orbiter alone, the DOD has agreed to modify an existing upper stage for use until the Space Tug becomes available. DOD development of this interim upper stage will assure that DOD payloads will be able to make an early transition to the Shuttle with minimum impact.

I'll have more to say about the IUS Program later. Lastly, the DOD will acquire and operate a west coast Shuttle launch and landing facility at Vandenberg Air Force Base. Both NASA and DOD payloads requiring high inclination orbits will be launched from there. We are planning for the first Shuttle launch from Vandenberg in December 1982. At that time we will begin transitioning the remaining DOD satellite programs to the Shuttle, gradually phasing out existing expendable boosters except for a few that may be required to provide an emergency Shuttle backup.

These then, are our major tasks in the Shuttle Program, and we have directed our planning and budgeting activities toward these objectives. We are committed to these plans, and have been since the Director of Defense Research and Engineering informed the NASA administrator in April 1972 that the DOD concurred in the selection of Vandenberg Air Force Base and Kennedy Space Center as launch and landing sites for the Shuttle. He further agreed that the DOD would provide Shuttle facilities for all users at Vandenberg on a time schedule compatible with the progress in the Shuttle development.

The Deputy Secretary of Defense, Mr. Clements, in an August 1974 letter to the NASA Administrator, Dr. Fletcher, stated that the DOD is planning to use the Shuttle to achieve more effective and flexible military space operations in the future, eventually phasing out the inventory of current expendable boosters. He also assured Dr. Fletcher that the DOD is committed to the development of the interim upper stage and to the acquisition of a Shuttle launch capability at Vandenberg. Funds are provided in the Air Force budget for this purpose.

As the pace of the DOD participation in the Shuttle Program accelerates high level management reviews are being conducted to assure that the Air Force Program is being structured to meet these long term objectives in the most efficient manner possible. The first Space Shuttle Program Review by various senior DOD Management officials was held in April 1974, and another program review is scheduled for November of this year. The November review will be conducted concurrently with a review by the Defense Systems Acquisition Review Council (DSARC) of the Air Force's IUS Program. Before that program proceeds into the validation phase of its development effort, the DOD participation in the Shuttle Program is extensive, and requires that a close working relationship be maintained with NASA at all levels. On a formal basis there are a number of DOD and NASA coordinating committees that embody this close working relationship.

The Aeronautics and Astronautics Coordinating Board is the main body responsible for interagency policy guidance and cooperation of aerospace matters. Dr. Currie, the DOD Director of Defense Research and Engineering, and Dr. Low, the NASA Deputy Administrator, are co-chairmen.

The Space Transporation System Committee is the main joint agency coordinating and planning committee for the Space Shuttle and is co-chaired by Mr. Yardley, NASA Associate Administrator for Manned Space Flight, and Dr. LaBerge, the Assistant Secretary of the Air Force for Research and Development. Under this committee a steering group for Space Transportation System operations has been recently established to formulate joint agency operations and management concepts and to make policy recommendations to Mr. Yardley and Dr. LaBerge. This group is co-chaired by Mr. Bill Schneider, NASA Deputy Associate Administrator for Manned Space Flight, and myself.

On the DOD side of the house, the transitioning of critical military payloads to the Shuttle without loss of operational capability is of paramount importance. For this reason, the Deputy Secretary of Defense established a DOD Space Shuttle User Committee which I chair. It includes representatives from DDR&E, the Joint Chiefs of Staff, and the military departments. NASA is represented by an observer, the Director of Mission and Payload Integration, Mr. Phil Culbertson. The committee provides high-level focus for the broad user interest within the DOD and assures additional planning emphasis on DOD mission applications and transition plans. Using inputs from the various DOD program offices, the committee has developed an approved DOD Mission Model which identifies DOD payloads and their initial Shuttle launch dates in the period 1980 through 1991.

For those of you who may not be familiar with the Air Force management organization on the Shuttle, let me briefly discuss this topic. As I mentioned previously, the Air Force is DOD's executive agent on the Shuttle, with primary staff responsibility residing in my office, the Directorate of Space, through the Shuttle Program Element Monitor, Lt. Col. Chuck Tringali. Air Force Systems Command has been designated as the implementing command for the DOD Shuttle Program, with a System Program Office established under Lt. Gen. Schultz of the Space and Missile Systems Organization in Los Angeles. The SAMSO Deputy for Launch Vehicles, Col. George Murphy, manages the Air Force program activities. You will have an opportunity to meet members of the Air Force Shuttle management team in the technical sessions during the next few days.

In order to continue our close cooperation with NASA at all operating levels, Air Force SPO Representatives are co-located with the NASA Centers at Johnson and Kennedy Space Centers. They participate directly in various Shuttle Design Reviews and planning meetings. In addition, within the past month the Air Force has begun to detail 28 officers for three year periods to work directly for NASA supervisors on various aspects of the Shuttle Program. We expect this arrangement to be beneficial to both agencies, not only for the experience these Air Force personnel will bring to the NASA efforts, but also the valuable experience they will gain in working on the Shuttle Program during the early development phase. We expect many of these specially selected officers to form the initial cadre of the Air Force organization required to conduct Flight Operations from Vandenberg.

I would now like to describe briefly the current management and operations concept under discussion for the Space Transportation System, and in doing so, I will briefly summarize the DOD role in the Shuttle program. It's important to remember that NASA is developing the Space Shuttle as a National Space Transportation System; the DOD role in that development effort is one of support -- ensuring that the Shuttle will meet DOD requirements. The Air Force and NASA are currently evaluating options for near term, and far term operations.

The near term is defined as that period from the first operational Shuttle flight at KSC, until after the Shuttle facilities at Vandenberg become operational. In a preferred near-term option, the DOD would make maximum use of the NASA developed KSC launch and JSC orbiter mission control facilities. Orbiter command and control of Shuttle launches of DOD payloads will be exercised from JSC by an integrated flight team under the overall control of an Air Force mission director. Flight crews will be provided by NASA using Air Force pilots detailed to NASA as part of a common NASA-DOD flight crew pool.

In the far term, NASA will be responsible for orbiter launch and recovery operations out of KSC, and the Air Force for like operations at VAFB. Orbiter mission control for NASA missions will be exercised out of JSC for Shuttle launches from either KSC or VAFB. Orbiter mission control for DOD launches will be exercised by the Air Force from an Air Force MCC in the far term. For multi-mission launches carrying both DOD and NASA payloads, mission control responsibility will be based on which agency payload is primary. Overall management responsibility during this far term period is yet to be determined; however, some possibilities for the future are: A joint NASA/DOD Management Partnership, NASA or DOD as the overall manager, or a new agency. It is still too early, however, to predict the outcome of discussions relating to these long range policy decisions on STS management and budgeting. In the meantime, we continue to work very closely with NASA to formulate hardware procurement strategies, develop alternative user pricing policies, and define logistic, operations, and maintenance programs necessary to provide all Shuttle users with a low cost alternative to expendable boosters.

As previously described, the Interim Upper Stage program will be a DOD development, and will be used until the NASA-developed full-capability Space Tug becomes available.

The DOD is committed to acquire the general purpose Shuttle launch and landing facilities at Vandenberg. The planned operational date for this capability is December 1982.

We believe that the program the DOD is pursuing allows us to accomplish an early and orderly transition of all vital DOD payloads to the Shuttle and makes maximum use of common DOD/NASA equipment, logistics and training.

Let's turn now to the two major areas of DOD participation in the Shuttle program: The Interim Upper Stage (IUS), and our Vandenberg Program. First, the Interim Upper Stage. I mentioned before that, although the Shuttle will be operational from KSC in mid-1980, the full capability Space Tug will not be available until 1984 or later. Since almost all of the DOD payloads scheduled for launch from Kennedy on the Shuttle will require an upper stage, the DOD has agreed to develop an Interim Upper Stage (IUS) for use until the Tug is available.

The IUS, as presently envisioned, will be a minimum modification of one of the five existing upper stages. It will be developed to satisfy principally DOD requirements, and will not have an in-orbit servicing capability or will it be reuseable unless there are clear life cycle cost advantages in making it reusable. The intent is to accomplish a minimum cost development program that will allow early transition of DOD payloads. The IUS, will be, of course, available for other payload users.

The IUS concept will be selected from competing designs which have been submitted by the five existing upper stage contractors: Agena - Lockheed Missiles and Space Company; Burner II - Boeing Aerospace Company; Centaur - General Dynamics, Convair Division; Delta - McDonnell-Douglas; and Transtage - Martin Marietta. These five contractors completed study contracts on their respective stages on 30 June. Since the IUS is currently in source selection, with an Air Force announcement of the selected contractor expected shortly, I won't comment further on any source selection aspects of the IUS.

As is the case with the other aspects of the Shuttle Program, the DOD activity on IUS development and the transition of payloads to the Shuttle are keyed to the NASA development schedule. Critical design on the IUS will not commence until after a successful approach and landing test. If delays in the NASA schedule occur, the IUS schedule can be adjusted accordingly.

The other, and largest portion of DOD participation in the Shuttle Program, is the acquisition of Shuttle facilities at Vandenberg. A joint study was conducted by NASA and the Air Force in 1972 which considered the feasibility of a single launch site for the Shuttle. Of the 51 potential locations considered, none proved suitable, primarily because of safety considerations resulting from launch trajectories which would involve the overflight of populated areas. No single site examined offered the range of launch azimuths required to launch into both equatorial and polar orbits without overflight of populated areas or the imposition of severe payload weight reductions in performing avoidance dog legs. The study concluded that the use of two Shuttle launch sites would be required, and from the economic and technical viewpoint, modification of existing launch facilities at Kennedy Space Center and Vandenberg Air Force Base was recommended. At this point the DOD agreed to acquire general purpose Shuttle launch and landing facilities at Vandenberg for use by both agencies. The DOD acquisition will be completed on a schedule that allows utilization of the experience gained from NASA's development, test, and evaluation of the KSC Shuttle launch capability. Much of the Shuttle hardware, software, operating procedures, and ground equipment will be common and interchangeable to the maximum extent possible between KSC and VAFB.

This is not to say, however, that the operational concept at Vandenberg will be identical to that of Kennedy. The NASA Kennedy concept for Shuttle uses an integrate - transfer - launch sequence similar to that used by current Titan launch vehicles. The Air Force Vandenberg baseline uses an integrate-on-pad concept which makes maximum use of existing facilities - in this case the use of Space Launch Complex 6 constructed in the 1960's for the Manned Orbiting Laboratory Program which has since been cancelled. By adopting the integrate-on-pad concept and modifying excess Saturn IB and Saturn IC transporters to move the orbiter and huge external tanks, a new highway and railroad system will not be required for launch operations. Let's discuss our planned launch processing sequence at Vandenberg in more detail.

The first step in the launch process is to stack and align the solid rocket boosters or SRB's, whose segments will arrive at Vandenberg via existing rail facilities. The next step in the launch assembly is the mating of the large external fuel tank to the assembled SRB's. The external tank will arrive at Vandenberg via ocean going barge and most probably off-loaded at a marine facility which will be specially built for this purpose. Next, the orbiter itself is readied and mated to the external tank-SRB combination. After all necessary interface checks are accomplished, the payload will be installed in the payload bay, the propellants will be loaded, the flight crew will enter the orbiter cockpit, and the final launch countdown will begin . Use of this concept and existing Apollo equipment will save over $50 million from previous baselines which used the integrate-transfer-launch mode of operation.

Initially, only a single launch pad will be built, with provisions for a second pad two to four years later if mission traffic or operational considerations warrant. Design definition for these facilities will continue through fiscal year 1977 with the actual construction planned to begin in FY 1979 after review by the DOD decision makers in the fall of 1978. The timing of the construction start will allow any necessary changes in this planning baseline to be made as a result of the NASA Shuttle approach and landing tests that will begin in 1977.

That completes my discussion of the involvement of the DOD in the NASA Shuttle Program; let us turn now to the question of how does the DOD plan to use the Shuttle once it is operational. While not overlooking the tremendous new capabilities offered by the Shuttle, we in the DOD can answer this question by describing a phased approach. The first phase is primarily devoted to minimizing the cost of transitioning, coupled with strong support of the NASA development effort, and leads to the second phase where we expect to exploit all the new capabilities which a manned system affords.

In the near term, we are looking at the Shuttle as primarily a reusable replacement for expendable boosters. We do this for two reasons. In the near term, which for convenience let's say through 1985, we will be transitioning critical and very expensive existing DOD payloads to the Shuttle. As I've mentioned before, the transition of our payloads to the Shuttle, while still maintaining an expendable booster backup capability, is a very complex and expensive task. It involves redesign of these existing payloads to be compatible with the Shuttle, the initiation of phaseout on the procurement of expendable boosters, and the phase out of expendable booster launch complexes. Let's just consider the phaseout of existing booster systems. It currently takes us 42 months from the initiation of procurement action on a Titan booster until the vehicle is actually delivered. In phasing out the Titan booster, we must carefully plan for some expendable booster backup as a hedge against unforeseen Shuttle schedule delays or performance probelms. With a 42 month lead time on Titan procurement, we must plan this exercise with the greatest care to avoid either the lack of sufficient boosters to meet emergency situations or a surplus of boosters which would ultimately be wasted. Almost all the DOD payloads that will be flying on the Shuttle during the transition period already exist or are in advanced development states. **Changing** these existing designs, which are constrained by launch on expendable boosters, to take advantage of the Shuttle's relaxed weight and volume capabilities or capability for on-orbit retrieval or repair, would be prohibitively expensive. On the other hand, we do plan to make maximum use of the Shuttle in its "Cargo Hauling" sense. Since all the payloads we'll be flying in the early years of the Shuttle will not take up the entire payload bay because they are all designed to be flown on existing boosters, the DOD and NASA are actively investigating multiple payload launches to decrease the number of Shuttle flights required. In our joint investigations we are considering combined launches of two or more like communications satellites, as well as mixing payloads betwen programs and between agencies. Let me develop this latter point, because it is important to our current thinking on the Shuttle. Right now we're looking at the Shuttle in the early years in much the same manner as a loadmaster looks at one of today's jumbo cargo jets. It's relatively rare when one shipper has enough cargo to fill one of these planes, so other compatible cargo, and this is the key of course, is loaded with the prime shipment to make the flight economically attractive. In a similar way, we plan on making maximum use of the Shuttle's capability to place twice the cargo weight and roughly three times the cargo volume into low Earth orbit on a single launch than is possible today. I mentioned cargo compatibility before as the key to this exercise. To be compatible, cargo or, in this case, payloads, must satisfy the following criteria: First of all, they must be going to approximately the same destination. In the case of satellites, destination might be more appropriately termed orbital inclination (the angle between the plane of the satellite's path and the earth's equator). Inclination must closely match because of the relatively large amounts of energy required to make even small plane changes. Then, there are the physical constraints of the orbiter payload bay itself. Since existing DOD payloads are all designed to be flown on current expendable boosters, they vary in diameter and length, but in only a few cases do they completely fill the payload bay. The weight of some payloads, however, combined with their length and orbit requirements, will preclude sharing with other payloads. There may be cases, also, where DOD payloads will be able to share the Shuttle payload bay with other payloads. It is in these cases where we are trying to determine the ultimate benefit. Our planning is not parochial. That is, we are not prohibiting DOD payloads from mixing with NASA payloads or the payloads of commercial users, for that matter. NASA, of course, will be the mission agent for payload suppliers from the civil, commercial, and international fields. Right now we believe that about 40% of the DOD payloads are inherently capable of being mixed.

I'm sure you're asking yourself - what about security - the classification of DOD payloads? In our planning we hope to use the same concept we now use in placing classified material in the U.S. mails where the material is properly protected from the disclosure of classified information to unauthorized persons by the registered mail system. The postman does not have a security clearance. All he knows is that he's carrying a registered letter or package. We hope to be able to utilize this same concept with DOD payloads, which brings us to how we plan on structuring the interface between DOD payloads and the Shuttle orbiter itself.

We plan to keep this interface as simple as possible, not only from the security aspect, but also to minimize the impact to DOD payloads. The interface today between DOD payloads and their expendable boosters is minimal. We hope to continue this interface in the early Shuttle era, again to avoid the extremely costly spacecraft redesign that a more extensive interface would require. In our current thinking, the interface between the orbiter and the payload would be limited to monitoring payload telemetry which affects the safety of the crew only and limiting commands to those required to safe the payload in event of abort. Payload checkout, in one plan under investigation, will be accomplished after deployment from the orbiter bay, using the existing DOD ground stations. The orbiter will stand by and stationkeep at a safe distance after release while checkout occurs. If a problem is detected, the orbiter may be asked to assist in resolution, or possibly retrieve the payload for return. I should emphasize that this is our near-term planning baseline and more detailed analysis may change some current ideas.

Now, what about the future -- the far term -- that period of time after the initial transition period. Lest we be accused of myopic thinking, I wish to assure you that the DOD is actively investigating using the unique capabilities of the Shuttle to revolutionize current thinking on spacecraft design. A top level panel, the Defense Science Board DOD Space Shuttle Utilization Task Force was established last fall under the Air Force Undersecretary, Mr. Plummer. Their charter is to examine how the Space Shuttle with its new capabilities can lead to more effective military space operations in the future. The task force was charged to review all past and present studies for Shuttle utilization, suggest innovative spacecraft designs for Shuttle use, and consider entirely new space system operating concepts which may be practical and desirable with the Shuttle. The membership of the task force is composed of creative, experienced national leaders in spacecraft design well acquainted with current military space programs. I would like to mention these gentleman in recognition of their contributions: Mr. Sam Araki from Lockheed; Mr. Jack Bell, Rockwell International; Mr. Bill Hamilton, Boeing; Mr. Charles Hunt, General Electric; Mr. Fred Kaufman, TRW; Mr. Ed Offenhartz, Grumman; and Dr. Hal Rosen, Hughes. In addition to Mr. Plummer, the DOD members are the Air Force Chief Scientist, formerly Dr. Mike Yarymovych who recently left us -- his place being taken by Dr. Bob Naka; and Dr. Bob Cooper from the Space and Advanced Systems Division of the Office of the Director of Defense Research and Engineering. **Undersecretary** Plummer has encouraged and supported many innovative ideas which were originated or amplified by his study group. For example, his committee proposed that a Titan booster with the IUS and a shroud the same diameter as the Shuttle payload bay could be made available prior to the Shuttle IOC as a means of releasing constraints on the design of new spacecraft such as the third generation vehicle for the Defense Communication Satellite System. Other near-term recommendations stressed the importance of the loadmaster concept and the simplification of the orbiter/payload interface and development requirements.

In the far term, the task force recommended using the Shuttle capability to retrieve high value spacecraft and return them to the earth for refurbishment. It also suggested the design of cheaper, shorter life satellites specifically designed for in-space servicing using the Shuttle and full capability upper stage. Another suggestion was the use of the European developed Spacelab for DOD research and development projects which require earth orbit. Discussed also is perhaps one of the biggest future payoffs of the Shuttle - its unique capability to deploy large structures, and then use its manned capability in the assembly of even larger structures. With this capability, new worlds are opened up. No longer are designers constrained by satellites which fit into relatively small confines. Huge structures can be assembled in the weightlessness of space -- structures using solar arrays or nuclear fuel sources with power outputs many, many times that which is available today. The application of this technique to military defense systems is obvious and will receive very thorough study. In addition, the pursuit of space technology will be given a "shot-in-the-arm" with the advent of numerous opportunities to make "space available" flights on partially loaded Shuttle launches. The Air Force Space and Missile Systems Organization is currently planning on developing a standard satellite which would provide the bus to be used by space technology experiments supported by the DOD Space Test Program. The Air Force has also completed a long-range planning study, called New Horizons II, which considers the role of the Shuttle in the future of the Air Force in space. This study was initiated at the request of the Air Force Chief of Staff, General Jones, and highlights the growing need for a more comprehensive definition of the role of space in future Air Force activities. The space portion of New Horizons II projected the likely thrust of space operations of the 1990's and provides the Air Force with a perspective of the value of space missions in the future. Specific mission and organizational recommendations for future Air Force use of the Shuttle were included.

New Horizons II is being paralleled by a study being conducted by NASA called "Outlook for Space." This study, which was headed by Mr. Donald Hearth of NASA Goddard reported to the NASA Administrator only a few weeks ago. It identified sixty-one desirable and practical civilian space objectives that could be carried out by the United States in the 1980-2000 time period, many of which are only feasible with the advent of the Shuttle.

In addition to these studies, a special Air Force study is being conducted by Col. Joe Stegall of Air Force Systems Command with an objective similar to that of the NASA "Outlook for Space" study.

I appreciate very much the opportunity you have provided me by your invitation to participate in your meeting today and I'd like to conclude my remarks with the quote from Oliver Wendell Holmes that the President used in announcing his decision to proceed with the Shuttle: "We must sail sometimes with the wind and sometimes against it, but we must sail, and not drift, nor lie at anchor." When the President spoke those words over five and a half years ago, we didn't know which way the wind would blow on the Shuttle program. Today it appears that we're running with the wind, and on our way to a goal by the end of this decade every bit as climactic as the goal we achieved in the last decade.

I hope my remarks have enlightened you on the DOD role on this voyage. We may not be steering the boat, but we're not just along for the ride either.

AAS 75-124

SPACELAB PROGRAMME - STATUS REVIEW

Bernard Deloffre*

Spacelab will probably be the first operational payload to be carried by the Space Shuttle, and it is anticipated that perhaps 30% of all Shuttle flights will employ Spacelab. This system, which will provide laboratory and observatory facilities on orbit, is currently being developed in Europe under the management of ESA acting on behalf of its member-States who provide the development funding. This paper presents a management overview of the Spacelab Programme and reports on recent progress achieved. The standing of Spacelab in the context of the overall Space Transportation System is first explained and the chosen concept and its mission flexibility is described. Funding, management, and scheduling aspects of the programme are then discussed and the paper concludes with some comments on European-related aspects of Spacelab operations.

INTRODUCTION

A particularly attractive feature of the Space Shuttle orbiter is its ability to remain on-orbit for periods up to 30 days. On such missions, the inherent capability of the orbiter to provide life support and living quarters are complemented by Spacelab which will furnish a working environment for the payload specialists. Spacelab, in fact, provides for a "shirt sleeve" environment laboratory (the module), and an unpressurised space observation platform (the pallet). The Spacelab concept will permit scientists and engineers to accompany

_____

* Director, Spacelab Programme, ESA, Neuilly-sur-Seine, France.

their experiments into space. It will also permit direct access to space for observational facilities that may be controlled from the orbiter cabin, from the Spacelab module, or from the ground.

Spacelab is therefore, a very particular cargo of the orbiter, and must be regarded as an integral part of the Space Transportation System. It is being developed to meet, on the one hand, the interface requirements imposed by the Shuttle, and, on the other, the needs of a varied user community. This task must be performed within a fixed financial envelope.

The Spacelab Programme entered phase C/D just over one year ago. In essence the programme remains as described in Ref. 1. However, certain refinements have been introduced since development go-ahead, and it is the intent of this paper to present a management overview of the current programme status.

BACKGROUND AND HISTORY

European participation in NASA's Space Transportation System formally commenced in mid-1972 with the signing of the Memorandum of Understanding between NASA and ESA (then ESRO). This act constituted a firm European commitment to the Spacelab project. After competitive definition studies and tender actions, the main hardware development contract was awarded to an industrial team led by the German company VFW-Fokker/ERNO as prime contractor. This sequence of events together with the two principal programme reviews (PRR Preliminary Requirements Review, SRR Subsystem Requirements Review) that have been conducted to date are depicted in Fig. 1. The two afore-mentioned Reviews have served to ensure that the Spacelab design is based on requirements that truly represent the needs of ESA and NASA. The resulting requirements baseline will serve as a firm foundation for the detailed design that is to be finalised over the next few years.

PROGRAMME REQUIREMENTS AND CONSTRAINTS

Spacelab was conceived, and is being developed, to meet the requirements of a user community drawn from many disciplines in the fields of science, applications, and technology. In addition, it was deemed essential that the Spacelab design and eventual operation be based on the same low cost principles as the Space Shuttle itself and should use to the full the benefits promised by the Shuttle. Among the latter, the most important tenets for Spacelab are: low cost transportation to and from orbit, reusability, short turn-around time, and relaxed design constraints arising from the relatively large cargo volume and weight capability of the Shuttle.

With these goals in mind, together with the recognised needs of the users, and the known European motivation for involvement in an advanced manned space programme, the major programme requirements given in Fig. 2 were formulated.

Compatibility with the Shuttle has certain advantages and disadvantages for Spacelab. It can draw on the many resources available in the orbiter as explained later, but it also has to conform to certain constraints imposed by the orbiter geometry and operating performance. Some of the more important constraints are summarised in Fig. 3. Most significantly, the Spacelab and its payload are influenced by the landing weight limitation and the CG constraint. The former calls for a light-weight structural approach and eventually limits Spacelab's payload capability, and the latter results in Spacelab being placed well aft in the cargo bay and the resulting requirement for a long tunnel for crew access between the orbiter cabin and Spacelab.

Additional factors currently influencing the management of the Spacelab Programme arise from funding aspects of the final hardware and its operation. Hence, one can cite - cost-to-completion, production cost, committed delivery dates and deliverables, operational cost - as important considerations that must be, and are being, considered in the conduct of the overall programme.

SPACELAB CONCEPT

Spacelab consists of two basic elements - a pressurised module and an unpressurised pallet - which can be used separately or in combination. The module is composed of two identical cylindrical shells enclosed by end cones. One of these shells contains a core of basic subsystems but also provides for several cubic meters of rack installed experiment equipment. The additional shell ensures room for additional experiment equipment if needed. The pallet can be made up of as many as five segments, each of 3 m length. They are of aeronautical shell-type construction and may be individually attached to the orbiter or mounted in series (of up to 3 pallet segments) when they are referred to as a pallet train. Fig. 4 illustrates the principal external features of the Spacelab module and pallet and indicates some design changes from the original phase C/D concept that are now in progress. A view of Spacelab in the cargo bay of the orbiter is given in Fig. 5.

The Spacelab subsystems provide basic services for running Spacelab itself and for the payload. As a result of cost/performance trade studies, a mix of orbiter-provided and Spacelab-provided resources are used. Orbiter support to Spacelab is summarised in Fig. 6. It must be borne in mind that even though a raw resource is provided by the orbiter, the Spacelab complement of equipment ensures that the resource is conditioned so that it is available in a form compatible with its needs and those of the experiment equipment.

The total Spacelab capability in terms of payload weight and volume, and services for the user, are summarised in Fig. 7 for four basic Spacelab configurations.

MISSION FLEXIBILITY

To ensure that Spacelab can be adapted to a wide spectrum of possible experiments and missions, a modular approach has been employed in the Spacelab design. Both module and pallet can be varied in size and length, as illustrated in Fig. 8. Additionally certain common payload support equipment, such as airlocks and viewports, that are indicated in the figure, may be flown if the experimenter so desires. Further, certain elements of the subsystems, such as racks and cold plates, have been identified as mission-dependent equipment and are carried only if needed by the mission. Additionally, it is likely that an Instrument Pointing Subsystem (IPS) providing $\sim$ arc sec pointing accuracy for payloads up to 2000 kg (nominal), will be available. Using this modularity of design, it is possible to select a total Spacelab system that best fits the needs of a particular mission and maximises the weight available for the payload. The configurations shown at the head of Fig. 7 are considered basic but, employing the techniques explained above, many more variations are possible.

Considerable work has been done by ESA and NASA on the accommodation of various typical payloads by Spacelab. The results in terms of services needed, crew size, volume and weight requirements and special needs (such as airlocks or windows) dictate the configuration to be used. Some typical results for six disciplines are illustrated in Fig. 9, where various combinations of module only, module plus pallet, and pallet only modes have been exercised.

A detailed account of possible Spacelab configurations, the available mission-dependent and common payload support equipment, and the corresponding performance data, is contained in the Spacelab Payload Accommodation Handbook (Ref.2).

PROGRAMME FUNDING

The funding envelope established for the Spacelab Programme in Europe is 380 million Accounting Units. The Accounting Unit (AU) is a monetary quantity used by ESA for international financing and, based on current official rates of exchange - 1 AU = $ 1.302 US. This cost envelope covers all development activities and the delivery to NASA of one engineering model and one flight unit. Subsequent Spacelabs will be procured by NASA in Europe. In fact, a firm order for a second Spacelab has already been placed, and additional orders are expected. The funding quoted above is provided by 10 ESA member States on the percentage basis shown in Fig. 10. ESA, acting on behalf of its member States is charged with the management responsibility for the project. The agreed financial envelope must not be exceeded, and, although a large margin is provided for technical changes and modifications that might arise from Shuttle or Spacelab variances, careful planning and use of the funds available for project completion is essential.

PROGRAMME MANAGEMENT

Although Spacelab is being developed and built in Europe, NASA will be responsible for its operation in conjunction with the Space Shuttle, and the US will provide the largest percentage of users. Thus, there are many areas of joint ESA-NASA interest, and Spacelab represents a truly international venture. The cooperative nature of the programme is evidenced by NASA involvement in the management at various levels from programme direction down to subsystem design.

The principles of the ESA-NASA agreement are contained in the Memorandum of Understanding between the two agencies. These principles are interpreted and expanded through a jointly-approved Spacelab Programme Requirements document and a Joint Spacelab Programme Plan. The key coordinating body that handles items and procedures of mutual interest is the Joint Spacelab Working Group (Fig. 11). Advice and recommendations on the needs of the user community is provided by another jointly convened committee - the Joint Users Requirements Group. Whereas JSLWG contains key members of the ESA and NASA management teams, JURG consists of representatives from all science, applications, and technology disciplines from Europe and the US.

A further example of the close ESA/NASA relationship on this programme is the planning for the first Spacelab flight. A joint planning group has already established the mission objectives, and the realisation of these objectives, in terms of choice of experiments and the execution of the flight itself will be very closely coordinated.

In addition to the trans-Atlantic features of the Spacelab Programme, the more conventional ESA-contractor arrangements must be managed. To carry out this function, the organisation shown in Fig. 12 has been established within the overall ESA structure. For simplicity, only the major elements of direct interest to the Spacelab development programme are shown in the figure. A Spacelab Programme Board, composed of representatives from the participating member-States, has been established, which advises the Spacelab Programme Director on policy and budget aspects of his duties. A low-cost approach to the programme execution has been fully implemented, and formal procedures for cost control, configuration and change management, and product assurance have been established.

As regards the status of the programme, system and subsystem design has been underway for about one year. System design activities have essentially been completed (see Fig. 4) and subsystem design is proceeding towards a state of accomplishment that will permit a full evaluation at the preliminary design reviews scheduled to start early in 1976. Some bread boarding of subsystem elements has already been initiated, for example, critical electronic circuitry for the data bus, input/output units of the command and data management subsystem, and the power bus of the electrical distribution subsystem. It is also worth noting that the computer that will be used for ground support activity of Spacelab has now entered pre-production testing. Metal for the main structure has been cut, and a full scale configurational mock-up has been manufactured (see Fig. 13). This mock-up has been used for for habitability and lay-out evaluations and as an aid to the Subsystem Requirements Review.

Certain major decisions affecting the Spacelab design have been taken during the past year. These are summarised in Fig. 14. A major cost item is the implementation of a remote control philosophy for Spacelab that will permit activation and control of subsystems when access to the Spacelab is not possible. Of particular interest, is the decision by NASA to permit emplacement of the airlock above the access tunnel during Spacelab missions, so that continuous access is possible to Spacelab even during extra vehicular activities (EVA). Another important action that affects both ESA and NASA is the agreement recently reached on the allocation of weights and reserves among the Spacelab subsystems, the tunnel and orbiter-support equipment. This agreement will go a long way towards guaranteeing payload weight capabilities and margins, and will permit realistic weight controls to be enacted.

Of course, a number of decisions remain to be taken that will influence the eventual Spacelab design. Some of the more important ones are listed in Fig. 15. Resolution of these issues will be made with the full knowledge of the needs of the user community, the design complexity involved, and the prevailing financial environment.

PROGRAMME SCHEDULE

In Fig. 1 a brief history of the Spacelab Programme in Europe was given. It is now relevant to review what is foreseen for the important next five years leading to the first flight. A tentative programme schedule of milestone events and actions is depicted in Fig. 16. It is likely that Spacelab will be flown on the first operational flight of the Space Shuttle and delivery of the flight unit to NASA will be one year earlier (mid-1979). A fully functional engineering model will be provided to NASA by mid-1978 so that compatibility testing, experiment integration and crew training may be carried out prior to the arrival of the flight unit. A system of formal reviews of requirements definition and design progress has been initiated that will permit ESA and NASA management to evaluate contractor achievement against the planned performance. The principal reviews are indicated as milestones in Fig. 16, but it must be realised that, in some cases, reviews pertinent to particular subsystems may be carried out over a given period but at different times. The review names are self-explanatory. The CDQR is an outstanding milestone since it formally establishes the final production baseline for the Spacelab system flight and ground hardware, and related software. It occurs towards the end of the qualification test phase of the integrated engineering model.

Overall, the schedule exhibits parallel subsystem and system development and makes maximum use of early test data obtained at subsystem level. Although flight verification is foreseen as a partial objective of the first Spacelab flight, no full scale flight testing of Spacelab will be required. This cost effective approach to development has been approved by ESA and NASA as being responsive to meeting the required delivery dates and assuring the necessary high standards of product assurance and flight safety.

Among the responsibilities listed for ESA in the Memorandum of Understanding is that for the delivery of certain Spacelab hardware and support engineering as summarised in Fig. 17. In addition, ESA will ensure that a follow-on capability is preserved in Europe for subsequent flight units, components, and spares. This will ensure that the future procurement needs of the US are fully satisfiable.

SPACELAB PAYLOAD OPERATIONS

A typical operations cycle for the Shuttle/Spacelab system is illustrated in Fig. 18. The profile is repeated from flight to flight, but with a different Spacelab payload complement. It is stressed that the overall responsibility for these operational procedures rests with NASA. However, ESA and the European users have important roles to play when European experiments are included in the payload. This is particularly true for the first Spacelab flight which is being planned jointly by ESA and NASA.

In general terms NASA visualises the following four levels of integration activities associated with Spacelab flights:

- Level IV, intra-experiment integration at the user site.

- Level III, inter-experiment i.e. payload integration at central facility with experiments installed in Spacelab racks (and floor segments as required).

- Level II, integration of payload with Spacelab flight system.

- Level I, installation of Spacelab in Shuttle orbiter.

The level II integration procedure is facilitated by the roll-out design concept adopted for Spacelab and illustrated in Fig. 19. The total payload is contained in the rack and floor combination which is literally rolled into the Spacelab shell by means of a roller-rail system. Fig. 19 also depicts how hardware elements of the Spacelab system might be distributed. Fortunately, convenient air transportation is available to effect a timely integration of the total payload even though it may be composed of elements originating in widely scattered locations, and may often be international in nature. As can be seen, it is anticipated that the basic Spacelab module and its subsystems will always remain at the launch site where refurbishment can be performed speedily.

For the first Spacelab flight, it is visualised that European experiments will undergo all level IV and partial-level III integration in Europe. To ensure efficient integration of these activities, ESA proposes to establish a small technical group to coordinate these activities (SPICE - Spacelab Payload Integration and Coordination in Europe). This group will answer directly to the Director of Spacelab Programme, but budgeting of their activities is not provided for in the previously described cost envelope. The functions of SPICE will be to: plan and coordinate all payload realisation activities, prepare interface specifications and supervise their implementation, approve compatibility and safety aspects of experiments, monitor all payload-related schedules, and coordinate payload specialist training.

Coordination with NASA (see Fig. 20) will be effected through a Joint NASA/ESA Planning and Integration Committee (JPIC), and a payload accommodation team that will be composed of NASA and ESA personnel. The former group is concerned with the planning of the payload and establishes and reviews, from time to time, the mission objectives, while the latter team will evaluate all aspects related to the accommodation of the selected payload hardware.

CONCLUDING REMARKS

Spacelab is now very much an "on-going project" in Europe with a total of approximately 1400 persons - at ESA and in industry- actively involved. Peak manpower and funding is foreseen for 1976 when the development programme funding level will reach approximately 90 MAU. The concept has been fixed and detailed design is well underway with actual hardware already being generated. Much remains to be done, but it is expected that any forthcoming technical, financial, or schedule challenges will be met and solved in the same closely controlled cooperation as those of the past. ESA and its member-States welcome this opportunity to participate in an advanced technological development and look forward to its successful completion. An active and fruitful participation in the operational phase of Spacelab is also enthusiastically anticipated, thus ensuring active involvement of Europe in future manned Spaceflight.

REFERENCES

1. D.J. Shapland "SPACELAB - Europe's Orbital Laboratory", Eleventh Handley Page Memorial Lecture, Aerospace, vol. 1, No 5, Dec 1974, pp 10-21.

2. Spacelab Payload Accommodation Handbook, ESA Report No SLP/2104, May, 1975 (Periodically updated).

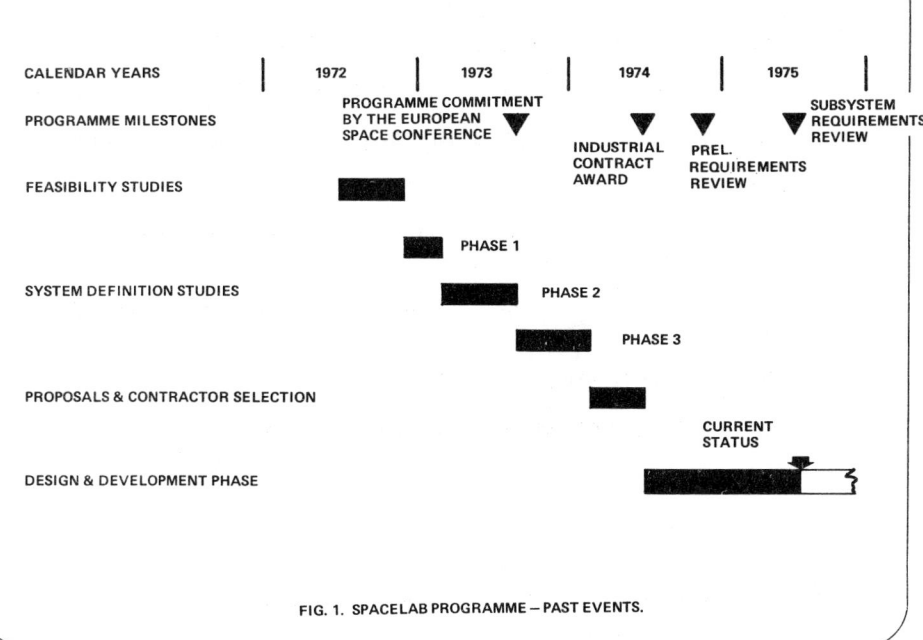

FIG. 1. SPACELAB PROGRAMME — PAST EVENTS.

- PRE-DETERMINED FUNDING CEILING
- DELIVERY OF FLIGHT UNIT EARLY-1979, ENGINEERING MODEL EARLY-1978 TOGETHER WITH ASSOCIATED GROUND SUPPORT EQUIPMENT
- LOW OPERATIONS COSTS TO BE ENSURED
- USER FLEXIBILITY TO BE PRESERVED
- EXPERIMENT PAYLOAD WEIGHT 5000 TO 9000 KG
- PROVISION FOR FOLLOW-ON PRODUCTION
- FLIGHT DURATION 7 TO 30 DAYS
- DESIGN LIFE 50 REUSES OR 10 YEAR LIFETIME
- CREW OF 1 TO 4 PAYLOAD SPECIALISTS
- COMPATIBILITY WITH SPACE SHUTTLE

Fig. 2. Spacelab Programme Requirements

- OVERALL DIMENSIONS GOVERNED BY SIZE OF CARGO BAY (18.3 M x 4.6 M DIA)
- COMPATIBILITY OF SPACELAB ATTACH FITTINGS WITH ORBITER ATTACHMENT PROVISIONS
- AVIONICS, FLUID, AND OPERATIONAL INTERFACES TO BE MAINTAINED WITH ORBITER
- MAXIMUM ALLOWABLE LANDING WEIGHT CAPABILITY (14,500 KG)
- CG OF SPACELAB AND ITS PAYLOAD MUST BE MAINTAINED WITHIN SPECIFIED ENVELOPE FOR ALL 3-DIMENSIONAL AXES
- ACCESSIBILITY TO AND SUPPORT OF SPACELAB PAYLOADS TO BE ENSURED AFTER MATING WITH ORBITER
- CONTAMINATION GENERATED BY ORBITER
- ORIENTATION RESTRICTIONS ON VIEWING ANGLE AND TIME

Fig. 3. Spacelab Design Constraints Introduced by Space Shuttle

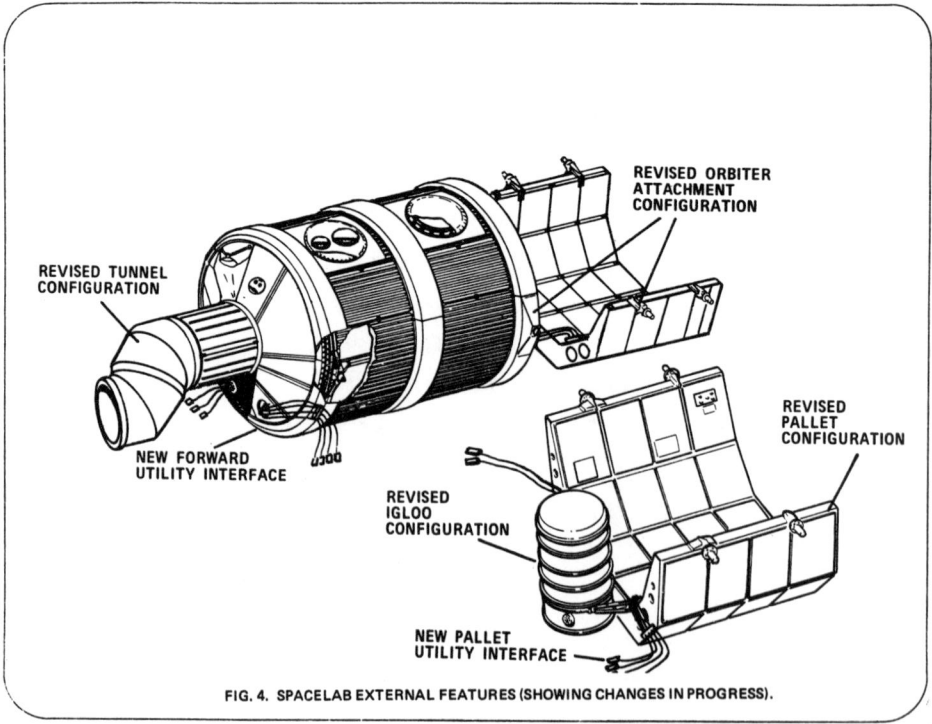

FIG. 4. SPACELAB EXTERNAL FEATURES (SHOWING CHANGES IN PROGRESS).

FIG. 5. SPACELAB IN ORBITER CARGO BAY.

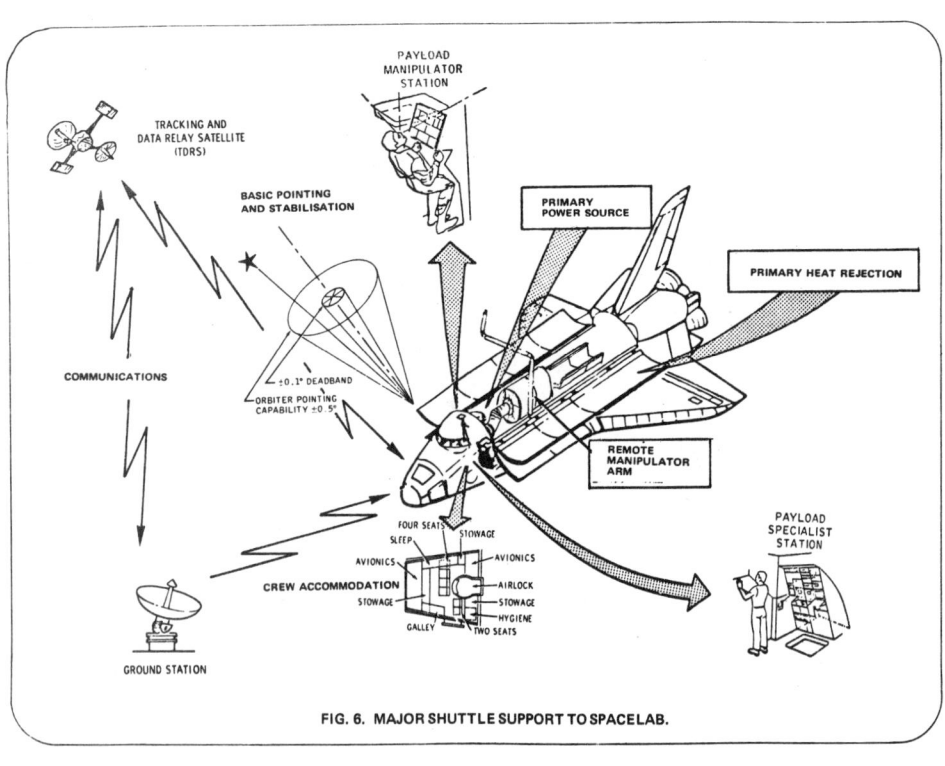

FIG. 6. MAJOR SHUTTLE SUPPORT TO SPACELAB.

SPACELAB CONFIGURATION AVAILABLE TO USERS	SHORT MODULE + 9 METER PALLET	LONG MODULE	15 METER PALLET	INDEPENDENTLY SUSPENDENT PALLET
• PAYLOAD WEIGHT (KG)	5500	5500	8000	9100
• VOLUME FOR EXPERIMENT EQUIPMENT				
PRESSURIZED (M³)	7.6	22.2	–	–
NON-PRESSURIZED (M³) (NO OVERHANG)	99.8	–	167.4	97.1
• PALLET MOUNTING AREA (M²)	51.3	–	85.7	53.6
• ELECTRICAL POWER (28 VOLTS DC/ 115/200 VOLTS AT 400 HZ AC)				
AVERAGE (KW)	3 – 4	3 – 4	4 – 5	4 – 5
PEAK (KW)	8.8	9.0	10.0	10.0
ENERGY (KWH)	> 400	> 400	> 600	> 600
• EXP. SUPPORT COMPUTER WITH CENTRAL PROCESSING UNIT	← 64 K CORE MEMORY OF 16 – BIT WORDS →			
	← 320 000 OPERATIONS PER SEC. →			
• DATA HANDLING TRANSMISSION THROUGH ORBITER	← UP TO 50 MBPS →			
STORAGE DIGITAL DATA	← UP TO 30 MBPS →			

FIG. 7. SPACELAB SERVICES FOR USERS.

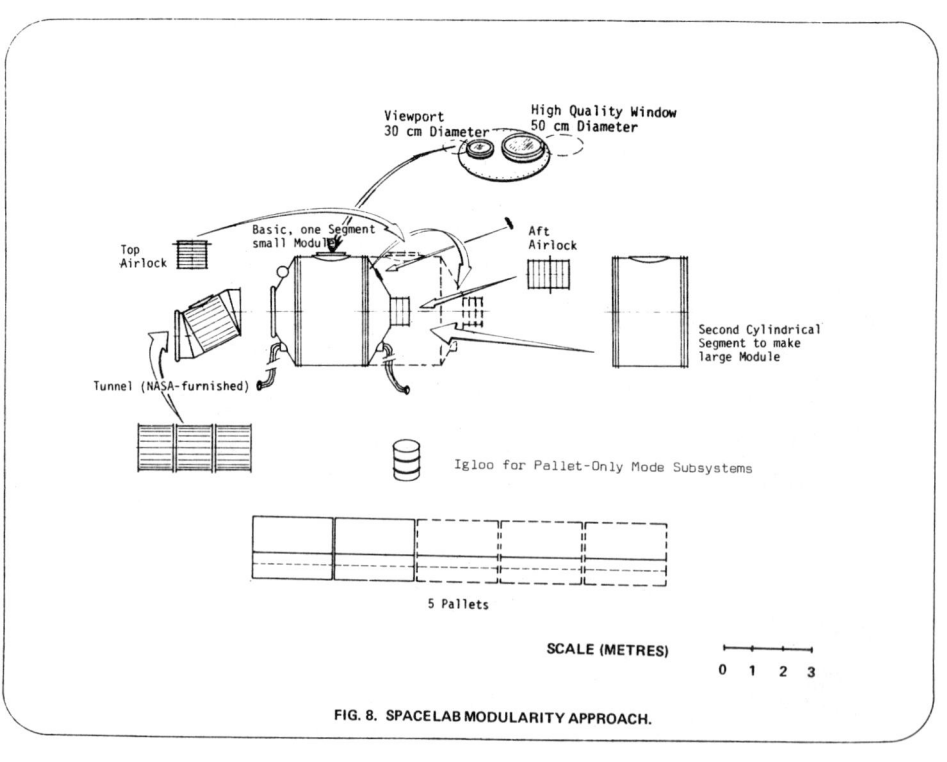

FIG. 8. SPACELAB MODULARITY APPROACH.

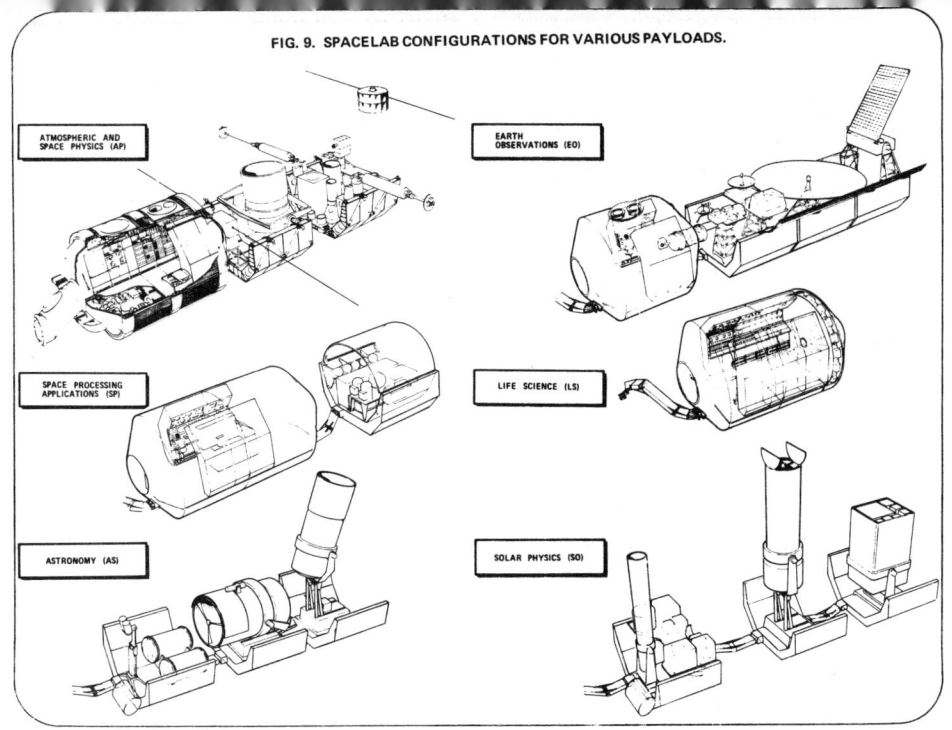

FIG. 9. SPACELAB CONFIGURATIONS FOR VARIOUS PAYLOADS.

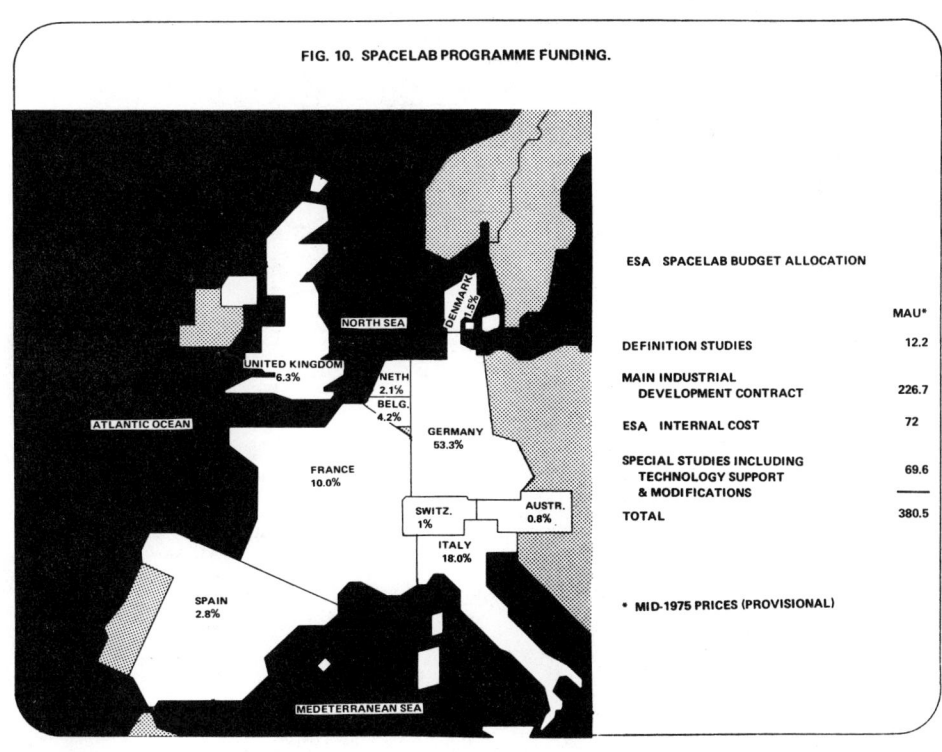

FIG. 10. SPACELAB PROGRAMME FUNDING.

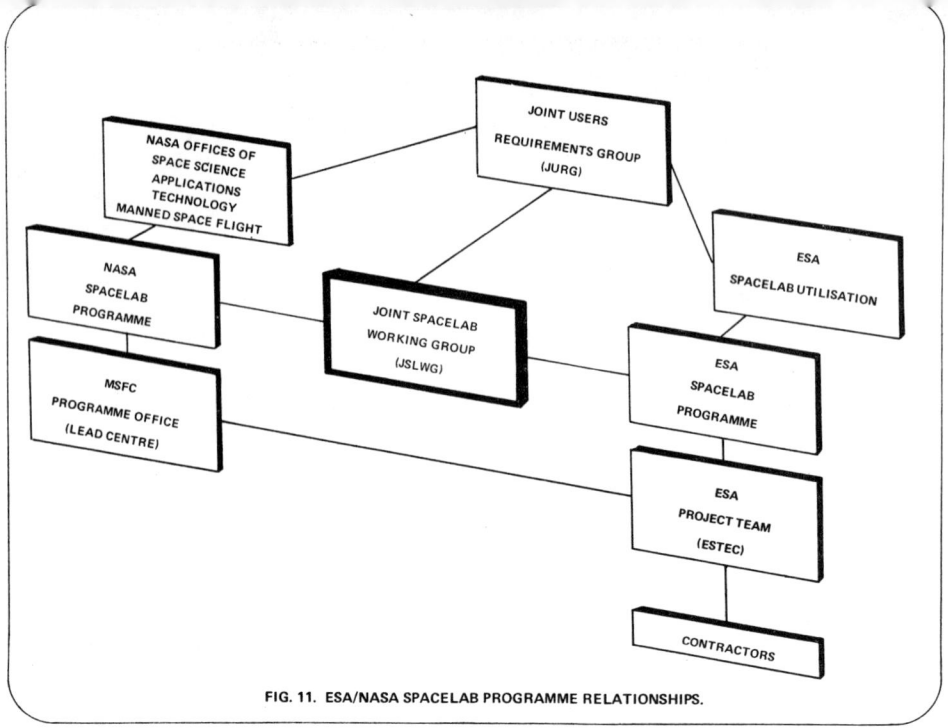

FIG. 11. ESA/NASA SPACELAB PROGRAMME RELATIONSHIPS.

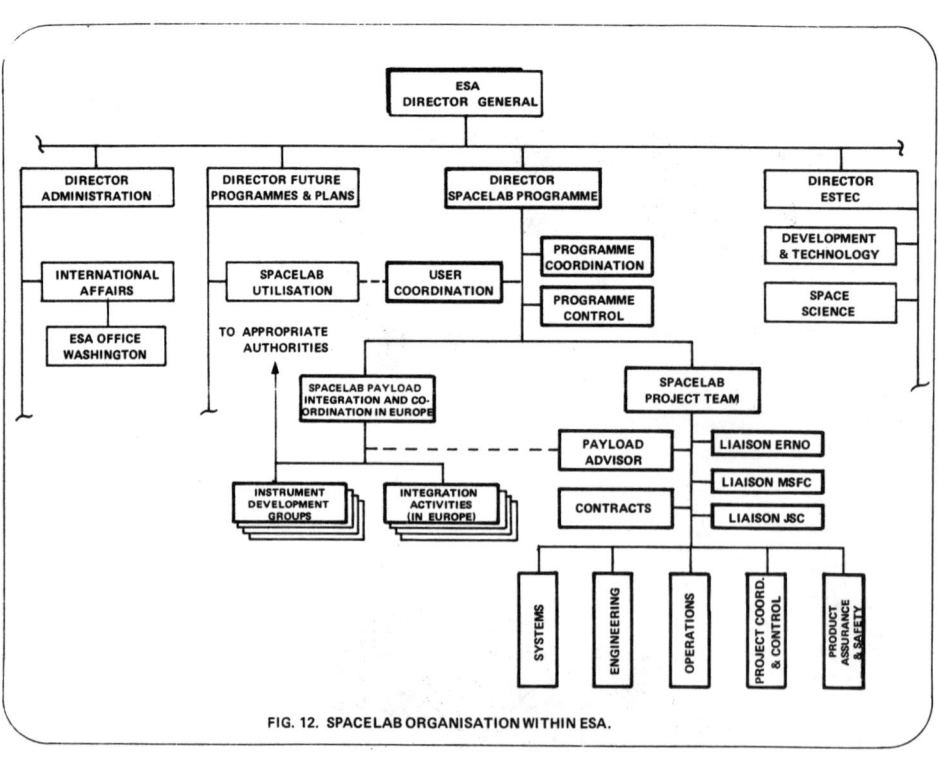

FIG. 12. SPACELAB ORGANISATION WITHIN ESA.

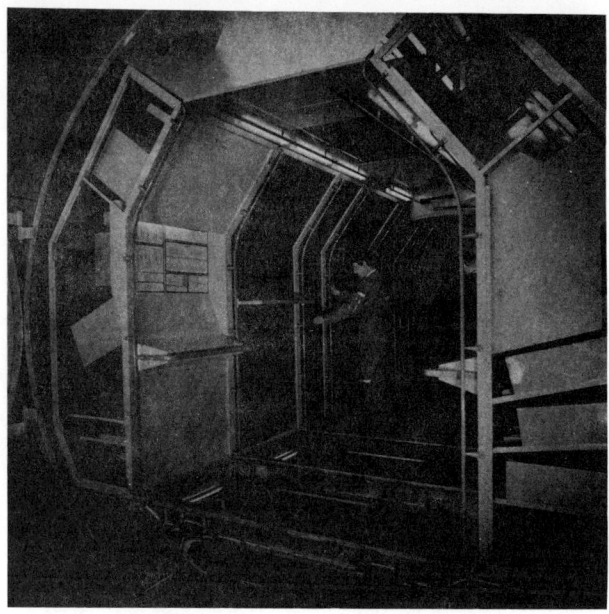

FIG. 13. SPACELAB FULL-SCALE MOCK-UP.

Fig. 14. MAJOR DESIGN DECISIONS

Design decisions have been taken and design actions implemented in the following areas of the Spacelab programme:

- Weight allocation
- Remote control of Spacelab subsystems
- CII computer for flight and ground segments and improved CDMS performance
- Orbiter to Spacelab Power Supply Concept
- Orbiter/Spacelab data transfer concept
- Orbiter/Spacelab caution and warning concept
- Orbiter mechanical attachment concept
- Orbiter airlock and tunnel concept
- Elimination of Spacelab crew transfer forward hatch
- Oxygen supply from orbiter / elimination of Spacelab O_2 bottles
- Contingency air revitalisation via tunnel air duct
- Fill and drain of N_2 bottles of Spacelab on pad
- Relaxation of road transportation constraints for pallet
- Payload dedicated heat exchanger
- IPS concept selected

Fig. 15 MAJOR PENDING DESIGN DECISIONS

The following Spacelab programme issues are currently under study and will be resolved in the near future:

- Extent of Ground Support Equipment
- Late access for life sciences experiments
- Spacelab power and heat rejection resources during Level I integration, prelaunch, ascent, descent and post launch phases
- Spacelab high speed digital data multiplexing and recording
- IPS Development Programme
- Routing of utilities between orbiter and Spacelab
- Peaking battery
- Rack design concept
- Venting facilities for space processing payloads
- Film vault design (size + cooling)
- Removal of H_2O loop in pallet-only mode

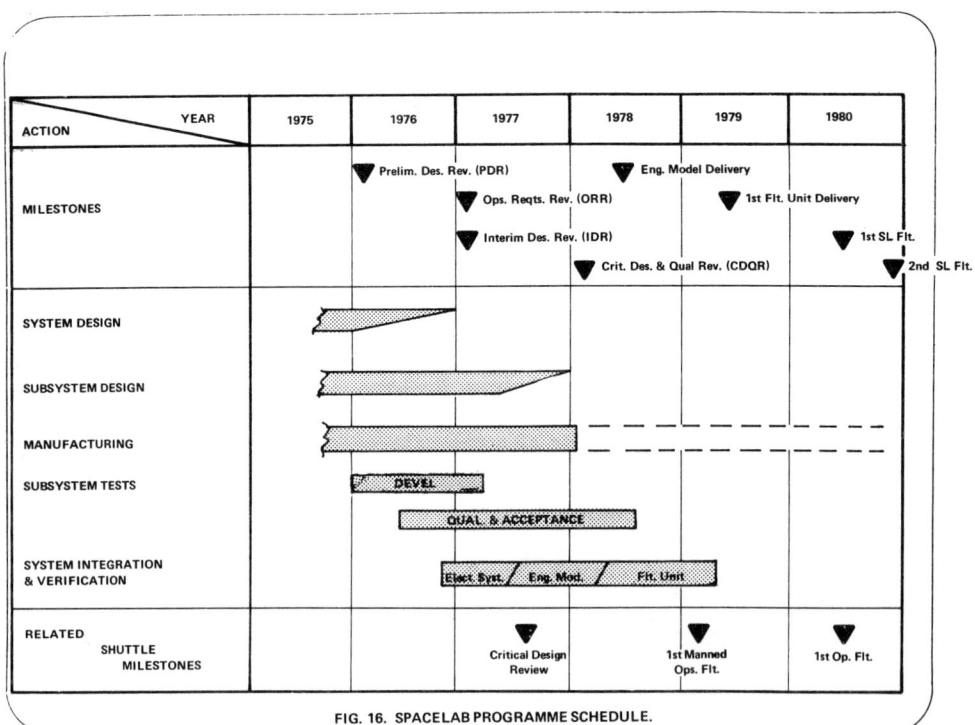

FIG. 16. SPACELAB PROGRAMME SCHEDULE.

Fig. 17 SPACELAB DELIVERABLES

In accordance with the Memorandum of Understanding and to ensure a viable Spacelab operational programme, ESA will deliver to NASA the following items of Spacelab equipment and services:

- One flight unit (basic module, extension module, subsystems, common payload support equipment)
- Engineering model
- Initial spares
- Ground support equipment (2 sets)
- Support software (non-applications)
- Documentation
- Sustaining engineering through first 2 flights
- Preliminary integration of ESA-supported experiments

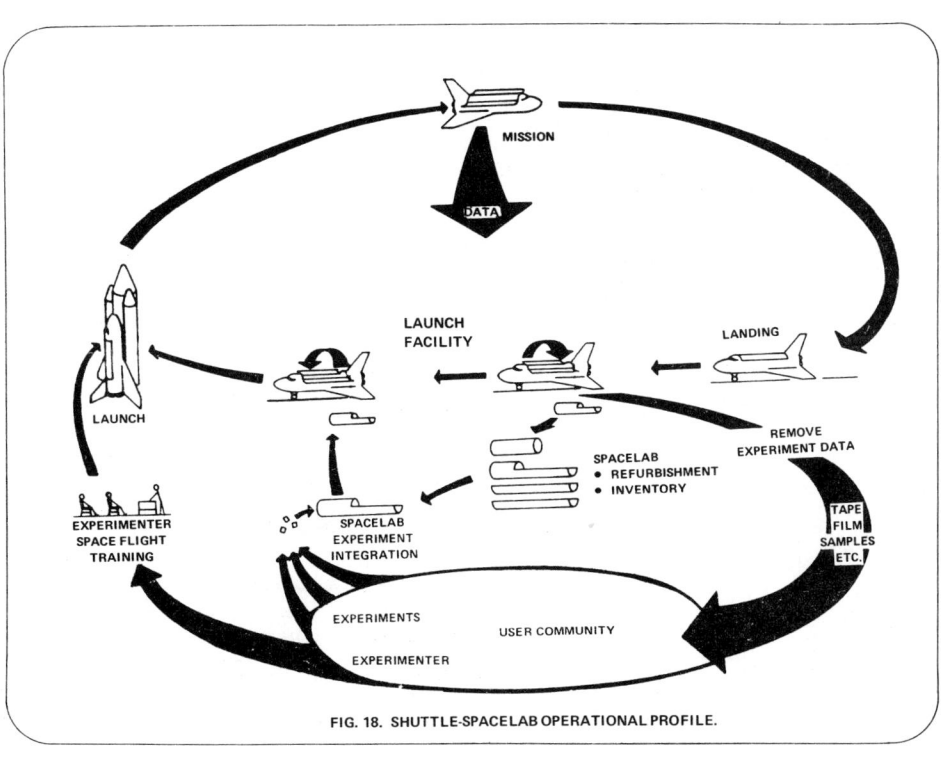

FIG. 18. SHUTTLE-SPACELAB OPERATIONAL PROFILE.

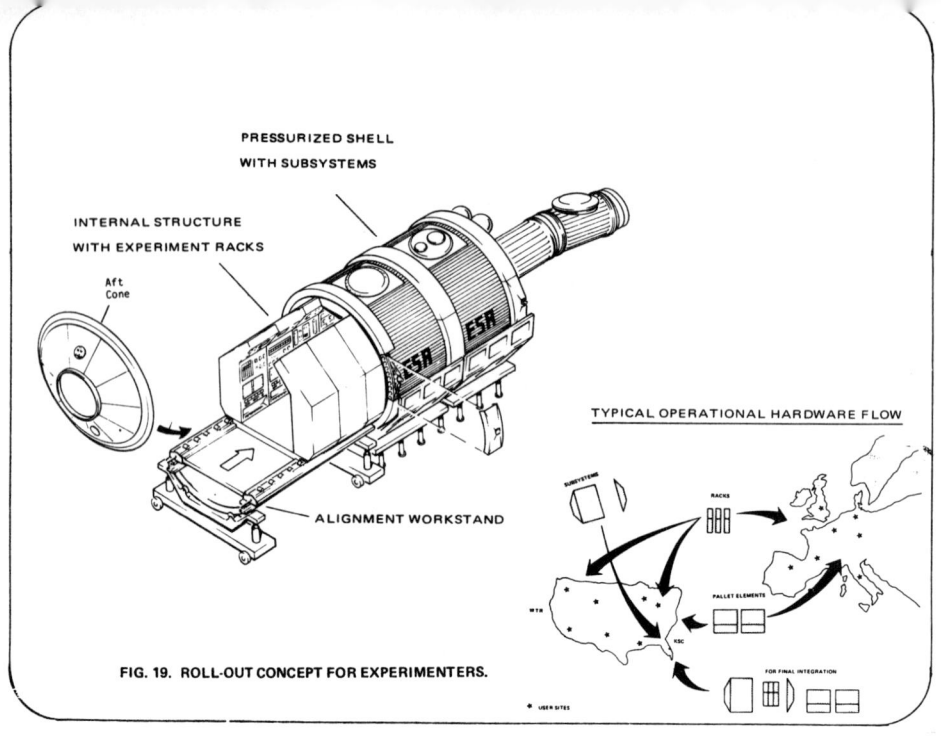

FIG. 19. ROLL-OUT CONCEPT FOR EXPERIMENTERS.

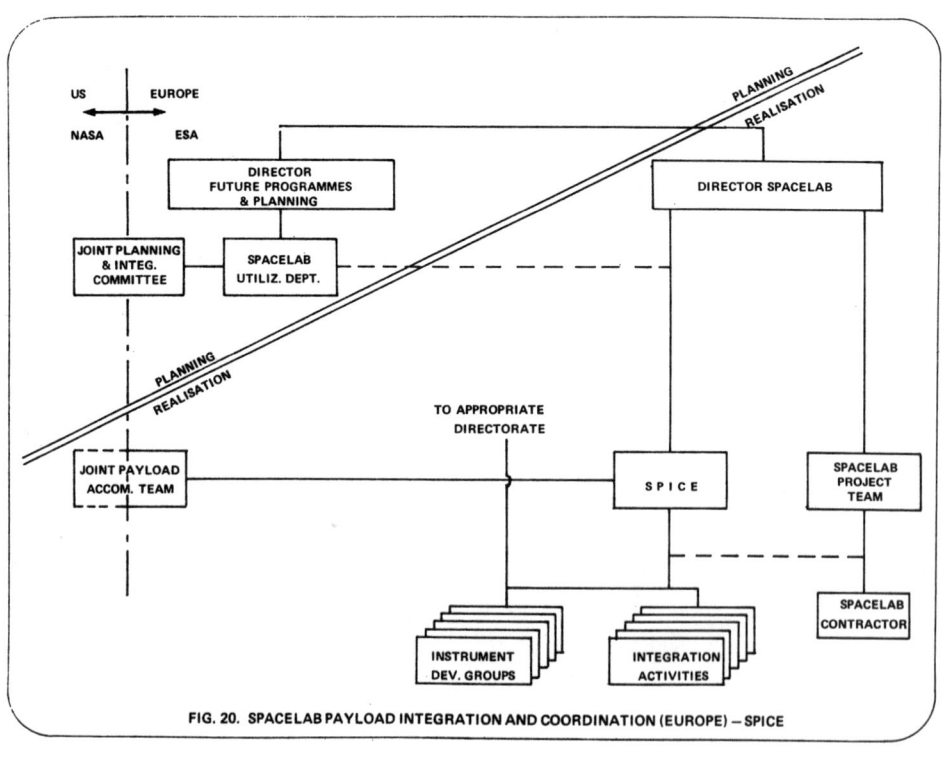

FIG. 20. SPACELAB PAYLOAD INTEGRATION AND COORDINATION (EUROPE) – SPICE

AAS 75-123

SPACELAB PROGRAM - TECHNICAL STATUS

AND SYSTEMS CAPABILITIES

Hans E. W. Hoffmann[+]

From mid '72 to mid '74 in an iterative study process Europe prepared for the development of Spacelab as a contribution to the US Post Apollo Program. In this study process industrial teams were formed, a technical concept for the Spacelab was worked out and the project was determined as a design-to-cost. In June 1974 VFW-Fokker/ERNO was selected to be Primecontractor for the Spacelab Project in Europe, leading a team of industrial firms as cocontractor within the ten European countries, which are financially contributing to this program. The paper will further describe the technical status of the project which has been reached after more than one year of the development and the near completion of the paper phase entering into the hardware phase. Program milestones achieved so far and those program milestones lying ahead will be described on the way to deliver the first hardware in the early part of 1979 to the US. The major technical interface existing between Spacelab and the Shuttle will be described. They are governing the present activities as Spacelab will not detach from the orbiter and thus will be technically dependant in many aspects from the orbiter.

INTRODUCTION

Spacelab represents for the European aerospace industry a unique project. Besides the fact that it is one of the three

---
[+]Spacelab Project Manager, VFW-Fokker/ERNO, Bremen, Germany

major parts of the new European long term space programs it is the only and first manned space project.

The Spacelab Project has been introduced in Europe by a systematic and carefully planned iterative study process, which provided a sound technical and planning basis for the actual development of this system. During this preparation process and up to this very day another unique feature of this program had to be observed: Spacelab in Europe is a design-to-cost program. The ten European countries, which are financially supporting Spacelab, have made their decision for participation in the understanding of the fact that a manned, non-detachable multi-purpose laboratory for the orbiter can be developed within a financial frame.

Another special feature of our program is the fact that during the development phase only one flight unit of hardware will be provided. This fact means that the development program does not allow for a phased introduction of improvements, modifications and changes or a planned stepwise introduction of performance increases into more than one flight unit.

Unique are, of course, also the interfaces with the orbiter which are of quite a different dimension than those of which we have known so far by our satellites, which we developed in Europe and launched on top of US rockets.

Finally it is important to mention that the selection of one clearly defined industrial Primecontractor for the leadership on the industrial side of the program represents for a project of this size in Europe also a new fact.

After more than one year of development work one can state that the program is running stable and according to the plan. However, the aforementioned unique features have given the program its particular characteristics and will influence the further course and the final successful completion of the development to a large extent.

THE INDUSTRIAL TEAM

The industrial team, which executes the development of Spacelab, is composed of companies of all ten countries which contribute financially to the program. The leadership is in the hands of VFW-Fokker/ERNO which has been selected after a competitive proposal phase. The selection of the Primecontractor was facilitated by the fact of the large German contribution of about 53 % to the program.

VFW-Fokker/ERNO which was awarded the Primecontract on June 5th, 1974, is the first European aerospace company which has tried through a merger between the former Fokker in Holland and the former VFW in Germany on the aircraft side to create an international industrial entity, which goes across the European borders (Fig. 1). The Prime-Contractor therefore possesses the very valuable experience of an international European cooperation within a merged industrial company. From the top management down to all working levels, people who work on Spacelab, have not only the experience in international cooperation through the projects but also through their daily internal company activities.

Technically Spacelab represents features of a spacecraft and an aircraft. Man on board and the space environment require reliability and standards which we are used to from our space work. Multi-purpose and re-usability require operations, flexibility, simplicity, standardization and easy handling, which we are used to from aircraft developments. To meet these requirements VFW-Fokker/ERNO, the Spacelab Primecontractor, and the cocontractors of his choice draw their engineering and facility resources from their two branches of space systems and aircraft development (Fig. 2). VFW-Fokker/ERNO has been developing civil and military aircraft since a very long time and through its space division ERNO is involved in the national and international development of automatic spacecrafts in Europe since the beginning in the early sixties.

Successful passenger aircraft like F-27 Fokker-Friendship, F-28 Fokker-Fellowship and the more recent VFW-614 in the civil field, the development of the Transall military transport aircraft and of the vertical take-off VAK 191 B as well as the contributions to the Airbus-development and production and to the Multi-Roll Combat Aircraft (MRCA) have given VFW-Fokker/ERNO a very good experience on how to handle large aircraft systems which involve man in all aspects and which also meet the requirements of easy and handy operation as well as in the economical point of view. Other partners in the consortium like AERITALIA who are responsible for the structure of Spacelab module, and Hawker Siddeley Dynamics who are responsible for the structure of the pallet, have a similar experience in aircraft development which is flowing into the Spacelab work.

In the space field the industrial consortium which has received from ESA the responsibility to develop Spacelab, has been involved in the development of spacecrafts in Europe in the past 15 years within the national programs as well as in the ESA programs. In the context of these programs Europe has contributed to the exploration and the use of space a number of remarkable projects, starting with simple earth-orbiter satellites carrying scientific instruments, to the rather sophisticated satellites and probes of the last days as the German Solar Probe HELIOS which has approached the sun to one third astronomical unit, and the Franco-German Telecommunication Test Satellite Symphonie. Another example is the development of ESA OTS Satellite which is the advanced standardized telecommunication payload platform of the future, and its maritime navigation derivative MAROTS. The teams who have worked on these projects and the facilities which have been constructed to test and qualify the European space projects, exist and are available for Spacelab. Adding these necessary space specific elements of experience, facilities and manpower to the described capacity in the aircraft field this has been the basis of VFW-Fokker proposal for the execution of the Spacelab development. Today we proceed very successfully along these lines of solving our task.

(Fig. 3) On the industrial side in Europe there is one contract receiving company vis-à-vis the customer. It is the VFW-Fokker company. This company has entrusted the execution of the program management of Spacelab to its space division ERNO, located in the northern part of Germany in the city of Bremen. Within ERNO one single project team for Spacelab has been built up, consisting of the technical departments Engineering, Operation and Integration, the service departments for Product Assurance, Project Control, Contract Administration and Configuration Control and a special department for Cocontractor Management (Fig. 4).

Roughly 70 % of the total work is executed by our cocontractor companies (Fig. 5). The composition of this team has been made in a way of an optimum compromise between satisfying the requirement of a financial return corresponding the contribution of the respective country, creating the least number of difficult technical and managerial interfaces and meeting at the best existing capacities and experiences in the companies. The core of this industrial consortium is the MESH satellite group which has already developed successfully so far the largest ESRO satellite TD-1A and at the end of 1973 has been chosen to develop the OTS Communication Satellite and the MAROTS Navigation Satellite. Through the iterative study and selection process for the Spacelab contract by the ESA organisation, some companies who are not member of the MESH consortium, joint our Spacelab team. Backbone are, however, the MESH companies, which since 1964 worked together very closely on different projects and which are by now very well known to each other.

In order to have a top-level platform above the project management level the Spacelab consortium, led by VFW-Fokker/ERNO, has created the Board of Directors, in which one member of the top management of all companies of the consortium participates. This Board of Directors meets every three months or prior to a major program milestone in order to discuss the general problems of the functioning of the consortium and to deal with questions which are beyond the responsibility of the appointed Project Manager. With this tool Spacelab possesses on the industrial side an institution which ensures at all times the complete involvement and the up-to-date information of the top management. It has helped so far considerably to make Spacelab inspite of its challenge such smooth an operation.

The technical status of the Spacelab project can be described best by a discussion of the schedule and the changes.

(Fig. 6 and Fig. 7) The first 12 months of Spacelab

development were to their largest extent a paper phase with the climax of two documentation reviews, the Preliminary Requirement Review in November 1974 and the Subsystem Requirement Review in June 1975. This period covers basically the preliminary design of systems and subsystems. At its end now is the start of the procurement of the bought out parts as well as of manufacturing and testing. Soon on subsystem level the preliminary design reviews will begin, leading into the qualification phase. 1977 will be the year of the integration, firstly of the Engineering Model No. 1 which will be delivered to the United States in the first part of 1978, and then the Flight Unit will follow, which is planned to be delivered in 1979.

(Fig. 8 and Fig. 9) In connection with the Subsystem Requirements Review in June we have completed the so-called Soft Mock-up, a scale 1 Mock-up of the Spacelab module including the internal equipment, which is supposed to solve as a design and development tool for interface problems. There will be two more Mock-ups in the program. The development fixture will be used to develop harness and plumbing layout and the hard Mock-up will serve as the design verification tool.

(Fig. 10 and Fig. 11) The external dimensions and appearance of Spacelab with the tunnel, the module and the pallet are very well known in the meantime. However, in the detail the interfaces with the Shuttle are solved finally in these days. We have a revised tunnel configuration. The utility interfaces between the module and the orbiter and between the pallet and the orbiter have been reconsidered as well as the mechanical orbiter attachment has been revised. In the case of the pallet only mode the proposed igloo to house the necessary equipment has been revised also in its configuration and its position.

The inboard profile of the module is also unchanged in its principles. The preliminary design phase of the subsystems, however, has brought up a number of small changes and improvements as it is very natural in such a state of a project (Fig. 12 and Fig. 13). These changes concern mainly questions of interfaces and of better economical solutions. In this sense improvements for the rack design and for the subfloor and the overhead secondary support structure were introduced. A special ceiling structure was added and the definition of the main floor was finalized.

With the deletion of the possibility of a road transportation the design of the pallet could be modified such that at the upper extremes no interruption in the structure is any more necessary. Besides the interface questions concerning the orbiter attach fittings, utility services connections with the orbiter and the positioning of the igloo the whole problem of the thermal environment of the pallet within one mission cycle of the orbiter is at present under investigation. (Fig. 14 and Fig. 15) In the area of common payload support equipment no major changes have occured. Here as well as in the basic Spacelab design refinements have been introduced and interface problems have been clarified. The attempt is made to design and manufacture these items in as close in contact and agreement with possible users as it can be done at this time.

For the environmental control subsystem the decision to use the orbiter oxygen supply was very important and can be considered to be the major change. The cabin temperature minimum has been fixed to be $19.5^\circ C$. There will be only two cabin fans available. (Fig. 16 and Fig. 17) For the thermal control system the addition of an experiment dedicated heat exchanger can be considered as a major change as well as the change to goldized capton insolation for the module. In most cases the component procurement specs are completed and a procurement process has been initiated.

With the selection of the CII computer for flight and ground segments the major decision for the command and data management subsystem has been taken earlier this year. After this change and its peripheral consequences have been implemented now further improvements of the CDMS performance are under discussion. Among other things considerable performance increases of the remote acquisition units are considered (Fig. 18 and Fig. 19).

(Fig. 20 and Fig. 21) The concept for an Instrument Pointing System has been studied, discussed and finally decided by the customer. The final proposal for the development of this subsystem has been worked out and the final decision for development is expected within the next weeks.

For the integration of Spacelab in Bremen a dedicated building is prepared which will allow from the 1st of January 1977 on the parallel integration of two Spacelabs at one time (Fig. 22 and Fig. 23). The building will include a storage facility and a room for a Mock-up as well as office facilities for the integration team, a large meeting room and laboratories and workshops for different technical subsystems and for our individual cocontractors. This special Spacelab facility will allow a continous Transfer of the present development program into a follow-on-production program of further Spacelabs and, of course, the integration of any payloads into Spacelab and the integration of payloads themselves. Beside this largest visible sign of Spacelab in Europe a lot of hardware, materials and components exist already. In fact, the first hardware was cut in February this year at AERITALIA in Torino for a test flange of the airlog.

(Fig. 24 and Fig. 25)
(Fig. 26 and Fig. 27)
(Fig. 28 and Fig. 29)

CONCLUSION

In concluding it can be stated that on the industrial side an enthousiastic team has been built up to conduct the development of Spacelab. A clear organisational form of cooperation has been found and an excellent relation with the customer ESA exists. Between the NASA Organization in the United States and the Shuttle industry and us in Europe working relationships have been built up, which should be sufficient to solve the difficult interface problems between the orbiter and its major payload. In realizing our task we should, however, not forget the constraints imposed on us in Europe by the conditions under which our ten governments decided to participate in the Post-Apollo-Program with the Spacelab. We hope and will do everything possible that this program will be so successful that our governments will be encouraged to continue with a strong support of future space programs in the era of manned space flight and in the closer future in the field of the utilisation of Spacelab.

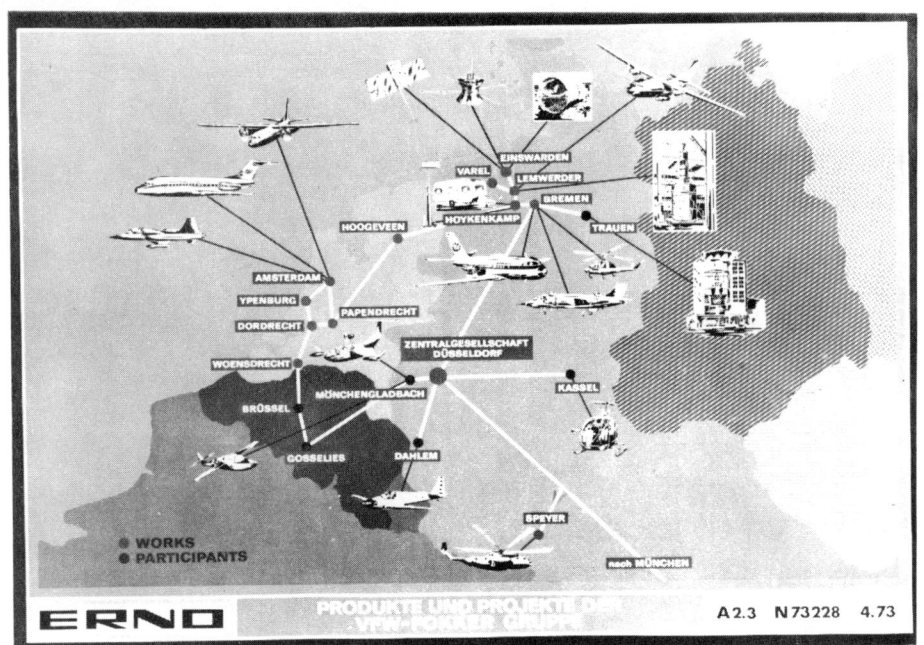

Figure 1

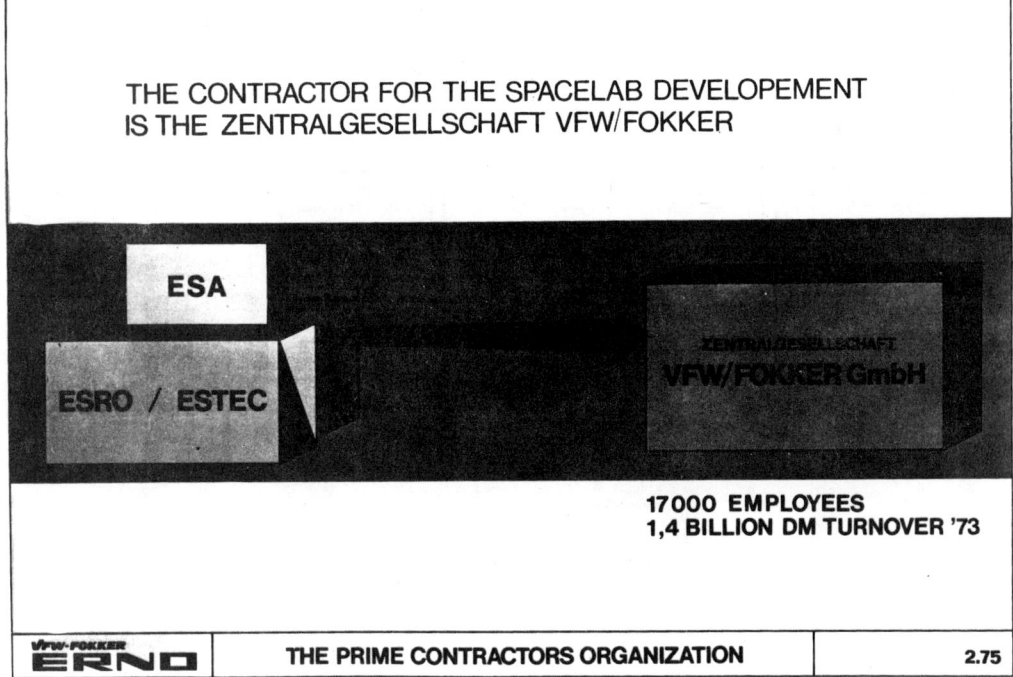

Figure 2

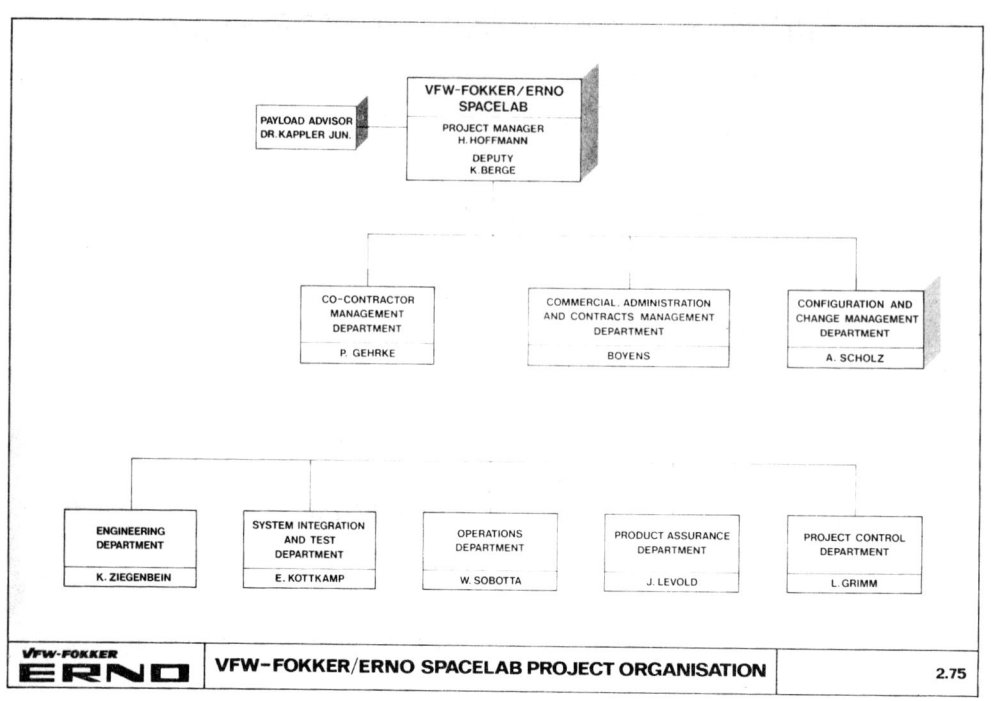

Figure 3/4

THE INDUSTRIAL SPACELAB CONSORTIUM

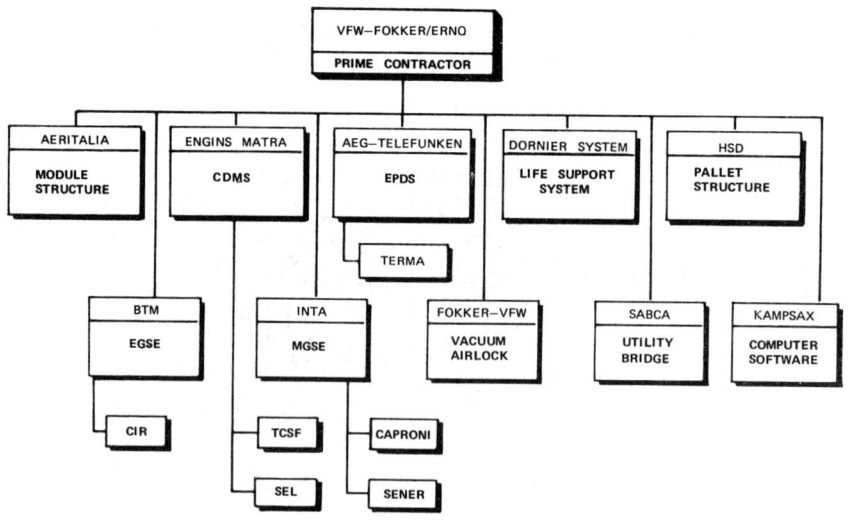

AEG-AERITALIA-BTM-CIR-DORNIER-FOKKER/VFW-HSD-INTA-KAMPSAX-TERMA-MATRA-SABCA-SEL-SENER-TCSF

Figure 5

NEAR TERM DEVELOPMENT AND TEST

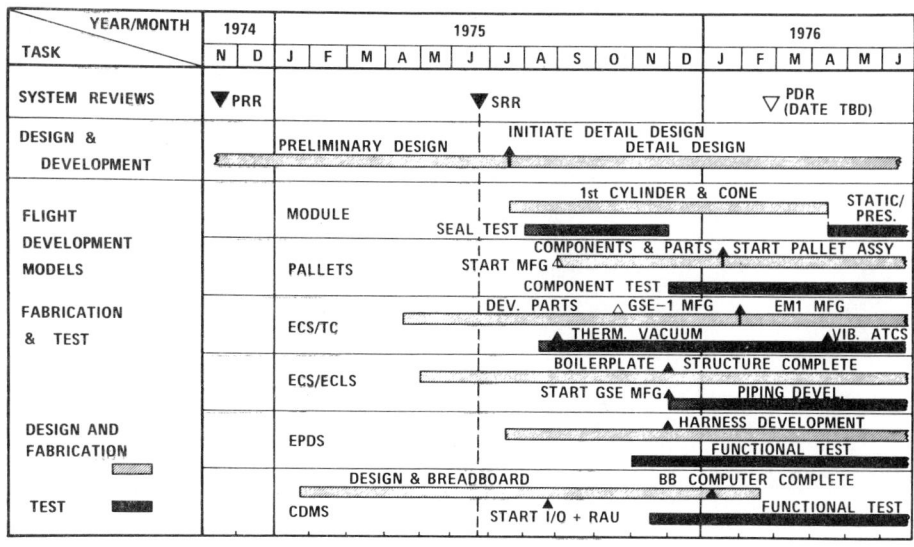

AEG-AERITALIA-BTM-CIR-DORNIER-FOKKER/VFW-HSD-INTA-KAMPSAX-TERMA-MATRA-SABCA-SEL-SENER-TCSF

Figure 6

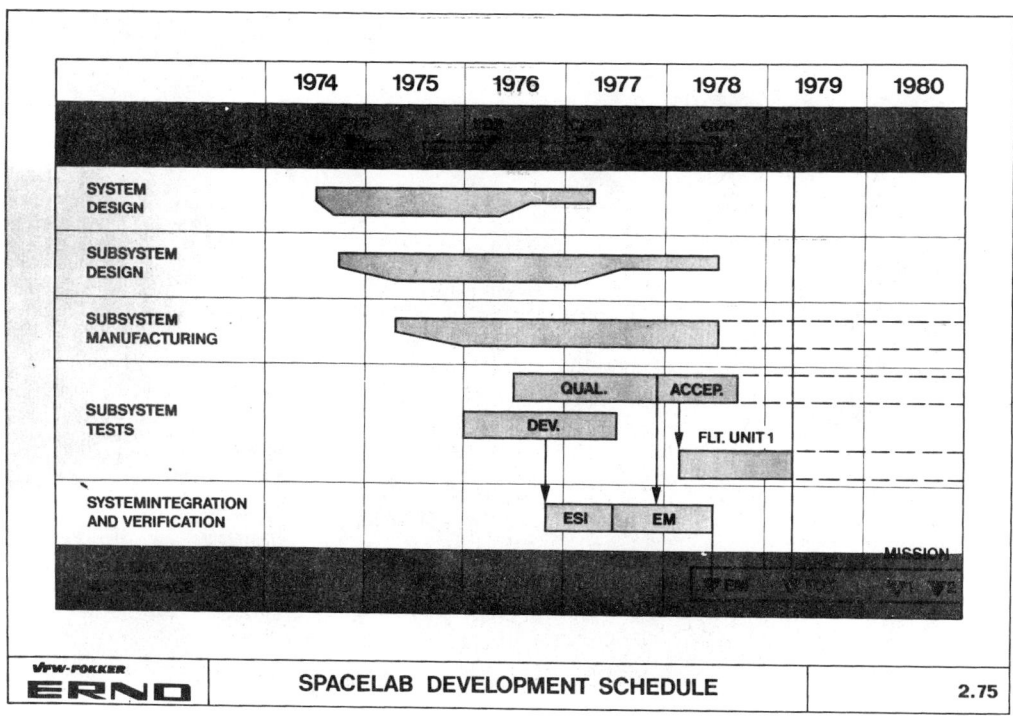

SPACELAB DEVELOPMENT SCHEDULE

Figure 7

SYSTEM DESIGN AND DEVELOPMENT

MOCKUPS
- SOFT MOCKUP
    - DESIGN AND DEVELOPMENT TOOL TO SOLVE INTERFACE PROBLEMS THREEDIMENSIONALLY
    - IN OPERATION SINCE 1 APRIL
- HARD MOCKUP
    - DESIGN VERIFICATION TOOL
    - HARD MOCKUP PLAN SUBMITTED
    - DETAILED DESIGN HAS STARTED
    - SCHEDULED TO BE AVAILABLE MID 76
- DEVELOPMENT FIXTURE
    - NEW PROGRAM ITEM
    - USED TO DEVELOP HARNESS AND PLUMBING LAYOUT
    - SCHEDULED TO BE AVAILABLE DECEMBER 75

AEG-AERITALIA-BTM-CIR-DORNIER-FOKKER/VFW-HSD-INTA-KAMPSAX-TERMA-MATRA-SABCA-SEL-SENER-TCSF

Figure 8

Figure 9

SPACELAB EXTERNAL FEATURES
(MAJOR CHANGES IN PROGRESS)

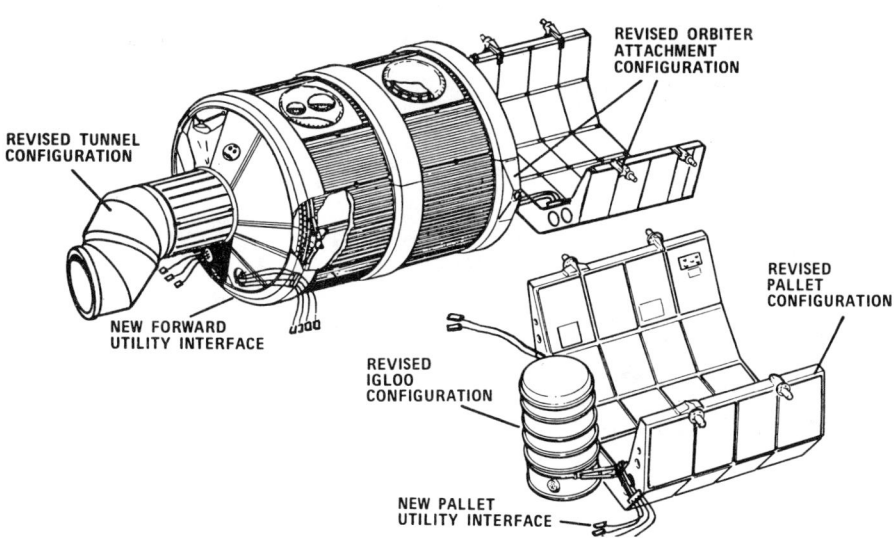

Figure 10

INBOARD PROFILE

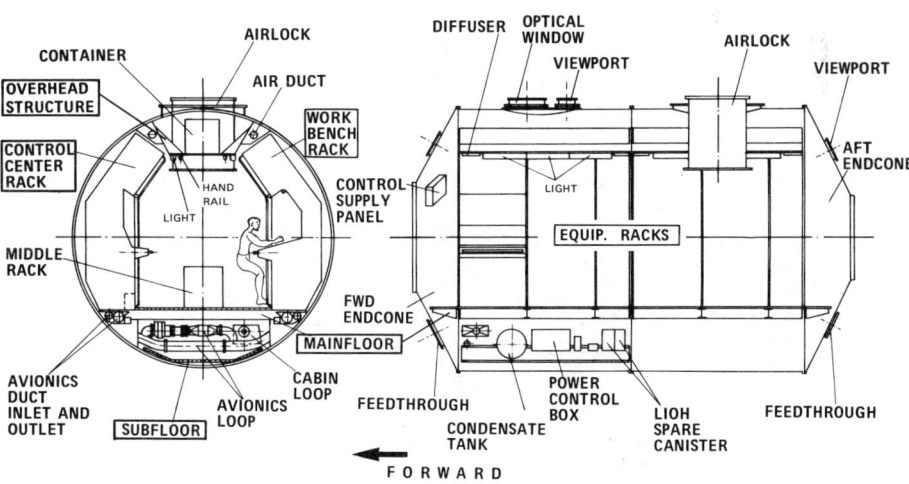

Figure 11

MODULE STRUCTURE

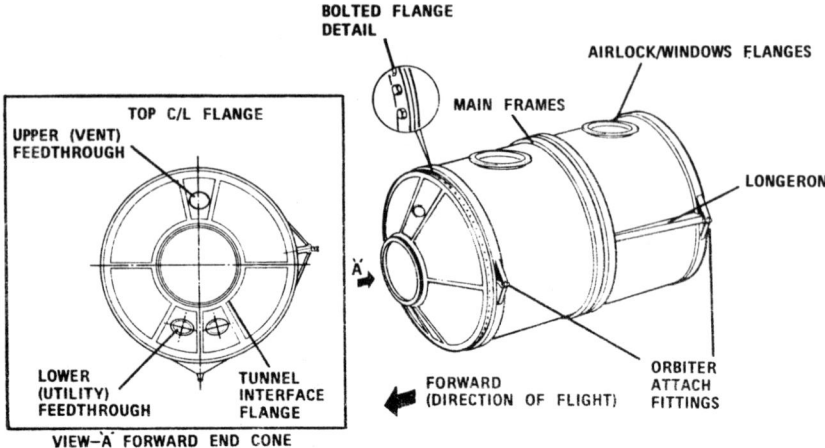

Figure 12

PALLET ASSEMBLY
REVISED DESIGN CONCEPT

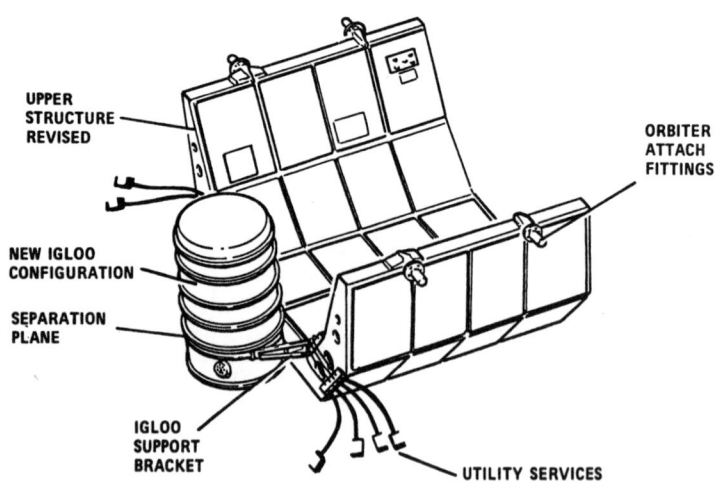

Figure 13

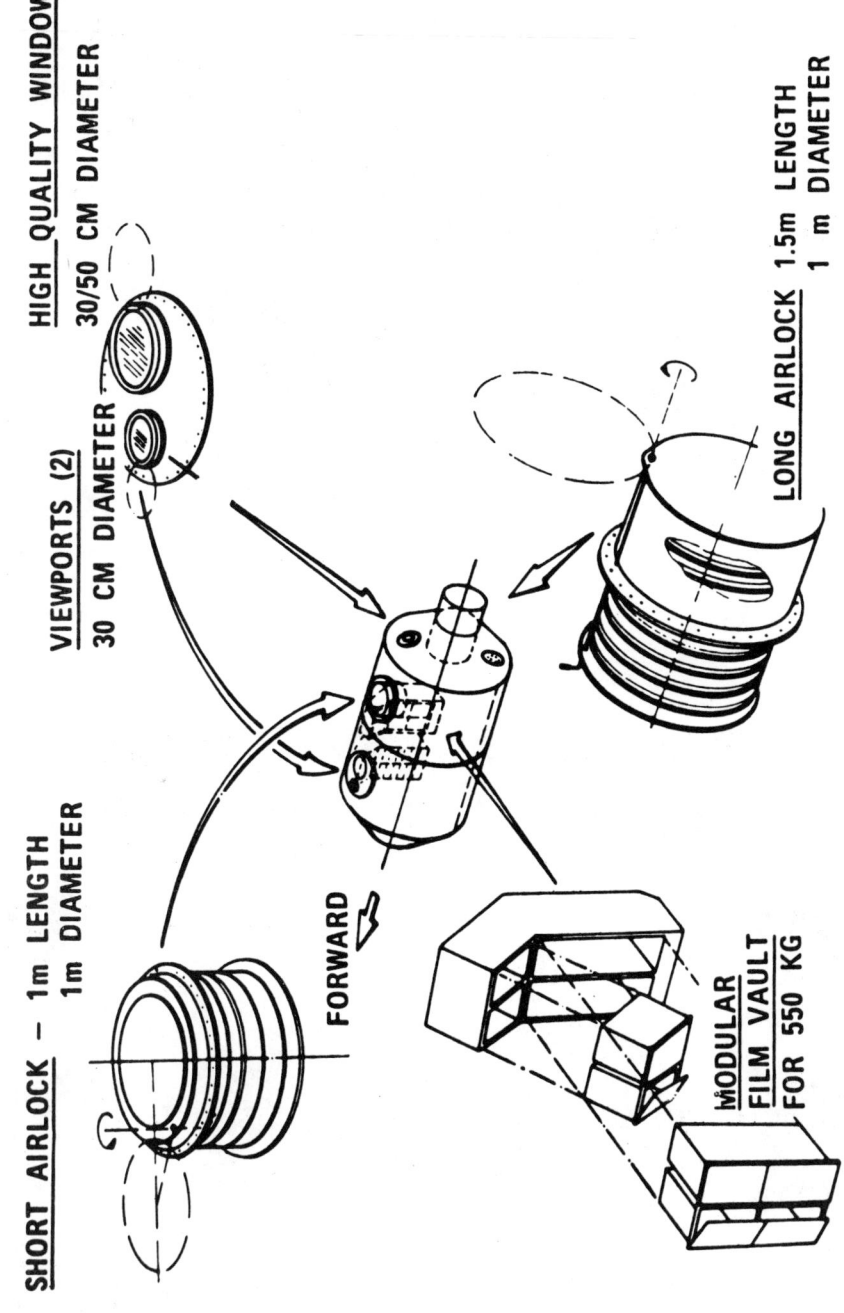

Figure 14/15

Figure 16

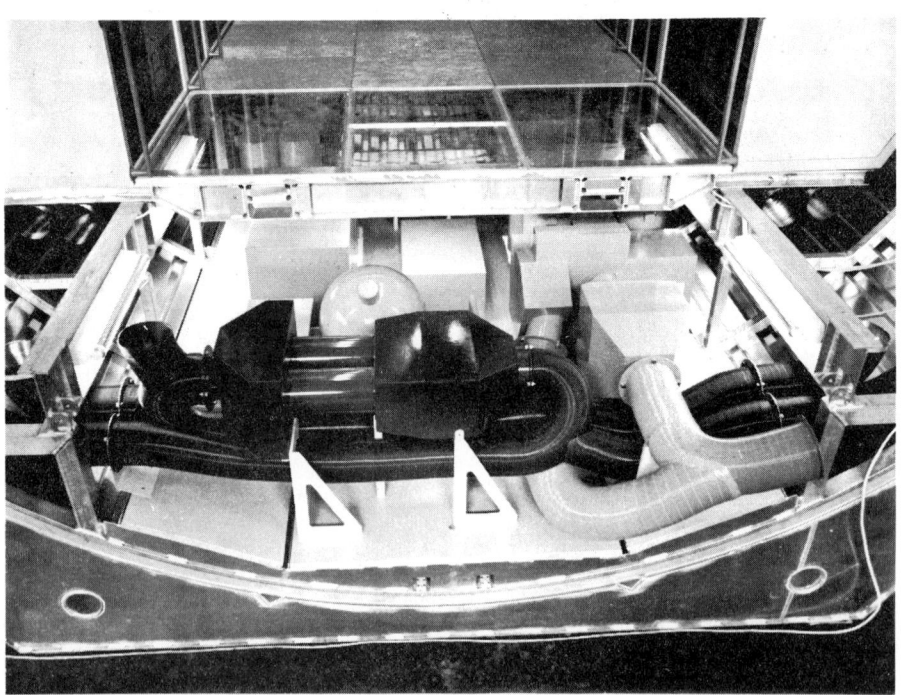

Figure 17

COMMAND AND DATA MANAGEMENT SUBSYSTEM

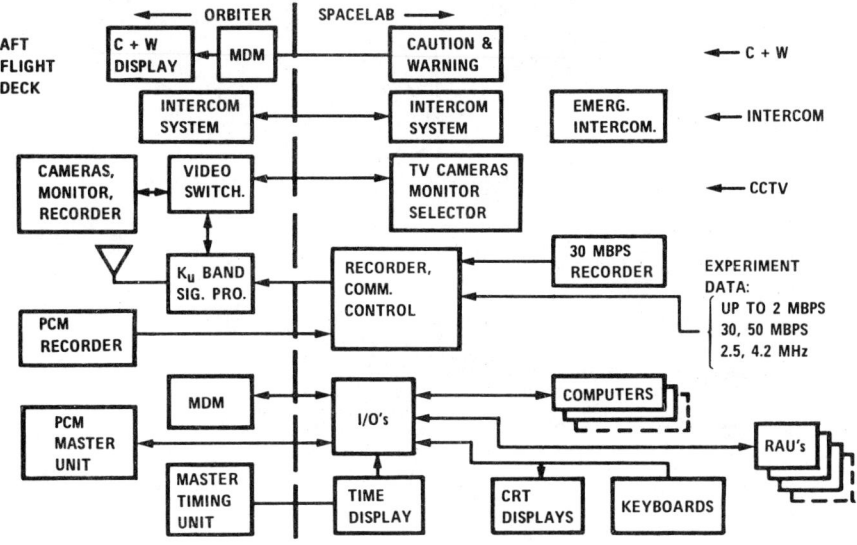

AEG-AERITALIA-BTM-CIR-DORNIER-FOKKER/VFW-HSD-INTA-KAMPSAX-TERMA-MATRA-SABCA-SEL-SENER-TCSF

Figure 18

CDMS DESIGN PROGRESS

- CHANGED TO ALPHA NUMERIC KEYBOARD
- CHANGED TO 12 INCH COLOR CRT
- IMPLEMENTED DIRECT INTERFACE BETWEEN SPACELAB I/O's AND ORBITER MDM AND PCMMU
- ADOPTED COMMON AUDIO NETWORK WITH ORBITER
- IMPLEMENTED INTERFACE WITH ORBITER TV
- ADOPTED AMERICAN TV SIGNAL STANDARD

- ADDITION OF MAINTENANCE RECORDER – BEING CONSIDERED
- IMPLEMENTED INTERFACE WITH ORBITER CAUTION & WARNING
- ORBITER TIMING AND SYNCHRONIZATION INTERFACE WITH S/L COMPUTER– BEING IMPLEMENTED

AEG-AERITALIA-BTM-CIR-DORNIER-FOKKER/VFW-HSD-INTA-KAMPSAX-TERMA-MATRA-SABCA-SEL-SENER-TCSF

Figure 19

INSTRUMENT POINTING SUBSYSTEM

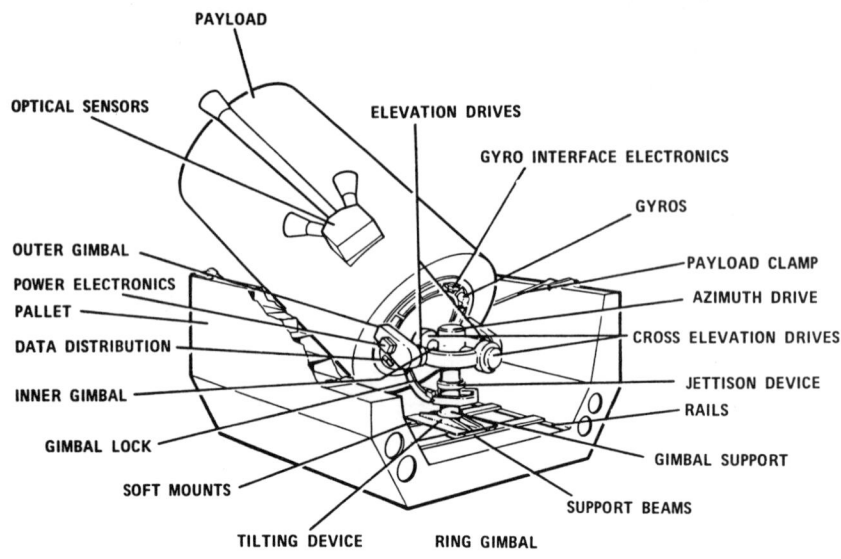

AEG-AERITALIA-BTM-CIR-DORNIER-FOKKER/VFW-HSD-INTA-KAMPSAX-TERMA-MATRA-SABCA-SEL-SENER-TCSF

Figure 20

IPS PERFORMANCE CHARACTERISTICS

VALUES ARE IN ARCSEC UNLESS OTHERWISE SPECIFIED

ERROR TYPE	REQ'M'T	DESIGN GOALS	NOTES
BIAS:			
LATERAL AXES	2	1.1	ASSUMES CALIBRATED
ROLL AXIS	40	18	SENSORS
QUIESCENT STABILITY ERROR			
A. LATERAL	1.6	0.8	VALUES ARE RMS.
A. ROLL	25	16	A = SMALL PAYLOAD
B. LATERAL	1.2	0.6	B = LARGE PAYLOAD
B. ROLL	25	12	
DISTURBANCE RESPONSE ERROR			
A. LATERAL	5	3	VALUES BASED ON
A. ROLL	30	20	3–AXIS SIMULATION
B. LATERAL	3	1	
B. ROLL	25	15	
LIMIT CYCLE	5	–	THRUSTER ON TIME = 200 m sec.

AEG-AERITALIA-BTM-CIR-DORNIER-FOKKER/VFW-HSD-INTA-KAMPSAX-TERMA-MATRA-SABCA-SEL-SENER-TCSF

Figure 21

Figure 22

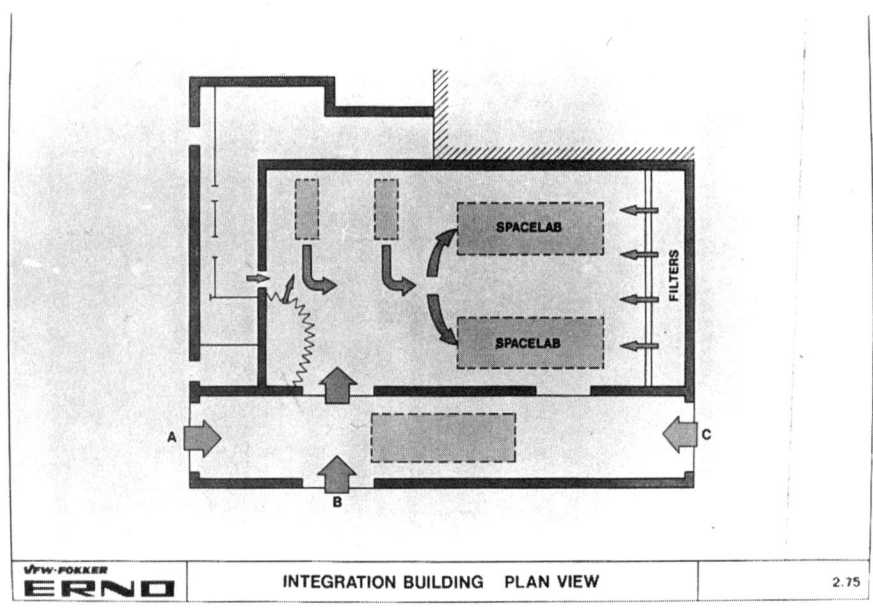

Figure 23

Figure 25

Figure 24

Figure 27

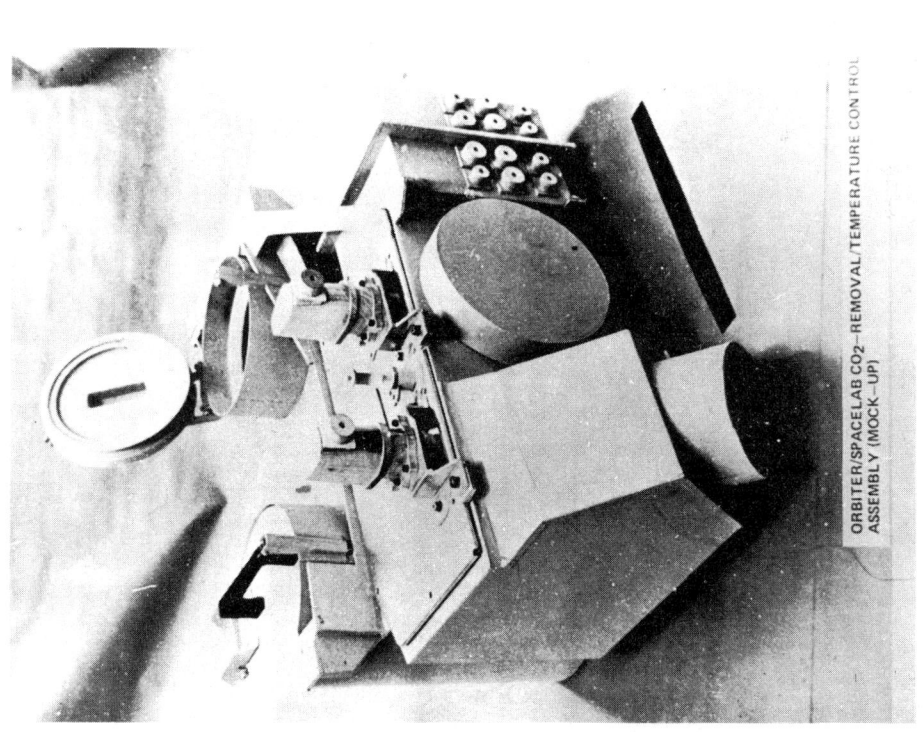

Figure 26

Figure 28

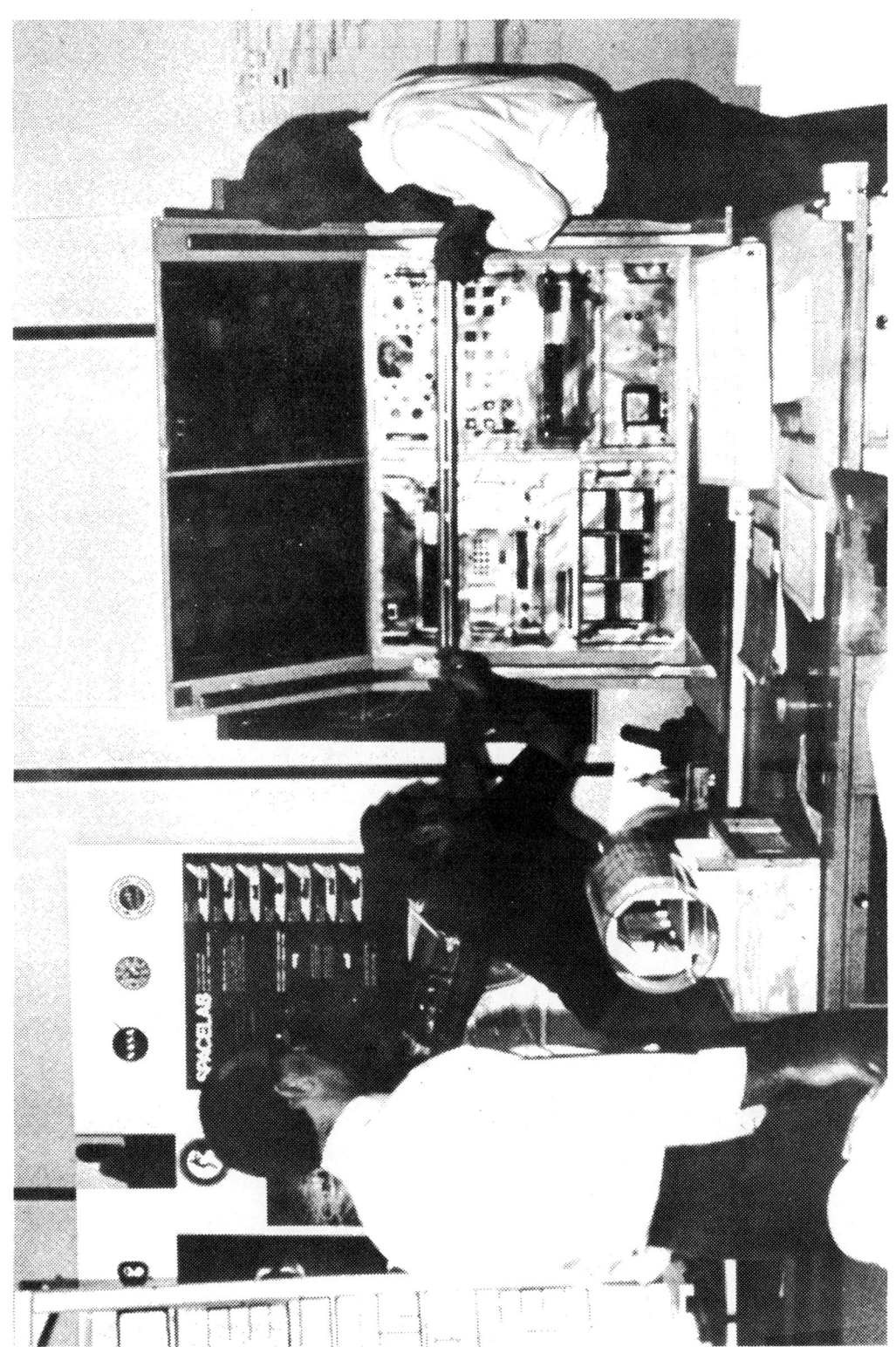

Figure 29

AAS 75-125

SPACELAB PAYLOADS

G. W. Sharp
NASA Headquarters
Washington, DC

Feasibility studies have been conducted for a substantial number of scientific and technological payloads for Spacelab. Payloads are being studied to use the full range of Spacelab configurations and available resources. The intent of this paper is to describe the various payloads that are being studied by the National Aeronautics and Space Administration for the Spacelab in the areas of life sciences, technology and space science.

INTRODUCTION

To date, spaceflight research payloads have been constrained by weight, volume, power, and telemetry limitations imposed by the spacecraft. Space Shuttle (orbiter) borne Spacelab will allow a significant relaxation of these constraints as well as bring new, unique, and exciting dimensions to the old ways of conducting space experiments and investigations. In the Spacelab, the researcher will find a work environment similar to his familiar ground-based facility where he can freely interact with his equipment and subjects.

Spacelab missions are nominally designed for seven days on-orbit but missions of up to 30 days on-orbit are possible. For the orbiter to supply resources for an extended Spacelab mission, it must carry additional power kits and other equipment that can significantly reduce the weight available for payload equipment.

The modular design of the Spacelab allows a great deal of flexibility and variety in the make-up of payloads. Payloads are envisioned that utilize the long pressure module only, long module with one or two pallet elements, short module with two or three pallet elements, and which use only pallet elements (from one up to five elements). Being defined are single discipline dedicated payloads with facility-class instrumentation, mixed discipline payloads with both principal investigator and facility-class instruments having compatible orbital and altitude requirements, and small rocket or balloon-class instruments as well as "suitcase" carry-on instrumentation which are to be flown on a space available basis.

The intent of this paper is to describe the various payloads that are being planned by the National Aeronautics and Space Administration (NASA) for the Spacelab in the areas of life sciences, technology and space science. Applications payloads are discussed elsewhere. Included in these descriptions will be the experimental objectives of the payload, the state of planning or development, and the intended flight schedule and mode for the payload.

LIFE SCIENCES

Scientific Objectives

For the purpose of this paper, it is considered that life sciences include all disciplines which seek to understand, measure, or control life processes at the subcellular, cellular, systemic, organism, or multiorganism levels. Thus, they include biology and physiology, medicine, behavioral and social sciences, biomedical engineering, life support systems engineering, human factors engineering, and all allied disciplines.

The two primary objectives of NASA's life sciences program are to enhance man's well-being and productivity in space, and to use the space environment to better understand all life processes.

Research to attain these objectives will include:

Biological research which uses the unique properties of the space environment to better understand life processes, as well as investigations which provide insight into the biological effects of long-term exposure to spaceflight;

Studies of the prevention, diagnosis, and treatment of diseases;

Research into the basic nature of physiological mechanisms and into means of enhancing man's well-being in space and on earth; and

Behavioral studies aimed at increasing man's effectiveness and efficiency in spaceflight, as well as other studies to measure the effects of spaceflight conditions on psychological and psychophysiological mechanisms.

Instrumentation

As a tool for accomplishing these objectives, NASA is considering development of a manned laboratory equipped expressly for life sciences investigations and carried into orbit aboard the Space Shuttle. This dedicated life sciences Spacelab can be staffed by scientists, engineers and technicians (men and women) who have received only minimal spaceflight training. This Spacelab will provide scientists with a "shirt-sleeve" environment. Furthermore, it will be outfitted to facilitate the use of off-the-shelf laboratory equipment, thereby reducing the cost of space missions and substantially reducing the time and effort required to prepare adequately for missions. To promote further savings, NASA will outfit the Spacelab with a battery of equipment which is frequently needed in life sciences flight experiments. Called Common Operations Research Equipment (CORE), these equipment items will be supplied on each Spacelab mission according to flight experiment requirements (Fig. 1). Mission capabilities

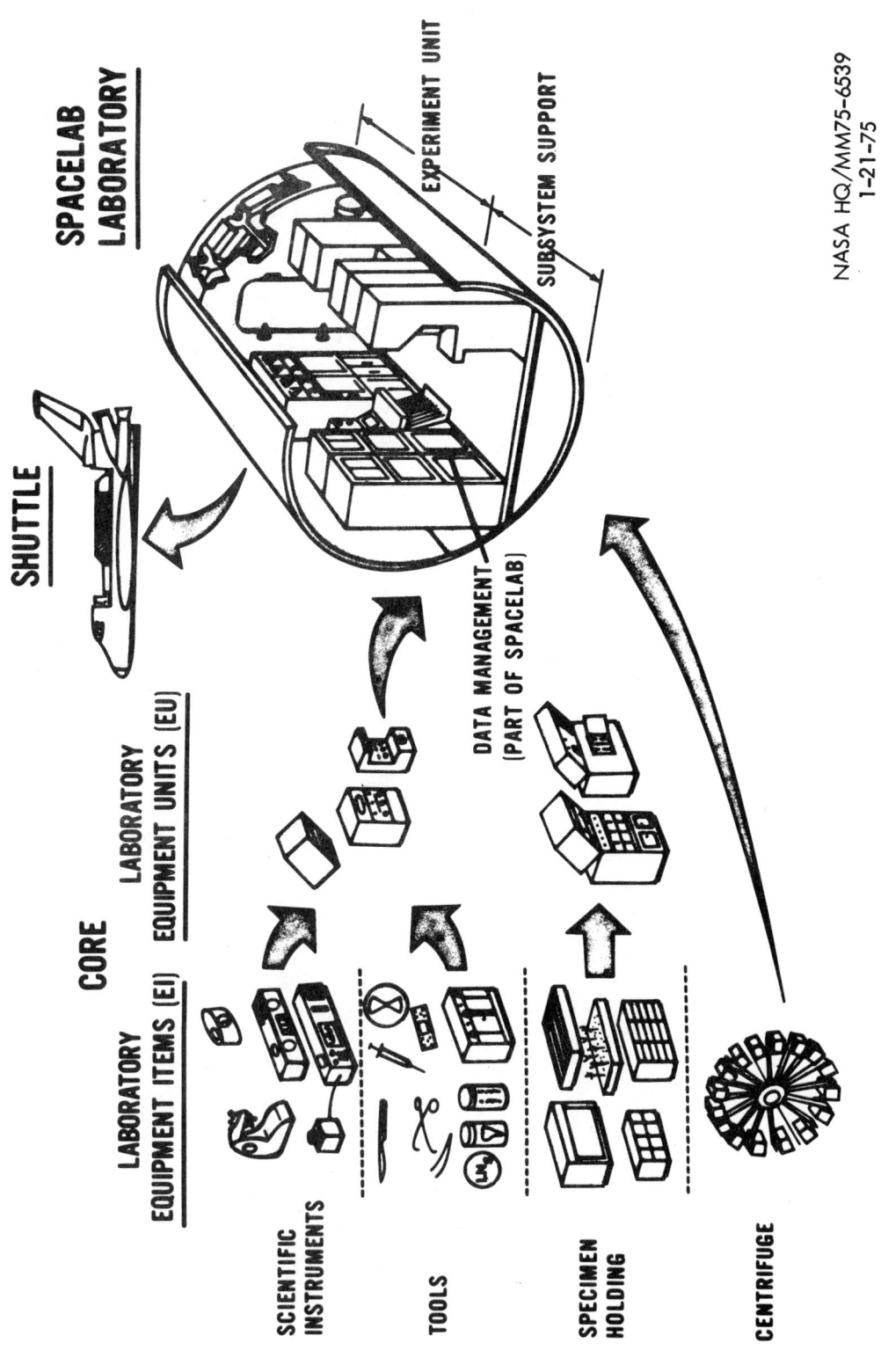

Fig. 1 Life Sciences Core Concept (Common Operations Research Equipment - CORE)

of the life sciences Spacelab are described below:

- Mission Schedule – One or two missions per year, beginning in 1981

- Mission Duration – Seven to 30 days on-orbit

- Crew Complement – Adequate working space for up to four payload specialists (men and women)

- Spaceflight Training – An estimated six months of specialized training for payload specialists during the year prior to launch.

- CORE – Based on typical life sciences experiments, NASA has developed a preliminary listing of CORE items which will be available in the Spacelab. This listing is shown in Table 1.

- Equipment Weights – Up to 454 kilograms (1000 pounds) of experiment-peculiar equipment

- Experiment Volumes – Up to 2.3 cu meters (80 cu ft) of experiment-peculiar equipment

- Temperature – 21°C (70°F)

- Communications – Television, voice and data telemetry will be provided as will voice communications between principal investigators located on-orbit and those on the ground

- Atmosphere – Two gas sea-level equivalent atmosphere, (760 mm Hg, 20% oxygen/80% nitrogen).

- Power – Sufficient electrical power (AC and DC) to operate experiments

- Test Specimens – Experiments may use virtually any test specimens up to and including humans, small primates, and small vertebrates. (Specimen holding facilities are part of CORE.)

Table 1

POTENTIAL COMMON OPERATIONS RESEARCH EQUIPMENT (CORE)

Sample Preparation & Preservation
─────────────────────────────

Invertebrate Anesthetizer
Glove box and liners
Refrigerated, High Speed Centrifuge (10 ml tube)
Chemicals
Radioactives
Cryogenic Freezer
Low Temperature (Holding) Freezer
Refrigerator
Bench Chemical Analysis Kit
Hematology Kit
Microbiology Kit
Microdissection Kit
Liquid Volumetric Measurement Kit
Macro Mass Measurement Device
Micro Mass Measurement Device
Staining System

Biochemical & Biophysical Analysis
─────────────────────────────

Air Particle Collector
Mass Spectrometer
Rh Spectrophotometer
Refractometer
pH Meter
Sound Level Meter
Surgical Work Bench

Internal Centrifuge
─────────────────────────────

14 Foot Diameter Specimen Centrifuge 0.5 to 1.5 G

Data Management
─────────────────────────────

Digital Computer
Computer Input:
  Keyboard
  Electronic Sensors (e.g. EKG)
Computer Output:
  Data Plotter
  Data Printer
  CRT - Type Display
  Digital (Electromagnetic) Data Tapes
Voice Recorder
Event Timer
Oscilloscope
Analog (Electromagnetic) Data Tapes

Biomedical/Behavior Research Support
─────────────────────────────

General Purpose Experimenter Control Console
Electrophysiology Back Pack and Receiver
Psychogalvonometer
Rotating Litter Chair

Biomedical Research Support
─────────────────────────────

Exercise Equipment
Doppler Flowmeter
Physical Examining Kit
Metabolic Analyzer
Limb Plethysmograph
Sono Cardigram
Lower Body Negative Pressure Device

Small Vertebrate Holding & Research Support
─────────────────────────────

Primate Holding Unit
Rat/Hamster Holding Unit
Small Vertebrate (General) Holding Unit
Water Vertebrate Holding Unit
Veterinary Kit

Plant Research Support
─────────────────────────────

Plant Holding Unit
Clinostat

Cells and Tissues Holding & Research Support
─────────────────────────────

Sealable Colony Chamber
Incubator Holding Unit for Cells & Tissues
Prepared Media

Invertebrate Holding & Research Support
─────────────────────────────

Invertebrate Holding Unit
Incubator Holding Unit for Invertebrates
Tool Kit for Insect Handling

Life Support Sub-System Test
─────────────────────────────

General Purpose Test Console

Human Factors Engineering Measurement
─────────────────────────────

Psychomotor Performance Console

Crew Mobility Measurement
─────────────────────────────

Anthropometric Grid
Mobility Unit Protective Corridor

Radiobiology Support
─────────────────────────────

Radiation Badges
Radiation Source (Prepackaged and Shielded)
Biochemical Sample Radiation Counter

Data Processing - On-board facilities for analog, digital, and television data storage and retrieval, and data processing computers will be available.

An important difference between Spacelab and previous missions is that the investigator can interact in real time with his experiment. Experiment procedures, equipment configurations, and measurement variables can be altered immediately to take advantage of novel observations, unexpected results, and insights gained during the experiment.

In summary, the life sciences Spacelab "... constitutes a major advance in the capability for definitive life sciences investigations in space. In addition to biomedical investigations relevant to man's well-being in space, basic principles of biology and medicine can be examined using the zero-G environment as a research tool. The laboratory will also provide the operational conditions necessary for the evaluation of components of advanced life support systems and of man-machine integration technology" (National Academy of Sciences, Scientific Uses of the Space Shuttle, 1974).

In addition to the above dedicated life sciences Spacelab, it is planned to piggy-back life sciences research on Shuttle flights dedicated to other disciplines on a space-available basis. For these piggy-back missions, selected life sciences CORE items will be integrated into small modules and data will be collected on a part-time basis.

SPACE TECHNOLOGY

The Office of Aeronautics and Space Technology (OAST) is planning a Spacelab program dedicated to the conduct of its research and technology activities in space. This dedicated Spacelab, known as the Advanced Technology Laboratory (ATL), will utilize the long pressure module of the Spacelab plus up to two pallet elements. Variations of this basic configuration will, of course, be possible.

An artist's concept of one version of the ATL is shown in Fig. 2.

The goals and objectives of the ATL are to provide the technical and programmatic support to enable OAST to use the Shuttle/Spacelab to extend its research and technology programs into the space environment where the unique properties of space are an essential element in a technology experiment, demonstration or test, and where the interaction of man is beneficial, if not essential. On the basis of the program goals and objectives, it is expected that the experiments flown on the ATL will be provided directly from on-going NASA research and technology programs. However, additional investigators whose technology research effort would benefit by association with the NASA payload and by access to space will be identified and accommodated.

The ATL is planned to be able to accommodate a wide variety of experiments and will essentially be reconfigured for each flight. The program goal is to develop plans and procedures that will facilitate rapid integration of individual experimental equipment into the Spacelab. Feasibility studies for this payload are completed. The present plan is to have the first flight of ATL in 1981 followed by a flight rate of about two per year beginning in 1982.

ATMOSPHERIC AND SPACE PHYSICS
Scientific Objectives

Atmospheric and space physics represent three fields of research: the atmosphere, magnetosphere, and plasmas-in-space. These are closely related and are combined here not only because common instrumentation is used for their study, but also because the couplings between different atmospheric regions are very important. Specific research objectives for each of these fields will now be discussed.

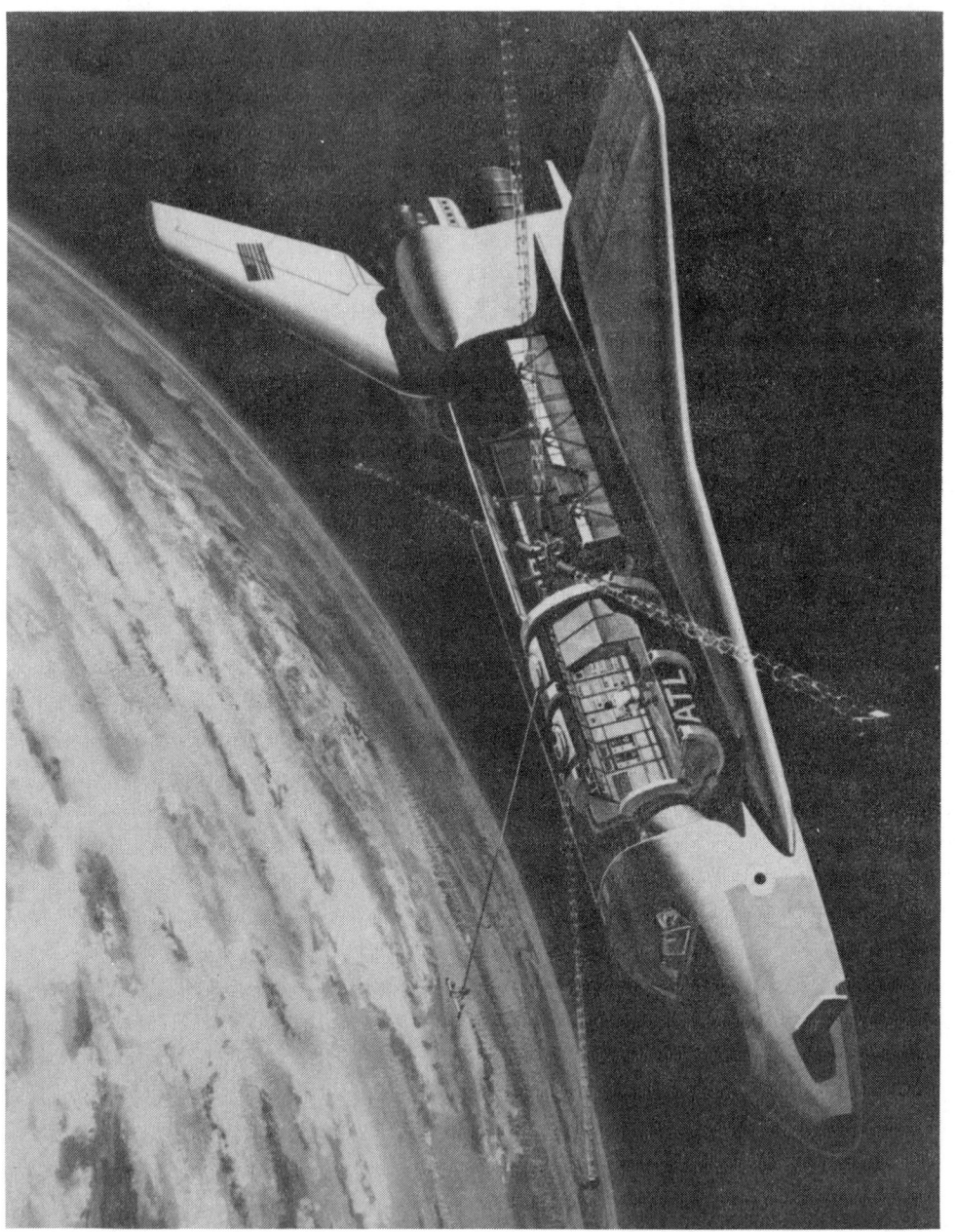

Fig. 2 Artist's Concept - ATL

Atmospheric Science. Particularly important regions of the atmosphere to be studied are: the stratosphere, mesosphere, and lower thermosphere up to 120 km height. While much of the interplay between optical radiations and chemical reactions is reasonably well understood in these regions, the dynamical properties and their interaction with the structure and chemistry are not. An important objective is to understand the general circulation of these regions and the effects of transport on chemical composition there. Particular emphasis is on how these effects influence the ozone layer.

Magnetospheric Dynamics. A prime objective is to understand the key physical processes associated with the energy, mass, and momentum transfer from the solar wind to the magnetosphere and atmosphere. The technique for meeting this objective will be to break into the chain of cause-and-effect relationships by introducing well-defined perturbations and measuring the response of the magnetosphere and atmosphere to these.

Plasma Physics in Space. The objectives here are to understand the basic plasma processes and instabilities which are believed to give rise to the release of energy in substorms, in solar flares, and perhaps pulsars. Additional objectives are to understand the ways in which particles obtain energy and produce aurorae and low energy cosmic rays (as from Jupiter) as well as the mechanisms whereby radio emissions are generated.

Instrumentation

To meet these objectives, the Atmospheric, Magnetospheric, and Plasmas-in-Space (AMPS) Spacelab has been studied as a versatile manned laboratory designed for dedicated studies of the near-earth space environment. An artist's concept of this is shown in Fig. 3. This facility, planned to be a short pressure module with three pallet

Fig. 3 Artist's Concept - AMPS

elements, is being designed to capitalize on the opportunities afforded the Spacelab system for the researcher to conduct his own experiments in space. The AMPS will permit him to react to unexpected observations and to modify his experimental parameters in a manner similar to those used in ground-based laboratories.

The AMPS facility will consist of a set of basic stimulation and diagnostic instruments as well as any required special purpose equipment which will be available to the experimenter upon request. It is anticipated that these instruments will be built for NASA by various universities and contractors both from within the United States and from foreign countries. They will then be integrated into the laboratory and will be provided to scientists as needed.

Major facility-type instrumentation presently planned for AMPS (See Fig. 4.) will include a laser radar system for probing the earth's atmosphere below the Shuttle orbit, chemical releases for measuring electric and magnetic field morphologies and investigating aeronomic reactions, particle accelerators and radio transmitters for stimulating the space plasma, and a complete array of optical and plasma diagnostic instrumentation. The total complement will probably be of the order of 50 different instruments, capable of operating together in various combinations.

The program will utilize the Spacelab reflight capability to allow a "building-block" approach to the development of a complete payload. Scientists and engineers will be able to develop the individual facility instruments and the total instrument complement as they learn with each mission. In this way they will become an intimate part of the experiments with the ability to react and modify experiments in flight.

It is expected that a cost effective project will result from the versatility of the facility which can be flown many times to perform different experiments with the same

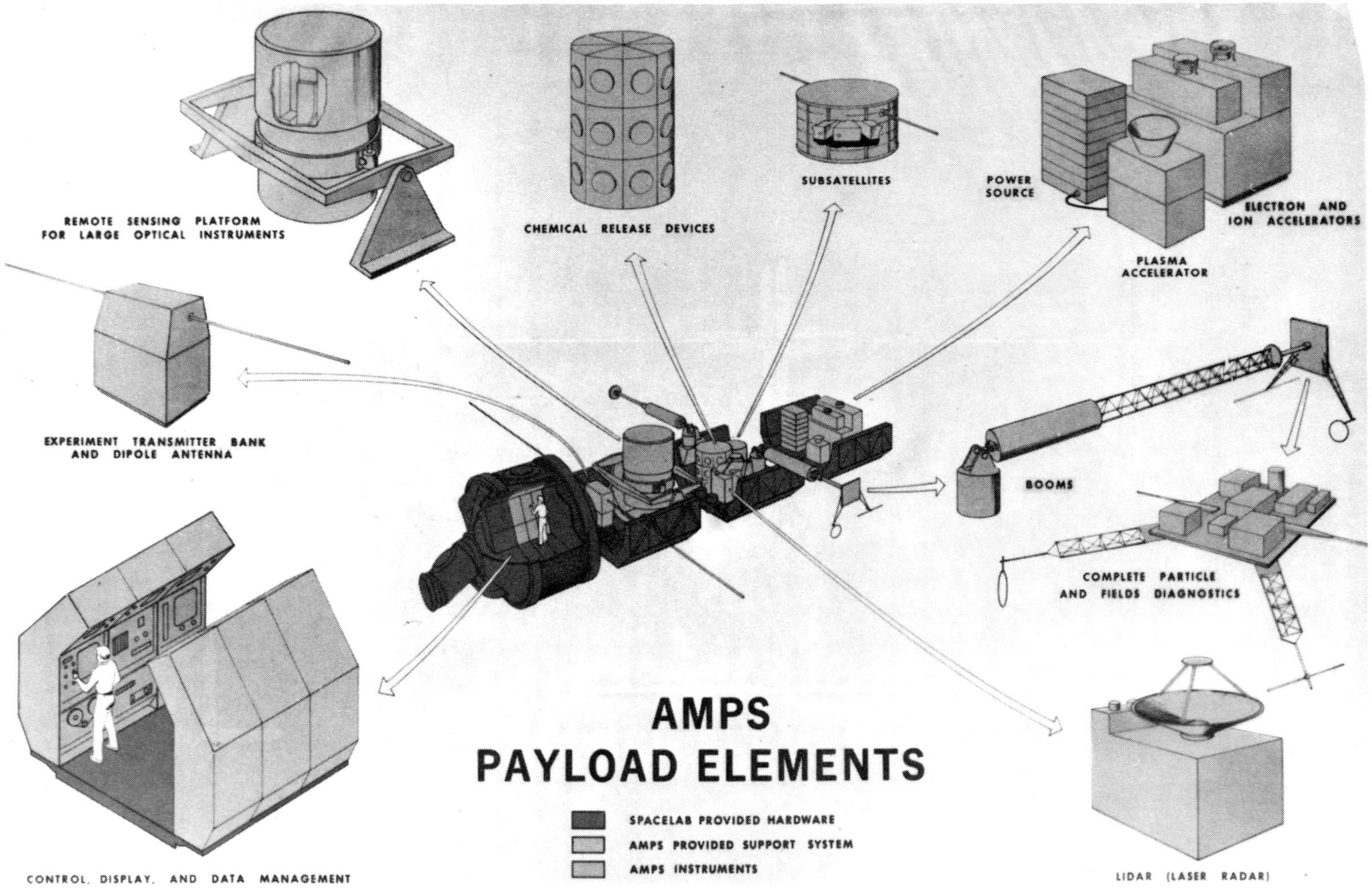

Fig. 4 AMPS Payload Elements

set of basic core instruments. These will be available to
a large constituency of users at relatively short notice.
A feasibility study has been conducted on the AMPS payload
and definition studies are about to begin. This payload
is planned as an early Spacelab payload, its first use
being anticipated in 1981. In addition, several of the
AMPS core instruments are candidates for the first Spacelab
payload to be flown in mid-1980.

INFRARED ASTRONOMY

Scientific Objectives

Infrared astronomical observations are crucial to the
understanding of nearly every fundamental question posed in
astronomy today. For example:

Solar System Formation. Broad band high resolution
spectroscopy of planets and comets give molecular and atomic
abundances, as well as isotopic ratios, which provide a clue
to the history of the objects within the solar system.

Stellar Evolution. High sensitivity spectroscopy provides
measurements on the composition and structure of the interstellar medium, stellar atmospheres, and circumstellar gas,
which are crucial to theories of nucleosynthesis, stellar
atmospheres, and evolutionary models of stars.

Galactic Structure and Evolution. High spatial resolution
maps of the galaxy will define the distribution of gas and
dust and its relation to the stellar content in the galaxy,
which with sensitive long wavelength photometry will permit
the study of the mechanisms that generate the enormous
infrared luminosities of many galaxies and cosmological
objects.

Cosmology. Good narrow-band measurements of the short
wavelength portion of the spectrum is of fundamental
importance to the full understanding of the cosmic background
radiation. Any spatial or spectral anisotropies in this

radiation would have great cosmological significance as it may well be the remnant of the hot early phase of our universe.

Instrumentation

At least two facility-type Spacelab telescopes will be required to make the measurements needed to meet the above objectives. These are:

- meter-class cryogenically cooled telescope
- 2.6 meter ambient temperature telescope

In addition to these, there will be a need to fly rocket-class special purpose telescopes and sensor packages on the Spacelab. These would be flown either in combinations as dedicated infrared astronomy missions or in conjunction with other instruments from compatible scientific disciplines.

NASA is just completing a feasibility study of the meter-class cryogenically cooled telescope, an artist's concept of which is shown in Fig. 5. The European Space Agency is studying the ambient telescope.

The configuration of the cryogenic telescope being studied is shown in Fig. 6 and has the following specifications:

System: Double folded Gregorian configuration
Aperture 1.16 meters
Focal length 7.97 meters
f/# 6.87
Area obscuration 25%
Field of view 15 arc min

Primary: Elliptical configuration
Aperture 1.2 meters
f/# 3.15

Secondary: Elliptical configuration
magnification 2.18

The second folding mirror is articulated to provide space chopping for background discrimination. All mirrors, baffels, and internal surfaces are cooled to $17°K$-$24°K$.

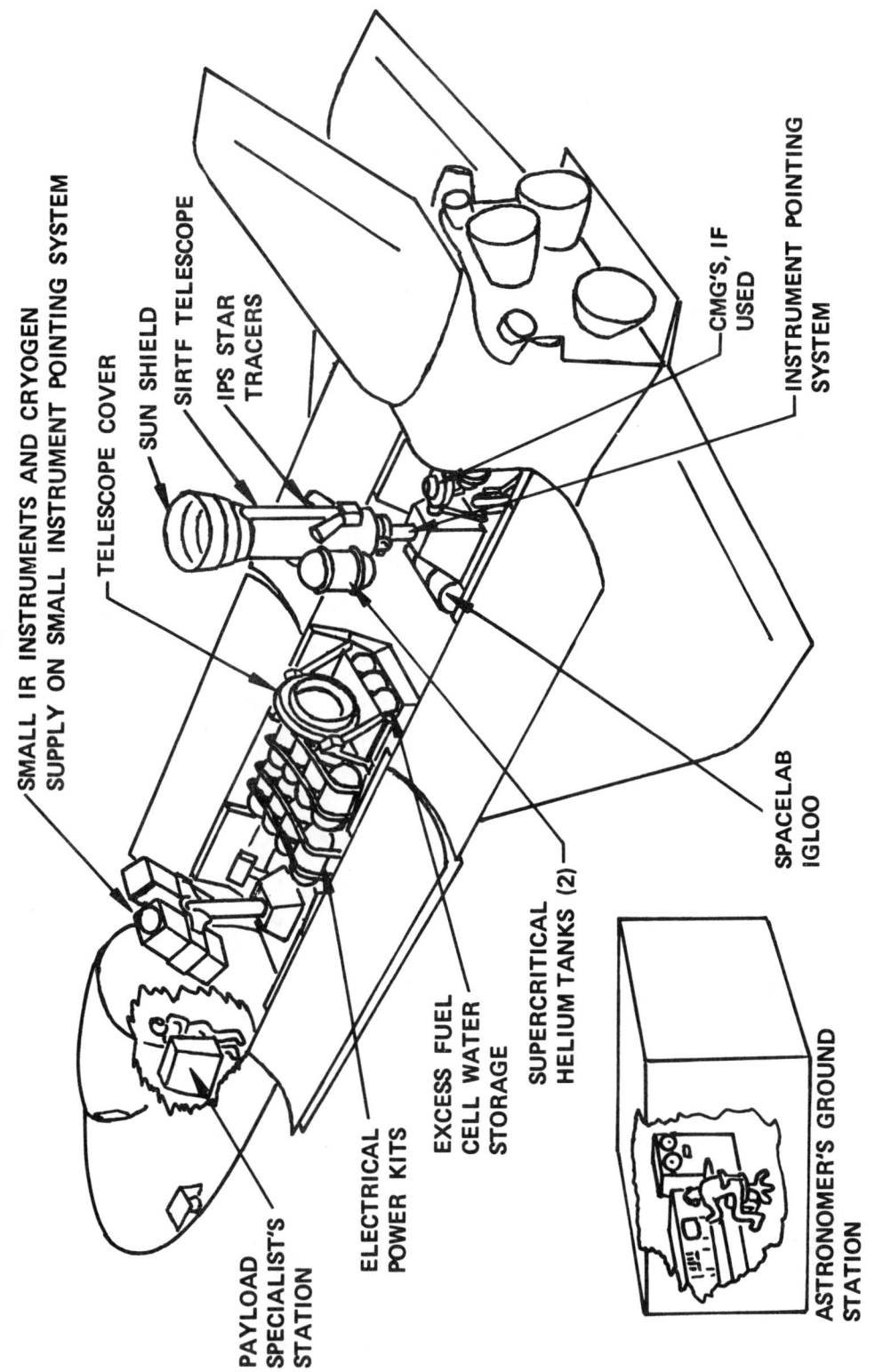

Fig. 5 Cooled IR Telescope as Part of an Integrated IR Payload

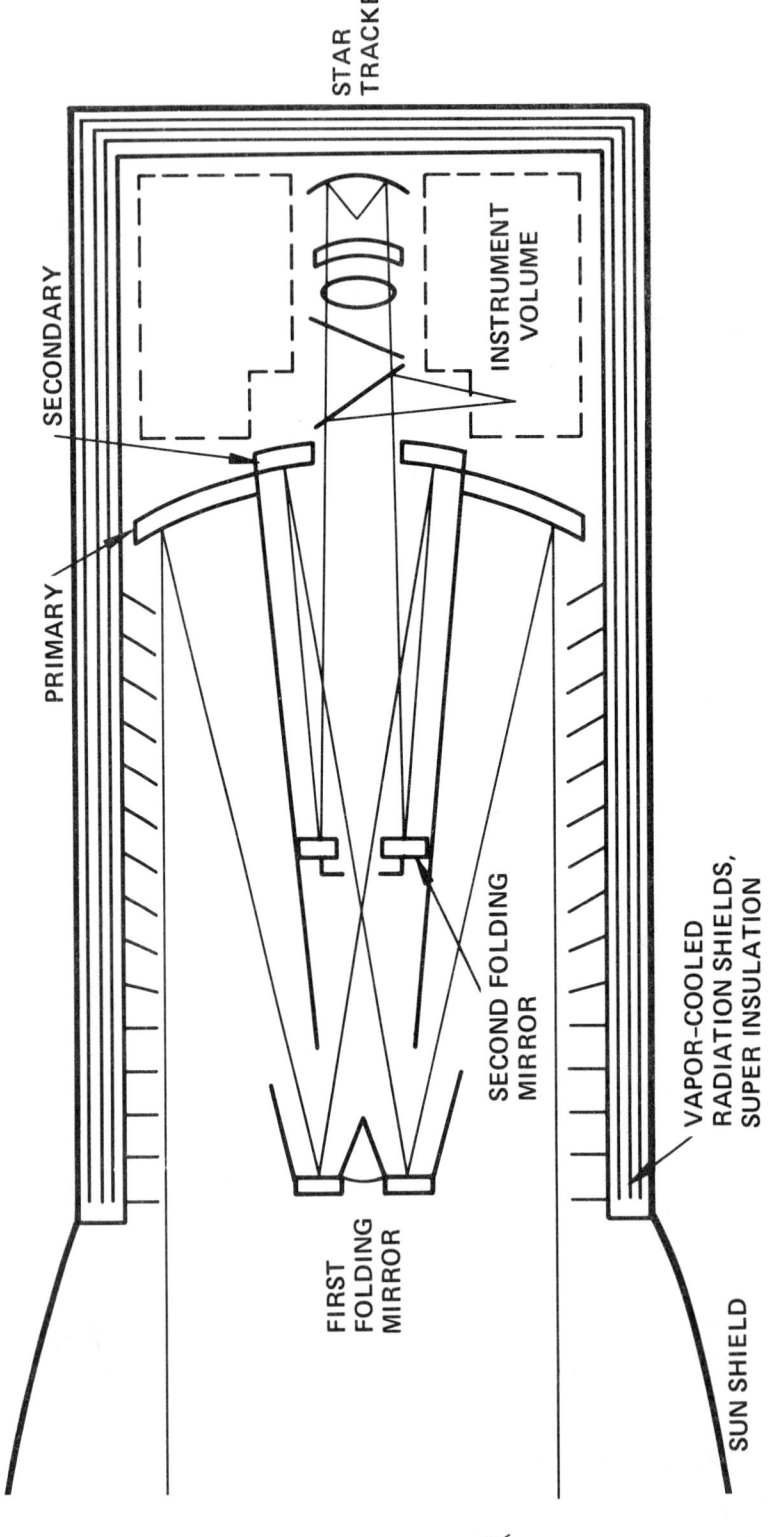

Fig. 6 Telescope Schematic

The development of such an infrared cryogenic telescope for spaceflight is an exciting prospect for astronomy. With this instrument in orbit, it will be possible to study the infrared radiation of galaxies billions of light years away from the earth. It will permit rapid, sensitive searches for sources in stages of currently unstudied stellar evolution. In the seven to 30 micron spectral region, this telescope has the spectroscopic and photometric sensitivity to make in hours observations which would require centuries for ambient temperature telescope of the same size.

Our present estimates are that the cryogenic telescope could be available for a 1981 Shuttle mission. Following this first flight, it could be effectively flown on one or two Spacelab missions per year.

SOLAR PHYSICS

Scientific Objectives

Solar physics research is pursued not only to understand the sun itself but to understand astrophysical processes which have application elsewhere in physics and astrophysics. The sun is an important source of disturbances to the earth's environment: the magnetosphere, the ionosphere, and possibly the atmosphere. The major goals of solar astrophysical research are to obtain detailed understanding of the basic mechanisms (nuclear processes, atomic processes, magneto-hydrodynamic processes, and plasma processes) as they occur on the sun. This understanding would then be applied not only to the sun, but to interplanetary studies, solar terrestrial relations and to the study of other astrophysical objects in which the same basic mechanisms occur.

The achievement of these objectives requires the coordinated use of a broad range of electromagnetic and particulate detectors in space and on the ground. The large weight carrying capability of the Shuttle/Spacelab will permit space observations using instrumentation with large collecting areas and high angular resolution.

Instrumentation

To aid in accomplishing these scientific objectives, we are studying the feasibility of several facility-class and a variety of balloon and rocket-class instruments for flight on the Spacelab.

Meter-Class Normal Incidence Telescope Facility. A preliminary design of this telescope has a Gregorian configuration with the following characteristics:

    Aperture   1.0 meter
    Focal Length   35 meters
    f/#   35
    Field of View   0.1 arc sec

This telescope is suited to studies of the heating of the solar chromosphere by searching for evidence for the formation and dissipation of shock waves, by studying the turbulence spectrum of the photospheric intensity and velocity fields, and by evaluating the relationships of magnetic fields to the structure and behavior of the chromospheric network.

EUV, XUV, and Soft X-Ray Facilities. A facility for the study of EUV, XUV, and soft X-rays would probably contain three instruments. These are:

- a normal incidence off-axis EUV telescope in the wavelength range 400 Å to 1500 Å,
- on XUV Wolter telescope in the 100 Å to 600 Å wavelength range, and
- a soft X-ray Wolter type 1 full figure of revolution telescope for the 2 Å to 100 Å wavelength range.

The EUV-XUV-Soft X-ray observations are directed toward understanding the processes of mass and energy balance and the transport and dissipation of nonthermal energy in the tenuous transition region and corona. The EUV/XUV/Soft X-ray facility will also study the large scale organization of the coronal magnetic field as revealed by magnetically confined coronal material in loop prominences, streamers and interconnections between active regions.

<u>Hard X-Ray Imaging Facility</u>. A hard X-ray imaging facility would include instruments to study X-ray, gamma ray and neutron emissions from the flaring and nonflaring sun, to study the triggering mechanism of flares, to determine the energy content of flares, and to observe the release of charged particles during flares. The spectral energy distribution of X-rays and gamma rays in continuum and line emission is needed as well as the temporal variations and morphology in this spectral region. This facility will consist of four instruments:

- a full sun 5-600 keV spectrometer with temporal resolution of 10^{-2} sec

- a hard X-ray imaging collimator operating in the 5-100 keV range with spatial resolutions of 4 arc sec full width half maximum

- a nuclear gamma ray spectrometer for the 50-100 MeV range, and

- an X-ray polarimeter for the 5-100 keV range located behind the hard X-ray imaging collimator, if possible.

Quick Reaction and Special Purpose Instruments. Quick reaction and special purpose instruments can be included with the facility instruments or on multidiscipline missions. The solar physics instruments being studied include gamma ray and neutron telescopes that will explore the processes of electron and proton acceleration in flares, and visible and near ultraviolet coronagraphs that will infer coronal densities, the temperature profile of the corona and solar wind, solar wind velocities, the hydrogen to helium abundance ratio as a function of position, and the structure of the coronal magnetic field. Rocket and balloon-class instruments which have traditionally provided great opportunity for innovative measurements are planned for inclusion in the program. Monitoring instruments for measuring the level of solar emissions and solar education experiments are also being considered.

Although the ultimate scientific objectives of the solar physics program employ facility-class instruments making simultaneous observations of the sun, early solar physics flights on Spacelab will probably make use of existing equipment that can be modified to upgrade its performance. One possibility would be the reflight of an Apollo Telescope Mount canister of instruments using some Spacelab hardware, illustrated in Fig. 7. Another potential early solar physics Spacelab mission is shown in Fig. 8. This mission contains two balloon-class instruments mounted on one pallet element, eight nonfacility-class and two facility-class instruments mounted on four pallet elements utilizing four Small Instrument Pointing Systems (SIPS) which provide the precision pointing control required. The SIPS canisters provide a controlled thermal environment for the instruments. Missions such as this one in Fig. 6 have been studied for feasibility and are being considered for a 1981 or 1982 launch.

Fig. 7 Apollo Telescope Mount (ATM) Reflight

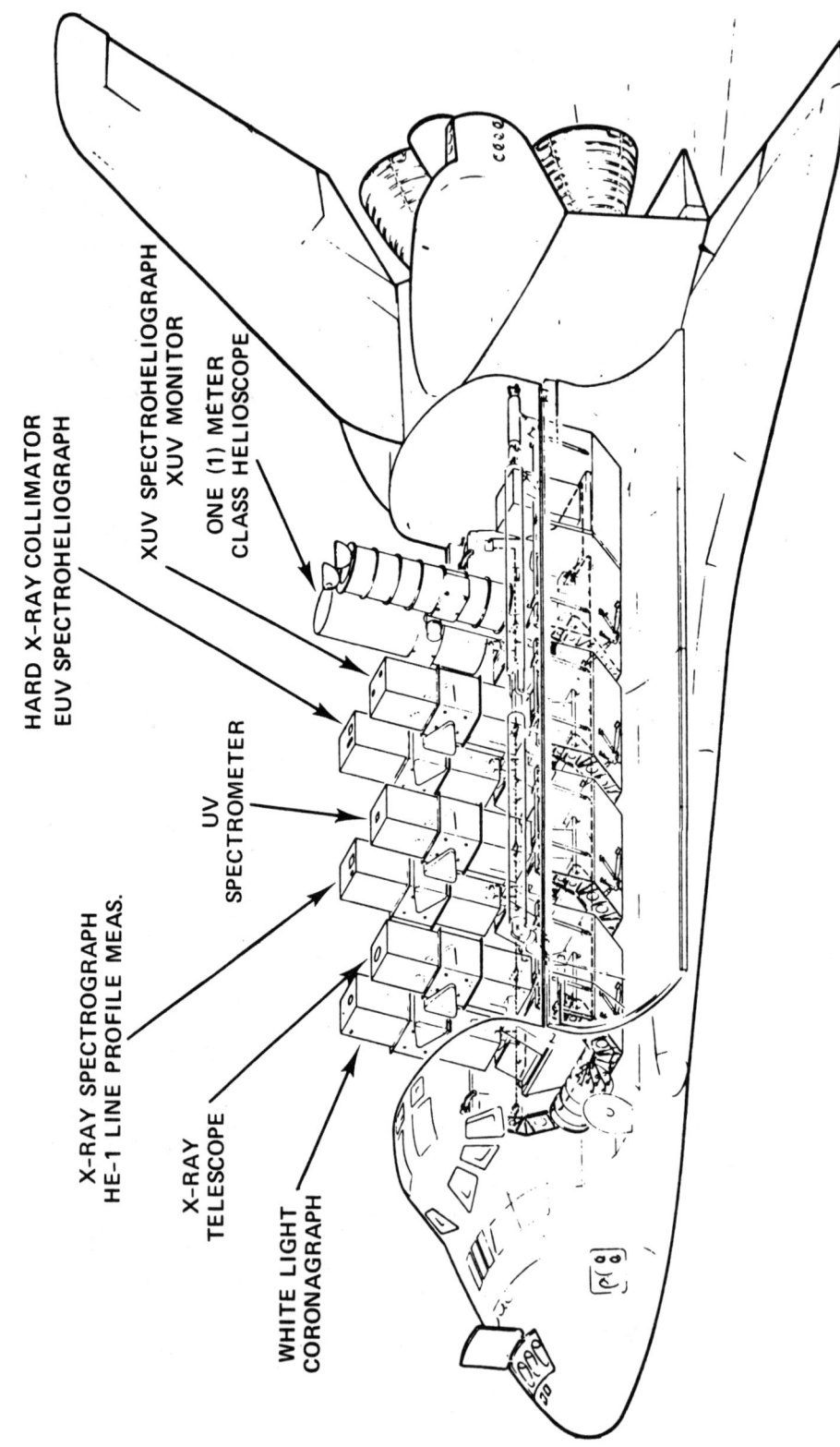

Fig. 8   Shuttle Solar Observatory

ULTRAVIOLET/OPTICAL ASTRONOMY

Scientific Objectives

The overall objectives of this branch of astronomy are to understand the origin and the nature of the universe. Understanding the problems identified to meet these objectives will require studies in areas such as the structure, composition and phenomenology of planetary surfaces and atmospheres; the composition and physical nature of the interstellar medium; the composition, structure and life history of stars, especially those in advanced stages of evolution; the stellar populations of other galaxies; such enigmatic phenomena as X-ray binary black holes, pulsars, active galactic nuclei and quasars; large scale interactions between galaxies and the nature of the intergalactic medium; precise calibration of the Hubble law for the expansion of the universe, the density of the universe, and the existence of extraterrestrial life.

A program for ultraviolet/optical astronomy for Spacelab is being considered which would complement the expected returns from the Large Space Telescope but, at the same time, would have strong scientific merit in its own right. This Spacelab activity concentrates on four observational programs:

<u>High Angular Resolution Imagery Over Fields Significantly Larger Than 2.5 Arc Min.</u> This will include studies of stellar evolution in globular and open clusters, the history of star formation in nearby galaxies and studies of intergalactic matter in clusters of galaxies.

<u>Far Ultraviolet Spectroscopy.</u> Here will be provided improved measurement of the present average density of the universe, and the rates of ion molecule exchange reactions in interstellar clouds, and investigations of the galactic halo of great distances from the galactic plane and the extent of ionized hydrogen and helium around stars.

<u>Precisely Calibrated Spectrophotometry and Spectropolarimetry
Over a Wide Wavelength Range</u>. This will make possible
extension of the interstellar extinction law into the far
ultraviolet; measurement of spectral energy distributions
for X-ray binaries, quasars, Seyfert galaxy nuclei, faint
blue stars, etc.; measurement of bolometric luminosities for
individually resolved globular and galactic cluster stars;
and measurement of polarization of planets, nebulae and
interstellar dust.

<u>Solar System Studies</u>. These could include mapping of
distinct geological provinces on Mercury; observations of
ultraviolet clouds on Venus; studies of the relation between
Martian water ice clouds and the large Martian volcanos;
establishment of heights and distribution of ammonia in the
Jovian and Saturnian upper atmospheres; to name just a few.

<u>Instrumentation</u>

To begin to realize the Shuttle Spacelab potential for
ultraviolet/optical stellar astronomy, two facilities for
the accommodation of scientific instruments are being
defined:

> A general purpose, one-meter class Spacelab Ultraviolet/
> Optical Telescope (SUOT) facility to be mounted on an
> instrument pointing system, which will provide
> wavelength coverage from 90 to 4000 nm and images of
> excellent quality (0.2-0.3 arc sec) over a wide angular
> field ($0.5°$) to interchangeable focal plane instruments
> carried in groups of two to four on each flight, and

> Small instrument pointing systems which will provide
> three-axis stabilization, standard instrument canisters
> for thermal control and contamination protection, and
> command, data and power interfaces for relatively small,
> autonomous instruments analogous to those currently
> flown on sounding rockets, balloons and Explorer-class
> satellites.

The feasibility of both facilities has been established by current NASA studies.

The current concept of the SUOT facility is based upon a one-meter, f/15, Ritchey-Chretien telescope which, with a Gascoigne corrector and a field flattener, will provide a flat field 0.5° in diameter with image diameters in the range 0.2-0.3 arc sec (70% encircled energy) at wavelengths > 2000 Å. Without refractive correctors it will provide similar image quality in a 0.1° flat field or a 0.2° curved field over the wavelength range determined by its optical coatings. The choice of f/15 is the best compromise between desired field size and the dimensions and linear resolution of currently envisioned electrographic or intensified photographic detectors. It is also dictated by the desire to provide full-field baffling, while still maintaining an obscuration ratio below 0.40, and by the difficulties of flattening the strongly curved field of a system as slow as f/30.

The 5 m long SUOT will occupy two 3 m pallet elements when stowed for launch and landing and will thus occupy 40% of the payload volume in a 5 pallet Spacelab flight configuration. The total estimated weight of the SUOT facility, a representative set of focal plane instruments, the IPS and other payload chargeable hardware is approximately 2900 kg.

Spacelab astronomy payloads analogous to current sounding rocket, balloon, airplane or Explorer satellite-class instruments will typically have a minor impact on the overall Spacelab system, a weight $\leq$ 450 kg, dimensions smaller than one pallet element (3 m length) and stabilization requirements in the arc second range. The support facilities for small astronomy payloads will provide a powerful extension of NASA's current sounding rocket program. By analogy with that program, payloads will be developed with relatively short lead-time to bring the most current

technology to bear on timely astronomical problems. A
Spacelab payload dedicated to ultraviolet optical astronomy
has been studied. As shown in Fig. 9, this payload would
include the one meter facility-class telescope (SUOT) on
two pallets using SIPS and three additional pallets of
research rocket experiments with three SIPS's for an
overall total of 14 different experimental instruments. The
smaller experiments, although able to satisfy the target
viewing requirements, are restricted in their flexibility by
having to share the SIPS canisters and thus the available
observation time. A variety of other possible payload
configurations is also possible. A Spacelab payload such
as the one shown in Fig. 9 could be available for flight by
mid-1981, modified, refurbished and reflown up to twice per
year for the next ten years.

HIGH ENERGY ASTROPHYSICS

Scientific Objectives

High energy astrophysics includes the studies of celestial
X-rays, gamma rays, and cosmic rays, observations that must
almost exclusively be carried out in space. The scope of
high energy astrophysics includes nearly all astronomical
objects ranging from normal stars (such as the sun and its
environment) to stars at the end point of stellar evolution
(such as white dwarfs, neutron stars, and possibly black
holes). It also includes the study of our galaxy, its
interstellar medium, other galaxies, clusters of galaxies
and the intergalactic media.

There are three basic scientific areas included in the study
of high energy astrophysics:

Cosmic Ray Astronomy.
The objectives of cosmic ray astronomy
are to determine the sources and acceleration mechanisms for
cosmic energetic particles by accurately measuring the
elemental composition, isotopic composition, and energy spectra
over a wide range of energies and for all known elements.

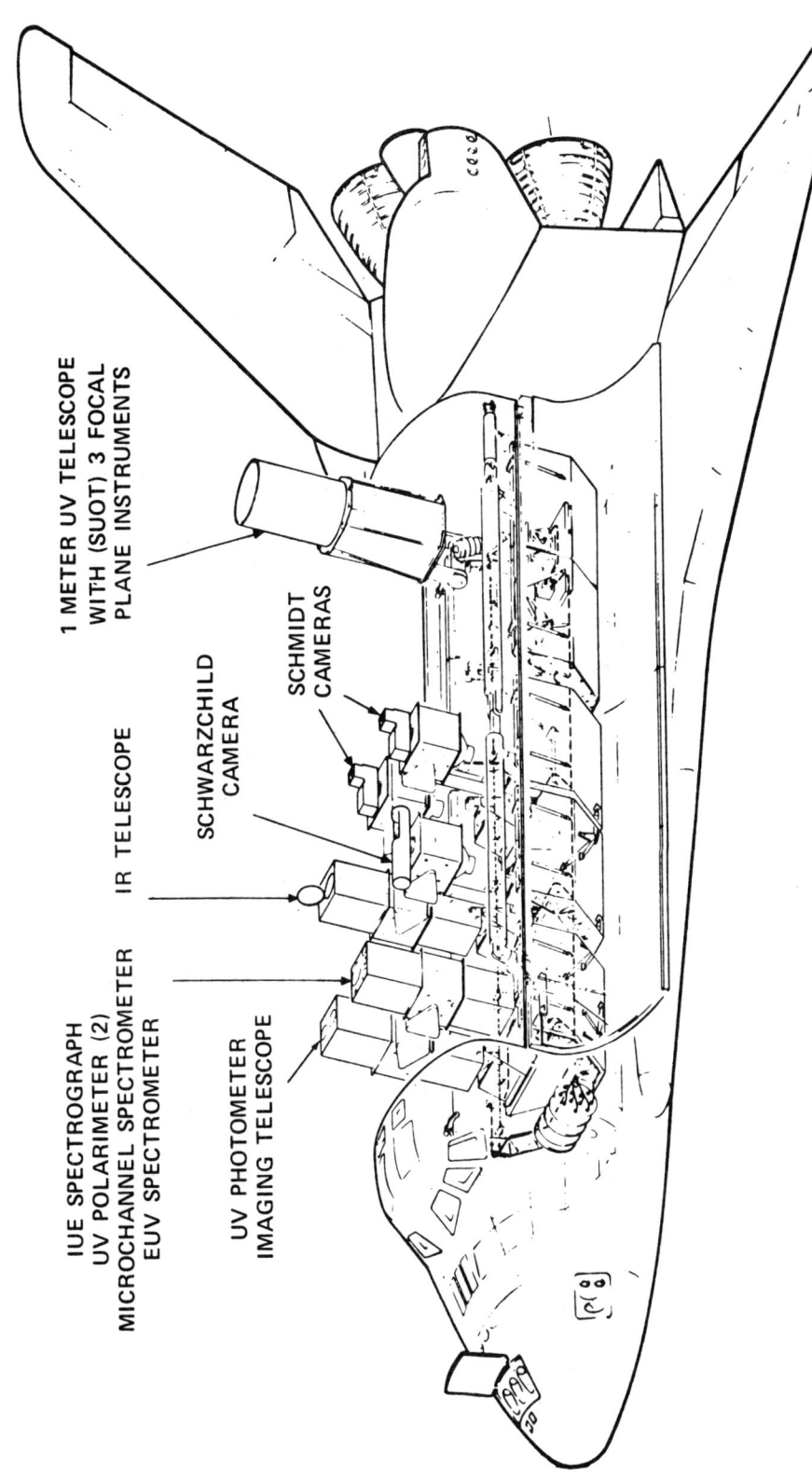

Fig. 9 Shuttle Stellar Observatory

This will contribute to recent developments in the understanding of explosive nuclear synthesis.

<u>Gamma Ray Astronomy</u>. A main objective of gamma ray astronomy is to determine the dynamic properties of our galaxy by measuring the directional and time features of antiparticles. This study provides knowledge of the dynamics and energetic processes in our galaxy and the universe.

<u>X-Ray Astronomy</u>. The objectives of X-ray astronomy are to understand the properties of stars near the end point of stellar evolution and to understand the physics of matter at extreme pressures, densities and magnetic field strengths. X-ray observations provide a study of the generation, containment, and dissipation of high energy particles and the mechanisms of energy transfer in the interstellar medium. They also could provide the key to understanding the fundamental processes that give rise to the enormous production of energy in active galaxies.

<u>Instrumentation</u>

The types of instruments to be flown on the shuttle Spacelab missions in high energy astrophysics have been outlined <u>Scientific Uses of the Space Shuttle</u>, published by the National Academy of Sciences, Woods Hole, 1973 and in more detail recently in <u>A Program for High Energy Astrophysics, 1977-1988</u> by the ad hoc planning group of the high energy astrophysics management operations working group. We have undertaken a Spacelab payload study which has included "typical" experiments which are basically a subset of those listed in the National Academy of Science (NAS) report. The subset was chosen with the goal of not only having representative experiments from each of the disciplines of X-rays, gamma rays, and cosmic rays, but also experiments which would present the more severe strain on Spacelab resources. Typical experiments for high energy astrophysics spacelab missions are listed as follows:

NASA Number	NAS (1973) Identification	Description
GI	SX-1	Large Area X-ray with Concentrator
GII	SX-3	High Energy X-ray Sources
GIII	SX-7	Bragg Spectrometer
GIV	SG-8	High Energy Gamma Rays
GV	SG 5 & 8	Low Energy Gamma Rays and Nuclear Lines
GVI	SC 1 & 4	Cosmic Ray Ionization Spectrometer
GVII	SC 1 & 4	Cosmic Ray Transition Radiation Spectrometer
GVIII	SC-4	Negatron Positron
GIX	SC-2	Isotope Abundance

In studying these experiments, several general concepts related to high energy astrophysics in the Spacelab era have emerged. First, most if not all of these experiments can be accommodated on a single pallet element. Second, in almost every case the technology exists and in many cases the experiments would be extensions of instruments which have been flown successfully on balloons or sounding rockets. Third, there is no single facility-type instrument which dominates the field; rather, there are a large number of generally quite different experiments with different objectives. Fourth, on the basis of past experience in high energy astrophysics, balloon, sounding rocket, and satellite experiments, the principal investigator concept is clearly the most appropriate one to adopt for Spacelab. In this concept, the principal investigator is responsible for the instrument, including its meeting the scientific objectives, quality control, and maintaining the cost within the budget guidelines. In the larger experiments, the experiment team would consist of members of several institutions. The scientist from the various universities and/or government laboratories would combine their talents to develop the experiment, but one scientist, the principal investigator, would have the primary responsibility.

In studying payloads assembled from several high energy astrophysics experiments, it was found that in general, with the exception of a few experiments, it was relatively easy to interchange instruments with little or no impact on the scientific objectives of the individual experiments. Further, most instruments fit efficiently onto a single pallet segment. The flexibility that is gained from these two features greatly facilitates the integration of high energy astrophysics instruments into missions and the interchange of instruments if one develops difficulty.

These nine typical experiments have been grouped into two high energy dedicated Spacelab payloads and analyzed. One of these is shown in Fig. 10. Pointing requirements are relatively coarse and most instruments are easily satisfied with pointing control available from the Shuttle orbiter. Most high energy experiments desire as much viewing time in space as possible and thus extended missions are highly desirable for this discipline. Some or all of the instruments shown in Fig. 10 could be flown on the first pallet-only Spacelab mission in 1980 and all could be available for missions beginning in 1981.

MIXED ASTRONOMY

The modular nature of the pallet-only payloads described for the above astronomy related disciplines suggests the possibility of mixing elements from each discipline to make up a Spacelab payload. The dedicated mission approach is the most desirable and scientifically efficient since the orbit, orientation, and mission sequence can be optimized for a particular discipline. Nevertheless, the three astronomy disciplines are mission compatible, to a certain degree. For example, the solar physics instruments can observe during the daylight side of an orbit and the ultra-violet/optical during the night side with high energy cosmic and gamma ray instruments able to collect data over the entire orbit, except during earth occultation.

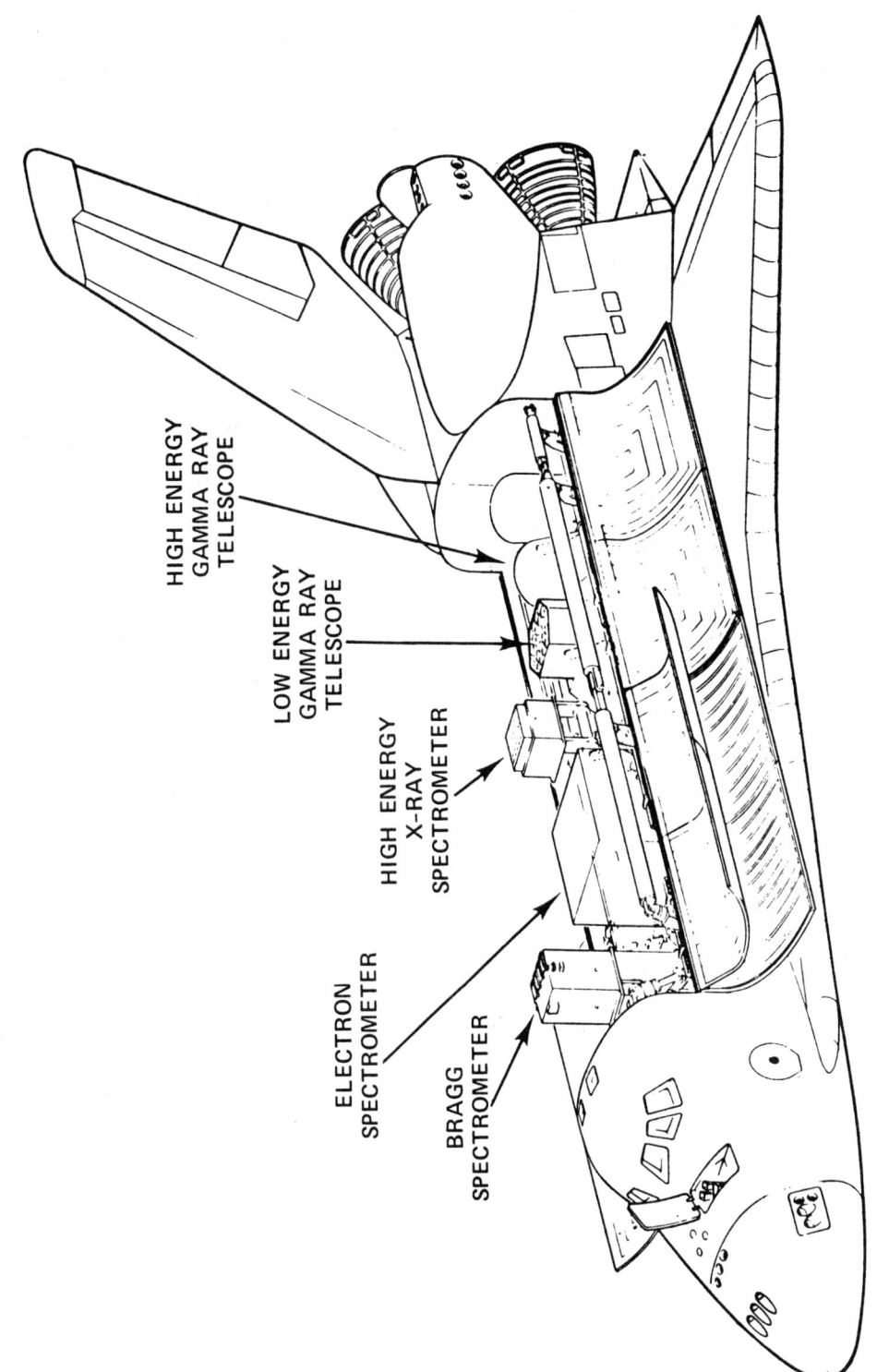

Fig. 10 Shuttle High Energy Astrophysics Observatory

One possible mixed astronomy payload is illustrated in Fig. 11. This payload consists of three pallet elements carrying rocket and balloon-class instruments in the high energy, solar physics, and the ultraviolet/optical areas, respectively, in addition to a high energy instrument too large for a pallet element. This payload is weight constrained but it would be possible to satisfy the operating needs of each of the seven instruments.

FIRST SPACELAB PAYLOAD

The first Spacelab mission will have two prime objectives: the first will be to operationally verify the Spacelab system and subsystems, and the second will be to demonstrate the unique capabilities of the Shuttle/Spacelab system for conducting experiments in science, applications and technology. Since the European Space Agency is developing the Spacelab system, it has been agreed that the experimental objectives for this first payload would be jointly determined and agreed upon. These objectives are:

Demonstrate the capability to investigate the fundamental science in vapor, liquid and solid phase interaction under gravity free conditions, observing among other things:

- crystal growth, metallurgical phenomena and separation of biological material
- cloud microphysics
- drop dynamics

To investigate key natural cause and effects relationships that exist in the near-earth environment by performing active and interactive experiments on and in the earth's atmosphere and magnetosphere.

Conduct investigations on the effects of the space environment (zero G and/or hard radiation-HZE) on body fluid redistribution, vestibular function, growth,

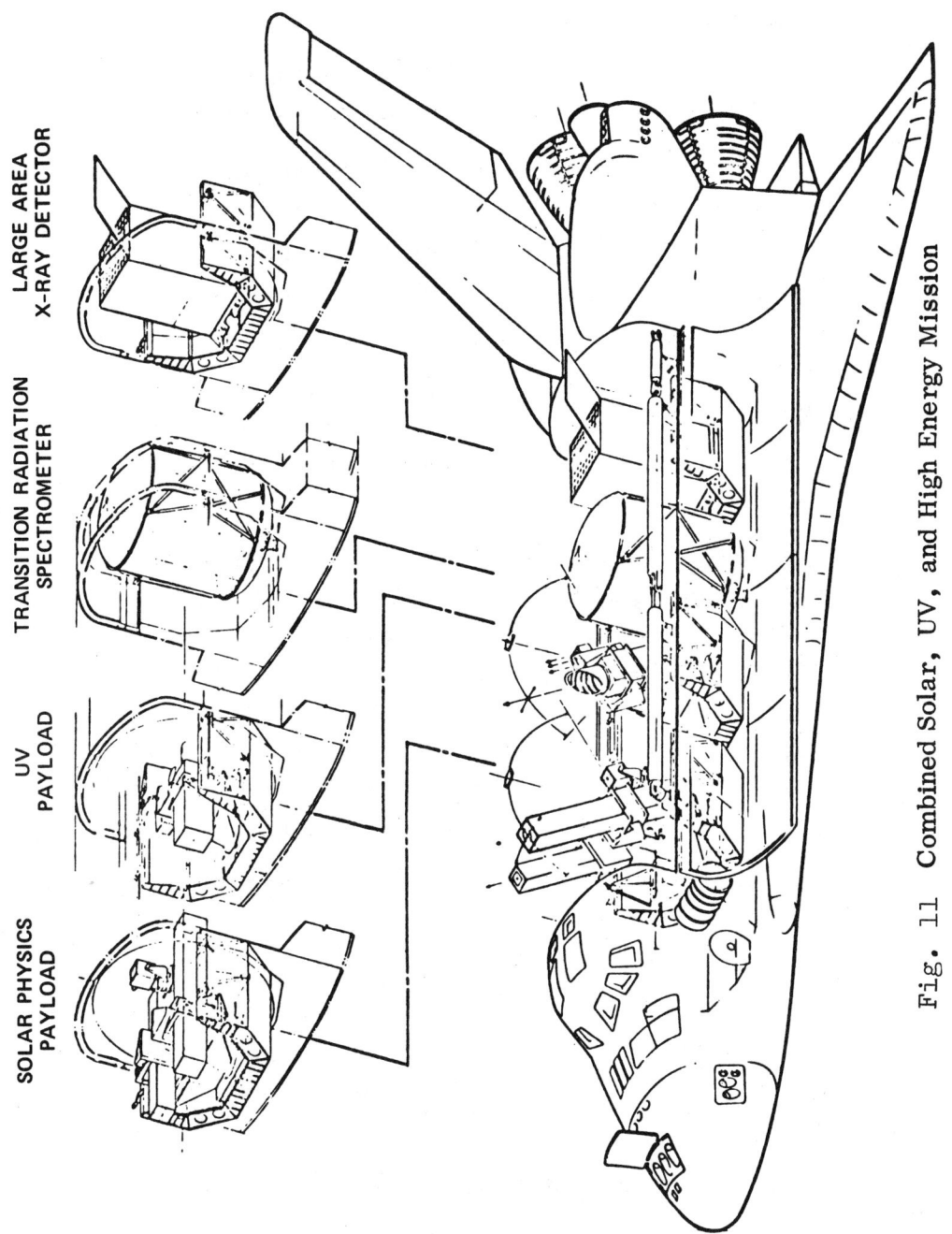

Fig. 11 Combined Solar, UV, and High Energy Mission

development and organization on living systems such as man, animals, plants, cells and tissues.

Demonstrate the capability to monitor the atmosphere and its effect on environmental quality by surveying the atmosphere for trace constituents, identifying their sources, flow patterns and decay mechanisms.

Demonstrate the capability to observe and monitor the earth's surface; in particular, to obtain high resolution, metric quality images, and to develop spaceborne all-weather remote sensing methods.

Observe extended sources of radiation in the visible, ultraviolet, and infrared spectra too faint for earth-based observations and possibly evaluate the effect of the Shuttle/Spacelab environment on such astronomical studies.

Demonstrate and use the capability of Spacelab as a technology development and test facility to perform experiments in the space environment in areas such as tribology and heat transfer.

In the field of communication, conduct investigations that will provide a basis for the efficient utilization of orbital spacing and frequency spectrum, including:

- studies of effects and anomalies of propagation from earth and space, and
- demonstrate the performance and operational capabilities of advanced satellite communications and navigation subsystems.

Certain configuration and resources constraints will be imposed by the verification requirements of this first mission. Key among these are that the configuration will be a long pressure module with one pallet element, the instrument pointing system will not be available, high data rate transmission via the Ku-band antenna will not be operational, and only one experiment airlock will be used. These constraints, along with the above experimental objectives, will form the basis upon which the experiments will be selected for this first mission. It is planned that the second Spacelab flight, also a verification flight, will be a pallet-only mission.

As seen from the experimental objectives, the first payload will be a multidiscipline payload. One guideline for payload selection will be to select experimental instrumentation that can and will be used on subsequent Spacelab flights. Accordingly, many elements of the Spacelab payloads described above, and elsewhere, will be candidates for this first mission. It is not planned to develop instrumentation that would have first mission use only.

CONCLUSION

Feasibility studies have been conducted for a substantial number of scientific and technological payloads for Spacelab. Payloads are being studied to use the full range of Spacelab configurations and available resources.

A number of very important scientific objectives in a broad range of disciplines can be effectively addressed by the payloads in planning. The potential for successfully solving vital scientific questions from Spacelab payloads appears very high. In addition, if the plans outlined for these payloads materialize as expected, there will be many exciting opportunities for researchers to participate in the solution of these problems.

AAS 75-126

DEVELOPMENT OF SHUTTLE PAYLOADS

James T. Murphy*

The development of payloads for the Shuttle is described in terms of two activities — one is oriented towards development of payload hardware, while the other is oriented towards the development of payload/Space Transportation System (STS) planning, scheduling, and integration techniques. This second activity is called STS Utilization Planning (SUP). In addition to describing the major aspects of SUP, a discussion of development activities on three payloads that typify direct Shuttle placement missions is included. These payloads are the First Spacelab Mission, the Atmospheric, Magnetospheric, and Plasmas in Space (AMPS) payload, and the Large Space Telescope (LST).

---

*Director, Program Development, NASA-Marshall Space Flight Center, Huntsville, Alabama.

INTRODUCTION

The Space Transportation System (STS) represents an entirely new approach to providing access to space. It offers a significantly larger return capability and more flexibility than any space vehicle previously available. The unique capability of the STS will open a new era in space, and we must strive to obtain maximum benefit from the national investment in this unique transportation system.

The development of payloads for future STS missions requires that the goals and objectives established by the users (DOD, scientific community, etc.) be transformed into the various instruments and equipment which can gather the desired data. Also the STS must be developed in such a way that it supports the needs of its "customers" (or users), who must plan their activities so as to capitalize on the unique characteristics of the STS. To ensure that the transportation needs of the users are satisfied by the STS as efficiently as possible and to provide to the user an overview of STS systems capabilities and progress toward these capabilities, NASA is actively engaged in the planning, scheduling, engineering, and integration needed to fulfill these objectives. This overall activity is referred to as STS Utilization Planning (SUP).

Since other papers and sessions at this conference are to be devoted to payloads, this paper will concentrate on the SUP aspects and will discuss briefly two payloads that typify the two classes of Spacelab payloads. These payloads are the First Spacelab Mission which is typical of "multi-user" missions and the Atmospheric, Magnetospheric, and Plasmas in Space (AMPS) payload which is a typical "exclusive-user" Spacelab mission. Also discussed is the Large Space

Telescope (LST) which is a large automated satellite. The LST is an example of a payload that can be placed directly in orbit by the Shuttle and subsequently maintained and upgraded by the Shuttle to extend the payloads's useful lifetime.

STS UTILIZATION PLANNING

By way of general introduction to this effort, an analogy can be drawn to the transportation industry. This industry has long been aware of the vital role that careful planning and scheduling play in its operations, and the cargo master or master scheduler, as he is sometimes called, is recognized as a vital link in the efficient operation of any carrier. Further, as the pressures for better efficiency have grown, more integration among all the elements involved (carriers, shippers, terminals, etc.) has been effected (e.g., containerized cargo), and a systems approach has emerged that is producing a more optimum solution to the "end to end" or total transportation problem. This has not only required improved planning and scheduling, but also greater overall integration of the diverse elements of the total systems. In the STS, SUP is analogous to a combined cargo master and overall system integrator.

For the STS, the operating problems are not a great deal different from other transportation systems, and the objective is the same — to obtain maximum efficiency from a transportation system — but the problem is significantly more complex. While much of the discussion often centers on the three basic pieces of flight hardware, the Shuttle, the Spacelab, and the Upper Stage, the total system is much more encompassing than this. It is a complex, "organic" system where many interdependent parts function toward a common goal. For the total system to function effectively, not only must the capability and availability of the basic STS flight hardware be assessed, but also the combined or total systems effect of mission objectives and schedules, compatibility of the various experiments with each other, experiment/carrier interdependence, ground operations, flight operations, post-flight activities, software capability, data handling, network availability/capability, facilities, support equipment, logistics timelines, etc., must be understood. While responsibility for the above functions is assigned to many different organizations, it is the role of SUP to provide the broad systems engineering, integration, and planning needed to "orchestrate" the many parts of the STS, so that they "play" together harmoniously. To accomplish this requires a level of sophistication in analysis, scheduling, and planning well above anything that has been used to date and an integration of "customer" needs and carrier accommodation to a degree not seen heretofore.

The basic activities in SUP are depicted schematically in Fig. 1. The genesis of STS Utilization Planning is shown in Box 1 of Fig. 1. It begins with liaison with potential STS users and an assimilation of their needs and requirements. As this work with the users proceeds, candidate experiments and payloads are identified and cataloged. This catalog is called the Payload Model. Once these potential uses of the STS are cataloged, work is done to characterize each of the payloads in the model and, in conjunction with the user, to develop technical descriptors of them. This work is an iterative process as new information is developed and as designs mature. This information forms the data base from which all of the analyses, planning, interface, and integration activities flow.

The baseline NASA Payload Model, released in October 1973, contains 986 candidate payloads. It covers a period of 12 years beginning in 1980. The Payload Model was discussed with Congress last year. As stated at that time, it does not represent Agency plans, rather it is a baseline to understand the intricacies of user needs and STS operations. The 986 payloads are distributed among the major users as shown in the diagram in Fig. 2 (51 percent NASA, 31 percent DOD, and 18 percent non-NASA).

A more recent Traffic Model has been established that requires a total of 572 Shuttle flights to satisfy the mission requirements. The 572 Shuttle Traffic Model represents a building of capability to 60 Shuttle flights a year. This provides a non-varying baseline model that can be used to determine the requirements for the major STS Program Offices, i.e., Space Shuttle, Interim Upper

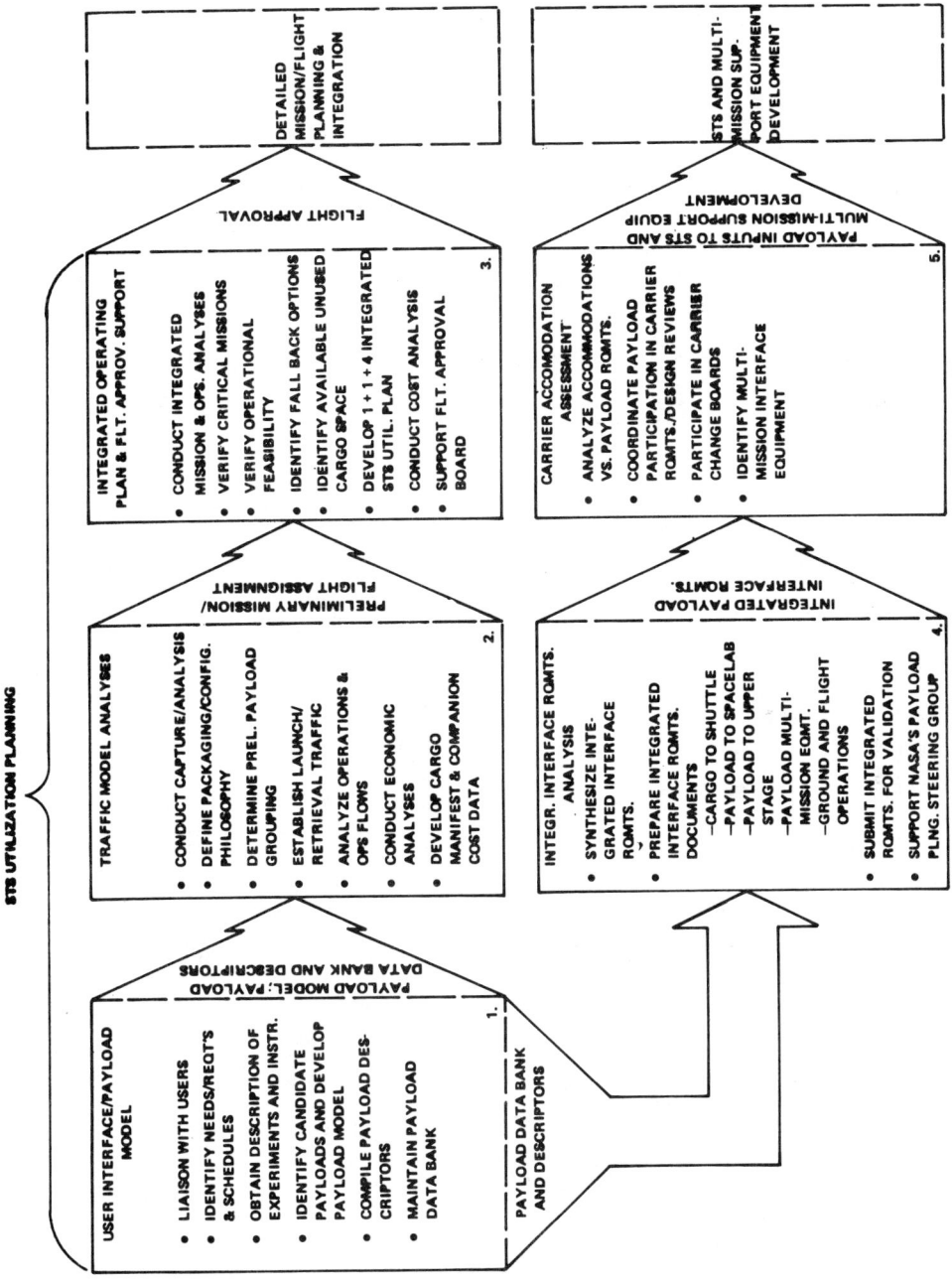

Fig. 1 Phases of STS Mission and Flight Planning

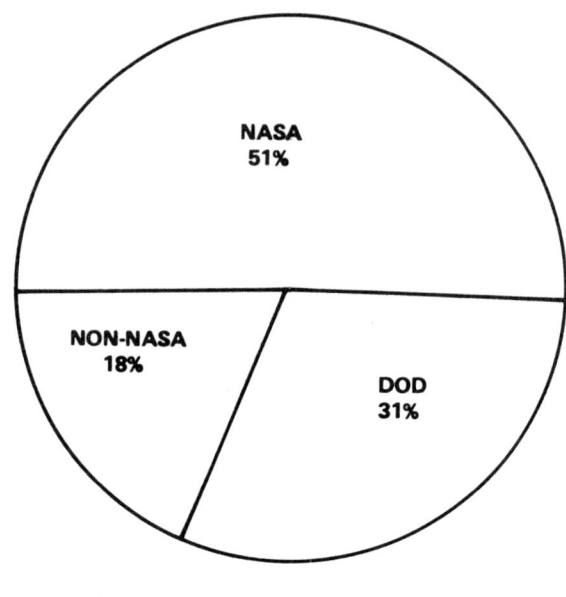

DISTRIBUTION OF 986 PAYLOADS

Fig. 2 STS Payload Distribution

Stage (IUS)/Tug, and Spacelab. The Shuttle operations traffic is divided into three major areas: Spacelab, IUS/Tug, and Shuttle direct placements. From Table 1, one can see that after the early years of initial operations and the activation of the Western Test Range, the traffic is distributed approximately equal within these three categories. Also, the model shows the anticipated distribution of traffic between the Eastern Test Range (ETR) and the Western Test Range (WTR).

Box 2 of Fig. 1, entitled "Traffic Model Analyses," depicts how the payloads are grouped and integrated with the STS to obtain maximum utilization from the carrier and supporting ground and flight operations equipment consistent with customer needs.

Beginning with the Payload Model and its accompanying data base, analyses are run on the various carrier and payload parameters to determine those

Table 1

572 SHUTTLE TRAFFIC MODEL

	79	80	81	82	83	84	85	86	87	88	89	90	91	Total
Shuttle R&D	3	3												
Shuttle Operations														
Spacelab		2	6	12	17	19	21	21	24	24	24	27	29	226
IUS/Tug		3	8	12	15	17	22	21	21	20	19	20	19	197
Shuttle Direct		0	1	0	16	24	17	18	15	16	17	13	12	149
Total	0	5	15	24	48	60	60	60	60	60	60	60	60	572
Total Shuttle Flights	3	8	15	24	48	60	60	60	60	60	60	60	60	578
ETR	3	8	15	23	36	40	40	40	40	40	40	40	40	405
WTR	0	0	0	1	12	20	20	20	20	20	20	20	20	173

payloads that are compatible with each other and with the STS capability. These analyses consider the weight of the cargo and the volume of the cargo bay, the center of gravity of the composite cargo, the influence of the STS on the payload, the influence of the payloads on each other, and the services required for the payloads on the ground and in flight. In addition, these groupings are analyzed for cost effectiveness and schedule lead times. A preliminary mission/flight assignment evolves from this activity. Several iterations are required to reach optimal groupings. Typical major activities supporting these analyses are shown in Table 2.

Table 2

TYPICAL MAJOR ACTIVITIES REQUIRED FOR INTEGRATED
MISSION PLANNING

- Mission Analysis
    - Orbit Selection
    - Flight Profiles
    - Event Sequence
    - Experiment Opportunities
    - Launch Opportunities
- Systems Engineering Analyses
    - Configuration Layouts
    - Systems Integration
    - Payload/Resources Compatibility
    - Mass Characteristics/C.G.
    - Structural Analysis
    - Thermal Analysis
    - Contamination Analysis
- Avionics Analysis
    - Electric Power
    - Instrumentation
    - Computer and Data Management
    - Guidance and Navigation
    - Command and Telemetry
- Ground and Flight Operations
    - Ground Flows and Schedules
    - Network Requirements and Scheduling
    - GSE/Facilities/Manpower
    - Handling, Transport Plans
    - Ground Software Sizing
    - Safety Plans
- Costs
    - Payload Costing
    - Spacelab Costing
    - Integration Costing
    - Transportation Costing

Before finalizing flight assignments, additional detailed work is accomplished, as reflected in Box 3 of Fig. 1, entitled "Integrated Operating Plan and Flight Approval Support." These activities are the in-depth doublecheck of the conclusions reached in arriving at the preliminary mission/flight assignments.

Currently, emphasis is being given to developing recommended groupings of payloads for the first 2 years of STS flights. A sample of some early payload groupings is given in Tables 3 and 4. Several iterations are required before the final assignments of payloads are made. You will note that there is a variety of missions being studied involving domestic, military, and foreign payloads. In-depth analyses of selected missions are in progress, whereby fuller understanding of the interaction between the payloads, and the carriers and payloads with each other will evolve. This information will flow back to improve both the carriers and the payloads.

The transition from expendable vehicle operations to the STS is also receiving detailed analysis. Some effort has been expended in this respect, and as the STS solidifies in the coming year, this problem will receive additional emphasis. In this case, STS Utilization Planning is coordinated with the Office of Space Science which is responsible for expendable launch vehicle services. The transition from expendable vehicles to the STS must be smooth and must include proper contingency planning.

Another significant portion of the STS Utilization Planning effort at MSFC involves the compilation, validation, and integration of payload requirements. The activity provides a focal point for the payload community to deal with the carriers on interface matters. Boxes 4 and 5 of Fig. 1 summarize the effort to understand the ramifications of customer or user requirements versus carrier accommodations. These two boxes illustrate the continuing effort in the evolution of system requirements, the review of the STS compliance with these requirements, the

Table 3

EARLY SHUTTLE MISSIONS PLAN, 1980 FLIGHT SCHEDULE

Flight No.	Date		Payloads	Inclination (degrees)	Altitude (km)
7	June	IUS	DOD Dedicated	—	—
8	July	L + P	NASA/ESRO Spacelab	57	470
9	Sept.	IUS	DOD, Space Technology Program Satellite Retrieve Long Duration Exposure Facility	—	—
10	Oct.	P	High Energy + Multi-Applications	28.5	300
11	Dec.	IUS	DOD Dedicated	—	—

IUS — Interim Upper Stage

L + P — Spacelab Lab + Pallet

P — Spacelab Pallet

IUS/KS — Interim Upper Stage/Kick Stage

Table 4

EARLY SHUTTLE MISSIONS PLAN, 1981 FLIGHT SCHEDULE

Flight No.	Date		Payloads	Inclination (degrees)	Altitude (km)
12	Jan.	L*	Life Sciences, Demo. Teleoperator	28.8	400
			Deploy — Explorer Space Technology Program		
			— DOD Space Technology Program Satellite		
13	Feb.	IUS	DOD Dedicated	—	—
14	Mar.	L + P	Multi-Discipline: Space Processing Communication, Navigation, ESRO	50	278
15	Apr.	IUS/KS	Deploy — Explorer — Lyman α		Escape
			— Explorer — Medium Altitude	28.8	1852 × 37,038
			Retrieve Solar Maximum Satellite	28.8	490
16	May	IUS	DOD Dedicated	—	—
17	Jun.	L + P	Advanced Technology Lab	55	370
18	Jun.	IUS	DOD Dedicated	—	—
19	July	L + P	Atmospheric and Space Physics	55	370
20	Aug.	IUS	Deploy — Foreign Synchronous Meteorological Satellite	0	Sync.
			— Geosynchronous Operational Environmental Satellite	—	—
21	Sept.	P	Infrared Telescope + ESRO	28.8	400
			Deploy — Biomedical Experimental Scientific Satellite	—	—
22	Sept.	IUS	DOD Dedicated	—	—
23	Oct.	—	Deploy — Earth Observatory Satellite — C	—	—
24	Oct.	IUS	Deploy — Foreign Communication Satellite	0	Sync.
			— Disaster Warning Satellite	—	—
25	Nov.	IUS	DOD Dedicated	—	—
26	Dec.	L	Life Sciences		
			Deploy — Solar Maximum Satellite	28.8	480
			— Space Technology Program Satellite		

*See Table 3 for key.

analysis of interfaces between the STS and its payload, and the definition of systems required to conform to these interfaces. This work assures compatibility of the STS and payloads.

During the past year extensive analyses were performed in this area, and a series of documents, which we call "Integrated Interface Requirements Documents," were published. These documents reflect the payload community's needs for carrier-supplied services. Typical of these documents are "Cargo to Shuttle Interface Data Book," "Spacelab Payload Requirements Document," "IUS Payload Requirements Document," and the "Ground Facilities Requirements Document."

These requirements documents state explicitly what the payload community desires from the carriers. The carriers prepare accommodation documents that state what they can efficiently provide. The payload requirements are then compared to the accommodations available and, through an iterative process, the needs of the payload community are reconciled with the STS accommodations. Included are land-based, as well as flight, systems. Where it is not practical for the carrier to accommodate the need, supplemental equipment is identified. MSFC is engaged in studies to standardize this supplemental equipment so that it can be used repeatedly. This standard equipment, called Multi-Mission Support Equipment, will be developed and provided to a variety of users from an inventory. This approach will lower equipment costs and simplify utilization of the STS by the payload community.

One final area is the planning of ways to better exploit the inherent capabilities of the STS. For example, studies are underway to examine the concept for servicing satellites on orbit. Early indications are that this can be an important element in reducing the overall cost of operating in space. This, together with our studies that explore areas of innovative applications for the STS, are being undertaken to ensure that the Space Transportation System yields maximum benefits to the nation.

In summary, the STS Utilization Planning is a key link in bringing the nation's Space Transportation System to maturity with high utility and cost effectiveness. It provides the mechanism for:

a. Assuring the user his requirements are accommodated.

b. Maximizing utilization of the STS while minimizing inventory.

c. Recommending payloads and compatible grouping of payloads for flight.

d. Maximizing STS utility and minimizing payload changes during integration with the carrier(s).

e. Minimizing total system cost.

PAYLOAD DESCRIPTIONS

Payloads for the Spacelab generally are classified "exclusive-user" or "multi-user" depending on the type of flight on which they are accommodated. "Exclusive-user" flights are those flights for which a single NASA center and/or organization provides and/or controls the entire payload. "Multi-user" flights are those flights for which more than one NASA center and/or organization provides the instruments that are integrated onto or into the Spacelab. The First Spacelab Mission could be either an "exclusive-user" or a "multi-user" payload. The AMPS payload is depicted in this paper as an "exclusive-user" payload. However, we believe that the initial payloads in this area will be flown on "multi-user" flights. Therefore, the AMPS configuration evolves only after precursor instruments or equipment have been flown on earlier missions. This evolutionary instrument development approach permits a comprehensive program of scientific investigations in various disciplines to begin early in the Spacelab/Space Transportation System lifetime without requiring large expenditures of discipline funds. Also, this approach allows a learning process to be used in the design and selection of instruments to be flown on later dedicated flights.

First Spacelab Mission

Although the primary purpose of the first two Spacelab flights is Spacelab systems verification, space and resources are available for a limited experiment program. During the past year, NASA and ESRO have engaged in numerous planning meetings to establish a set of experimental objectives to be accomplished on the First

Spacelab Mission. NASA and ESRO agreed that two of the more important criteria for selection of a candidate payload would be:

   a. The payload should be complementary to and consistent with future Spacelab missions, while emphasizing reusable hardware elements and techniques.
   b. Payloads should demonstrate to the user community and the general public the uniqueness of Spacelab and its broad potential for research and applications.

With these criteria and certain established mission guidelines and constraints, NASA and ESRO derived several sets of experimental objectives. MSFC accomplished system analysis studies on various experimental complements to ensure general compliance with all mission constraints. Based on the results of the systems analyses, NASA and ESRO agreed that the following experimental objectives should have prime consideration for the First Spacelab Mission:

   a. Investigation of key natural cause and effects relationships that exist in the near Earth environment by implementing active and interactive experiments on and in the Earth's atmosphere and magnetosphere.
   b. Demonstration of the capability to investigate the fundamental science of vapor, liquid, and solid phase interactions under gravity-free conditions.
   c. Investigations of the effects of the space environment on body fluid redistribution; vestibular function; and growth, development, and organization of living systems (i.e., man, cells, tissues, etc.).

The first experimental objective (item a) can be realized by flying several precursor versions of AMPS instruments, such as the lidar system, an electron accelerator, and/or a diagnostic package. In addition to providing valuable scientific information, data obtained from these instruments will be useful in configuring the AMPS payload and in accomplishing its scientific objectives.

The second experimental objective (item b) can be achieved by the flights of some precursor Space Processing instruments/experiments (e.g., crystal growth or separation of biological materials) and some early Zero-G Cloud Physics experiments (e.g., warm cloud processes).

Thus, the First Spacelab Mission (Fig. 3) affords the opportunity to accomplish a meaningful science program at a reasonable cost and at the same time provide the requisite knowledge for configuring more comprehensive Spacelab payloads for future flights.

Fig. 3 First Spacelab Mission

AMPS Payload

The Atmoshperic, Magnetospheric, and Plasmas in Space payload will be developed as an efficient reusable laboratory that can actively probe the coupled dynamics of the Earth's atmosphere and magnetosphere, while permitting a close interactive involvement of the scientist in each experiment. The laboratory environment provided by the Spacelab permits the scientist to control the experiments and to modify them interactively on the basis of the results obtained. The AMPS payload (Fig. 4), with its basic set of core instruments (Fig. 5), will be designed to accomplish a variety of different experiments during a series of recurring flights, which would probably include international participation.

Fig. 4 Lidar Probing of the Upper Atmosphere

As a result of the recommendations from the National Academy of Science's Woods Hole Summer Study 1973, NASA competitively selected scientists to participate in the detailed definition of the AMPS payload. The AMPS Scientific Definition Working Group began operation in August 1974 and to date has defined the scientific objectives, cataloged candidate experiments, identified required instrumentation, and provided the functional requirements for these instruments.

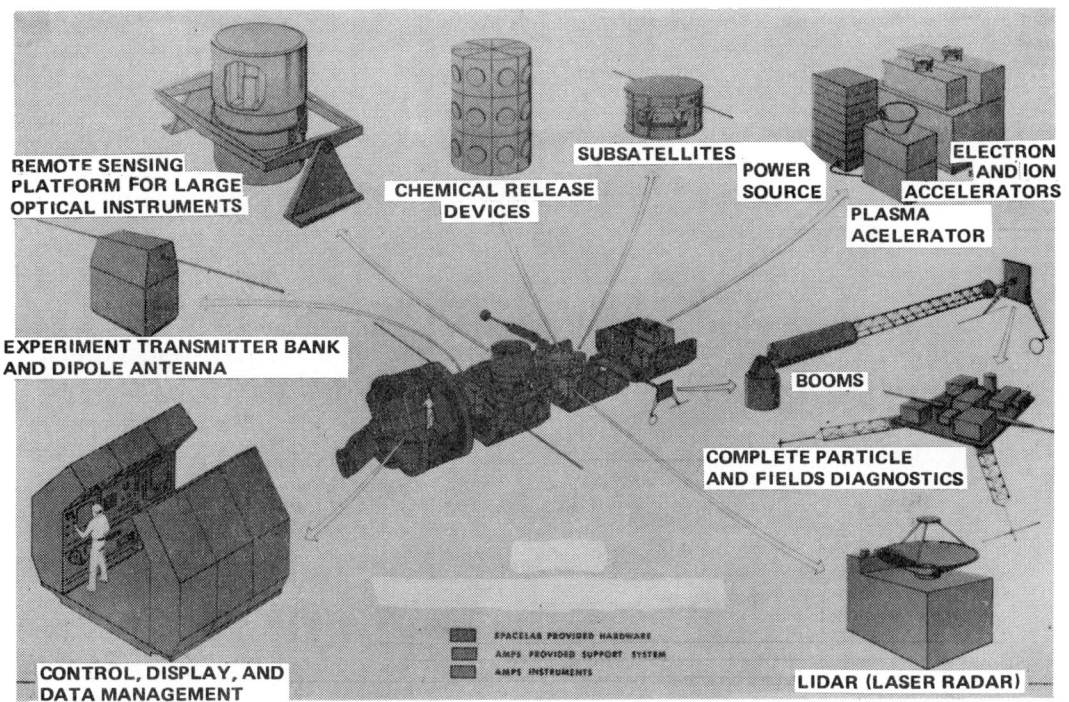

Fig. 5 AMPS Payload Elements

The Working Group includes members from eight foreign countries — Canada, England, France, Germany, Japan, Netherlands, Norway, and Sweden.

The initial phase of the Working Group's activities will be completed in the fall of 1975, in conjunction with the beginning of two parallel Phase B studies by outside contractors. In the second phase of activity, the Working Group will act in an advisory capacity to the contracted Phase B activities.

NASA will release an Announcement of Opportunity (AO) in 1976 for investigations and associated instrument development for the first AMPS mission planned for 1981. Subsequent AO's will solicit specific investigations for follow-on missions.

Several technology areas have been identified and are being studied in support of the AMPS Scientific Definition Working Group. These areas include the definition of scientific instruments and support systems. In general, technology activities on the scientific instruments are being carried out through contracted efforts, whereas several support system technology studies are being conducted in-house by MSFC. The technology studies typically include the following:

   a. Accelerators — These studies provide the basic criteria for the design of electron, ion, and plasma accelerators. They also include definition of functional requirements for the accelerator systems and an assessment of the need for and the sizing of support systems.

   b. Lidar — Studies in this area are concerned with defining the required laser output characteristics which include the power profile, beam resolution, operating profile, and collector optics.

   c. Transmitters and Antennas — These studies investigate the frequency ranges, power, and antenna types required to conduct a variety of wave phenomena experiments.

   d. Chemical Release Devices — A family of chemical release devices including canisters and modules is required to conduct many AMPS experiments. These efforts are concerned with establishing the canister sizes, gas release techniques, means of canister deployment and positioning, and general mission definition and analysis.

   e. Optical Spectrometers — Several studies are being carried out to define the requirements for optical spectrometers, interferometers, and photometers that the experimenters will need to conduct their investigations. These studies define the focal plane optics, cooling requirements, and pointing and stability required to conduct a variety of investigations.

The AMPS payload is oriented toward the study of those mechanisms that influence and control the Earth's environment. Using the results of the AMPS investigations, scientists will be able to understand the dynamic processes of the atmosphere and magnetosphere that can influence everyday problems such as communication blackouts, effects of contaminants on the Earth's atmosphere (e.g. ozone depletion), and solar-weather relationships.

LARGE SPACE TELESCOPE

The primary objective of the LST program is to develop and operate a large, high resolution optical space telescope system that is unique in its usefulness to the international science community and will significantly extend man's knowledge of the universe.

Like ground-based telescopes, the LST (Fig. 6) will be designed as a general-purpose instrument, capable of utilizing a wide variety of different scientific instruments at its focal plane. This multipurpose characteristic will allow the LST to be used effectively as a national facility, capable of supporting the worldwide astronomical needs of an international user community. By using the Space Shuttle to provide scientific instrument upgrading and subsystems maintenance, the useful and effective operational lifetime of LST will be extended to a decade or more.

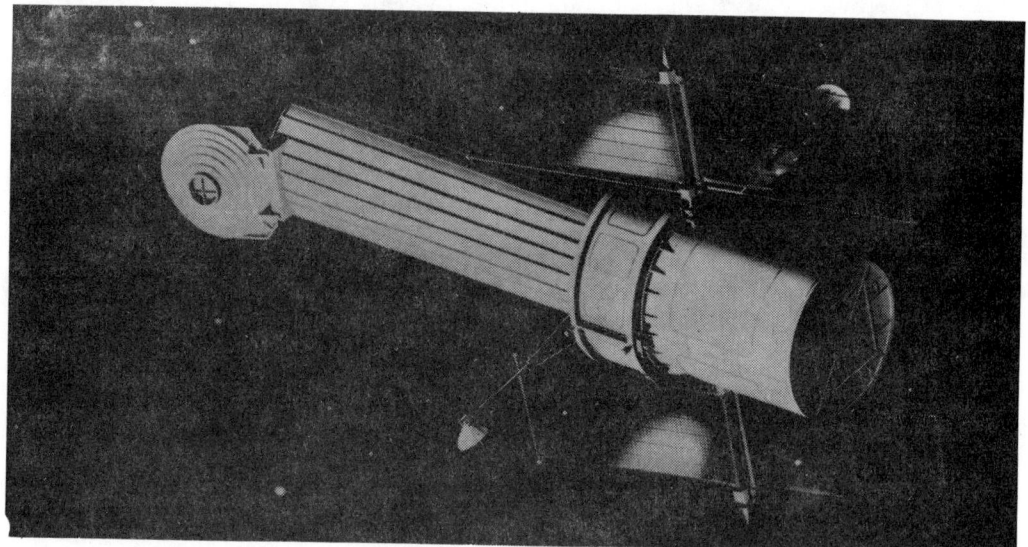

Fig. 6 Artist Concept of LST

NASA has for the past several years been evolving the technological and operational capabilities that are needed to place such a telescope into Earth orbit and to utilize it effectively. These technology advancements have occurred as a natural result of other orbiting astronomical satellite programs such as Orbiting Astronomical Observatory (OAO), Orbiting Solar Observatory (OSO), and Apollo Telescope Mount (ATM), as well as through supporting research and technology activities. In addition, a number of LST mission definition activities have been underway within the agency and by contractors for the past several years.

The LST consists of three elements, the Optical Telescope Assembly (OTA), the Scientific Instruments (SI), and the Support Systems Module (SSM).

In May 1972 the NASA Associate Administrator selected MSFC as the lead project management center for LST. MSFC, through contract with Itek Corporation, completed a conceptual design study of the Optical Telescope Assembly in 1972. GSFC, through contract with Kollsman Instrument Company, completed a similar study of the Scientific Instruments in 1972. MSFC carried out conceptual design studies of the Support Systems Module and an overall systems integration study in-house.

Definition studies for the combined OTA/SI were awarded in mid-1973 to Itek and Perkin-Elmer. These studies will be completed in February 1976. Definition studies for the Support Systems Module were awarded in December 1974 to Boeing, Lockheed, and Martin. These studies also will be completed in February 1976. These definition studies are making use of available designs, hardware, facilities, NASA standard components, and proven technology to the greatest extent possible to minimize risks and costs.

Current planning is to start the design and development of LST in Fiscal Year 1977, working toward a launch date of late 1982.

AAS 75-127

APPLICATIONS PAYLOADS
Charles W. Mathews*

INTRODUCTION

Space capabilities are extremely well suited to help solve many of the large scale and complex problems that plague today's world. The NASA program of applications of space technology and systems (see Figure 1) is aimed squarely at such help and, therefore, is most pertinent to the theme of this conference. I am sorry to say that insufficient time is devoted to this important subject.

Application of space should be a stimulating challenge for all of us in aerospace for three interdependent reasons. First, as I have already stated, space capabilities can make major contributions to the solution of very pressing earthly problems such as food shortages, scarcity of natural resources, and environmental pollution--problems of global scale. Second, aerospace-related R&D is, in many disciplines, the vanguard of advanced technology. To remain competitive in the world economy, the U.S. must maintain its technological leadership. Third, if the nation is to support the space program, the program must contribute in ways the American people can see as beneficial and thereby find important.

The problems the general public encounters are here on earth, and their solutions must impact activities, institutions, and systems here on earth. Thus, in addition to space flight missions and payloads, as important as they are, there are many ground-based activities that are key to the success of the space program and are absolutely necessary to effective use of the space elements. One current example is the Large

---

* Associate Administrator for Applications, National Aeronautics and Space Administration

Area Crop Inventory Experiment (LACIE) which NASA, the Department of Agriculture, and NOAA are mutually conducting. The objective is to use imagery from satellites and weather data to monitor the progress and predict the wheat crop production, first in the United States, then in other countries, and then worldwide. The immediate task is to perfect a practical processing system which provides timely output information for management of grain inventories.

Why is this information important? In many parts of the world, population is increasing as fast as, or faster than, food output. Starvation is an ever present danger, particularly in the less developed nations. World grain reserves have fallen from a 60-day supply in 1960 to a 20-day supply in 1974. This is a precarious situation as small changes in weather can produce large fluctuations in the availability of food.

On the economic side, the wheat sales to Russia provide a good example. In 1973, in the absence of information about their wheat harvest shortfall, the U.S. sold the Soviets about 11.5 million tons of wheat at about $91 per ton. As news of the sale became available, spot prices rose to about $172 per ton. Had the Soviet shortfall been forecast earlier, the price would have risen as the early purchases occurred. The final sales would have been made at close to the $172 per ton price. As a result, the U.S. would have gained an additional $500 million in foreign exchange.

Perhaps of most general importance is the ability to base policy and management decisions on pertinent and timely information and upon an adequate information base. This need applies across the board. Crop inventory management is just an example. Water resources management (i.e., watershed inventory, irrigation strategy and reservoir control) is another. Pollution control strategy is another. And there are many more.

But just acquiring space data is not enough. The principal task (see Figure 1) is to work with end users to develop appropriately matched systems and capabilities so the users can utilize space-acquired information along with other sources in their operational decision making

APPLICATIONS PROGRAM OBJECTIVES

- ESTABLISH USEFUL APPLICATIONS OF SPACE AND SPACE KNOW-HOW
    - DEVELOP USER RELATIONSHIPS
    - DEVELOP REQUISITE TECHNOLOGY
    - CONDUCT APPROPRIATE GROUND, AIRBORNE AND SPACE FLIGHT INVESTIGATIONS
- PROVIDE SUPPORT TO OPERATIONAL SYSTEMS
- CONTRIBUTE TO NATIONAL SPACE EXPERTISE

NASA E72-4020
1-3-72

FIGURE 1

processes. This is a very challenging and demanding activity. But the issues involved in such transfer activities must be dealt with and efficient transfer mechanisms must be developed if we are to realize the full potential of the vantage point of earth orbiting satellites.

I mention this because such ground-based activities must grow in the future--hopefully at least as fast as the space-based programs with which they are so strongly interdependent. I hope that industry and other elements of the technical community will become more closely involved. Perhaps a meeting such as this should be focused on the issues of space capability transfer.

Now to the subject at hand -- Applications payloads in the shuttle era. The increased capabilities and flexibility afforded by the new Space Transportation System, the Shuttle and Spacelab, offer very real opportunities to the Applications Program. In addition to its use as a launch and deployment system, we will conduct important experiments in the environment provided by Spacelab. Experiments in materials processing are natural to the Shuttle system where samples can be prepared on earth, processed in space, and returned to earth for analysis or application.

Shuttle and Spacelab also promise to accelerate technology and technique development. We will be able to deploy, retrieve, modify, and redeploy developmental systems. Even in the developmental stage, sensor systems aboard the Shuttle will allow the investigation of sensor performance parameters, information extraction methods, the checkout of ground data handling systems, and the study of operational techniques. This capability should speed the process of transfer to use. In the paragraphs which follow, I will cover examples of these uses of the Shuttle by reviewing its applicability to some of our planned payloads.

AUTOMATED PAYLOADS

The second and third figures list, in a planning sense, our payloads through 1985. At first sight, the program appears broad and diverse, and, from one point of view, that is true. There are, however, very strong interrelationships among the flight missions. SEASAT-A (1978), illustrated in Figure 4, is our first effort in gaining detailed physical oceanographic data, and Nimbus-G (1978), illustrated in Figure 5, is our first effort in comprehensive pollution monitoring from space. Nimbus-G will carry, among other things, a Coastal Zone Color Scanner whose synoptic data on water quality will complement the wind speed and wave spectra data from SEASAT. Jointly, these data should help us understand and predict the dispersion of estuarine and river discharges. Air-sea interface information obtained using SEASAT will greatly aid our efforts to understand weather and climate as an adjunct to our meteorological satellites. On the other hand, meteorological data from the operational NOAA and GOES satellites and from the improved instruments on TIROS-N will provide key complementary data for the crop production forecasts predominantly associated with the LANDSAT series. At about the same time period, we expect earth albedo and solar irradiance information from the Radiation Budget Satellite to contribute to long range (monthly to seasonal) weather and climate forecasts.

OFFICE OF APPLICATIONS FLIGHT MISSION MODEL

MISSION	75	76	77	78	79	CY 80	81	82	83	84	85
EARTH OBSERVATIONS											
LANDSAT	B		C			D	D[1]	E	E[1]	F	F[1]
NIMBUS	F										
AIRSAT				G			A			B	
TIROS				N		TF	N[1]	B	O		
STORMSAT						A			A		
SEOS											B
SMS	B										
AEM			(HCMM)		SAGE		RBS				
SPACELAB (∅-g FACILITY)						2	2	2	2	2	2
COMMUNICATION											
DISASTER WARN.							A	B			
EODAP											
GEOS	C										
SEASAT				A				B			
GRAVITY FIELD							A				
GRAV. GRADIOM.								B			
GEOPAUSE											B
VECTOR MAGNET.										A	
MAGNET. MONIT.										A	
LAGEOS (CLUSTER)									A		
AEM							MAGSAT				
LAGEOS		A									
SPACE PROCESSING						3	6	7	7	10	10
EARTH VIEW. APPLIC. LAB.						2	2	2	2	2	2
TOTAL MISSIONS (LESS SPACELAB)	4	1	2	3	1	3	5	5	5	5	4

NASA HQ E76-303 (1)
8-12-75

FIGURE 2

APPLICATION FLIGHT MODEL
(CIVIL, NON-NASA)

MISSION	CY75	76	77	78	79	80	81	82	83	84	85
EARTH OBSERV.											
ITOS		E-2, H		I							
NOAA				A	B	C	D	E	F	G	H
GOES	A	B	C		D			E			F
FOREIGN SYNC. MET. SAT.			A				B		C		
COMMUNICATIONS (SHUTTLE LAUNCHED BEYOND 1980)											
INTERNATIONAL	3	1	3	1	2	3	1	1	2	3	2
US DOMSATS	2	5	2	4	1	1	2	2	4	1	1
AERO/MAR. TRAF MGMT	3	-	-	3	1	2	2	1	1	1	-
FOREIGN	2	1	1	1	-	2	2	2	2	2	2
TOTAL MISSIONS (Less Spacelab)	11	10	8	11	6	10	8	8	11	8	7

NASA HQ E76-304 (1)
8-12-75

FIGURE 3

141

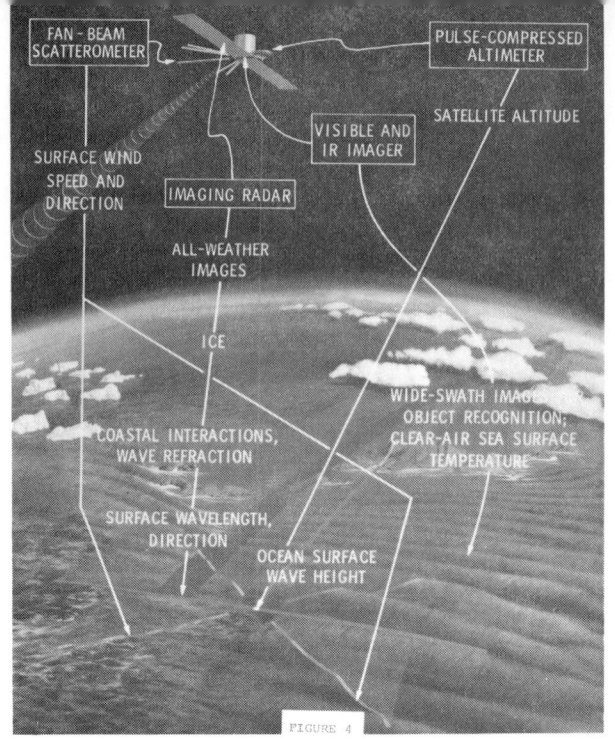

FIGURE 4

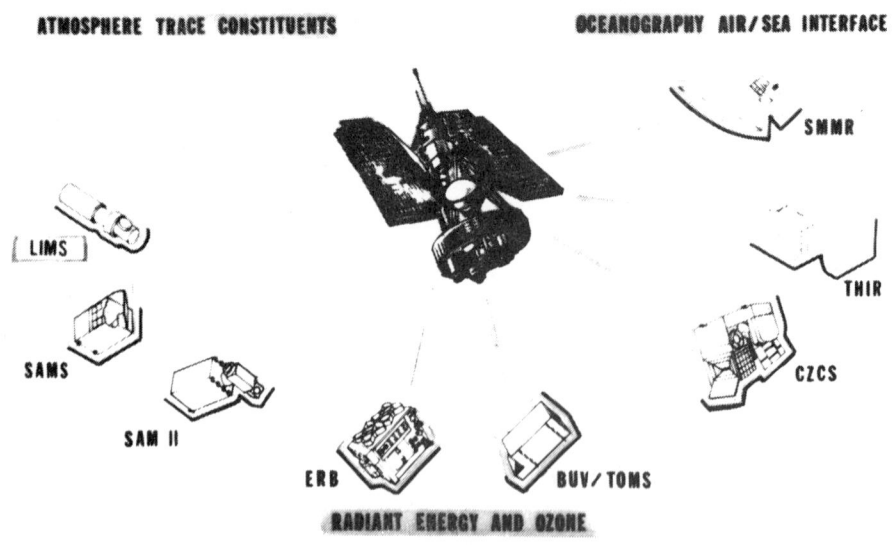

FIGURE 5

It is also noteworthy that our payloads generally cluster about geosynchronous and polar-sunsynchronous orbits (for near continuous observation of local scenes or for full earth coverage). These requirements will influence the ways that we utilize the Shuttle. The geosynchronous orbit capability provided by the Interim Upper Stage/Tug is very vital to us, and we are planning to launch our latest severe storm monitoring system, STORMSAT, on the first Interim Upper Stage (IUS) flight. As Shuttle-launched polar orbit capability will not be available until 1983, we will have to continue using expendables for several of our missions. However, we hope to have an early test of the Shuttle compatibility features of the Multi-mission Modular Spacecraft, formerly called the Earth Observatory Spacecraft. We will be looking at such things as teleoperator replacement of modules, spacecraft deployment, and spacecraft retrieval. We will also test an engineering model of a much improved Limb Scanning Infrared Radiometer (LSIR) for measurements on the nitrogen reaction side of the stratospheric ozone chemistry.

SORTIE MISSIONS

In addition to using the Shuttle as a launch vehicle for automated Applications satellites, we expect to make very substantial use of its capacity to operate as an orbital laboratory in conjunction with the Spacelab system. A space processing laboratory concept is illustrated in Figure 6. If it lives up to its early promise, the activity we call space processing is likely to become the most significant use of the STS as a laboratory spacecraft. Its possibilities are manyfold. The Spacelab is a natural way to work in this area and, in all probability, it will involve numerous and diverse privately-sponsored efforts even early in the flight program.

Space processing has to do with useful applications of space capabilities to materials science and technology, in particular, the weightless environment. In general, very large economic interests are bound up in materials technology, and some of the greatest commercial success stories of our times were made possible by advances in or control over the properties of materials and in understanding their behavior. The development of the jet engine and the solid state electronics industries came

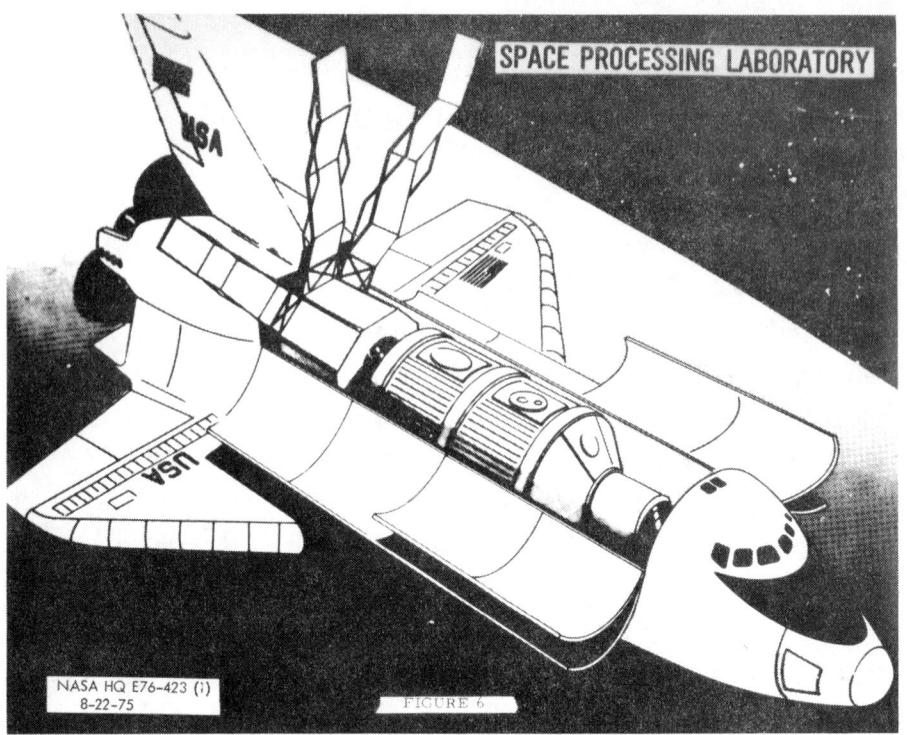

FIGURE 6

about in this way, and industry continues to devote large resources to materials research and development because of the tremendous leverage exerted by this technology.

We believe that processing in space can make very valuable contributions to many areas of materials R&D because the weightless conditions that are available in space flight make it possible to get highly improved results from processes that are subject to unpredictable and detrimental disturbances on the ground. Because of this, certain classes of experiments done in space can be more productive because they can be analyzed to a greater depth of detail than ground experiments, and processes can deliver results that are unachievable on the ground because they can be brought under unprecedentedly precise control.

For example, the space processing experiments on Skylab demonstrated that one can predict just how atoms of different species in a weightless melt will distribute themselves when the material freezes, and they also showed that crystals of electronic materials grow in a more orderly way

and achieve higher perfection when disturbances driven by gravity are
eliminated. We expect that experiment results from the Apollo/Soyuz
mission will confirm these findings and also demonstrate how space
techniques can be used to produce preparations of biological materials
that have unique value in medical research and applications. Early
observations of the returned samples appear to confirm our expectations.

These and other examples have persuaded us that space methods can take
some areas of materials research out of the cookbook cut-and-try category
and yield new understanding and new applications. Therefore, we believe
that an adequate technical incentive exists for industrial materials
research to enter space and that the entry of industry depends on space
being made accessible at a price that the market can bear. The Shuttle
should do this for us.

We plan to include space processing equipment in the payloads of at
least four STS missions in 1980 and 1981. This equipment will be used
for a fairly extensive NASA-sponsored experiment program in biological
preparations, electronic materials, metallurgical processes, glass and
ceramics, and other physical and chemical processes. We expect that the
experiment program will demonstrate that space methods can produce unique
and useful results in all of these fields and that such results can be
obtained at costs commensurate with their value to commercial industry.

Two of the first four payloads will be carried in the Spacelab pressurized module and configured for human operation. A typical layout made up
for an early Spacelab mission is shown in Figure 7. The apparatus
occupies a double rack and comprises two electrophoretic separation units
for work on biological materials and a group of small electric heat
treating furnaces which are similar in design to those used on Skylab
and the ASTP mission but somewhat higher in capability. Although its
layout is compact, we estimate that this payload will produce about three
times as many processed samples during a five-day sortie mission as the
Skylab space processing facility did in all three Skylab missions. It
will be critical to achieve high rates of productivity in our space
experiments because this is the most effective way to reduce costs to
the individual user.

EARLY SPACE POROCESSING PAYLOAD

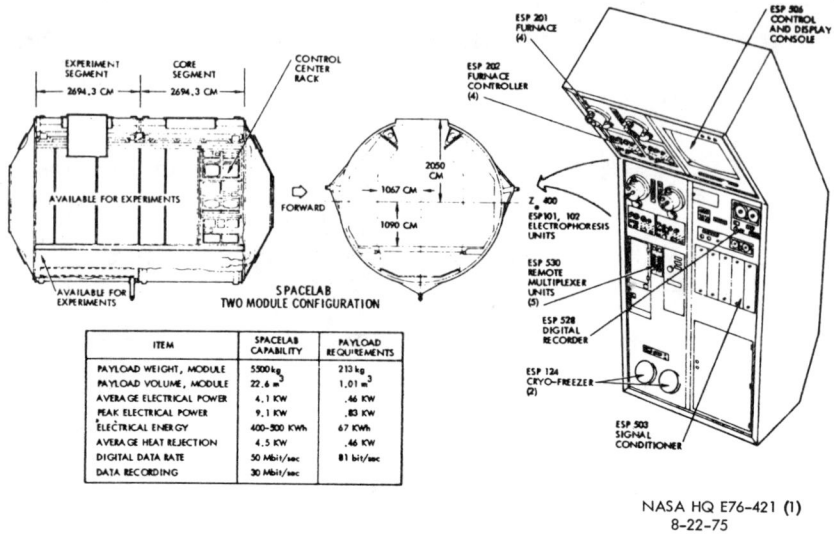

FIGURE 7

The other two space processing payloads to be flown in 1980-81 will perform experiments that can be automated and will fly as "partial payloads" in the shuttle payload bay. On these missions, the experiments will generally involve heat treating processes with power consumption levels that could create problems inside the pressurized module. Such processes lend themselves to automation, and, in order to take advantage of the full potential of automation for high productivity, we shall need to run all of the processing equipment continuously during all of the time it is in space. This mode of operation will require much larger power resources than the STS can be expected to provide, and, for this reason, the space processing program is planning to build its own Auxiliary Payload Power System (APPS).

An artist's sketch of the current baseline design for the APPS is shown in Figure 6. The system would include two Orbiter fuel cells, and all of the tankage, power conditioning, pumps, etc., would also be Orbiter hardware. As shown in the figure, it would have a deployable radiator that would use the existing ATM solar panel structure, and, in the concept, it would be mounted on a strengthened Spacelab pallet. With two fuel cells operating, this APPS configuration can deliver 20 kw continuously to the payload associated with it, and the structural envelope includes about 16 cubic meters of space for experiment equipment.

To the maximum possible extent, the actual processing equipment in these payloads will consist of general purpose apparatus that can be used for all types of experiments, and it will be modular in design so that each payload can be configured for the experiments it will perform. It will also be designed for easy maintenance and repeated reuse so that the share of capital costs associated with the average individual experiment will be minimized.

With this design philosophy, we expect that we can run a very flexible and cost effective experiment program. Moreover, we believe that it will be easy to accommodate private experiments in NASA payloads at low rates for equipment rental and for usage of mission resources. This will enable industry to enter on an economically-phased schedule without requiring large initial investments of private capital, and we feel that the prospects for early industrial participation on this basis are quite good.

ATMOSPHERIC CLOUD PHYSICS LABORATORY

A number of other significant activities involving use of the weightless environment of space have been defined, and I would like to briefly describe to you another important facility, the Atmospheric Cloud Physics Laboratory (ACPL). Some of the factors involved with the need for such a facility are outlined in Figure 8.

Local and regional weather results primarily from the redistribution and release of energy in the clouds. While we have made progress in understanding large scale atmospheric processes, our efforts to understand the cloud microphysical processes have been hampered by the lack of a long duration facility in which to study the processes. In the low gravity environment aboard Spacelab it will be possible to suspend

ZERO GRAVITY CLOUD PHYSICS LAB

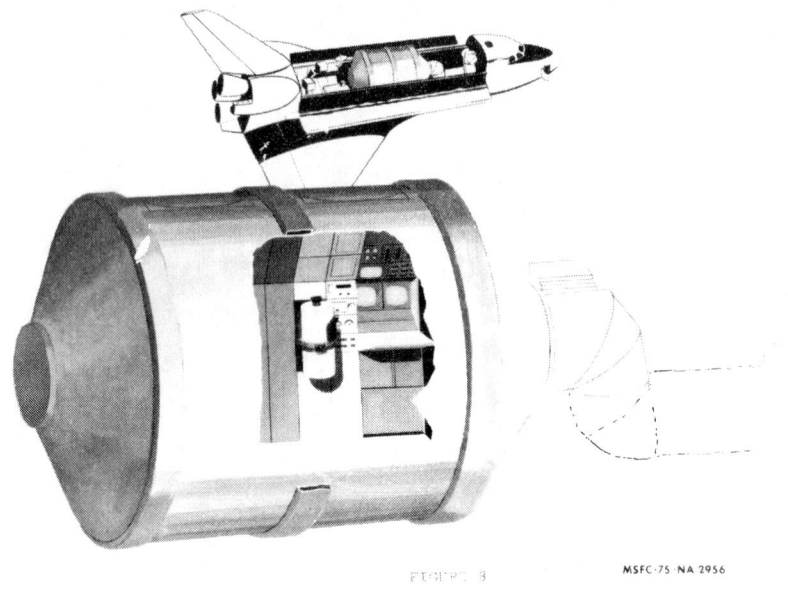

aerosol particles in important size ranges (0.1 to 10 microns) for test durations from 30 minutes to hours. On earth, these durations are limited to a few minutes at best.

An understanding of cloud microphysical processes has an impact on society's problems in several ways. These include understanding and depicting the self-cleansing mechanisms of the atmosphere, the potential of weather modification and the weather-modifying influence of airborne pollution, and the prediction of the formation of short-term, very severe weather events. The fact that cloud activity plays such a major role in atmospheric processes and cannot be fully understood without such a facility implies future scientific and applications benefits of great magnitude.

The objective of the ACPL is to provide a flexible research facility in which to conduct research that will increase our knowledge of these microphysical processes. The laboratory apparatus (Figure 9) utilized in conducting these experiments consists of an expansion chamber supplied

ATMOSPHERIC CLOUD PHYSICS LABORATORY

OBJECTIVES

PROVIDE A ZERO GRAVITY FACILITY TO OVERCOME LIMITATIONS INHERENT IN GROUND BASED RESEARCH

- SUSPEND AEROSOLS FOR TESTING TIMES OF HOURS COMPARED TO A MINUTE
- AVOID CONVECTION AND WALL EFFECTS

SIGNIFICANTLY INCREASE LEVEL OF KNOWLEDGE OF CLOUD PROCESSES

- NUCLEATION AND GROWTH
- SCAVENGING
- ELECTRICAL CHARGE

OBTAIN BASIC DATA ON

- STICKING COEFFICIENTS
- THERMAL ACCOMMODATION COEFFICIENTS
- MEMORY EFFECTS

PLAN

DEVELOP A PARTIAL SPACELAB FACILITY INCLUDING AN EXPANSION CHAMBER, GAS INJECTION SYSTEM, PARTICLE COUNTER, AND SUPPORT SUBSYSTEMS. THESE AND COMPLEMENTARY BUILDING BLOCKS MAY EVOLVE INTO ADVANCED CAPABILITIES AS REQUIRED.

FIGURE 9

by an environmentally controlled gas system. Aerosols will be obtained by having particles injected into the gas system; then concentration will be controlled through the use of particle counters, and the resulting phenomena will be observed in the expansion chamber and recorded on film. A scientific working group will be formed this winter to assist us in the final design of this equipment.

Analyses indicate that meaningful experiments can be conducted in nucleation, growth, scavenging, and electrification. A series of warm cloud processes experiments is planned for the initial flights in order to establish an experimental technique base in an area which has undergone extensive ground-based testing. Once the facility is available and checked out, it is planned to be utilized by a wide spectrum of scientific users and user organizations.

EARTH VIEWING APPLICATIONS LABORATORY

The space laboratory approach also encompasses the earth observations disciplines, even though much of the operational work will be accomplished by free-flying satellites. Earth observations from space perhaps have the greatest near-term value of all the applications of space, and we envision a heavy activity in this area in the Shuttle era.

Since a large variety of experiments are to be performed, the Earth Viewing Applications Laboratory (EVAL) is designed to contain general-purpose equipment that can be modified easily or adapted easily to accept new sensors in order to optimize a particular observation or a switch to a completely different kind of measurement. In such a laboratory, we can also bring many experimenters, instruments, and observational techniques to bear on the different facets of a single complex problem and thus accomplish much more than would be possible by a single experimenter working alone.

The Spacelab configured to carry out earth observations missions is shown in Figure 10. EVAL provides the capability to recover payloads, and by reconfiguring and refurbishing, it allows for multiple flights in a cost-effective manner. Overall costs will be lowered further by the large weight, power, and volume available for sensors and instruments: up to 8000 kg and 5800 watts, depending on specific configurations. EVAL provides an unpressurized instrument environment on its open pallets and a pressurized space inside its optional closed module. A crew will

EARTH VIEWING APPLICATIONS LABORATORY
(EVAL)

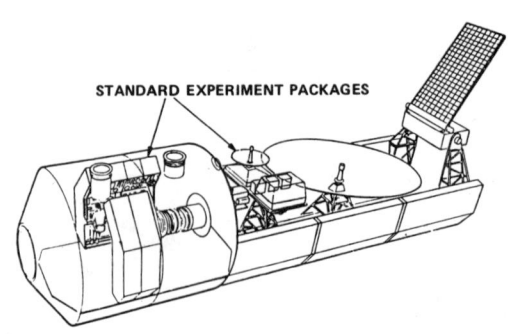

STANDARD EXPERIMENT PACKAGES

- MULTI-DISCIPLINE SPACELAB
  - EARTH RESOURCES
  - WEATHER & CLIMATE
  - EARTH & OCEAN PHYSICS
  - COMMUNICATION & NAVIGATION
  - MATERIALS TECHNOLOGY
  - SPACE PROCESSING
- SPACE APPLICATIONS MISSION

FIGURE 10

NASA HQ E76-305 (1)
8-12-75

be present to control sensor operation and to perform repairs as needed. A variety of orbit inclinations will provide mission flexibility, and the flexibility of Shuttle launch schedules should provide for investigation of special events or conditions. Mission duration can be anywhere from 7 to 30 days.

In the past, the NASA Applications Program has employed multi-purpose, experiment-carrying spacecraft, first to develop the needed technology and then to provide to the user community the necessary data to meet their various needs. But these spacecraft had major limitations. First, the number of experiments was limited by the payload weight limitations of the spacecraft and booster combination employed. Second, the size of the experiments was limited due to the restructions imposed by the launch vehicle shroud. Third, the experiments had to be designed for long life without the possibility of refurbishment or repair for they could not be returned back to earth and this necessarily kept overall costs high. The Spacelab overcomes all of these limitations, and, in addition, by common use of instrument support facilities for multi-disciplinary tasks and by providing recurring flight opportunities, will greatly enhance the accomplishment of NASA's Applications Program goals.

EVAL will fulfill a variety of needs, as listed in Figure 11. It will provide self-sufficient scientific measurements which require only a short period of observation; with proper coordination, its instrument complement will enable calibration of instruments on free-flying spacecraft already deployed in orbit; it will carry out signature studies of earth scenes required for the development of interpretive analytical models; and, it will test instrumentation or techniques as precursors to later free-flyers.

The potential for flying a number of diverse instruments at one time permits interaction of data in a synergistic manner and their use for various applications disciplines simultaneously. Thus, antennae and microwave receiving systems may be used for the evaluation and measurement of the electromagnetic communications environment on earth, but, in conjunction with other equipment, may serve as radiometers to measure

EARTH VIEWING APPLICATIONS LABORATORY

A DEDICATED SPACELAB FACILITY TO PROVIDE...

- SYNERGISTIC GROUPINGS OF INSTRUMENTS
    - --CROSS CORRELATION OF DATA
    - --SIGNATURE DEVELOPMENT AND EXTENSION
    - --PROOF OF INFORMATION EXTRACTION TECHNIQUES
    - --DIRECT COMPARISON OF PERFORMANCE

- FLEXIBILITY OF ACCOMMODATION
    - --REPEATED FLIGHT OPPORTUNITIES
    - --VARIETY OF ORBITAL INCLINATIONS
    - --RELAXED WEIGHT AND VOLUME LIMITATIONS
    - --RELAXED RELIABILITY REQUIREMENTS

FIGURE 11

environmental radiation from the earth surface and atmosphere, providing data to earth resources and meteorology discipline investigators. The infrared radiometer used to provide data for the interpretation and detection of geological phenomena also could evaluate the health of crops and the availability of liquid water for irrigation. A combination of infrared and microwave radiometers, a synthetic aperture radar, a film camera, and a thermal infrared mapper, when employed simultaneously, should provide synergistically for the measurement of most snow, soil moisture, and surface water parameters.

Another technical advantage of a spacecraft that has the capability to carry a number of instruments is that competing technologies may be carried into space, comparative tests performed, results analyzed under the same conditions, and decisions made as to the better item. This "fly before buy" concept has not been practical before the appearance of Spacelab. EVAL could carry a laser communications transponder and a millimeter wave communications system to test wide band high data rate telecommunications performance, or laser versus radar ranging for earth surface measurements could also be compared.

The shuttle bay's large area of 18.2 meters by 4.6 meters also provides EVAL with a capability to carry very large experiments that the automated spacecraft program has not been able to provide due to launch vehicle shroud limitations. Large antennas for communications with small ground stations, or for obtaining radiometric information, or for radars applicable to earth and ocean physics needs are examples of where large size structures can be employed.

Finally, we believe that the cost of placing space applications experiments into EVAL will be markedly reduced when compared with automated satellites. The experiment will operate for a short period of time, and, therefore, the need for long-life components and high reliable parts can be relaxed. The need for extensive ground tests may also be reduced. In fact, engineering models of the experiment could be considered as the type to fly. The presence of man near the experiment and his knowledge to modify and repair a faulty instrument should result in substantial reduction in hardware cost. A last, but important, cost reduction factor for EVAL is its payload return capability. This factor will permit reflies of the experiment at very low cost and at minimal cost with modifications.

The NASA Applications Office is looking very favorably on utilizing the Shuttle-launched Spacelab in an earth-viewing mode to develop the technology needed for our future automated spacecraft and to gain initial operating experience with the technical data obtained. There is every reason to believe that our user community will greatly benefit by taking part with us in EVAL experimentation.

COMMUNICATIONS LABORATORY

Obviously, Spacelab facilities can be effectively employed in other areas, such as in communications investigations (see Figure 12). Mr. Ehrlich of my office will describe this area in more detail in a paper to be presented later in this conference. For present purposes, I will discuss in a brief way two potential users in this area.

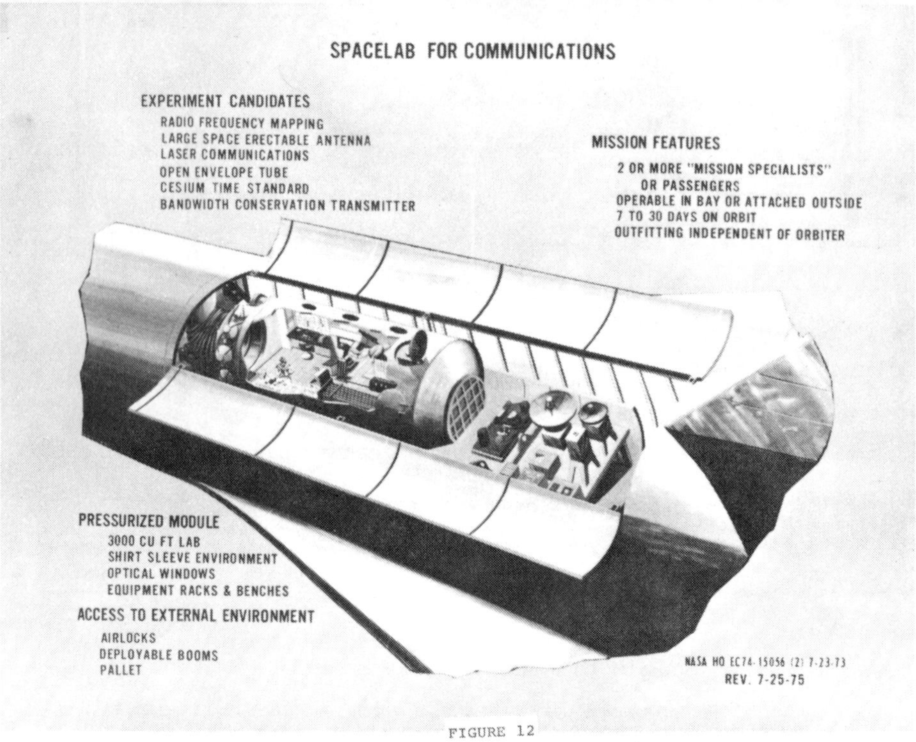

FIGURE 12

During recent years, increased radio frequency usage has created spectrum-resource problems for the various regulatory agencies. Since spectrum occupancy is expected to experience an ever-increasing growth in the future, a capability for providing continuous global spectrum monitoring and electromagnetic environment mapping is essential in order to avoid over-crowding and interference in particular bands. With the advent of the Spacelab program, there is increased opportunity to develop space technology to monitor the radio frequency environment.

The radio frequency survey program will establish both a system and a data base for determining limits on interference in crowded spectral bands and updated listings of those bands void of interference. This type of information will then become available to NASA and the various U.S. and foreign regulatory agencies. Based on this resultant data, more efficient spectrum usage should accrue in the future. The space-borne survey system will also provide mapping of certain geographical areas that would have been impractical to achieve via terrestrial methods and be useful in establishing levels or limits of earth-to-space radio frequency emissions on a global basis. Finally, this type of data will place the U.S. in a more informed position regarding negotiations for international frequency allocations.

BANDWIDTH COMPRESSIVE MODULATION EXPERIMENTS

In another approach to solving the problem of space communications traffic congestion, we are planning a Bandwidth Compressive Modulation (BCM) experiment. The experiment will evaluate bandwidth compression ratios of up to 6:1 through use of multiple phase and amplitude modulation techniques.

The experiment will establish a direct communication link between selected ground stations and the Shuttle using a separate pallet-mounted antenna and special equipment within the Spacelab module. The Shuttle communication system will not be required for the experiment. The experiment will be conducted over a succession of several missions. It will require establishing one-way links from the Shuttle to ground stations on early flights (and two-way links between ground stations via a transponder in the Spacelab on later flights).

The Shuttle/Spacelab provides an excellent vehicle for testing BCM performance in space prior to a commitment on automated satellites. Specific types of test data will be transmitted for the purpose of evaluating operational performance, measuring error characteristics, and assessing effects of propagation. Selected microwave frequencies will be examined. Use of multiple flights will permit repeated transmission under a diversity of weather patterns, restrict the equipment frequency range required in any one flight, and allow evolution of the experiment from one-way links to two-way links.

LARGE STRUCTURES IN SPACE

I have already alluded to the large cargo weight and volume capabilities of the Shuttle, and a broad range of important applications will flow from the ability to place and control very large structures in space. This effort is a natural consequence of the trend toward having a greater share of the system complexity and capability in the spaceborne portion of the system. For instance, the early communications satellites required large, complex, and expensive ground receiving terminals. Now, the current ATS-6 satellite with its much higher power levels and larger antenna (just over 9-meter diameter) is able to broadcast directly to simple, low-cost receiving antennae. We are continuing demonstrations in educational broadcasts to remote towns and villages using this system. Current studies foresee the test of a 30-meter class deployable antenna, Figure 13, in about 1982. Such a range of antenna size will have immediate application to communication and position fixing of mobile platforms (vehicles and sensors), to low-cost data collection and relay, and to direct broadcast with multiple, tailored beam patterns.

Other areas requiring large structures are passive optical systems and high powered laser systems. Lasers will be used for communication, for detailed high accuracy (order of centimeter) measurement of earth crustal and ocean features, for pollution monitoring (LIDAR techniques), and subsequently for energy transmission. At current and foreseen laser

LARGE DEPLOYABLE ANTENNA SHUTTLE EXPERIMENT

NASA HQ EC75-15814 (3)
FIGURE 13

efficiencies, such systems will require large surfaces to radiate away large amounts of waste heat.

There is also a great deal of interest in very large structures--the order of kilometers--for microwave energy transmission. In one method, the energy would come from conversion of incoming solar energy; in another, energy produced at remote sites could be transmitted to urban sites while protecting the urban environment.

There are, however, significant technical and social challenges associated with implementing such programs. On the one hand, to maintain the shape, attitude, and integrity of very large structures will call for major advances in controls, guidance, and materials capability. Furthermore, the erection of such structures, in some cases, will call for space transportation systems well beyond those currently planned. To avoid some of these problems, another approach calls for achieving the effect of a very large structure by means of an array of independently positioned, small structures. This would trade material and structure problems for position fixing and switching problems in arriving at, say, a desired composite wave form from the separate emitting structures. Again, though the basic technology may be available, there are significant system issues which remain.

SUMMARY AND CONCLUSION

In summary, the increased capabilities and flexibility afforded by the new Space Transportation System afford very real opportunities for the Applications Program. In addition to its use as a launch/deployment system, we will conduct important experiments utilizing the low gravity capabilities of Spacelab--particularly for materials processing and cloud microphysics studies. Shuttle/Spacelab will also be important in accelerating technology and sensor development. An applications mission involving both pollution and hydrologic sensors will be our first opportunity to test modular replacement features of the Multi-mission Modular Spacecraft and to deploy, retrieve, modify, and redeploy a developmental sensor system. The Earth Viewing Applications Laboratory concept should result in mutually beneficial and complementary sets of earth viewing instruments with which to cross-correlate data and speed up information extraction. Finally, the potential that the Shuttle affords for the use of large structures is an exciting trend with many very important possibilities.

Though the emphasis in this paper has been on flight hardware, the bottom line in our program is getting useful information into the hands of decision makers as a basis for action. The Space Transportation System will contribute importantly in the acquisition of space information and the provision of other space capabilities to do that job.

CONGRESSIONAL VIEWS ON LONG RANGE SPACE PLANS

Honorable Don Fuqua*

Ladies and gentlemen, participants in "Space Shuttle Missions of the 1980's" distinguished guests, it is good to be with you today. A little over a month ago, in remarks to the National Space Club, I said that if we were to pick only one subject from the multitude of issues in our space program today that issue would be payloads.

You have embarked this morning on an intensive three days of discussion, debate and analysis of some of the crucial payloads for the Shuttle era. The Large Space Telescope, Spacelab and Upper Stage for the Shuttle called "IUS," "TUG," or by some other name, and manned and automated payloads tailored to take full advantage of the Space Shuttle's potential, now do and will continue to occupy your attention in this conference. In reality this is a critique of near-term Space Shuttle opportunities. My encouragement to you during this time is to question every basis for these projects. Test again their logic. Determine what more can we derive from these efforts during the early missions of the Space Shuttle. We need to be our own strongest critics on payloads for the next several years.

My initial comment has been only to the near-term effort of this conference. It seems to me that this meeting can also help set the stage for amplifying two future oriented efforts recently concluded. The first of these were hearings by the Subcommittee on Space Science and Applications on future national space programs, and the second the NASA future programs study called the "Hearth Report."

---

*Representative, State of Florida, United States House of Representatives and Chairman, Subcommittee on Space Science and Applications

We are still in the process of evaluating testimony taken in the hearings and attempting to assimilate the many excellent contributed papers. I can state now that the quality is high and the content diverse. The charge we made to the many participants was a difficult one. On the one hand we encouraged the contributors to define in whatever way they wished the scope of their discussion of future space programs. We followed this general charter with the request to give us the benefit of their thinking on criteria by which any future space programs should be evaluated by the Congress and the Committee on Science and Technology.

It is already apparent from our review that we need to apply broader tests to each future program that is proposed. This should be beneficial since the Government-University-Industry mode of operation is already a phased study approach. If anything is apparent from our yet incomplete evaluation it is that more difficult questions need to be asked earlier in the study cycle. This also implies, at least to me, that we need to reach out even more to define as a minimum, and in a preliminary way, as many potential new space programs as possible, notwithstanding the real fiscal limitations of our year-to-year program. Our basket from which to make selections needs to be as full as possible. If we turn off innovative ideas too early in the study cycle, our shopping basket will diminish. Our hearings have once again demonstrated that a diversity of substantial ideas are available if we provide the climate of analysis and debate essential to developing these ideas in their early stages. Perhaps we have penalized ourselves in the past several years by not building our potential programs even though our total program opportunities have been limited by inflation and a reduced budget. The critical advanced planning has not been and is not, as you well know, the expensive part of the research and development cycle. The outcome of our hearings make it apparent that such "mental" limits to advanced planning for future space programs only delay our definition of the full potential of space.

It is popular today to say we have entered a new era in space utilization and exploration. Yet, I believe that when we have the advantage of time we will look back on the mid-1970's as a time when we really did not appreciate the dimensions of that new era in space. Informally over the past six months we have been examining with the executive branch what immediate utilitarian opportunities the various agencies should grasp which would employ space based systems to render less expensive, better, or heretofore unavailable federal services. Again the test applied should be strict to assure that we do not merely do a necessary job another way without accruing real benefits. Early experience with satellite remote medical services and remote education services for teachers and students offers promise. Earth Resource Surveys -- be it maritime, agricultural, geological or demographic -- are increasingly in demand. There is substantial progress being made to cause these things to happen. I don't think we can be complacent and expect these opportunities to just happen, but the momentum is increasing in the federal area. Another area of payload opportunity needs attention now. It is a difficult task, however, to achieve the attention needed. Industry use of space early in the Shuttle operational program is dependent on accepting some level of risk and proceeding to explore what space manufacturing opportunities are available that likely offer a good return on any investment which is made. Certainly, the early work in defining these opportunities is not the province of industry alone, but the impetus must come, just as in the user federal agency, from those who stand to benefit from the results. Our chemical, metalurgical, power and other industries are not yet sufficiently stimulated to take the business risks essential to making space a place of commerce. It is doubtful that the remainder of the free world will neglect this opportunity. If we wish to reap the ultimate benefits of our high technology space developments this is an area -- Space manufacturing and processing -- that needs more emphasis in the next several years.

We have ranged over the potentials and opportunities of our future space program for the last few minutes concentrating on what the industry and NASA can do. Congress, too, shares this obligation. My feeling is that, despite the problems cited daily in our news media, real progess in Congressional research and development overview has been taken by concentrating more of this jurisdiction within the Committee on Science and Technology.

Under the able leadership of our distinguished chairman, "Tiger" Teague of Texas, the Committee on Science and Technology -- and its subcommittees -- has moved rapidly and effectively to fulfill both its new and old responsibilities.

The Subcommittee on Space Science and Applications has assumed both legislative and oversight responsibility for all NASA programs with the exception of aviation. The subcommittee on Aviation and Transportation R&D, chaired by my colleague, Dale Milford, reviews NASA aeronautics effort with a perview of aviation on a multi-agency basis. We have two new energy subcommittees, an Environment and Atmosphere Subcommittee and a broadly based Analysis and Oversight Subcommittee.

Superimposed on this major Congressional reorganization, broadening the base of the Science and Technology Committee, has been the Budget Reform Act. Another indication of how rapidly the chairman and members of the committee have brought the committee up to full speed is that the committee, in authorizing both NASA and NSF funds for fiscal year 1976, met the budget committee timetable this year, although this is not mandatory until next year. Intensive action by the committee is now shaping a new science policy. The new Subcommittee on Science, Research, and Technology has moved quickly to fulfill its role as the replacement for the former Subcommittee on Science, Research and Development. A new authorization for a new agency, ERDA, has been hammered out by the committee. Our new Subcommittee on Environment and the Atmosphere has already reported out a bill authorizing legislation for the Environmental Protection Agency Research and Development Program and taken a leadership role in Congressional oversight on the potential problem of ozone depletion in the upper atmosphere. In sum total, this has been an active and, I believe, productive few months.

This work will need to be sustained by our committee over the next several years if the Congress is to play a proper and positive role in the evolution of our future space programs.

While we all plan for this future in meetings such as this one, maintaining the quality and commitments of current programs is essential. The acceptance that the Space Program enjoys, in my view, is based in its record of success. Every program has had major hurdles to overcome but sustained success has allowed the general public to accept the Space Program on faith during the past decade. We can fulfill most, if not all, of the expectations that have sustained that public faith by doing our planning homework.

I commend you on your activities here this week and look forward to seeing the results of your work reflected in our NASA authorization hearings in November.

Thank you.

AAS 75-136

THE EXPLOITATION OF SPACE FOR THE BENEFIT OF MANKIND

Honorable Frank E. Moss*

I've become very much interested in the Bicentennial celebration lately. I was reading a book called Democracy in America written by a young Frenchman named Alexis de Toqueville. He visited America in the early 1800's and described our young country in his book. In it he devotes a chapter to "Why American Orators Often Use An Inflated Style."

"I have frequently noticed, " he said, "that the Americans, who generally treat of business in clear, plain language, devoid of all ornament and so extremely simple as to be often coarse, are apt to become inflated as soon as they attempt a more poetical diction. They then vent their pomposity from one end of a harangue to the other; and to hear them lavish imagery on every occasion, one might fancy that they never spoke of anything with simplicity."

In deference to the Frenchman and to you, I shall try to get down to business and speak with simplicity tonight.

I was asked to talk about "the exploitation of space for the benefit of mankind." But I stumbled on the word "exploitation." I'm sure the phrase has always been "the exploration of space." Besides, "exploitation" has become an unkind word in our vocabulary. It brings to mind colonial empires, environmental destruction, sweat shops, and other things I assumed no Senator should advocate.

But, as it turns out, the definition of "exploit" is merely to turn a natural resource to economic account. That is precisely where I hope we are headed in our space program right now. And, I believe the Space Shuttle is the means by which we will bring space to an economic account.

But convincing the public of the worth of the Space Shuttle has not been easy. Convincing certain of my colleagues in the Senate has been downright difficult. After all, we start out with three strikes against us -- the Space Shuttle's name, its mission, and just the looks of the beast.

---

*United States Senator, State of Utah

Senators and their constituents can get excited about projects with a name like Pioneer. And Viking stirs the imagination. I think Apollo was an appropriately dignified name for its mission. And then there's Atlas and Titan. But what does Space Shuttle bring to mind? Perhaps a crowded subway car. Even Shuttle-Sat would have been more in keeping with other NASA program names.

As for missions, it wasn't so difficult for President Kennedy to stimulate popular support for going to the moon. And what could be a more dramatic mission than sending a spacecraft to scoop up a lump of Martian soil -- to examine it for the presence of life beyond Earth. But with the Space Shuttle the question I often get after explaining what it does is -- yes, but what does it do?

Yet there is an even bigger problem. We were all impressed I'm sure with the sight of the Saturn rocket lifting off the pad. I know that at the ASTP launch I was awestruck again at the beauty of that pointed white column easing into the blue sky. But let's face it, all artists' renditions aside, the Shuttle is a rather ungainly looking creature. It doesn't look like a thoroughbred, nor a pampered filly. More like the accidental offspring of an airplane mated with a missile. And yet, let's not forget for a moment that from the crossing of a horse with the donkey came the mule -- the hardiest draft animal of all. The Space Shuttle may not be a showhorse -- but it will be a work horse. There is nothing dramatic about having the mission of ploughing a field, but this country was built on such labor -- and as for fancy names, you never heard a mule named "Man of War."

The Space Shuttle will serve what sounds at first to be a rather mundane role -- economy. The Shuttle is aimed at economy -- its reusability, its huge payload capacity, its simplification of the launch process, and its retrievability capability. Yet it is the economy of the Space Shuttle that is going to slash the price of the admission ticket to space. As that occurs, I predict the benefits of mankind will pour out of the space program in an avalanche of new uses of satellites.

You know NASA worries a lot about understanding the potential uses of satellites, and I think they certainly should be doing that. But I'm reminded of the time two salesmen were arguing about how to sell more tractors to farmers. Tractors had just come on the market and sales were not very brisk. One salesman said, "First you've got to convince farmers that tractors can be used safely on steep hills." The other contended, "No, it's more important to show them how tractors can be used in mud." A man in overalls standing nearby was heard to mumble, "Cut <u>the price</u> and the uses will follow."

Look at the history of air cargo in the United States. In the beginning of aviation, the only cargo considered worthy of air travel was people and mail. As aircraft improved and the cost of air shipment dropped, then all sorts of perishable items, such as fruits and vegetables, found their way into cargo aircraft.

Now, with the advent of the jumbo transports, there even are discussions about shipping <u>automobiles</u> by air <u>because</u> it is economical.

When you start talking about the long-term consequence of an economical Space Transportation System (in other words the Space Shuttle) the <u>key word</u> is "payloads."

I see the Space Shuttle as being the basis of three new industries that are now on the horizon -- communications, remote sensing, and space manufacturing.

Already the communications satellite industry, though still in its infancy has been weaned from government assistance and has found a lucrative market. Comsat stock is one of those "I had the chance but I didn't" investments.

Look at the growth in the technology in this area! The first synchronous commerical satellite launched for Comsat in 1965 linked Europe and America with a mere 240 telephone channels. Still that wasn't a bad first step -- it nearly matched the combined capability of the four transatlantic telephone cables laid on the ocean floor 10 years earlier. But today, Intelsat IV-A is being equipped with 12,000 channels -- about 50 times as many as that first Early Bird satellite.

With the advent of the Shuttle the cost of communication satellites should drop significantly. Right now the launch costs about the same as the satellite itself.

Considering the past growth of this industry, it hardly seems possible that anything can further accelerate the pace, but if the cost drops, surely more developing nations will want to avoid the high cost of terrestrially based systems by starting with a satellite. A single rocket launch can tie together a small nation's entire communications network. Such a system is now under development by the Indonesian government.

Another facet of the emerging communcations satellite industry is ATS-6. Experiments with it showed the feasibility of bringing first-rate health care to remote areas, such as in Alaska, via television. A doctor thousands of miles away can diagnose a patient's disorder and supervise treatment. In another experiment with ATS-6, educational services were beamed to isolated communities in Appalachia. Now placed over East Africa for a year's use, ATS-6 will beam educational programs by direct broadcast to community television sets in thousands of villages in India.

I see that NASA is soliciting proposals for further experimentation with ATS-6. I found the announced order of priorities for experiment selection to be intriguing -- first, societal disciplines; second, communications; and finally, the technological disciplines. ATS-6 is indeed a remarkable instrument.

Casper Weinberger of HEW was so excited with the ATS-6 results that he has recommended launching a <u>social services satellite</u>. The problem of course is <u>cost</u>. When an hour of color TV communcation via satellite costs thousands of dollars, you can see why remote diagnosis of a single patient's illness via satellite is prohibitively expensive. But the Shuttle will decrease the cost and should bring some of these social services via satellite within the financial grasp of government and private industry. ATS-6 is furnishing the opportunity to test these ideas. The Shuttle could furnish a realistic opportunity to put these ideas into practice.

There is another type of communications satellite system that offers an unprecedented array of benefits to mankind. It's called the NAVSTAR Global Positioning System. The plan calls for 24 satellites to orbit the globe transmitting very precise positioning information. Any vehicle in the air, on land, or at sea, having a receiver, could determine its location and altitude to within 10 meters and its velocity to within one-tenth of a foot per second. A simpler version of the receiver could give a foot soldier or a surveyor his exact location on land and yet the receiver can be carried on his own back.

If the Department of Defense approves the program, the 24 satellites would all be in orbit by about 1984. Naturally, the Shuttle and probably the Interim Upper Stage will figure into plans to launch and maintain the 24 satellites. The beauty of the system is that because the Department of Defense would have our military forces depend on NAVSTAR, we can be sure that the military will pay for it. Thus, all civilian users need only buy a receiver, nothing more. The satellite system would be available to the civil sector <u>free</u>.

I have already sent a letter to Jim Fletcher requesting NASA to consider investigating the possible non-defense applications of NARSTAR, to include a national air traffic control system and an aircraft collision avoidance system. I certainly hope that other agencies, especially the Department of Transportation, will take a serious look at how NAVSTAR could revolutionize the method of finding the answer to an ancient question "Where am I?"

The potential economic benefits to mankind of such a system are bounded only by our imagination. The DOD decision to proceed with the program will come probably late in 1977. I think it is imperative that the non-defense benefits will have been investigated adequately enough to feed them into the decision-making equation in 1977.

The second industry which I believe will have its genesis in the space program concerns <u>remote sensing</u>. The weather satellites and Landsat, of course, come to mind.

The benefits to the world of our weather satellites are so apparent to this group that I won't describe them again. Suffice to say that we cannot control weather, yet we have virtually eliminated one of its ageless and most terrifying characteristics -- total surprise.

I want to rush on to one of my favorite topics -- Landsat. A watershed event occurred for the space program on May 17, 1975. Probably only a handful of you noticed it. On that day an editorial on Landsat appeared in the Washington Star entitled, "Satellites and Consumers." What was so notable about the article was its author, a consumer advocate named Ralph Nader. I noted that not once did he question the value of the Landsat program or the importance of the information received from Landsat. To the contrary, using his own well-known knack for imagery, he described the uses of Landsat imagery -- "Super-eye" he called it.

Nader's concern is that the consumer is not going to get his share of the Landsat pie. While the Landsat data is easily available, interpreting it does require expensive equipment and skilled personnel which only relatively large businesses and government can afford. Whatever you think of Ralph Nader, I think he has raised a legitimate point here that is worthy of further consideration. But, I brought this article up mainly to announce with pleasure that GM's nemesis is apparently on our side, with certain qualifications. I'd say that's the most novel endorsement yet of Landsat's multitude of potential benefits to mankind.

While we're on the subject of Landsat, I might mention that the Space Applications Board of the National Research Council just released a very interesting report called "Practical Applications of Space Systems" which makes a number of recommendations, some of which concern Landsat. The Board recommended that future Landsat satellites should have a "zoom" capability to focus on selected areas with a resolution of 3 meters to 10 meters, mainly for land-use planning. If such advice proves wise and is followed, we _may_ eventually see _many_ Landsat satellites in orbit simultaneously, instead of just two. That, of course, will mean more requirements for the Shuttle.

The third major industry that I believe will spring from the space program is space manufacturing.

That is literally the exploitation of a natural resource for economic gain -- but the natural resources have never been available before -- weightlessness, endless vacuum, and the absence of Earth's seismic vibrations.

There seems to be two schools of thought as to the value of space manufacturing. One group sees a host of new products on the horizon -- crystals of unprecedented size for the electronics industry, entirely new alloys of metals, the ability to separate biological materials such as vaccines into ultra-pure constituents, extremely accurate vibration tests of small electric motors, and on and on.

The second school believes that we needn't look ahead for new products; we should be content to unmask the effects of gravity on material processes. We should be content to take a gaint step in the direction of a more fundamental understanding of physical processes. I personally am of a third school of thinking -- that the first and second schools are both right and that we better be ready to use every opportunity that the Shuttle affords us to develop space manufacturing.

I believe these experiments may well evolve into the bread and butter payloads for the Shuttle in the initial years.

In the <u>long run</u> the Space Shuttle program will emerge as an immense success. It will eventually become saturated with business from three new industries.

However, the program may suffer initially from, oddly enough, a trend toward better satellites. It seems that the aerospace industry is now planning satellites with considerably better reliability and longer life than anyone anticipated several years ago.

Ultimately, I think this will prove a great boon to the Space Shuttle because <u>again</u> the cost of maintaining a satellite service will drop, which should attract even more users into orbit.

However, in the short run, earlier estimates of demand for satellites may prove to be a little high, and this includes the estimates of the Department of Defense.

That presents a problem for NASA because the agency has a new job in the Shuttle program. NASA is an old hand at being the pilot who flies his cargo into space. But now NASA is also the ticket agent who is responsible for booking flights. If we are going to realize the full opportunity for economy and new uses of space that the Shuttle promises, it must fly with a load factor of one. So the challenge to NASA as I see it will be to adopt a flexible, aggressive approach to payload planning. I think that NASA recognizes its new challenge now and is heading to meet it.

Some scientists and space enthusiasts have recently urged NASA and the Congress to consider creating some kind of institutional mechanism to assure user participation in defining new space applications. Former Presidential Science Advisor Edward David suggested the possibility of commissioning a "space launch corporation" or of assigning the role to an existing corporation. That way market forces would predominate in the policy planning.

The Space Applications Board of the National Research Council recently recommended formation of a National Space Applications Council to do the job of planning those new payloads.

I have not reached any conclusions as of yet. But I do think the Senate Committee on Aeronautical and Space Sciences should take this matter up and examine these ideas with care. It may be that we do need a new institutional arrangement.

In closing, I want to say that I see the unmistakable marks of maturity in the space program. In its adolescence the program emphasis was on exploration, as it should have been. When the subject of benefits to the American people came up we could point to national pride, stretching new technologies, spin-offs, and other indirect benefits.

Now the emphasis is shifting. We will still be looking outward at space with planetary probes of the solar system and hopefully with the Large Space Telescope, but our attention is turning back toward Earth as well. We are rapidly learning how to use space, how to exploit this infinite natural resource, for direct benefits to all mankind.

And if there is any single event that will demonstrate the space program's attainment of maturity, I think it will be the launching of the Space Shuttle.

SHUTTLE MISSIONS - RETURN ON INVESTMENT (ROI)
PANEL DISCUSSION

<u>MASTER OF CEREMONIES</u>

    Robert Gervais
      Executive Vice President
      American Astronautical Society

<u>CHAIRMAN</u>

    William E. Dean
      Vice President and Program Manager, Shuttle System Integration
      Space Division, Rockwell International

<u>PANEL</u>

    Dr. Richard Goody
      Chairman, Space Science Board
      National Academy of Science

    Dr. Preston Layton
      Aerospace Systems Laboratory
      Princeton University

    Capt. Walter M. Schirra, USN (Ret), Astronaut
      Director, Technology Purchases
      Johns-Manville Corporation

    Dr. Robert J. Anderson
      Vice President, Math. Tech. Division
      Mathematica, Inc.

    The Honorable Patricia Schroeder
      Congresswoman
      First District of Colorado

    Dr. Gottfried Greger
      Head, Orbital Systems Section
      Federal Ministry for Research and Technology
      West Germany

    Dr. Gerard K. O'Neill
      Department of Physics
      Princeton University

INTRODUCTORY REMARKS - Robert Gervais

It is my pleasure to welcome you to this luncheon panel session entitled "Shuttle Missions -- Return on Investment." The purpose of this session is to give perspective to the symposium. That is, what is the return on investment that will be realized from the Shuttle program that we have been discussing for the last few days? Initially, Governor Lamm opened the symposium with a challenge in this regard. During the symposium, you've heard the details of some of the possibilities for the Shuttle payloads program. They've been succinct and subtle, but you've been given an inkling. Those of you who are concerned with the space program on a day-to-day basis have heard from your counterparts on what are the potentials for the future.

We have convened an outstanding panel and, in a sense, are challenging the speakers, in their areas of expertise, to draw from them their opinions on what -- in fact -- may be the real benefits. I'm going to use a slip of the tongue made by Senator Moss last night to characterize the challenge. When he was trying to introduce the Space Shuttle, the slip of the tongue was the "subtle Shuttle" -- and that's what it is right now. It's subtle in the sense that we know there are benefits; the problem is to crystallize them and focus them.

OPENING ADDRESS AND INDUSTRY ROI - William E. Dean

As Bob indicated, we've had three days of papers on Space Shuttle Programs and Missions, and last night Senator Moss did describe the benefits which will accrue to mankind from this family of new capabilities, new spacecraft, and new missions. But, there are new areas, if you will, which are under study or barely into their development aspects. They are the missions of the 1980's. The question is asked whether we must wait until the end of this decade of investment, the 1970's, in order to begin receiving the returns from the investment. I feel very strongly the answer is no, and the panel will be bringing you some thoughts on that from their particular areas of expertise today.

The low cost transportation system which we call Shuttle, as you know, was begun several years ago. I'd like to recap quickly the status of some of the hardware in the program to provide a little background for the other discussions. Of course, the Shuttle has been established as one of the nation's major new space objectives. Development of the main engines was initiated by Rocketdyne Division of Rockwell in 1971, and the Orbiter Development and the Shuttle System Integration contracts were awarded to my Division, the Space Division of Rockwell, in 1972. Martin Marietta was selected for the External Tank and Thiokol for the Solid Rocket Motor, in 1973 and 1974.

This is the industry team that is committed now to develop the Shuttle within the cost and schedules set by the NASA in 1972. And happily, I can report that we are on cost and on schedule to those targets set in 1972. In fact, as I think most of you know, the carrier for the payloads which have been described and discussed in the last three days is well into the fabrication of the test and flight hardware. We're way beyond the paper stage with the Shuttle System. The main engine is now on the test stand in Mississippi and its first firing took place on June 24th. The structure for the first orbiter is well into the systems installation phase in Downey, and in Palmdale, California. By February, the first flight article will be assembled into a complete unit at Palmdale. The External Tank Critical Design Review will occur in December, and hardware for the Structural Test Articles is already being fabricated. The tooling, of course, for the flight hardware is under way. The solid rocket motor has been through the Preliminary Design Review. Solid rocket motor case, the nozzle metal components, and flex bearing subcontracts have been awarded. The first case segment has been roll-formed.

Modifications to the facilities at Cape Kennedy to create the first reusable space launch and landing port are under way. Those of you who have seen any photographs of the new runway, for example, just to the north and west of the Vertical Assembly Building will see a lot of concrete being laid. There is a lot of new activity at the Cape on the facilities.

From now on, the progress being made is going to be more visible, and the reality of your payload carrier will become increasingly evident. In 13 months the first orbiter will roll out of the hangar at Palmdale. In about 18 months, the first captive flights on the 747 will begin, and within 24 months, the orbiter will be into the approach and landing test phase. All are aimed, of course, at the first manned orbital flight in the early part, or first or second quarter of 1979.

Over half of the development decade, counting just the 1970 decade for the moment, has passed. As the industrial member of the panel, however, I feel that I can answer the question "What is the return to industry from this effort?", and I would like to cite a few examples. Within the context of their social responsibility, of course, business organizations have always had two primary objectives: the earning of a fair return on the stockholder's equity, and the development of new business in order to perpetuate the enterprise. Participation in the Space Shuttle program has provided for attainment of both of these objectives in large companies as well as small, and I will try to give you some examples. It has provided other unique returns which enhance the long-run performance of the business firms. I'm going to be speaking of Rockwell and some of our subcontractors in the examples, but I would like to ask each of you from industry to be thinking of your own companies in the same context as the examples from Rockwell. Each of you who have participated in the space program understands the continuous investments which have to be made -- first, from your own research and development and proposal efforts and also from the dedication of your most creative talent in making your company competitive in today's market place. Now, this investment in itself provides the company with technical capability to compete in other business areas. Too often we think of ourselves in one business. Think also of what your company has been able to do in other business areas because of the investment in the space business. Suppose there is a competition in the space business. Once having won that competition, the company must continue to make investments needed to fulfill the contractual responsibilities. For the Space Shuttle team, a foremost requirement has been the increased investment to assure quality, because quality is the key to safety and success in space flight programs. This means quality is imposed as a company philosophy from top management down to the individual member in the production line. It means the same quality investments required by each prime contractor are also required of all their subcontractors and similarly from the subcontractors to their suppliers. It has been this unceasing attention to improving quality of design, quality of testing, and quality of production that has been the hallmark of the manned space program.

A second major investment has been occasioned by the funding limitations imposed on the Space Shuttle contracts as a result of the fixed cost and schedule commitments that the NASA had made to the Congress and to the President. We have had to invest in developing new and more rigorous management systems and techniques for adherence to cost and schedule, for judging risk, for weighing changes. This has been referred to as "constraint management" which is a process of continuously balancing performance, risk, cost, and schedule. The new element here is the risk schedule, risk in the sense of determining the parts of a program to be dropped, or deferred, or done in a different way in order to continue the total program in the context of the schedule and dollar commitment. It has taken money and time on the part of each industrial member of the Shuttle team to develop these skills. It requires a program management team that is able to evaluate the impact of progress and problems from all facets of the program, not just from the part of the program under their jurisdiction. This is another return which has resulted in the formation of a very close-knit and cooperative customer/contractor team. Coordination between contractors is unlike any that I've seen before, and we, therefore, have a team with the capability of adjusting program content while maintaining the overall goals. The management capability developed for project Apollo has been improved and made more responsive to the new priorities and national needs. This has been a very unique, very helpful, very useful, long-term benefit from participation in the space program.

We have also been attentive to our social responsibilities, and we have major investments in providing equal opportunity for minority employees and subcontractors in accordance with formal Affirmative Action Programs. The returns in local economic stability and the developing of human resources are very immediate. The Shuttle team is committed to assuring nation-wide distribution of the contractual efforts in order to achieve the widest possible participation in what is truly a national program. The Orbiter contractor alone has awarded subcontracts and purchase orders in 47 of the 50 states. Some 53 percent of the total dollars are out-of-house, if you will, in subcontract and purchase orders.

In times of economic uncertainty and rapid change such as we've all been experiencing over the past year, programs such as Space Shuttle provide a financial stability to a business. It gives us a basis for weathering the storm, you might say, and provides for earning the investments in tools that are needed to be competitive in the next round of good times. In Rockwell's case, the financial returns from the Space Shuttle program have not nor ever will equal the good years from the commercial side of our corporation. But, for the past six months, it has been the Space Shuttle in California that has provided the company profit, and hence the capital, for new industrial machinery of our weaving machinery plants in Massachusetts, for example, and some help to the automotive plants in the mid-west. The investments in quality and management techniques and attention to detail result in the attainment of profit incentives on Award Fee contracts. Our financial objectives are being met and bettered as a result of the investment that we have made in order to perform well on the Shuttle program.

The Space Shuttle requires advances in technology as well as in management. These advances are being developed by our subcontractors for new composite materials, for completely new insulation systems, and for new types of avionic systems and equipment. Advances which are made on this program become a national asset in that they are available for commercial application and exploitation with the encouragement and with the support of the NASA. This creation of new technology is a continuing return on our investment in the space program. It is the application of this technology on a wide scale that has made terms such as "spin-off" and "technology-utilization" and "space-rated" household words, but it is more than just a spin-off. The development of advanced technical equipment for our manned-space systems has become the product-development activity for a number of our smaller subcontractors. This is the means by which they develop the business to succeed the start-stop sequence that we sometimes find in aerospace. I would like to cite a couple of examples. A small company by the name of Tylan in Torrance, California developed mass-flow meters for the Apollo Command Module to measure the flow-rate of astronauts' breathing oxygen. Today, these mass-flow meters are the key component in mass-flow controls and gas-blending panels that are used extensively throughout the world for process control in the manufacture of semiconductors. As a result of this direct, commercial application following the completion of Apollo, the annual sales and jobs of this company have grown from $300,000 and 20 people to over $3,000,000 and 80 people -- a small company but four times it size now. A business has grown and an industry benefitted from the quality products resulting from the space program.

Another example of the industrial benefit from space hardware development is ELDEC of Washington. ELDEC Corporation was a supplier of instrumentation for the Saturn Rocket and the Apollo Spacecraft, and they provided such items as electronic signal conditioners and water-heater electronic controls. To meet our requirements, it was necessary for them to develop space-rated methods for welding cordwood modules, for potting electrical connectors that would withstand launching and space use, for the manufacturing of miniature transformers, and for improving the packaging density of their electronic components. Commercial products first made possible by these space hardware manufacturing techniques began with a load-cell system for the C-5 and then the L1011 aircraft weight and balance systems. Then, a simplified version of the system was developed for logging trucks in the states of Washington and Oregon. This weighing system allows the truck driver to get a digital readout of the exact weight of the logs being loaded on his truck, and thus he is able to maximize the number of logs on his truck yet stay within the legal weight limits for that state. Finally, ELDEC has convinced the state of Washington that they need a load cell system to measure the weight of those huge logging trucks that rumble over their roads. Guess who is supplying that system to the state of Washington -- ELDEC!

In both these instances, the companies were able to significantly expand their business operations and to provide new and improved products for wide-spread usage because of both the advancement in space technology and the accessibility of these advances to America's industries -- the very key which we have to remember as we look at return on investment.

The investments in quality bring about a different type of return to companies that participate in a space program. We have seen the pride and the workmsnship which go into articles that are to be used in a manned spacecraft. Knowing that only the best can be used and will be used results in people deliberately trying to do better work, to pay more attention to the hardware which will go into space programs such as the Space Shuttle. This pride is contagious. It rubs off onto the other products of the company and increases the value and the quality and the competiveness of everything the company does.

The application of constraint management techniques, I believe, is the answer to being able to produce products of constant value for both government and commerical programs. It is this combined value of technology plus quality plus cost control that makes companies competitive for both U.S. and international markets. It is recognition of this long-range return, this increase in productivity which makes foreign countries and companies anxious to participate and willing to invest their own resources to become members of the space program team.

If we are to achieve our other humanistic and social goals we must have an economic base to provide the national wealth with which to help those who are victims of poverty, or prejudice, or ill health. For the foreseeable future, it is industry and business that are the source of this wealth. Improvements in products and efficiency are what make industrial growth possible. The application of the management, quality, and technology advances from the investment in our national space program, and in particular in the Space Shuttle and its missions, will contribute to, indeed will, accelerate National economic gains. An on-cost, on-schedule operational Shuttle will benefit the process of establishing achieveable National goals and policies, will increase mankind's understanding of his Universe and of his place within it, and will give us the tools for monitoring and managing our earth's resources. Nevertheless, it begins with investments from industry and it provides continuous, on-going returns. We are receiving them now -- even before the first flight, or the first payload, or the first mission.

SCIENCE RETURN ON INVESTMENT -- Richard Goody

My subject is science -- science on the Space Shuttle. Science is knowledge, and my part of knowledge is not just the conventional science, perhaps, but also that associated rather directly with applications through their contact with the earth sciences. This is the broad area that I want to make a few remarks about in the minute or two that we have. The Space Science Board did a fairly thorough blue sky study in 1973 of what you might do on the Shuttle with science and produced an extensive document. Much has been added to this in the meantime. Obviously, I can't even attempt to describe to you the various findings along the lines of the ways in which Shuttle would benefit science. Following that period of identification of valuable things that could be done on the Shuttle, the NASA has gone into a lot of Phase A studies trying to look at some of these concepts like LST and AMPS. NASA is now coming out with complicated studies which are being returned into the system, and that's where we are in this whole business of assessing what you can do on the Shuttle when you have got it.

Now, in the midst of all the complexities which I find very difficult to bring into focus -- even when I have an unlimited time to try and do it -- I'm going to try to distill my experience of that process down into two points, if I may, as a highly over-simplified version of how we would view the use of the Shuttle with respect to its advantages and value as a science tool. The first, and I think obvious point, is that it immediately becomes clear that the size, the weight, the revisiting capability, the possible existence of an operator on the Space Shuttle are all features that can be used to great advantage for scientific purposes. You can do very much more than you otherwise could with other available systems. One particular example is semipermanent observatories in space (such as LST) which are operated from the ground, thereby extending your ground observatories up into space. These would scarcely have been feasible before the Shuttle, but with the Shuttle they have become an eminently reasonable and practical thing to do. Now, I add, just so that one should not be solely euphoric about this point of view, that for a very great deal of science -- all of extraterrestrial science and all free fliers, which is most of terrestrial science -- the Shuttle is only a booster, and the question of which booster you use is a matter for proper evaluation at the appropriate time.

The second point I want to make is that the advantages are certainly exceedingly strong. The possibility of running a large astronomical telescope in space is obviously a tremendously valuable and exciting prospect. However, let me say that when we come to the actual exploitation of those advantages, I have no clear conception of what they mean and how they may be realized. I have a feeling that others do not either. Whenever I go in any depth on the exploitation of these remarkable advantages of this big tool one sees management problems -- science management problems. One also sees problems which almost involve restructuring NASA, and I do not in a minute see that the proper attention is being given to them. This, then, is one of the points I want to make to you. Let me take as an example the Astronomy Observatories which are being proposed and suggested. These are supposed to last for some 15 years, but if you give them the proper ground base science support, computers, software, and all the necessary types of support that we know are necessary for that kind of operation, you are talking, in fact, about a program which costs at least two or three times more than the initial cost of the LST itself. This, in fact, would probably be the Nation's main astronomy program. Therefore, you bump into questions on what is the right way to run the Nation's main astronomy program. Is it a part of the agency? How does it work with other agencies like NSF? There is a vast array of exceedingly difficult problems which I am now concerned that we need to be able to try to answer a little more carefully if we are to go ahead. There are many other examples of this kind of thing and I draw them to your attention because we are even now considering the necessary new starts for the things we will actually do on the Shuttle. We are talking about a concept with which we have no practical experience, and we are going to have to do a lot of very careful and thoughtful work if that planning is going to be fruitful. Now, we can of course limp into the 1990's, you may say, when we will have gotten the necessary experience on operating the Shuttle. Or, we can get into intensive examination of straw models to see how the whole thing works and thereby become sensitive to the nature of the use of this new capability. I believe that this latter course in absolutely essential. I can't see just experimenting on its use. I think we have to think our way into it. In order to do that, I would like to make as a comment that I think one has to have, when talking of 5 to 10 years from now, a very clear idea of where one is going. I also think one has to have a clear idea that we are talking about a National commitment to a central purpose for the NASA. This, in fact, is to use, employ, and to operate the Space Shuttle as a natural operating basis for the purpose of doing -- of doing what? Well, as I see it, it is of doing applications to man's needs and scientific exploration. It is a case here, that Shuttle was, in the area of applications and in the areas of science, a promise deferred. Many of these programs could have been done by more conventional means. We deferred these programs to the point of getting, hopefully, a better transportation system with which we could, in fact, do them better. The actual problem of doing these missions better on the Shuttle is that we have to make the commitment, I believe, in the period five years ahead. This is the position in which we are now relative to deciding where the future NASA programs will have to go. Once we have that commitment, I don't see any particular difficulties in solving these problems that I have been raising. With such a program of utilization of the Shuttle to carry out the already defined and perceived programs that are available in the applications area and those in scientific exploration, I have no doubt that the NASA program will be something of which the Nation can be extremely proud.

EDUCATION RETURN ON INVESTMENT - Preston Layton

As Chairman of the AIAA Committee that assessed the need for a new space transportation system, I became convinced that the NASA Space Shuttle System represented a necessary next step in the United States Space Program. While I have some real doubts about the realism of the mission models that have been used, in terms of both the present payload concepts that they include and the traffic build-up, it is my opinion that NASA's patterning of the Shuttle development and the planned introduction to operational use have distorted the remaining space program somewhat unnecessarily. It seems to me also, that the economic payoff may come later -- rather than earlier -- although I'm sure its there. As to the overall necessity of the Shuttle to man's future endeavors in space, there should no longer be any rational doubt.

In addition to the quantifiable returns on investment that the Shuttle permits and that will be described by the others, there is a prime opportunity to use space in novel ways with increased levels of activity. It seems to me that is its strongest recommendation. From the educational standpoint, the Shuttle/Spacelab combination permits the undertaking of research in space and the exploration of new kinds of knowledge that could never be accomplished on earth. Both basic and applied research will be undertaken, and a professor, for example, may take a student into orbit for the conduct of experiments. All manner of other industrial and scientific endeavors will be, I believe, carried out. The opportunity for education by direct broadcast to all areas of the world is facilitated by the Shuttle and the space stations that it will serve. I believe that much can be learned about a multitude of earth-surface phenomena, including man himself, by direct personal observation that will supplement the observations from automated spacecraft. Art forms, as evidenced by the show here, are already being affected by this space vantage.

Best of all, the Shuttle and the capabilities derived from it will permit man to learn more about himself. This has already happened more substantially than is commonly realized from the Apollo and the unmanned spacecraft results. Remarkable photographs taken of the "big blue marble" by Capt Schirra and his fellow astronauts have underpinned the idea of spaceship earth and the ecological efforts that are now being made to the advantage of us all. That's a return on investment. We no longer have to worry about the little green men from Mars. We have apparently been given the solar system to make use of as we see fit. The Viking and its follow-on planetary and lunar spacecraft will continue to educate us about this unfamiliar real estate and the space between. Also, think of the benefits of putting some politicians, perhaps even a statesperson such as Representative Schroeder, into orbit. Of basic importance is man's idea of his place in the Universe as it will be strengthened by the use of the Shuttle. For example, in servicing the

LST and other automated observatories and by direct observation, our basic philosophy of life will be much affected and our religious views will be surely modified. Maybe we should put some philosophers and religious dons into orbit too. A new moral ethic would be a real return on investment.

MAN ONBOARD THE SHUTTLE AND ROI - Walter M. Schirra

As you know, I'm in Denver now. I've been here since 1969 but have been watching the space program -- with another Walter. On those occasions I've seen some rather changed attitudes in the public's opinion of what's happening to the space program. Before I get to that, though, I think its appropriate to reminisce a little bit. When I had my first flight scheduled in 1962, somebody handed me a bag of rocks and a pile of books about geology. One week before launch he said that he hoped we could bring back some good data on the surface of the earth out of the little tiny window that we were supposed to look out of half of the time while we were in orbit. Acknowledging that the other half is night time, it is kind of hard to look at the surface of the earth. We progressed from that point to some rather good experiments in the Gemini program. I really think, looking back on Gemini, that we accomplished two things. The first was the capability of performing rendezvous in space and the second man enduring in space for two weeks. That is all that you can say that the poor crew on Gemini 7 did -- endure for two weeks. If you recall, Tom Stafford and I rendezvoused with them and saw the dirty laundry hanging out.

With the Apollo series, of course, we put both of these techniques to work. That was to go some place else, do a rendezvous, and quite often people leave out the phenomenon of coming back to earth. They always talk about going to the moon, but we've always tried to book a round trip. With this, of course, we explored another planet, and I think most of us in this room look at the moon as another planet rather than as an associate of earth. We are learning things about our celestial system by having observed the moon.

We have also learned how man can perform, in that we had men walking on the surface of the moon, picking up selected samples -- now having had years to study geology rather than the week that I had with a bag of rocks in front of me. We came back with some good data, and some of the data proved that our opinions were correct. I can recall our initial discussions with the people who worried about volcanoes being on the moon in contrast to meteor deposits or meteor impacts. We'd get both teams to argue, because neither one of them believed the other was right. We would take notes, and that way we learned to pick the brains of both sides. But, in all three of these missions, and then, of course, we had the follow-on with Skylab and finally our Apollo-Soyuz mission, we were trying to find out how well man could perform on these missions. What could man essentially do to add to the number of nines in the success reliability of the mission. I think it is time now that we also face up to the fact that the world and the nation, not most of the people in this room because we are all fans of each other, but the rest of the world wants to know, what is in it for me, what's the pay-off? A man has proven, and I would defer to Mrs Schroeder that womankind as well can do this task, that mankind can go into space, do a task, and come back. Now we have a task ahead of us which is to do something with a payload, a transportation system as Dr. Layton mentioned, that will carry not only experiments into space but practical equipment to look back on earth and report on earth in a way where we can profit from the venture. The payloads we are talking are, in many cases, the unmanned satellites which so many people accuse astronauts of not appreciating. If you don't think we didn't appreciate what Surveyor did when it dug around on the moon (thinking it might have been just a big bag of dust and we would have gone through it) you have missed a big bet. If you don't think all astronauts aren't interested in what Viking is going to report back on Mars, I think you have missed the fact that we like to know what we are landing on. Whether Viking will land safely or not is something we are looking forward to. So, there is a mixture of manned or unmanned that we are looking for in the Shuttle. I do believe that we have to start selling the Shuttle as a vehicle that can provide services, and then prove its economic point -- not prove its economic value by the fact that we are doing work making Shuttle but doing work using Shuttle.

ECONOMIC BENEFITS AND ROI - Robert J. Anderson

I have argued in various places that it is difficult, indeed impossible, to apply rigid economic criteria to the various facets of the space program. This isn't so surprising, since, as economists, we can't look ahead six months and say with any great confidence what is going to happen to the gross national product. So, why should one suspect that we could look into the 1980's and say what kinds of good things will emanate from the Space Shuttle program? Nonetheless, it would be bad manners to stand up here and say what we can't talk about. Indeed, over perhaps the last 4 or 5 years, economists have made some headway in measuring, and I think rather well, some of the returns of the space activity.

I would like to talk just a little bit today about two different kinds of studies the economists have done and the results of those studies. The first sort of study is a very general statistical relationship between R&D efforts in the space program and the gross national product. There have been a number of such studies. The first one was by the Midwest Research Institute under contract to NASA, and then Chase Econometrics has followed up with several others under the sponsorship of both Rockwell and NASA. As a newcomer to space economics, a fellow who was more interested in the value of clean air, the results of these studies were rather incredible to me as I first learned about them. In particular, almost every study, no matter who did it, showed something like a 35 percent internal rate of return to R&D in aerospace. So, I looked somewhat more closely at these things. They were all done by people of good reputation, honest and intelligent, and although there are some minor points to quibble with, I think the conclusion seems to stand. That is, if one looks at the rate of return that equates the stream of increases in gross national product with the outlays for R&D, something like a 35 percent rate of return emerges. These studies, by the way, have often been summarized by Dr. Fletcher somewhat differently. Many of you have heard the 7 to 1 number -- 7 dollars in additional gross national product for each dollar of expenditure on general R&D. These are interesting studies, nonetheless, because they are statistical and they talk about aggregates like gross national product. They are also perhaps somewhat mystical. In consequence, we of Mathematica were commissioned to do another study using a case study approach. We looked at four particular aerospace technologies and very selected applications of these technologies. We concluded that, over a twenty-year period, the kinds of advances that have been made since 1958 have resulted in a present value of economic benefits of something on the order of 28 million dollars. Of course, the technologies that we looked at were things like integrated circuits, and people other than people in space have interest in this technology. Some of the benefits may derive, therefore, from things other than our interest in space. When we account for this, we come up with something like 7 billion dollars, figured conservatively. Unfortunately, it wasn't possible for us to isolate R&D spending in these technology and applications areas, so we couldn't come up with any kind of good hard internal rate of return estimate. That's one area -- the general returns from space program activity and space program technology.

The second kind of study, and one that is just now getting under way, includes studies of the benefits of particular applications of space technology. Space communications satellites and sensing satellites are two of the more obvious examples. In general, we as yet have estimates of only a portion of these benefits. In particular, it is relatively easy to quantify cost savings where a system in space does the same thing that a system on earth does, or cost deficits if it breaks the other way. This is not an area in which I have worked very much so I'd rather pass from the applications area.

There is a third kind of benefit that we never really have addressed, probably because we don't know how to. That is the benefit associated with finding a new place for man to do something -- a very difficult problem. This is certainly the kind of benefit that we refer to when we talk of benefits being in the distant future. Just for the heck of it, while I was sitting out by one of the mountains somewhat north of here, I decided to do a little intellectual exercise. I tried to cast myself back into the 1500's or 1600's, and, with the benefit of hindsight, reasoned about what the benefits, the internal rate of return might have looked like in thinking about exploring, colonizing, and developing the

New World. I looked up at the sky, and with some knowledge of the economic data from the ensuing period calculated a rate of return of between 20 and 30 percent. This is quite literally a blue sky estimate, since I was looking up at the time. Even if its a perfect estimate, realize, of course, that it may have absolutely no relationship to what the value of finding and charting new places in space are, or developing them for man's use. Nonetheless, in the process of doing the exercise, I began to get the feeling that we could begin to think about this problem. What is the value of a new place, and the use of a new place to mankind over the far distant future?

Finally, I would like to conclude by sharing with you a little anecdote that Professor Morgenstern passed along to me. It concerns Faraday who was apparently being pressed by a politician concerning what the worth of his discoveries in electricity were. He thought about it for a moment. Then he said, "I don't know yet, but someday you'll tax it!"

NATIONAL POLICY AND ROI - Patricia Schroeder

I must admit that I worked my way through school adjusting aviation losses, so I always got to see the failures, but I must admit it had great economic return for me. I was able to get through. Nevertheless, I looked at this session and saw that you were going to ask us questions. I really wish I could ask you questions, because, I think what happens so often is that we all come in with our own little perspective and it is very hard to get the overall picture. I think if I were to ask this group how many people were against deficit spending that most of you would be with it. I think we could ask people how many thought we were spending too much on food stamps; or how many people thought we were spending too much on education. Or, we could look at welfare; we could look at unemployment; we could look at child-abuse programs; we could look at foreign aid; we could talk about aid to the cities as we see New York City coming very close to going belly-up. I would like to ask all of you, and we could just go through each little Federal slot in the budget, to check yes or no on which you thought we had too much and which we had too little. Then, we could do the same thing with the Space Program and put it all in perspective. I think you would begin to see some of the problems that we, as politicians, have when you start from the given point that nobody wants a deficit. I've always wanted to conduct a poll on what Federal programs directly benefit people and where they think there is any waste that can be cut. Every time I've asked that of anyone who comes into my office, I find that there's no waste in their program. In fact, they need more. Well, that's very difficult to accomplish if you start with the idea that nobody wants a deficit.

The other point is that everybody says that either you are 100 percent for us or you are against us. You know, there is no way to understand the give and take. Now, I don't think there is anyone in the world who doesn't think that there is some value in the Space Program and the Shuttle Program. I really think that everybody agrees. The question is -- What kind of value?--as we are weighing all the different priorities, as we are balancing everying. My topic, of course, is what does the Shuttle Program do for our National image and what kind of international relations returns do we get on it. How do we justify spending approximately 10 billion dollars of taxpayer's money, which is probably where we'll be -- and it'll probably be more than that by 1979 for the Shuttle Program? I'm including the Tug and things like that, and I think with inflation and the way we're going that it is probably a very low figure but we'll start with that. What do you say to the third-world nations and to people where half the world is going to bed hungry? They are saying to us, "You're the haves, what are you doing for us?" That's heavy. Or, you go into Africa where they say the Japanese are building a sub, a dam, the Chinese are building us railroads, what are you doing for us? Now, we may not like that, and we are saying everyone has got their hand out. They are all saying we are all in this thing together and united we make it but you're going to split everyone off. Its tough! How do we respond to people in the United States who are saying, that's very interesting that you are putting all that money into the Shuttle but I'd like to have you a little more concerned about how I get to work in the morning -- our ground transportation. Or, they ask if anyone has looked at railroad beds in future years and what are you going to do about the railroad beds. Someone else says, well, what are you doing about pollution and what are you doing about alternative sources of energy and they mean in a consumer way. They don't mean in research. They mean -- look I'll put a solar furnace in my house, but I want to be able to go down to the store and buy dampers off the shelf so I can afford it. They say, what are you doing to get industry moving in there, and you say well -- have patience.

We're a very impatient nation, let me say, and these are all very difficult options to balance. We have had a little problem with atomic energy Project Plowshare. There were many people who felt that that was technology in search of a justification for a long time and that we spent a lot of money on it. I think we have to be sure that we're geting something back. Now, I think the LST is probably going to offer us a lot, but I also hear considerable other dialogue about other things and I say, fine! But I think we can also get that from other programs, because I have been aware of other programs when they come and tell me how I ought to back them too. They tell me the same thing you're telling me and so, I want to know whether we can get these things in other ways. Someone said "what is the value of colonizing space?" I think if we ask the British and the Spanish and the French we find out that they're probably going to revolt after you spend thousands and millions of dollars settling them. Look how disrespectful we were two hundred years ago when we said to the King we don't care, we're on our own. So, I'm not sure we can really count on having colonies up there that are going to send us money. It would be nice, but I'm not sure we can count on that anymore. What we want to be careful about, and I think one of the bad images that has come upon us, is to make sure that the space program isn't a Roman Circus. We want to make sure it isn't something we are doing to take people's minds off earthly worries. We have to make sure that it is in balance. We have to make sure that yes, we are creating jobs with the space program, but the purpose of the space program is not just to create jobs. If the purpose is to create jobs, I think anyone would admit that we can create more jobs with less money if we are just out to create jobs, because you have to have incredible capital-intensive industry to go through this.

Granted that the Shuttle Program is in an early stage, so it is a little hard to see the returns, but, we have all these competing interests, and I think we have to keep that in mind. I think we have to be really prepared to give and take and to try to find some way that Shuttle becomes relevant to the average world citizen -- in some way other than being just an observer or participator when the thing happens on a television screen. Unfortunately, it's the kind of world now where everybody wants to know what's in it for them. We may not like that, but that's where we are. I think we have to keep that in perspective as we discuss this program, and we have to talk about the total costs versus the cost of other programs and its real return to the country. If you have put in ten billion dollars and go out and look for oil and find oil, then we are in better shape than we were before. I think that there are certainly a lot of things that we can do in research, and we may be in better shape than we were before, but we've got to start pinpointing and be a little more specific about those benefits.

SPACE SHUTTLE AND THE EUROPEAN COMMUNITY - Gottfried Greger

In trying to assess the impact of Space Shuttle on the European community, I must kick off with the statement that Europe is not engaged directly in the Shuttle development. You may remember we tried several years ago to do so but were not successful. So, I want to refer to the impact of the overall transportation system, which includes Spacelab as a substantial part. I need not repeat the impacts on the different areas of science and technology which have been mentioned already. They can be considered more or less equal for the European community. You have only to delete Shuttle and insert Spacelab, and you have to insert, instead of one national group of industry, an international grouping of industries from ten European member states.

I want to pick up three categories of problems which we see, and I think these are of real importance to us. First is the improvement of cooperation with the United States on an advanced technological field. Second is the possibility for Europe to enter manned space flight and for reorientation of the European program toward applications. Third is the improvement of the common use of investments, especially in the field of advanced space transportation and orbital systems.

The first point, transatlantic cooperation in one of the most exciting areas of technological development is substantially being improved. You know that ten countries are working together, and with the U.S.A., within clearly defined financial, technical, organizational, and schedule constraints. All levels of governments in the European Space Agency and the industry in ten European countries are equally involved in this venture. The importance of this cooperation, however, is reaching far beyond this specific development task. It is a useful exercise for tackling even more complex and demanding world-wide problems that we might be forced to overcome jointly in the future.

Regarding the second point, the Shuttle Spacelab program is THE step into manned space flight for Europe but under a special circumstance. The understanding of the real aims of space flight in Europe is considerably changing now. Space flight must prove itself as an effective tool to solve problems on earth. Its unique possibilities must be integrated into the overall technical infrastructure of our economy. Shuttle and Spacelab seem to be a key to reach this ambitious goal in Europe. How well we can reach this goal will determine the future of European space activities.

The third point, which is a very important one, is that transatlantic cooperation is devoted to the development of a transportation system which might be in the long run, the only basis for future space activities and commercial utilization. It's the real aim of this system. The The junior partner, Europe, will depend strongly on the availability of this system operated by the United States for the execution of European tasks which might have economic impact. I need not stress here that the European efforts are based on a strong confidence in this availability of launch services. I think cooperation is always an obligation for all the partners. I consider this growing, mutual confidence as a good and encouraging step forward.

Summing up from the European side, the Shuttle Program is really not underestimated as a significant pace-maker for cooperation in technology, economy, and last, but not least, in political levels. We regard, just as you do, this program as a challenge and we are ready to take the necessary steps to make the best use of it. In this context, if the chairman will allow me a few additional seconds, I would once more express my appreciation for the opportunity of the Europeans to take part in this meeting. It was a very interesting session for us. I hope you got the impression that Europe is trying to do its job honestly concerning Spacelab development and preparation for utilization. We, on our side, thank you for the confirmation that, concerning utilization, we are marching in the same direction and straight forward.

SPACE SHUTTLE AND SPACE UTILIZATION - Gerard K. O'Neill

Although I had agreed some months ago to take part in this panel discussion, I very nearly cancelled out at the last minute because in the final letter from the organizing committee, I found myself referred to (as just now) as a Space Visionary. I'm really not interested in visions -- I am interested in some good practical engineering and what can be done with it on a short-time scale. The name, space colonization, which has been mentioned once or twice today, has a touch of sensationalism about it, and I hope we find a better one -- perhaps space manufacturing facilities or high orbit manufacturing will serve. What we mean by it though, is four essential points. The first point is that of using non-terrestrial materials (initially from the lunar surface) to minimize transport costs for getting materials into high orbit. Second, getting up a manufacturing facility in free space where solar power is available full-time. The third point I'll term the boot-strap process -- bringing only equipment, essential elements, and a workshop from the earth to high orbit and there building a habitat for people, a first beachhead in space. Then, that work force will be applied to the task of building additional manufacturing facilities as well as products for use in deep space or in high orbit. That way, production goes exponentially with time rather than just linearly. I should emphasize that the economic reasons for construction of a space community all relate to products whose end use would be in free space or high orbit. Fourth, and a very important condition is the restriction to near-term technology -- that is, engineering and not science fiction.

The Space Shuttle is vital to the development of high-orbit manufacturing in three ways. First, Shuttle would be used for physiological testing and pilot plant checkout that could lead to great economies in the program. The second point is that the STS is the only people-rated system we will have in the early 1980's. In deference to Congresswoman Schroeder, I should say that I am trying to educate NASA out of two words that it is very fond of -- one is "man" and the other is "crew." I like to talk about people. Third, the Shuttle system main engines are exactly what we need for a near term heavy lift freight vehicle.

Now, the physiology tests, with the potential to permit economy in habitat design, are mainly on gravity and rotation rate. You drive the design very far if you set very tight conditions on gravity and rotation. Two other parameters, the physiology and fire control aspects of low pressure, oxygen-rich atmospheres, can, I think, be tested more cheaply on the ground. Manufacturing, metals processing, evaporative formation of metal shells, large scale assembly, and probably demonstration tests of satellite solar power systems are all within the capability of the STS.

The return on investment from a program of high-orbital manufacturing should be, we think, in two forms, energy and dollars. The projections made by the NASA Ames/Stanford Summer Study, which concluded last week, are that energy costs from satellite power stations (built at the libration point from lunar materials) should be low enough to undersell every competing form of baseload power. They may or may not be right. The important thing is that these numbers have to be checked. The amount of energy available is certainly so great that the build-up of space facilities, in the scenarios that were worked out this summer, appear to be limited only by market saturation. As for dollars, the Summer Study concluded that with 10 percent discounted economics and a 100 billion dollar cost for the first permanent facility (that is roughly 15 percent of Project Independence) the benefit to cost ratio will be somewhere between two and three. That is, the program ought to pay for itself and earn profits in the many tens of billions per year. The most critical factor is speed. As I think people in the aerospace industry can understand very well, in a lackadaisical program which extends over many decades, the interest charges just eat you alive and kill the benefit-to-cost ratio. Many people think that there is an equally important but non-tangible return on investment, and that is hope. The interest and enthusiasm shown in these possibilities by the general public during the last year almost surely, I think, derive at least in part, from a sense of relief that we may not be at the end of the line now but just at the threshold of a new frontier on which this country is very well equipped to make its mark.

LARGE SPACE TELESCOPE (LST) PROGRAM

LARGE SPACE TELESCOPE (LST) PROGRAM

Program Chairman   A. Hoag, Kitt Peak National Observatory
Assistant   J. C. Spencer, Martin Marietta Corporation

Mirror Development
     Chairman   Thomas Vogt, Itek Corporation
     Co-Chairman   James Raymer, Beech Aircraft Corporation

Telescope Performance
     Chairman   D. McCarthy, Perkin-Elmer, Incorporated
     Co-Chairman   H. Gier, Beech Aircraft Corporation

Instrument and Detector Development
     I   Chairman   S. Sobieski, NASA/GSFC
        Co-Chairman   A. Delamere, Ball Brothers Research Corporation

     II   Chairman   H. Lassen, TRW
        Co-Chairman   J. Hinricks, Ball Brothers Research Corporation

Precision Pointing and Control Systems
     Chairman   S. Seltzer, NASA/MSFC
     Co-Chairman   T. Spencer, Ball Brothers Research Corporation

Data Management
     Chairman   Fred J. Hudson, IBM
     Co-Chairman   John A. Dougherty, Martin Marietta Corporation

Mission Analysis and Operations
     Chairman   Marcel J. Aucvemanne, NASA Headquarters
     Co-Chairman   Mike Andrews, Colorado State University

Maintainability and Operations
     Chairman   Bert Bulkin, Lockheed Missiles & Space Co., Inc.
     Co-Chairman   Arthur A. Rosener, Martin Marietta Corporation

POTENTIAL FOR ADVANCEMENT OF SPACE ASTRONOMY

Arthur D. Code*

The opportunity to carry out astronomical studies in space has been the realization of a dream nurtured by astronomers for centuries. It has already provided many exciting discoveries and surprises and has significantly advanced our understanding of the universe in which we live. The Large Space Telescope represents a quantum jump in this intellectual adventure.

Space astronomy provides distinct advantages over the traditional techniques of ground-based astronomy. First, it has greatly expanded the utility of the experimental approach. It is now possible to carry out *in situ* measurements with space probes and to obtain and return samples of planetary and interplanetary material. Previously we had to be content with analysis of those few meteorites or cosmic particles that came to us from outer space. Today we can go look. For the most part however, experimental astronomy is restricted to the solar system and astronomy remains basically an observational science. However, observations from above the terrestrial atmosphere provide several distinct advantages. Firstly, the absence of atmospheric absorption greatly expands the available electromagnetic spectrum. Only in the extreme UV does the interplanetary and interstellar hydrogen interfere with our clear view of stars and galaxies. Secondly, the sky brightness is significantly reduced due to the absence of night sky emission and scattered light (the reduction of sky brightness depends upon the wavelength but is the order of a factor of 3 or 4). Finally, in orbit, astronomical instruments are no longer plagued by the erratic fluctuations of the sea of air above a ground-based observatory. Absent are the debilitating effects of atmospheric tremor, scintillation, and differential refraction that astronomers refer to as "seeing" disturbances. The spatial resolution is no longer limited by the earth's atmosphere but rather by the optical system

---

* Dr. Code is affiliated with Washburn Observatory, University of Wisconsin, Madison, WI 53706.

itself. Above the earth's atmosphere the stars are sharp, steady pinpoints of light, the sky is dark, and the spectrum is free of atmospheric absorption. To date space observations have primarily capitalized on this freedom from atmospheric absorption. Rocket and satellite measurements in the X-ray region have opened up a new and exciting chapter in astrophysics by providing a view of the high energy processes occurring in space. Many X-ray sources are compact objects, some of which may be the ultimate collapsed objects--black holes. Satellite observations in the ultraviolet have impacted all areas of stellar astronomy. The great hydrogen halos discovered surrounding comets have changed concepts of the structure of these objects, indicating that H_2O is the primary icy constituent. The extended constant luminosity behavior of the nova, FH Serpentis, which was discovered by ultraviolet observations, has significantly modified the interpretation of these cataclysmic events. The observations of a sharp peak in the interstellar extinction curve at 2200 Å has provided new insight on the nature of interstellar dust grains, while measurements of many interstellar resonance lines in the ultraviolet reveal the essential features of the dynamic interplay between stars and the interstellar medium. The extension of measurements over most of the electromagnetic spectrum has made it possible for the first time to determine empirically fundamental parameters of main sequence stellar structure, such as effective temperatures and luminosities; many other fascinating problems are under investigation using satellite observations of variable stars, clusters, and galaxies. These and other observations will be continued by a variety of techniques such as the International Ultraviolet Explorer, sounding rockets and Shuttle sortie missions. The LST will promise to extend these absorption-free measurements to much fainter limiting magnitudes and yield higher precision than so far possible. It is, however, in the area of high spatial resolution that the LST will make its most profound impact.

Let me illustrate the dramatic improvement in spatial resolution obtainable by the following comparison. It is possible by sophisticated techniques such as speckle interferometry to achieve angular resolutions of the order of two Airy discs with a ground-based telescope, despite atmospheric degradation; that is, it is possible to approach diffraction-limited performance. In Figure 1 the signal-to-noise ratio for interferometric observations is shown to be proportional to the observing time, τ, inversely proportional to the square of the diameter of a point image in the focal plane, θ, and to the square root of the number, Q, of individual exposures of length τ. Now in ground-based observations the maximum exposure time is limited by the length of time for which the stellar image or speckle pattern is stationary, and this does not exceed more than 0.1 of a second. The angular size of a typical stellar image observed from the ground is the order of 1 second of arc. On the other hand a space telescope can carry out observations for most of an orbit without significant temporal changes in the image distribution. If we assume that the optics are capable of putting most of the light from a point source within 0.2 arc-second diameter, which is rather conservative for LST, then the signal-to-noise ratio with the space telescope is approximately 5×10^5 greater than the signal-to-noise ratio achievable from the ground. It would therefore require some 2.5×10^{11} individual exposures (or 800 years of exposure time) with the ground-based telescope to achieve the same signal-to-noise ratio obtainable in space during a single orbit.

Before reviewing some of those areas in which it is expected that the LST will make its most significant contributions, let me briefly describe an astronomer's view of the universe. In the beginning, some 10 billion years ago, there was an expanding hot dense condensation of matter and radiation. As this primordial mass expanded, the simplest of all atoms, hydrogen, became the primary constituent along with a lesser amount of helium. Later density fluctuations within the primordial medium caused condensations of massive clouds of hydrogen and helium, which on further collapse and fragmentation under the forces of self gravitation resulted in condensations of galactic masses. As individual galaxies formed, perhaps with the violent events we see now as quasars, further condensations occurred, driven by shock waves produced in the collapsing mass. Thus stars were born, stars not too unlike our sun but with at least one very essential difference. These stars were composed of only hydrogen and

$[S/N] \approx Q^{\frac{1}{2}} \tau \theta^{-2} \cdot f[\lambda, D, \varepsilon, F]$	
Space Telescope	Ground-Based Telescope
$\tau \sim 30$ min. $\sim 2 \times 10^3$ sec	$\tau \sim 0.1$ sec
$\theta \sim 0.2$ sec of arc	$\theta \sim 1$ sec of arc
$\tau \theta^{-2} = 5 \times 10^4$	$\tau \theta^{-2} = 0.1$

$[S/N]_{space} / [S/N]_{Gnd} = 5 \times 10^5$

If Q exposures are obtained on the ground to achieve equal S/N then

$Q^{\frac{1}{2}} \tau \theta^{-2} = 5 \times 10^4$

$Q = 2.5 \times 10^{11}$

$Q\tau = 2.5 \times 10^{10}$ sec ≈ 800 years

Figure 1 Signal-to-Noise for Interferometry

helium, with no traces of carbon, nitrogen, oxygen, iron and other heavy elements present in the sun and here on earth. As these first stars contracted the temperature increased until thermonuclear reactions set in. First hydrogen in the deep interior was converted to helium, providing the light we see radiating from stars. As the central hydrogen fuel was consumed the temperature rose and helium nuclei began to fuse to form carbon, while in the final stages of the star's life heavier elements were built up. For those stars more massive than the sun, a significant fraction of their mass was then returned to the interstellar medium, in some cases by violent explosions producing what astronomers call supernova, or in other cases by less energetic losses of mass such as those resulting in planetary nebula. By these processes the interstellar medium became enriched in heavy elements and new stars were formed and the cycle repeated. The very atoms of which each and every one of us are made were formed in the center of some star in the distant past, and that these atoms should collect into the complex amino acids and proteins that form living organisms is not nearly so unlikely as was once thought. In the interstellar medium from which stars form, atoms condense into solid particles to form dark clouds of dust and in these dust clouds modern radio astronomy reveals organic molecules not too different from these essential amino acids. The formation of complex organic compounds appears to be a natural and likely event that has been duplicated in simple laboratory experiments. It is believed that in the formation of stars like the sun, matter condenses and forms planets which orbit about the central star, and that on those favorable planets, through the womb of time, self organizing systems have evolved by the process of mutation and adaptation and thus we are assembled here today. This scenario describes the framework of our general view of the origin and evolution of the universe. In detail much remains to be learned and each step is subject to uncertainty and some to controversy; it is however a fair description to say that his is what astronomy is about and it is in this arena that the LST displays its potential.

The scientific potential of the LST was first explored in depth in the National Academy of Sciences publication "Scientific Uses of the Large Space Telescope" (1969). Each of the current LST Instrument Definition Teams has prepared a compilation of the scientific objectives, and many of the exciting research programs were presented at the Washington meeting of the AIAA (1974). Let me survey these areas by simply listing the titles of the papers presented at that Aerospace Sciences Meeting. They read as follows: Cosmology with the LST, Allan Sandage; Contents and Structure of Galaxies as Observed with LST, Ivan King; Observing Quasars with LST, Margaret Burbidge; A Breakthrough for Astrometry, Laurence Fredrick; Infrared Capability with LST, Gerry Neugebauer; LST - A Window on Stellar Nurseries, George Herbig; LST Looks at Stellar Death, John Bahcall; and Solar System Astronomy from the LST, Harlan Smith.

I commend these papers to your attention. Had I the time, I could hardly do better than each of these imaginative authors have already cone. Rather let me single out one specific problem to illustrate the point that the LST does indeed represent a quantum jump in astronomical instrumentation.

The area I have chosen is the cosmological problem, a field that captures the imagination of all inquiring minds. If the LST is capable of making significant contributions to the cosmological problem, it then is a tool that necessarily has the capability required for most of the other investigations.

The cosmological problem in its most restricted sense is the problem of determining the appropriate geometry for describing the Universe. The more ambitious program of describing the present structure and origin and evolution of the Universe might more properly be called the cosmogonic problem. For the present let us consider the restricted problem of choosing the most appropriate world model. Is the universe finite or infinite, bounded or unbounded, closed or open?

The geometry may be completely specified by the metric. The metric is usually described in terms of the coefficients for each coordinate in space as they appear in the line element or differential distance between two points. For example if we write $ds^2 = g_{xx} dx^2 + g_{yy} dy^2 + g_{zz} dz^2$, in a simple three dimensional space, then ordinary flat cartesian coordinates are given by $g_{xx} = g_{yy} = g_{zz} = 1$. In general these coefficients, called the components of the metric tensor or the metric, are not constant. For example in spherical coordinates $ds^2 = dr^2 + r^2 d\theta^2 + r^2 \sin^2\theta d\phi^2$. In discussing the geometry of the universe we are interested not only in where an event occurs but at what time it happened; hence a four dimensional space is used where time is the fourth coordinate. The line element is written in a short hand form as $ds^2 = g_{\mu\nu} dx_\mu dx_\nu$ where μ and ν take on the values 1, 2, 3 and 4; hence in general we may have terms like $g_{23} dx_2 dx_3$ as well as terms like $g_{11} dx_1^2$. In principal the metric may be found from examining those simple consequences of the prevailing geometry. For example, in flat space the sum of the angles of a triangle equal 180°, while in a positively curved space, like the surface of a sphere the sum of the angles is greater than 180°. In a negatively curved space, like the open space of a saddle surface the sum of the angles is less than 180°. Astronomical observations do not directly yield simple accurate distances and angles, and one must examine the more complex relations between observed quantities such as brightness, doppler shift and number counts, and the geometry of space. The approach taken is to predict the observational relations that would exist for a given world model and choose the appropriate world model on the basis of the success of such predictions.

The construction of world models within the framework of the General Theory of Relativity follows the broad format shown in Figure 2. General Relativity is based upon two fundamental principles. The first is the principle of covariance, which simply states that the laws of physics should be independent of the observer and hence expressible in a form that is invariant to the particular choice of coordinate systems. The second principle, that of equivalence, also recognizes that there is no particular unique coordinate system or absolute frame of reference when it states the equivalence of inertial and gravitational mass. It is not possible to distinguish between a force and a coordinate acceleration. This means that by a suitable coordinate transformation, a force field can be transformed away. In a free-falling elevator we could say that we are falling under the attraction of the gravitational force of the earth or equivalently that we are moving uniformly in a coordinate system in which the grid spacing becomes progressively closer as we approach the earth. That is, we can choose complex laws of physics and a simple geometry or a simple law of motion, namely that all particles follow a simple path, a geodesic, and employ a complex geometry. The presence of matter has curved space and determined the appropriate metric. Planets move in near-elliptic orbits simply because the solar mass has curved space and the shortest path in this geometry is an ellipse. The equivalence or relation between the underlying geometry or metric and the distribution of mass and energy in the universe is the field equation. One of the beauties of this field equation as formulated by Einstein is that it contains implicitly all the conservation laws of physics and yet is the very simplest equation connecting the metric, described by the little $g_{\mu\nu}$'s and the mass of the universe. Other cosmological formulations such as the Brans-Dicke cosmology or the steady-state universe modify the field equations, but the recipe remains the same. Given the distribution of mass and energy in the universe we can

> Field Equation: $f(\text{metric}) = F(\text{mass energy})$
>
> Line Element: $ds^2 = g_{\mu\nu} dx_\mu dx_\nu = \text{invariant}$
>
> Geodesic Equation: $\delta \int ds = 0$
>
> Cosmological Principle: Universe is isotropic and homogeneous, i.e.
>
> $g_{\mu\nu} = 0$ if $\mu \neq \nu$
>
> $ds^2 = c^2 dt^2 - R(t)^2 [du^2]$

Figure 2 World Model Building

determine the appropriate geometry or metric, or conversely the determination of the metric specifies the large scale physical structure of the universe. From the metric or $g_{\mu\nu}$'s we can determine the invariant line element or interval between two events. The equation of motion of a particle in the universe is then described by the simple path of a geodesic in this space. The equation of motion is the equation of the geodesic, which says that the path is a minimum. Now when applying these concepts to the gross structure of the universe it is both necessary and reasonable to make a basic simplification. When sawing a board it is not necessary to consider the detailed atomic and molecular structure of the wood, rather it is sufficient to consider the board as relatively uniform. Over a sufficient volume of the universe it is reasonable, although not necessarily true, to assume that irregularities average out and a large scale the universe is homogeneous and isotropic. That is, the universe looks the same at any place and in any direction. This is the cosmological principle. It means that the line element can not contain any cross-product terms such as dxdy, otherwise the universe would look different depending upon whether we looked in the plus-x direction or the minus-x direction. The line element can therefore be written in a form in which the three space-like coordinates, du^2, simply vary in scale as a function of time by the factor $R(t)$. In other words, the universe is radially symmetric. The history of the universe is then described by the scale factor $R(t)$. In a closed universe this can be viewed as simply the temporal behavior of the radius of the universe.

Without prejudicing the discussion by a choice of a particular world model let us consider the local behavior of R(t). By expanding R(t) around the point in four-space here and now, that is, the value of $R = R_o$ at time t_o, we obtain the result illustrated in Figure 3. The linear term given by the Hubble constant is found observationally to be positive. The universe is currently expanding at a rate somewhere between 50 and 100 kilometer second^{-1} per megaparsec. The Hubble constant is determined from the slope of the velocity-distance relation in the near vicinity of our own galaxy, where the difference between different models is insignificant. The chief uncertainty in the determination of H_o is the accuracy of the extragalactic distance scale and whether or not the measurements have been extended far enough to be in the region of "pure Hubble flow," that is to large enough velocities that local velocity fields are unimportant. The Hubble constant establishes the time scale τ while the acceleration parameter q_o is a measure of the curvature of space.

$$R(t) = R_o + (t - t_o)\dot{R}_o + \tfrac{1}{2}(t - t_o)^2 \ddot{R}_o + \ldots$$

Let $H_o = \dfrac{\dot{R}_o}{R_o}$ = Hubble's Constant

and $\tau = (t_o - t)H_o$ = Look back time

Then $\dfrac{R(t)}{R_o} = 1 - \tau + \tfrac{1}{2}\tau^2 \dfrac{1}{H_o^2}\dfrac{\ddot{R}_o}{R_o} + \ldots$

if we let $q_o = -\dfrac{\ddot{R}_o}{H_o^2 R_o}$ = acceleration parameter

Then $\dfrac{R(t)}{R_o} = 1 - \tau - \tfrac{1}{2}q_o \tau^2 + \ldots$

Figure 3 The Scale of The Universe

Now if we know R(t) we can determine the behavior of all pertinent observational parameters. For example the matter density of the universe is simply $\rho = \rho_o (R_o/R(t))^3$ which is related to the number of galaxies observed to different magnitude limits. It is rather easy to demonstrate that the very important red shift parameter, $z = \Delta\lambda/\lambda$, is given by $(1 + z) = R_o/R(t)$ in any cosmological model. To the first order we can see from the expansion given in Figure 3 that $z \approx \tau_o$, and since τ_o is the length of time in units of the Hubble time for light to travel a distance, d, which is to this order the distance of the galaxy, $\tau_o = (d/c)H$. The expansion velocity cz is therefore, $cz = dH$, the linear velocity-distance relation appropriate for small z's.

The field equations of general relativity yield a set of simple solutions if the pressure, p, and the cosmological constant, Λ, are set equal to zero. Now at the present epoch the pressure term is exceedingly small compared to the gravitational interaction of masses, since collisions between galaxies or clusters of galaxies are very rare; thus p = 0 is an excellent approximation. The cosmological constant is a term occurring in the most general formulation of the field equation representing some large scale interaction not known in terrestrial physics. We do know that if there is a cosmological constant it is exceedingly small since it does not manifest itself on the scale of the solar system. Models based on the assumption $p = \Lambda = 0$ are referred to as the Friedmann Universe for which the solutions of the field equations are given in parametric form in Figure 4. The constant A appearing in these solutions is a function of H_o and q_o only. Thus we see that the model is completely specified by H_o and q_o, hence the importance of these two parameters. For a closed universe, $q_o > \frac{1}{2}$, the solution is a periodic function. In Figure 5 the behavior R(t) is shown for Friedmann models. A flat universe continues to expand proportional to $t^{2/3}$, q_o remaining equal to $\frac{1}{2}$ at all times. If the density of the universe is sufficiently high, gravitational interactions will decelerate the expansion and the universe will ultimately contract again as shown by the cyclical curve for a closed universe. The total energy of the universe, in this case, is negative and the system is bound; it did not have escape velocity. The critical density above which the universe is closed is given by inserting $\frac{1}{2}$ for q_o into the equation for the density of a Friedmann Universe given in Figure 4. For current values of H_o this density is more than

$p = 0; \Lambda = 0$ $\rho_o = \dfrac{3}{4\pi} q_o H_o^2$		
Closed $q_o > 1/2$ $R(t) = A(1 - \cos \psi)$ $ct = A(\psi - \sin \psi)$	Flat $q_o = 1/2$ $R(t) = K\, t^{2/3}$	Open $q_o < 1/2$ $R(t) = A(\cosh \psi - 1)$ $ct = A(\sinh \psi - \psi)$
$A = \dfrac{c}{H_o} \dfrac{q_o}{|2q_o - 1|^{3/2}}$		

Figure 4 Friedmann Universes

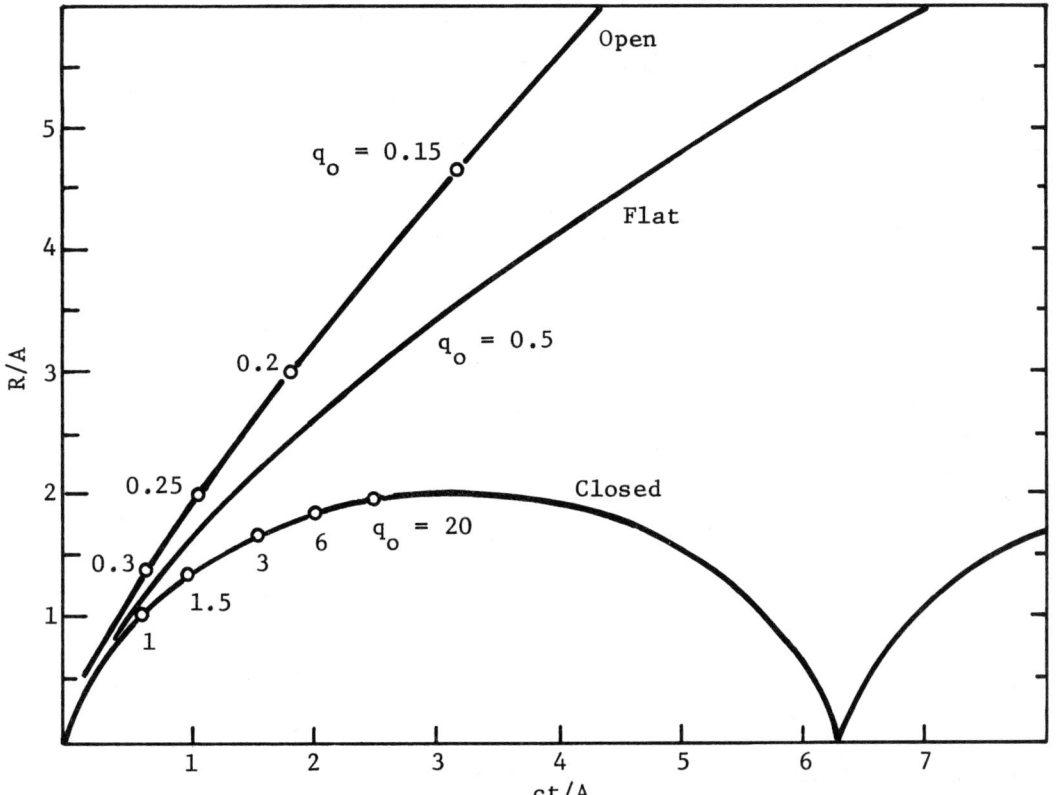

Figure 5

an order of magnitude larger than the observed density of the universe, and unless there is appreciable "missing" or unobservable mass the universe is open. It is possible, however, that locally the density exceeds the appropriate critical density for which space is closed; we then say we are dealing with a "black hole."

Now while this closed model gives a periodic or pulsating universe the Friedmann conditions are certainly violated as $R(t)$ approaches zero and the density approaches infinity. In a high density state it is no longer valid to neglect the pressure or the energy density of radiation, and thus unless very special conditions prevail the universe will not bounce and repeat its history as suggested by the curve.

The open universe indicated by the curve with the systematically decreasing value of q_o continues to expand indefinitely. If we knew observationally the value of q_o, we would then know our present location on one of these sets of curves. If in addition we knew the value of the Hubble constant, we would then know the scale of these curves and hence the age of the universe and its dimensions. Now let us return to the observational evidence available for relating these model universes to the "real" physical universe. We know that currently all distant galaxies give red doppler shifts and hence indicate that we are presently in an expanding universe. Of the variety of observational relationships that can be used to test the space metric the most powerful utilizing ground based data is, in fact, the relation of this doppler shift to the observed magnitude of galaxies. As mentioned previously the doppler shift $(1 + z)$ is given by $R_o/R(t)$. The total energy received from a galaxy decreases as the fourth power of $R(t)/R_o$ or as $(1 + z)^{-4}$. The energy is reduced by one factor of $(1 + z)$ due to the reduction in energy of a photon and by another factor of $(1 + z)$ resulting from the decreased rate of arrival of the photons. A factor of $(1 + z)^2$ accounts for the change in solid angle in which the photons are observed as a result of the geometry.

In Figure 6 the relation between the total power received, expressed as the bolometric magnitude, m_{bol}, and the logarithm of the doppler shift, cz, is shown for different values of q_o for Friedmann models. The curve

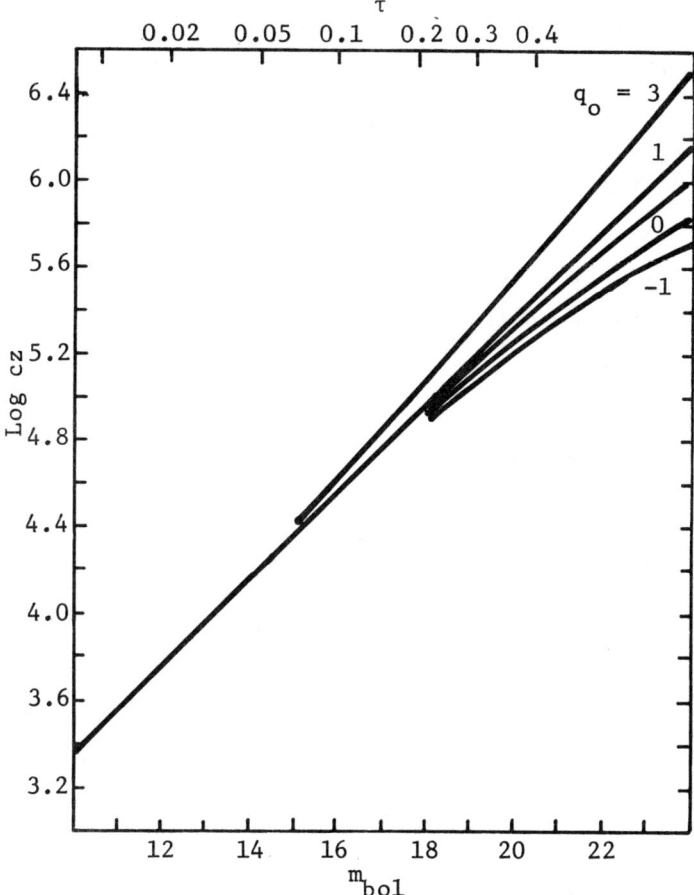

Figure 6

labeled $q_o = -1$ represents the results for the steady state cosomology. Significant differences between models set in fainter than 18th magnitude or a log cz greater than 5 $\left(\text{corresponding to } z > \frac{1}{3}\right)$. The look back time indicated on the upper margin corresponds to the times for $q_o = 1$. At about 19th magnitude one is looking back through nearly half the history of the universe. The curves plotted in Figure 6 refer to the variation in observed bolometric magnitude of an object of fixed luminosity, namely to a "standard candle." This standard candle is usually taken as the brightest galaxy in a cluster. The determination of the luminosity of this standard candle is intimately connected to the determination of the extragalactic distance scale and the Hubble constant, while the extension of the curve to larger red shifts involves the extent to which the brightest galaxies in clusters are really the same and to a variety of other corrections.

Figure 7 shows the results of a discussion by Sandage and Hardy (1973) which summarizes the data as of that time. The data presented are for 97 clusters or groups of galaxies and the straight line corresponds to $q_o = 1$. The magnitude given is the total corrected V magnitude. We do not observe the bolometric magnitude but only the magnitude of the object as observed over some selected bandpass, in this case the V magnitude centered near 5500 Angstroms. The V magnitude of a nearby star of known spectral type differs from the bolometric magnitude by a simple determinable constant. The V-magnitude of a distant galaxy is not so simply related to its bolometric magnitude, and the corrections applied are indicated along the abcissa of this figure. The magnitude has first been corrected for the fact that observations with a given focal plane aperture may *not* include all the light from a galaxy. The second correction called the K-term takes in to account the fact that for a red shifted galaxy one is not observing the same spectral region with the fixed band pass as for an object in the rest frame. The third term, A_v, corrects for the decrease in brightness due to extinction produced by interstellar dust grains in our own galaxy. The last two corrections attempt to correct for differences in the brightness of the brightest galaxies in clusters depending upon the kind of cluster and the richness, or population, of the cluster. It is clear that the step from observation to our theoretical models is not a simple one and each correction is subject to uncertainty. Moreover, one very essential consideration has been neglected. It may very well be that, due to evolution, those distant galaxies sampled back in time are significantly brighter or fainter than the nearby galaxies of the present epoch. We shall return to the question of evolution presently, but for the moment let us see what can be said about the choice of metric if evolutionary corrections are small.

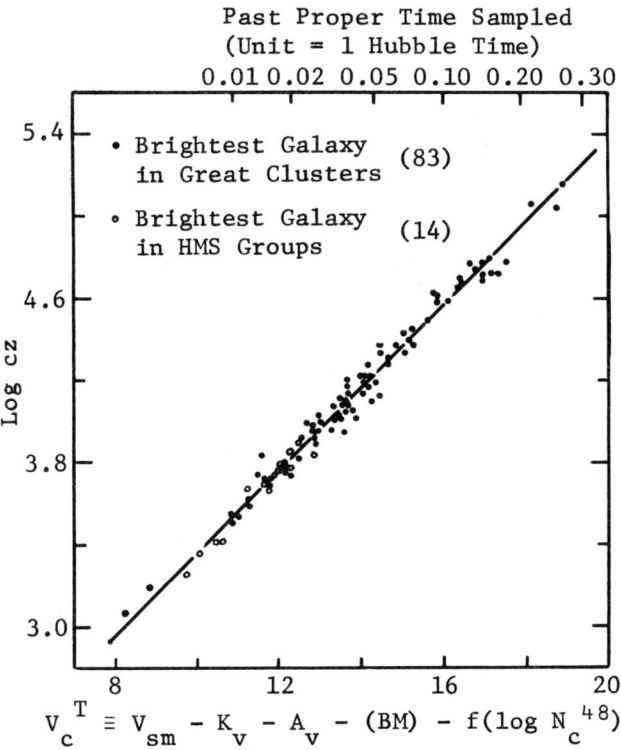

Figure 7

As noted before, the difference between different world models does not become significant until at least the 18th magnitude and thus Figure 7 gives little information on the value of q_o. In Figure 8 I have plotted the results for the most distant clusters for which we had observational data in 1973. These data are taken from Sandage and Hardy (1973). The deviation in magnitudes from the curve $q_o = 1$ is plotted against the red shift parameter z on a linear scale. The range of magnitude differences for the closer clusters is indicated by the arrows in the lower portion of the plot. While a formal least-squares solution could be made for the best value of q_o, it appears clear from the diagram that q_o is really undetermined. It is probably inconsistent with the steady state cosmology or a value of q_o much in excess of 2, but otherwise is uncertain.

Obviously what is required to derive information on q_o from the Hubble diagram is to improve the precision of the magnitude determination and to extend the results to large values of z and therefore to fainter objects and to increase the total number of clusters observed.

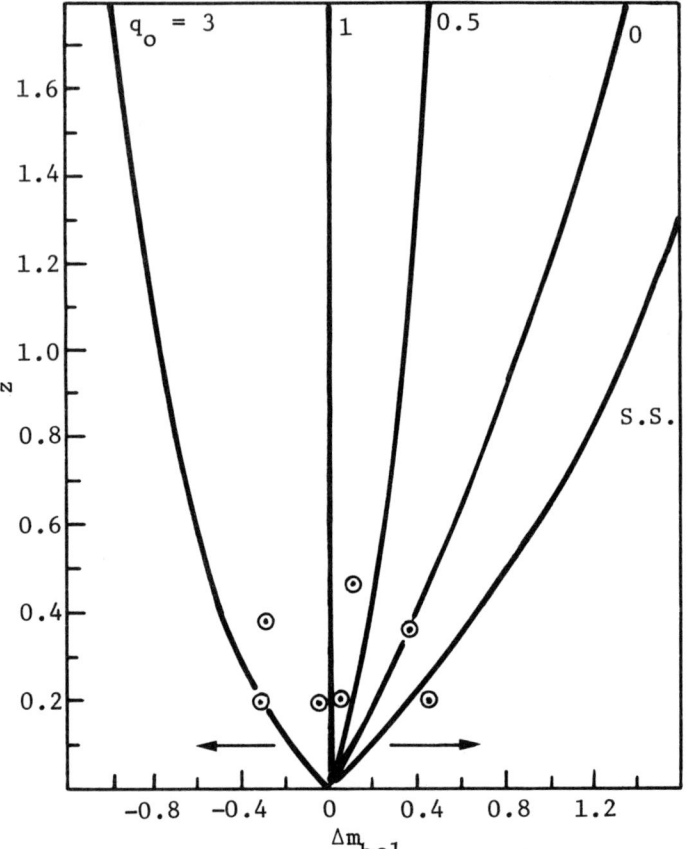

Figure 8

The LST should be capable of obtaining spectra for red shifts to $z \sim 1$ while the photometric precision at this limit would be appreciably better than current ground based photometry. At larger redshifts the ability of the LST to carry out observations in the ultraviolet may prove advantageous. The most important aspect of ultraviolet observations, however, will probably be the leverage provided to allow for galaxy evolution. This is very difficult to determine in the spectral region available to ground based telescopes. The extension of spectral energy distributions to the far UV, however, would give a very strong hold on the hot star component in the galaxy, which exhibits the most rapid evolution. Observations would be needed of both small red shifts and the largest values of z possible to provide the sensitivity required to calculate the evolution of galaxies with time. The LST could also provide information on evolution by extending galaxy counts to much fainter limits, since the effect of evolution on the apparent value of q_o has the opposite sign for number counts than for the red shift-magnitude relation.

The most fundamental contribution of the LST to the cosmological problem, however, will be in the refinement of our knowledge of the distance scale and the brightness of the brightest cluster galaxies. The extragalactic distance scale is based upon the observation of individual objects of various kinds in nearby galaxies. The cepheid variable stars, e.g., are luminous pulsating stars whose absolute magnitudes can be determined from their period of pulsation. The very brightest stars in our own and nearby galaxies also provide a valuable distance indicator that can be utilized to somewhat greater distances than the cepheid variables. A different type of distance indicator is provided by the angular diameter of the largest HII regions or ionized hydrogen regions. At the top of Figure 9 the distance to which these types of distance indicators are applicable are indicated schematically for typical ground based observation. Here the limiting B-magnitude for suitably accurate photometry is taken to be 22nd magnitude and the angular resolution to be of the order of 1 second of arc. Immediately below the figure some of the galaxy groups and clusters are given. It is only from these few nearby groups and the Leo and Virgo cluster that we can apply these basic distance indicators and thus connect the brightest cluster galaxies to these distance determinations. To increase the available statistics on the brightest cluster members one must resort to a number of indirect arguments of considerably lower weight. The bottom of Figure 9 shows the domain of the cepheids, brightest stars and HII regions accessible to LST if the performance is conservatively taken to be $B \approx 26$ magnitude and $\theta \approx 0.1$ seconds of arc. Distance indicators in thousands of galaxies become possible and the number of clusters that can be studied is substantially increased.

In this discussion I have restricted myself to only a few classical approaches to the cosmological problem. LST offers other alternatives. In addition I have confined the discussion only to the restricted problem of selecting the appropriate geometry. The LST can make very substantial

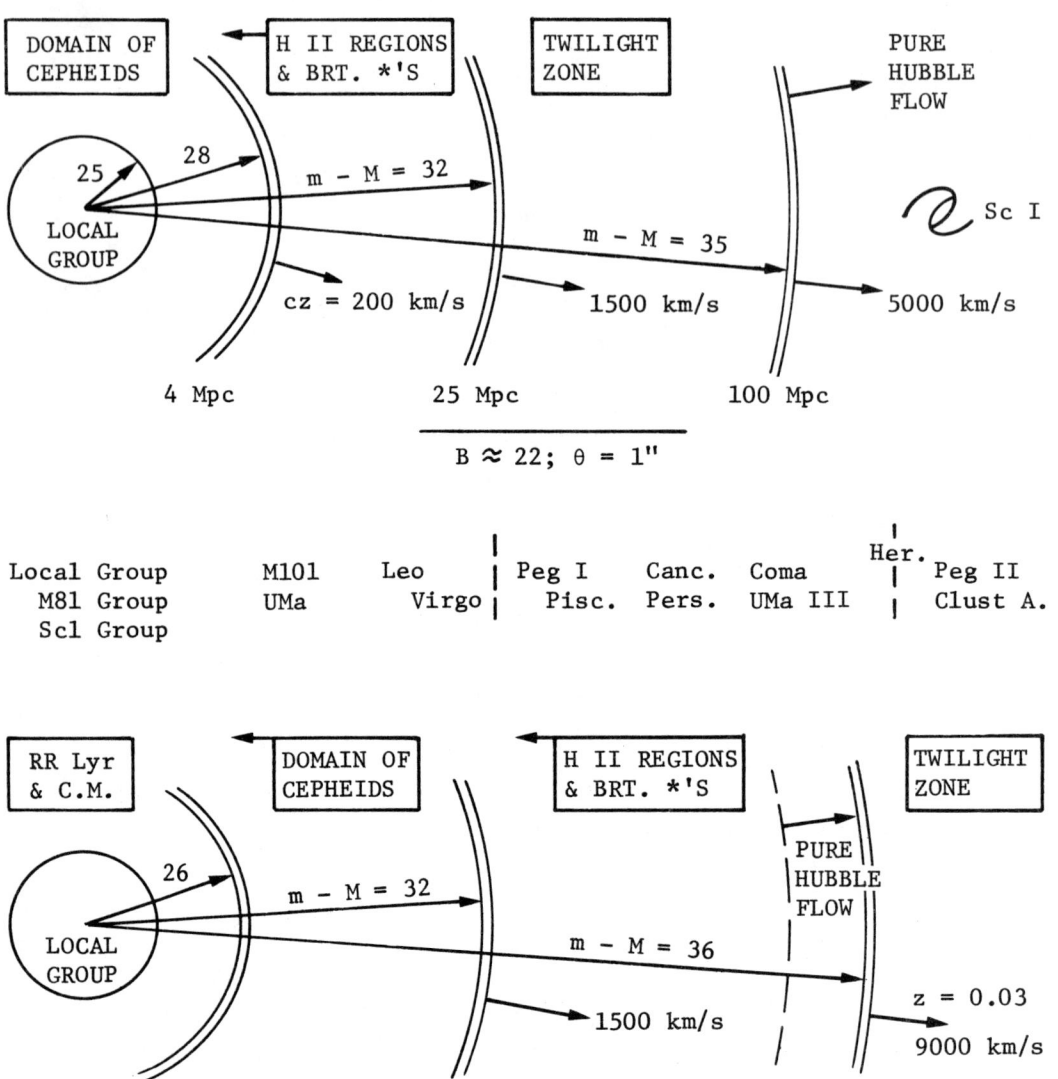

Figure 9

contributions to the broader questions of the origin and evolution of the Universe. I hope, however, I have convinced you that the solution of the cosmological problem is within our grasp.

The LST is an exciting adventure and whether it comes to pass here and now, it will come to pass. For thus it is that the primordial hydrogen atoms, of which I spoke, have, over 10 billion years of dynamic history, arranged themselves in such as we, so that they can ask from whence they came.

REFERENCES

AIAA 10th Annual Meeting. January 1974, *Large Space Telescope - A New Tool for Science* (New York: American Institute of Aeronautics and Astronautics).

Sandage, A., and Hardy, E.: 1973, *The Astrophysical Journal*, 183, 743.

Space Science Board Ad Hoc Committee on the Large Space Telescope. 1969, *Scientific Uses of the Large Space Telescope* (Washington: National Academy of Sciences).

AAS 75-129

CONCEPTS OF LST OPERATION
C. R. O'Dell *

As the first true astronomical space observatory, the LST will be faced with the operational requirements of both the traditional ground-based observatories and its predecessor satellite programs. Moreover, it will have its own set of new constraints and requirements. These all combine to make the LST a challenging system to operate efficiently. Operate efficiently we must, for the scientific demand for the use of LST will be very high and the cost per year of observing time will significantly exceed that of traditional telescopes.

What I shall address in this paper is the subject of operations in the broadest sense, and put down what I believe are the relevant points in moving from idea, through observation, to interpreted result. The actual control of spacecraft operations will be through a Mission Operation Center. Although it will be working with the TDRS System, this aspect of operations will not be particularly different from earlier astronomical satellites and will not be discussed in detail here. There are, however, two requirements that are special for LST.

Object acquisition and guidance are new problems for LST because of the precision required. Acquisition presents a new challenge because of the faintness of the objects being observed. Although methods using precise gyros are sufficient for most of the imaging experiments, these are insufficient for locating the correct star into a Scientific Instrument entrance aperture that is comparable in size to LST's minute images. The problem is further

---

*Dr. O'Dell is Space Telescope Project Scientist, NASA Marshall Space Flight Center

compounded by the need to often work on faint objects in crowded fields of view. When the object is isolated and bright, we can use a signal from the source itself to do the fine centering. However, in the faint object-crowded field case we must obtain an image of the field, then use that information to make the final setting. Almost all of the LST observations will demand an essentially continuous guidance with a stability only a small fraction of the size of the stellar image. Since we will be observing faint stars, the reference signal for guidance will have to come from guide stars, some 10 arcminutes away from the object field. The guidance system must know where to look for these stars, acquire them automatically, then initiate the guidance signal. This required preplanning and automation are among the real challenges of LST operations.

SCIENTIFIC REQUIREMENTS-IN

After the scientist establishes his desired observing goals, in terms of a limited or multi-faceted approach, he must formulate the observing needs to reach these goals. Using information of LST systems capabilities and procedures, he can turn these observing needs into a proposed observing program. These observing proposals will be solicited continuously and normally collected periodically at times 6-12 months before the observations would begin. Technical evaluation follows. This evaluation is combined with the scientific justification to allow a peer group review by other scientists, prior to selection. This peer group will select, according to scientific merit and feasibility, a set of observing programs to fill the time period in question.

The nature of these programs will vary enormously. Some will be one-shot measurements while others will require multiple observations. Some will require intensive observations during one part of the year when a particular part of the sky is optimally located, while others will require occasional observations throughout the year. Many different brightness levels will be addressed and some programs may even wish to use more than one Scientific Instrument.

The observer must then work with a specialist staff of scientists working for the LST program to prepare a list and sequence of specific Scientific Instrument operations to be executed in carrying out his program. These several requests for use of that Scientific Instrument and the other Scientific Instruments, must then be coalesced into an optimum observatory observing schedule. This optimization will consider the requirements for turning individual Scientific Instruments on and off, calibration needs, demands for observations at specific times, solar, lunar and terrestrial glare, communications requirements and slewing and setting rates for the telescope as a whole.

The final, optimized observing program will in large part be carried out without immediate scientist-user participation. However, in two areas he will have to work closely with the Mission Operations Center staff. The first of these will be in the quick look at some of the data, where changes in the observing program may be demanded by the nature of the source at that specific time. The second will be in the acquisition of faint objects and/or of objects in crowded fields. In the latter case, the observer-scientist will work with an acquisition-scientist who does the actual field recognition and spacecraft commanding, if necessary, with the observer-scientist beside him.

SCIENTIFIC DATA-OUT

Once the observation is made, then the process of turning data into information begins. After temporary onboard storage and transmittal to the ground and the Mission Operation Center, the scientifically relevant data (which can include spacecraft status data) is stripped out and put into a form to allow straightforward data reduction. Data reduction is expected to be a detailed process. Since the data processing requirements will be essentially identical for various observations with one Scientific Instrument, this process will again be centralized, with the user scientist working with a data scientist. During this process, field distortion corrections will be applied, photometric calibrations determined, spectral response calibrated, etc. All of these processes will be applied and then the scientist user can select several formats in which he can collect his reduced data. Basically the intent is to centralize those functions that are common to many observing programs and carry the data reduction up to the step of data analysis. The data then is the scientist's to do with as he pleases.

This data will also be stored in a readily accessible form. Experience amply shows that proprietary use of his own data is an essential ingredient in obtaining the best science. Observance of this tradition of proprietary rights is balanced by the fact that this is a nationally funded facility. We expect to balance these two opposing pressures by keeping the data in storage until one year after data reduction, after which time it can be distributed upon request.

All of the above describes how a scientist will work in carrying out the operation of LST. There are several plans for how such a plan should be implemented; however, the preferred ones must recognize that scientific excellence flourishes only under a free enterprise system while LST itself is a complex system demanding continued involvement of the agency developing it.

AAS 75-130

LARGE SPACE TELESCOPE PROGRAM STATUS

James A. Downey III*

INTRODUCTION

The concept of placing as astronomical observatory in Earth orbit is not a recent one. Scientific advisory groups, individual scientists, and various persons of vision have recognized the exciting possibility of putting a major astronomical facility in space above the obscuring effects of the Earth's atmosphere. The idea of a large optical observatory in orbit pre-dates the space era. Today a number of telescopes already have already been flown in space and observed celestial objects in various regions of the electromagnetic spectrum. These present systems along with the advance of aerospace technology in general and the coming availability of the Space Shuttle have paved the way for accomplishment of the Large Space Telescope (LST) Project. The LST will be an astronomical observatory that will operate in space for more than a decade.

LST DESCRIPTION AND STATUS OF DEFINITION

LST is an automated satellite which will be delivered to orbit by the Space Shuttle. Scientific data from the LST Scientific Instruments will be transmitted to Earth via telemetry through the Tracking and Data Relay Satellite in geosynchronous orbit. The LST differs from automated satellites of the present era in that it can be retrieved from orbit and returned to Earth for refurbishment by the Space Shuttle. Also, the LST will be equipped to permit on-orbit servicing by a space-suited astronaut. As the LST operational era evolves, a high degree of on-orbit maintenance and servicing will be possible. The LST is a relatively safe and inert payload from the standpoint of

---

*Manager, LST Task Team, NASA Marshall Space Flight Center

presenting a hazard to the Space Shuttle and its crew; LST does not have any on-board propellants and will not carry pyrotechnic devices.

The current over-all status of the LST Program is that more than one-half of the definition effort (Phase B) has been completed. Early definition was based on an LST configuration with a primary mirror of 3 meters (118 inches) diameter. However, since May 1975 all effort has been concentrated on a 2.4-meter (94-inch) LST. The events and rationale leading to this change are discussed below. The remaining part of the definition contracts will be completed by March of 1976, and the program is aimed as a new development start in FY-77.

The FY-77 new start plan reflects release of the RFP's for the development contracts in June of 1976, leading to the award of contracts early in 1977. Development of the LST system will require a period of approximately 6 years with launch in late 1982 or early 1983.

There is an intrinsically long development time associated with certain elements of the LST, e.g. the primary mirror. The LST Project Plan has approached the development schedule on a basis of minimizing costs. Obviously there are no launch window considerations in the case of LST. Planning of various events in the development schedule can be optimized from a standpoint of cost effectiveness.

As a result of recent reviews of the LST definition contracts, development cost estimates have been refined. These cost estimates have been submitted to NASA Headquarters for review and evaluation. LST development costs are not discussed in this paper because of the current review activity in NASA. The next key milestone from an over-all LST program standpoint is inclusion of LST development as an FY-77 new start in NASA's FY-77 budget plans. This decision will be made next month by NASA top management. Submission of LST as a new start in NASA's FY-77 budget will be the starting point in review and approval of initiation of the LST execution phase by the Office of Management and Budget (OMB) and the Congress.

The balance of this paper discusses some of the early history of the LST Program and concentrates on the recent decisions and work that remains to be done in the definition phase. By spring of next year, LST definition will have been completed. More time and resources will have been expended on LST definition than in the planning of any previous NASA scientific mission. Consequently, LST definition will be exceptionally thorough, providing high confidence that LST can be developed to achieve required scientific performance within predicted costs.

LST feasibility studies were completed in 1972, and project responsibility was assigned to the Marshall Space Flight Center. Goddard Space Flight Center also plays a significant role in the project, having responsibility for the scientific instrument definition and development. Mr. George Levin, the GSFC LST Study Manager, is providing a paper on the LST Scientific Instruments. Therefore, the status of the Science Instruments is not discussed in detail in this paper.

In late 1972, NASA decided to proceed with the definition and development of LST by letting separate contracts for the definition and development of the telescope portion of the LST, which is referred to as the Optical Telescope Assembly (OTA), and the LST spacecraft, called the Support Systems Module (SSM). The LST Science Instruments will be developed by Principal Investigators (or investigator teams) from the astronomical community. (In the present definition phase, the OTA contractors have responsibility for the combined definition of the OTA and Scientific Instruments, based on requirements established by the scientific community.) We refer to this contracting approach being pursued in the LST Program as the Associate Contractor approach. The NASA will have over-all LST systems engineering management responsibility, but will receive support in LST systems engineering and integration by the Support Systems Module contractor. LST definition effort was initiated in 1973. Scientists were selected in the summer of 1973 to participate in the LST definition. Their guidance and assistance have been invaluable. Contracts for the definition of the LST telescope and scientific instruments (OTA/SI) were awarded in August 1973 to Itek and Perkin-Elmer Corporation. Originally these contracts were for 17 months' duration, but extensions have been implemented to carry the work through February 1976. The initial work on the Support Systems Module (SSM) was started in-house by Marshall Space Flight Center in August 1973. The in-house definition of the SSM was completed in the fall of 1974. In December 1974, contracts for further definition of the SSM were awarded to Boeing, Lockheed, and Martin. These contracts are for 15 months' duration and will be completed early next year.

FY-75 was the first year that LST was identified as a specific and separate line item in the NASA budget submission to Congress. After passage of the FY-75 Authorization Bills, the House Appropriations Subcommittee recommended the funding requested for LST be denied. The Senate subsequently favored reinstatement of funding for LST. Ultimately, as a result of the actions and resolution of the Joint House/Senate Conference Committee, NASA received 3 million dollars in FY-75 for LST definition. This was approximately one-half the amount of funding requested by NASA for LST. Furthermore, this funding was appropriated with the stipulation that NASA pursue definition of a lower cost program involving substantial international participation. Definition of the 3-meter LST system had not matured to the state that accurate cost predictions were available. However, the intent of Congress was clear and action was taken by NASA to readjust the program objectives to reduce over-all project cost and complexity.

In the fall of 1974, the OTA contractors, who had been defining a 3-meter telescope system, were asked to recommend a lower cost system. What was being sought was a smaller size system that would cost less but still provide acceptable scientific performance. Both contractors accomplished studies which indicated that significant cost savings and increased confidence in cost and schedule estimates could be achieved by reducing the aperture of the telescope to 2.4 meters. When the SSM contracts were awarded in December 1974, Boeing, Lockheed, and Martin were requested to do 4-month studies to evaluate the cost and performance characteristics of three LST systems having primary mirror sizes of 1.8, 2.4, and 3.0 meters. The result of these studies showed that significant cost savings could be achieved by reducing the size from 3.0 meters to 2.4 meters. However, further reduction to 1.8 meters did not result in significant savings. Also, the 1.8-meter system degraded science performance to the extent that the class of scientific investigations envisioned for LST would not be achievable. Above the 2.4-meter size there is somewhat of a "step function" in higher cost and complexity. Therefore, in May 1974, NASA Headquarters decided that all further LST effort should be based on the 2.4-meter size system. This decision was a sound one, primarily because of the engineering margins available with the 2.4-meter system and the higher over-all confidence that cost targets can be met and program objectives attained.

During the earliest NASA studies of the LST, the principal efforts were directed at basic technology and subsystems performance. A high quality optical system is necessary to take maximum advantage of the viewing potential in space. Questions were raised about the ability to figure the primary mirror of an LST to the accuracies required to achieve near diffraction limited performance. Furthermore, the structure of the telescope would have to be very stable dimensionally in the orbital environment to keep the mirrors aligned in proper focus during scientific observations. Additionally, the higher the potential optical performance of the telescope and the finer the detail of the optical image, the greater the demands on the spacecraft stabilization system to hold the image steady during long term observations. A number of advanced development activities have been accomplished which studied and resolved these and other technical areas of concern to LST.

NASA sponsored an advanced technical development effort with Itek to figure a 1.8-meter (72-inch) mirror to the accuracies required by LST. This contract with Itek was successfully completed in the middle of last year. The mirror was figured to better than 1/60 of a wave (measured at 6328Å), exceeding the LST design goal for optically figuring the primary mirror. This 1.8-meter mirror was made of the same low coefficient of thermal expansion material that will be used for the LST primary mirror. The 1.8-meter mirror admittedly is somewhat smaller than the 2.4-meter primary mirror for LST. However, we now have utmost confidence that the 2.4-meter mirror can be figured to LST requirements.

To build a telescope mirror support structure using traditional structural materials would put severe demands on the thermal control system to achieve telescope performance and focus during scientific observations. Fortunately, the technology of graphite-epoxy materials has progressed rapidly. Graphite-epoxy has attractive properties for a variety of aerospace applications. A property that is of particular significance to LST is that graphite-epoxy structural members can be fabricated in a manner to provide an extremely low thermal coefficient of expansion. (Also, graphite-epoxy is a stiff material and has exceptional strength/weight properties.) Making structural components of the telescope portion of the LST, or optical bench, out of graphite-epoxy with its low coefficient of expansion relaxes significantly the performance requirements of the spacecraft thermal control system. NASA has accomplished in-house work and sponsored advanced development efforts to investigate the basic materials properties of graphite-epoxy for space telescope applications. Also, there were concerns about the fabricability of graphite-epoxy structures for use on LST, particularly techniques to make appropriate joints at the intersection of structural members. However, a full scale graphite-epoxy truss for a 3-meter-type LST telescope has now been completed recently by Boeing. This truss presently is being tested. Also, General Dynamics/Convair has fabricated an LST half scale structural shell out of graphite-epoxy. The cylindrical shell design is an alternate approach to the truss for forming the principal structure of the telescope.

The requirements on the LST spacecraft stabilization and control system are demanding. We have specified a design goal of maintaining stabilization of 0.007 arc sec during a scientific observation. This might appear intuitively to be an unachievable goal. However, the Orbiting Astronomical Observatory spacecraft has demonstrated stabilization performance of 0.01 arc sec (even better than 0.01 arc sec for short periods). OAO achieved this stabilization using a reaction wheel actuated control system, and this type is planned for use on LST. The Martin Company has developed a fine pointing simulation facility which is used to investigate pointing control and motion disturbance effects on a simulated LST system. This simulator, which is in a seismically quiet location, can resolve motions of a few ten thousandths of an arc sec. Lockheed has built a structural simulator of the LST system and is investigating induced vibrations in the simulated LST structure using components similar to those which will be used on LST. Of particular interest are vibrations from reaction wheel imbalance and the propogation of these vibrations to the focal plane region of the telescope.

The areas of investigation discussed above and other advanced development effort, sponsored both by the government and industry, have provided the assurance that LST development is achievable and practical. Certainly the technology exists today for development of LST. In addition, a significant part of the LST effort to date has been devoted to the testing and breadboarding of certain critical subsystems and components. Demonstration of various OTA and SSM subsystems performance in my opinion has been a most important contribution of the NASA- and company-sponsored work accomplished to date. The principal challenge that lies ahead is not further technological effort and advanced development work, but application of engineering design principles and systems engineering practices to implement LST as a complete flight hardware system.

The LST system is designed in a modular fashion with respect to the Scientific Instruments. The instruments are not interdependent in any way, and each on-axis instrument is designed separately to a common interface specification. Individual instruments may be added or removed without affecting the other elements of the system. This modularity in instrument design represents a principal difference between the early LST designs and the current approach being pursued during the present definition effort. The easy interchange of instruments during the operational life of LST provides the ability to update instruments to address the problems of greatest scientific interest during the long projected lifetime of the LST Observatory. Also, as the instrument and detector technology evolve, LST can be updated to benefit from the future state-of-the-art instrumentation. There is a strong analogy in this respect to the operation of the great

Earth-based astronomical observatories. The basic telescope optics and structure remain essentially unchanged; however, the scientific problems being addressed and instrumentation used at the large ground observatories have evolved with time.

The scientific participants in the LST Phase B definition activity have established minimum LST scientific performance requirements. It will be necessary for the LST system to provide performance equal to (or exceeding) these requirements in order to effectively address the class of scientific problems envisioned for LST. The over-all challenge of the LST is to meet these requirements in a most cost effective manner. A current listing of some of the over-all parameters of the LST system is indicated in Table 1.

Table 1
LARGE SPACE TELESCOPE CHARACTERISTICS

Primary Mirror	2.4-meter diameter
Primary f Number	2.3
System f Number	24
Angular Resolution	0.1 arc sec
Spectral Range	1150Å - 1 mm
Encircled Energy	60% at 0.15 arc sec dia.
System Wavefront Error	$\lambda/20$ at 6328Å
Spacecraft Stability	0.007 arc sec
Weight (typical)	6800 kg (15 000 pounds)
Length (typical)	13 m (44 feet)
Diameter (typical)	4.3 m (14 feet)

With the current designs, we are able to maintain a healthy engineering margin without incurrence of significant additional cost. For example, we have a design target of $\lambda/20$ for the over-all wave front error at the focal plane of the OTA. The image quality available to the LST Scientific Instruments is affected both by the intrinsic optical quality of the system and the accuracy the spacecraft control subsystem can stabilize the optical axis of the system during an observation. We have set a design target of 0.007 arc sec stabilization for the LST spacecraft. With an OTA wave front error of $\lambda/20$ and stabilization of 0.007 arc sec, the minimum scientific performance of encircled energy of 60 percent at 0.15 arc sec diameter (which relates to the profile of the point spread function of a stellar image) is met with reasonable margin. The encircled energy criterion still can be achieved if the telescope performance is reduced to $\lambda/14$ and the stabilization is reduced to 0.010 arc sec. Furthermore, if the optical performance is maintained at $\lambda/20$ the spacecraft stabilization could drop to 0.02 arc sec and the encircled energy performance criterion is still met. Margins such as this are necessary in definition of the system to provide assurance of cost and schedule predictions. During the execution phase of a program, when hardware is being manufactured and tested, margins can prove to be the salvation in preventing delays and associated cost overruns when the manpower being applied to the program is near its peak.

We have been working actively with the European Space Agency (ESA) concerning their potential commitment to LST. They have expressed potential interest in providing an LST Science Instrument and associated detector, furnishing the solar array and possibly other elements of the SSM power subsystem, and in supporting LST operations. Also, other possible areas of involvement are under consideration by ESA. The Europeans will make a preliminary decision next month concerning their participation in LST. ESA will be compensated for their contributions to the LST Program by a share of total observing time no less than the fraction of the cost of the total project which ESA contributes in areas of mutual interest. A detailed working agreement concerning NASA/ESA Large Space Telescope cooperation has been drafted.

During the balance of the Phase B effort, particularly in the second half of the SSM contracts, interfaces between the various elements of the LST system will be firmed up. There is a number of interfaces to be resolved including the LST-to-Shuttle interfaces, the LST-to-TDRSS communications interface, the OTA-to-SSM interface, and physical/optical interfaces between the Scientific Instruments and the OTA, and power and data interfaces to the SSM. The LST Project will be making a number of interface decisions by the first of October, based primarily on the technical material provided in the recent design reviews by the two OTA/SI Phase B contractors and the three SSM Phase B contractors. The interfaces discussed above are defined presently in general terms, but they will be "hardened up" in preparing final specifications for the development phase.

We do not anticipate that the basic designs of the OTA or SSM will change significantly during the balance of the definition phase. The LST Systems Engineering and Integration will be emphasized. Particular attention will be devoted to test and integration planning. Of course, costs will continue to receive prominent attention. Costs will be of key importance in all decisions that are made. The balance of Phase B will be mainly a "fine tuning" of presently available information and updating of cost and schedule estimates. We do not believe that the LST design, program plans, or cost estimates will change significantly.

It is our plan to initiate the LST development phase with performance-type specifications indicating requirements, not design solutions. However, the Requests for Proposals for the development phase will include detailed interface specifications which will form the boundary conditions for the various elements of the LST system. It is our intention to procure the lowest cost system which will provide acceptable performance. Furthermore, we will be willing to trade performance for cost, particularly in areas where significant cost savings can be realized. We are wary of proposed sophisticated solutions and complex designs. If one is considering a trade between sophistication and performance versus simplicity and somewhat reduced performance, it is usually prudent to adopt the simpler approach. Also, flight proven components and subsystems and standard components will be used whenever practical. Use of space proven equipment is doubly beneficial by reducing both development cost and performance risk.

CONCLUSION

The technology and advanced development work to assure the success of the LST Program has been accomplished and demonstrated. The LST system, as presently defined with the 2.4-meter primary mirror, meets scientific requirements. Appropriate engineering margins with this design are available. The exceptionally long and thorough definition phase, which will be completed next spring, will provide high confidence in the reasonableness and appropriateness of design approaches, specification requirements, and cost estimates. We are certain that all program objectives can and will be achieved.

AAS 75-131

THE LST SCIENTIFIC INSTRUMENTS

George M. Levin*

Seven scientific instruments are presently being studied for use with the Large Space Telescope (LST). These instruments are the F/24 Field Camera, the F/48-F/96 Planetary Camera, the High Resolution Spectrograph, the Faint Object Spectrograph, the Infrared Photometer, and the Astrometer. These instruments are being designed as facility instruments to be replaceable during the life of the Observatory.

INTRODUCTION

The LST Scientific Instruments represent a unique design challenge. The LST is to be a permanent National Astronomical Space Observatory. The Scientific Instruments (SI) for the spacecraft are being designed as facility instrumentation which will be replaceable during the life of the Observatory. To reduce the down time of the telescope, the instruments are being designed for manned on-board serviceability. The LST Scientific Instruments must be capable of serving the broader needs of the astronomical community.

These basic mission requirements have been translated into specific design approaches. The LST Focal Plane Assembly is configured around standard SI Modules. Each module contains one scientific instrument. The thermal, electrical, mechanical, and data interfaces between all modules and the focal plane assembly are identical. Figures 1 and 2 show the two concepts of the LST Focal Plane Assembly under study. Each concept has four standard axial scientific instrument modules which share the light in the focal plane.

At present, there are seven scientific instruments under design for the first LST mission. These are the F/24 Field Camera, the F/48-F/96 Planetary Camera, the High Resolution Spectrograph, the Faint Object Spectrograph, the Infrared Photometer, the High Speed Point/Area Photometer, and the Astrometer.

---
*Goddard Space Flight Center, National Aeronautics and Space Administration

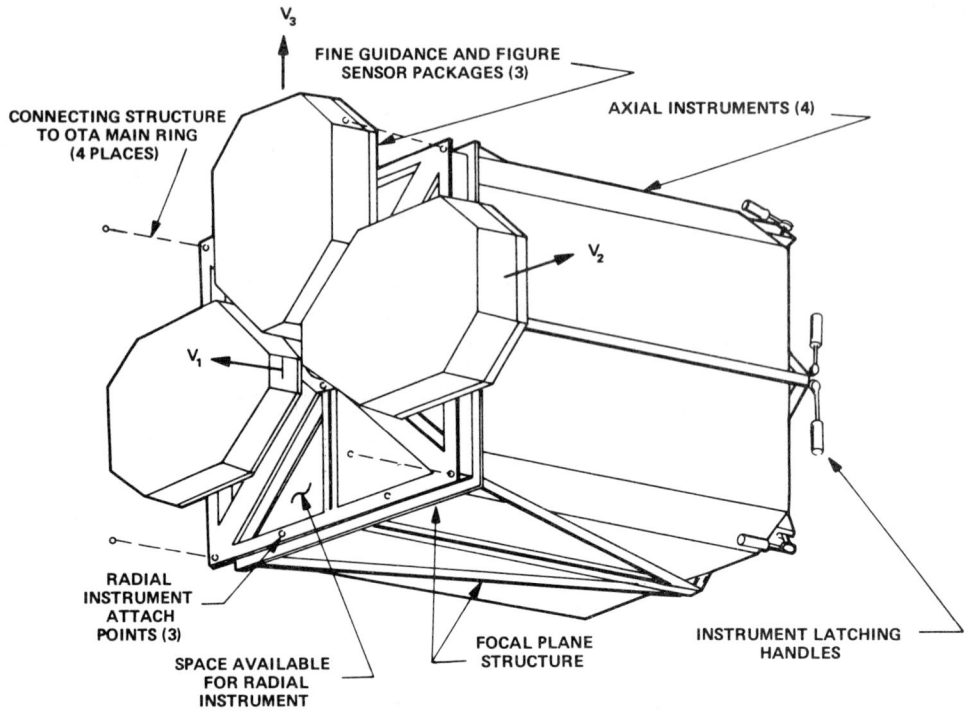

Fig. 1 Focal Plane Assembly

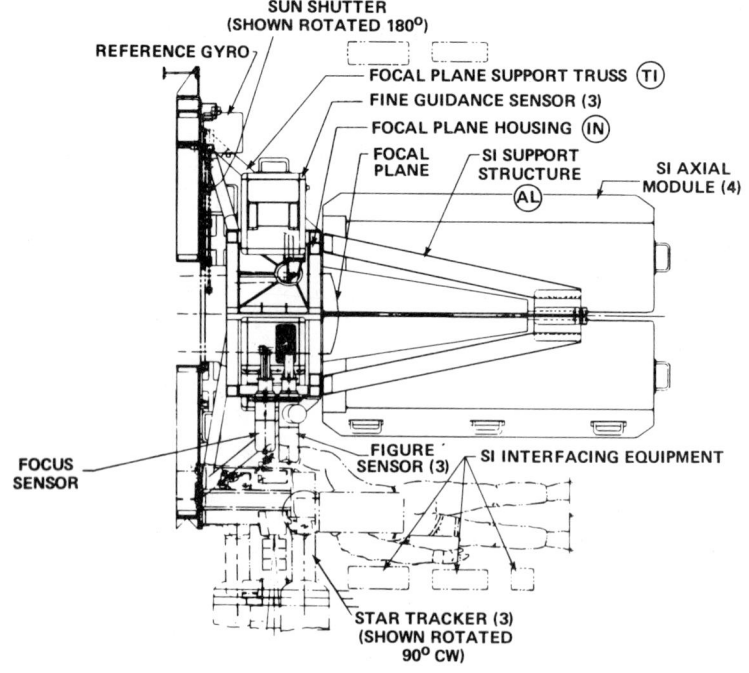

Fig. 2 Focal Plane Assembly

F/24 FIELD CAMERA

The design of the F/24 Field Camera is geared toward the study of faint extended and point sources, and for search and survey work. The instrument system is designed so that the F/24 Field Camera can be operated simultaneously with the other scientific instruments.

The F/24 Field Camera (Fig. 3) consists of the 70 mm SEC Orthicon (SECO) Camera submodule, a 1:1 relay, filter assembly, standard star calibration system, internal source calibration system, and capping shutter assembly. These subsystems are mounted on an optical bench in the SI module. The optics are housed in a forward bay which is enclosed with a thermal shroud whose temperature is thermostatically controlled. The SECO is mounted to the optical bench in an aft bay. The aft bay has a thermal reflector to increase the view factor of the SECO submodule to the outer wall of the instrument module.

F/48-F/96 PLANETARY CAMERA

The design of the F/48-F/96 Planetary Camera is primarily for the study of imaging photometry at high angular resolution for high surface brightness or multiple bright sources. The Planetary Camera (Fig. 4) has an optical system with two, selectable focal lengths imaging on a Charge Coupled Device (CCD) detector cooled to $-40°C$. The optical system consists of two independent, two-mirror relays nested together in such a way as to have two separate entrance apertures in the focal plane and a common output focal plane. Switching between the relays is accomplished by means of a multiposition capping shutter.

The wavelength range of the Planetary Camera is from 180 to 1200 nm. The field of view at F/48 and F/96 is 17 x 17 arc sec and 8 x 8 arc sec respectively. The Camera is to have 0.1 arc sec resolution at F/48 and to be diffraction limited at F/96.

HIGH RESOLUTION SPECTROGRAPH

The High Resolution Spectrograph (HRS) is designed for use in imaging spectroscopy of point or extended sources. The instrument (Fig. 5) consists of interchangeable slits, a collimator, interchangeable echelles and first-order gratings,

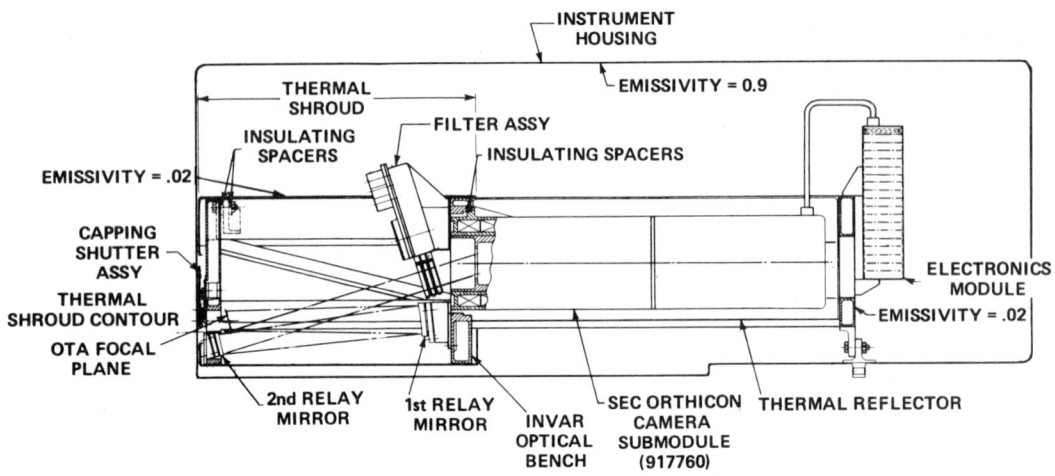

Fig. 3 F/24 Field Camera

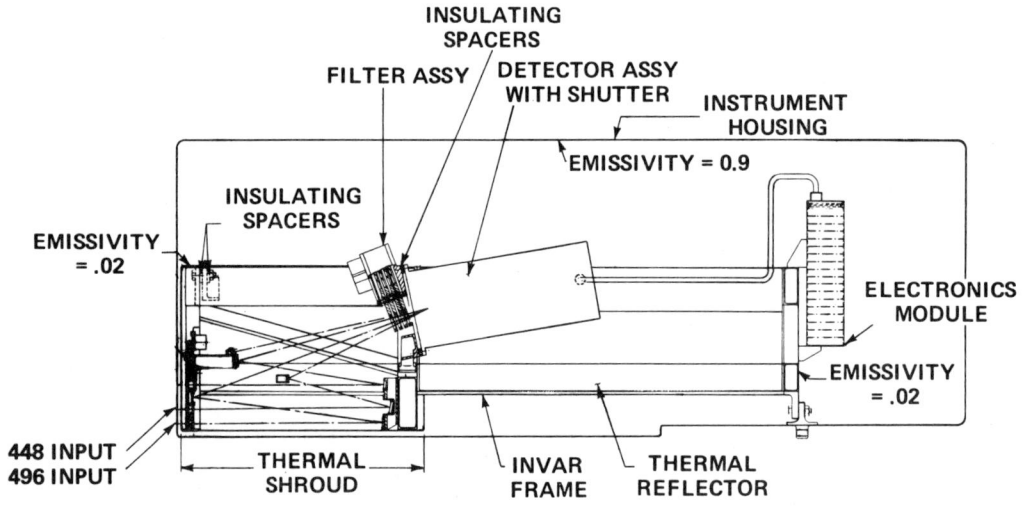

Fig. 4 F/48 - F/96 Planetary Camera

interchangeable cross dispersers, detector selector and compensator for orbital-velocity doppler shifts, and two SEC Orthicon Camera submodules. The principal operating mode of the HRS is as an echelle spectrograph with spectral resolution of 3×10^4 and 1.2×10^5. A secondary spectral resolution of 10^3 was also specified. The wavelength range of the HRS design is from 115 mm to 410 mm. This wavelength range is divided such that the first SECO Camera (with a CsI photocathode and a MgF_2 window) covers 115-170 mm and the second

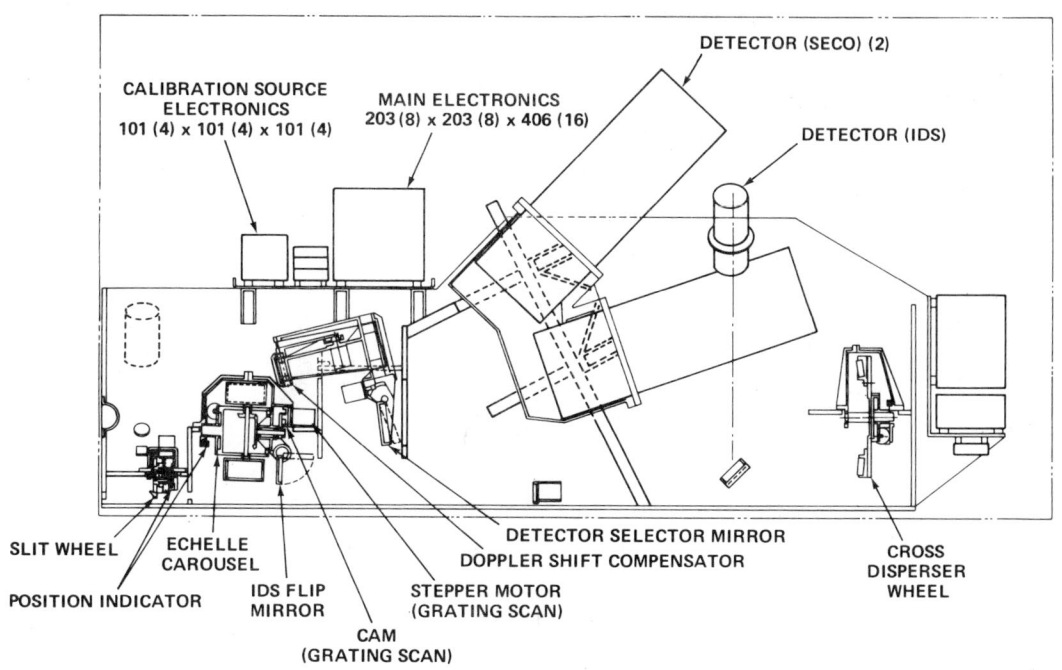

Fig. 5 High Resolution Spectrograph

SECO Camera (with a Bi-Alk photocathode and an SiO_2 window) covers 170-410 mm.

FAINT OBJECT SPECTROGRAPH

The Faint Object Spectrograph (FOS) is a composite instrument covering a broad spectral range and intended for imaging spectroscopy at moderate resolving power. Both point, or stellar, and extended sources are to be observed. Time variations, spectral profiles of broad emission and absorption features, and continuum flux distributions are to be measured for primarily faint targets.

The main components of the FOS (Fig. 6) are gratings used in a Wadsworth mount, three Intensified Charge Coupled Device (ICCD) camera submodules, a collimator, an image slicer, a slit jaw camera, and a spectral reference source. The wavelength range of the FOS is from 90-800 mm. It is divided into three regions. FOS 1 is from 90-190 mm. Its ICCD Camera has a CsI photocathode and a MgF_2 window. FOS 2 covers the range from 180-400 mm and has

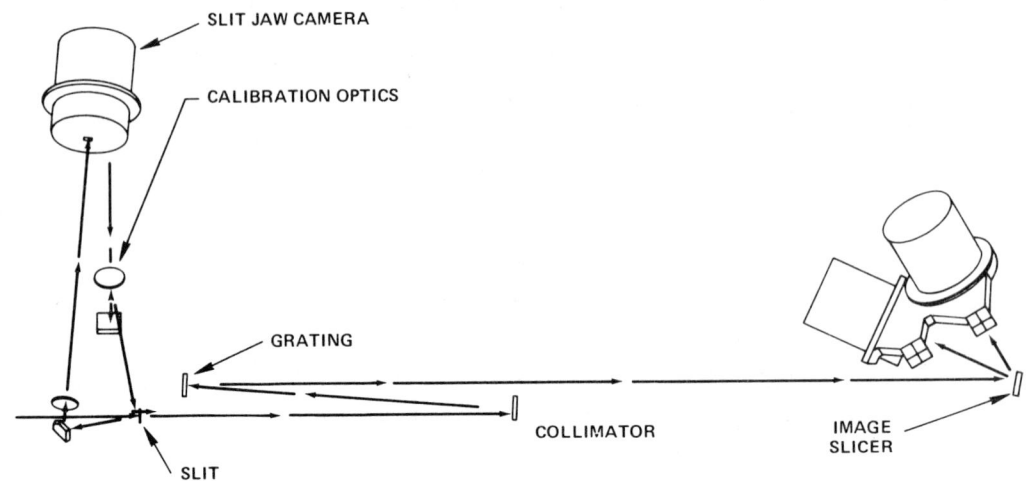

Fig. 6 Faint Object Spectrograph

a BiAlk photocathode and SiO_2 window. FOS 3 covers the range 380-800 mm and has a S-25 photocathode and an SiO_2 window.

INFRARED PHOTOMETER

The Infrared (IR) Photometer is primarily intended for use in the photometry of point or extended sources in mid and far infrared wavelengths. By commanding the telescope to move in a prescribed systematic manner, the photometer can scan a particular area of scientific interest. The photometer when scanning, can determine the surface brightness of IR sources. In addition, the photometer can accurately determine the position of new IR sources by scanning the area surrounding a nominal spacecraft pointing and determine the location of the peak signal.

Background limited performance requires detector cooling to less than the lambda point of He, i.e., $2.2°K$. In addition, offset signals due to background must be suppressed. Cryogen life and dewar design are significant design drivers of the IR Photometer.

The IR Photometer concept (Fig. 7) is based on the use of superfluid Helium to achieve the necessary detector cooling. A focal plane chopper provides the required background suppression. Cooled apertures matched to the diffraction size

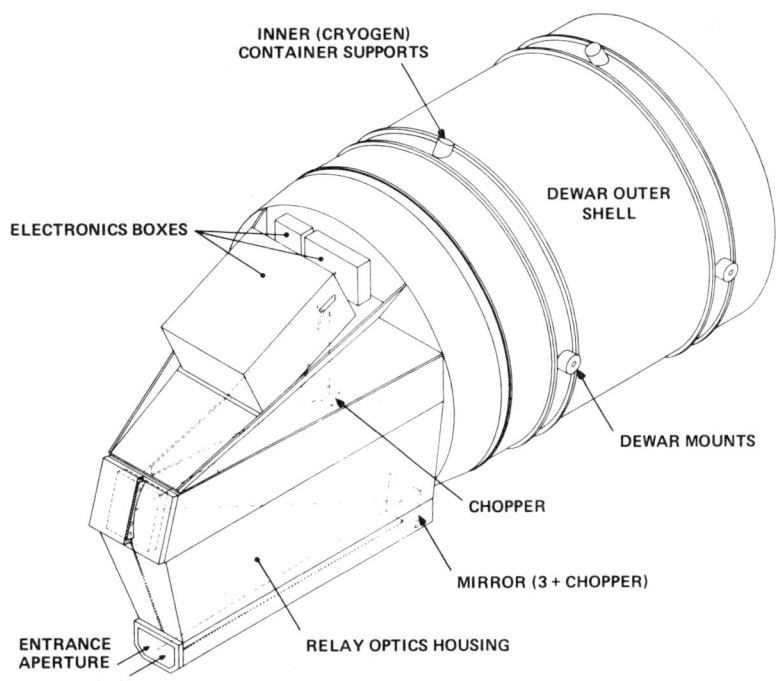

Fig. 7 Infrared Photometer

of the beam and cooled field mirrors provide the remaining background suppression. Multiple apertures and filters along with a beamsplitter provide instrument flexibility. In certain target cases, acquisition is handled in a special peak-up mode to determine the largest signal magnitudes encountered during the scan. Using the results of this determination, the telescope is automatically commanded to the new pointing.

The general arrangement and construction of the dewar and relay optics are shown isometrically. The system consists of three separable modules; the dewar, the instrument mounted within the dewar, and the external optics assembly. The electronics are mounted on the optics housing but are thermally isolated.

The dewar outer shell and relay optics will be at, or near, ambient container temperatures. The chopper and electronics will be above ambient. The optical relay system will be sensitive to temperature gradients that may cause bending misalignment; however, it will not be very sensitive to the uniform temperature level. To preclude temperature gradients it will be necessary to thermally insulate the heat producing elements such as the electronics from the structure and provide means to transfer this energy to the SI container. Low power heaters and insulation may be necessary on the optical housing to control gradients under various operating conditions.

The entrance aperture will require a cover to preclude contamination. As a minimum a removeable aperture cover for use during ground handling will be necessary.

HIGH SPEED POINT/AREA PHOTOMETER

The basic utility of a UV-visible photometer operating in a photon counting mode with an optional analog mode capability, is its capability to obtain precise measure of the constant or time variable intensities over a wide dynamic range of astronomical signal strengths. The data collected can be from either point sources or celestial fields of small angular size. Such a photometer makes feasible a variety of observing programs extracting the maximum information content possible. In addition, high time resolution and flexible sequencing allows observations of a variety of unusual time related phenomenae.

The first element of the High Speed Point/Area Photometer (HSAP) is the contamination door which is used to exclude contaminants during pre-launch, launch, return and maintenance operations, and as backup to the shutter which is next in reaction time and can quickly shut out unwanted incoming light in the case of inadvertent acquisition of a source whose intensity is bright enough to damage the sensors (Fig. 8).

An aperture wheel assembly is mounted at the telescope F/24 focus and provides selectable field stops. The two filter wheels follow the aperture wheel assembly. These filter wheels provide spectral band pass discrimination. The next element is the primary mirror carousel. The carousel contains space for mirrors on each of six sides of the turret. Four of these are used, one each for the two point detectors, and one each for the F/24 and F/96 images. Each of the mirrors is figured for the optical system in which it functions and is so placed on the carousel to direct the beam to the desired detector or secondary mirror.

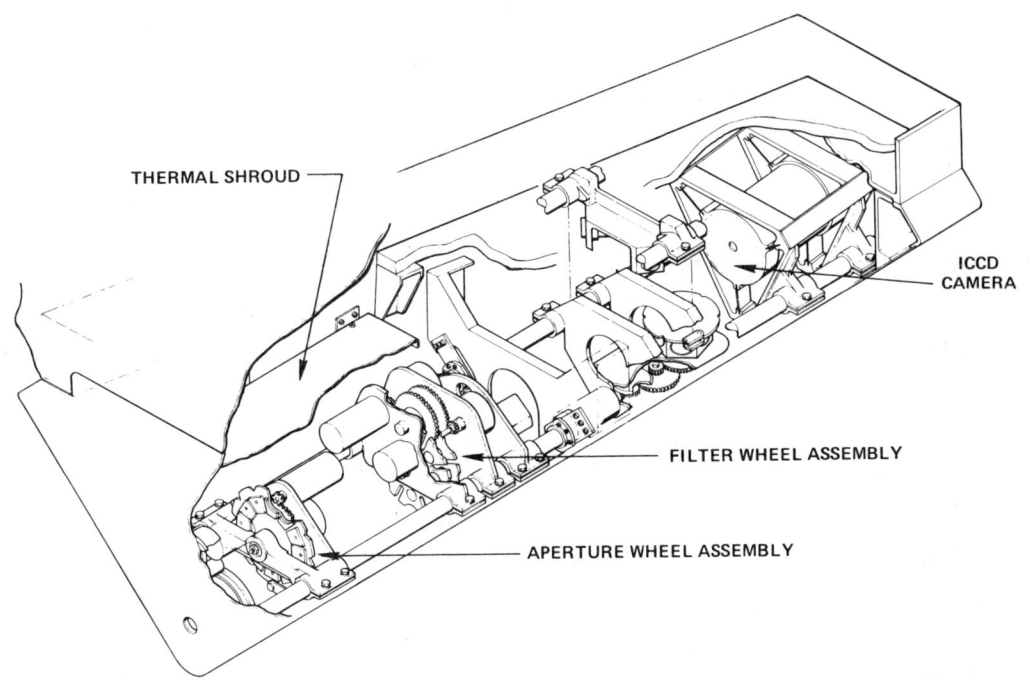

Fig. 8 High Speed Point/Area Photometer

The two point detectors receive their input from primary mirrors on the carousel, and are mounted to the bench-rods, receiving their input from the primary mirrors on the carousel. They, in turn, direct the beam to the area detector forming images at F/24 and F/96.

The final element is the Intensified Charge Coupled Device (ICCD) area detector. Mounted to the optical bench rods, it connects through a flexible thermally conductive copper web strap to a radiator.

Graphite epoxy is used for the optical bench rods to minimize thermal deformation. The rods are supported at mid point and aft end on flexures to allow only minimal stress to be induced in the rods from deformation of the shell.

ASTROMETER

The astrometer is designed to measure parallaxes and proper motions ten times more accurately than presently possible, to measure angular diameters for stars and the nuclei of galaxies, and to determine individual masses through observations for about 100 spectroscopic binaries.

Wide field astrometry for studies of parallax and proper motions requires a wide field of view (effective area equivalent); uniform image quality (cover that effective area); stable positional accuracy within the field of view (±0.002 $\widehat{\text{sec}}$ relative to other objects in the field of view); and occasional serendipitous operation regardless of which instrument is prime. Narrow field astrometry for single stars and close double stars requires high angular resolution (diffraction limited) and high time resolution (100 μsec ±2%).

Two different concepts of astrometry instrumentation are under study. In the first concept a modified multiplex area scanner AMAS (Fig. 9) is employed for the wide field studies. The high resolution camera and the area photometer may also be used to supplement and expand the capabilities of the AMAS for astrometric measurements.

In the second concept, the fine guidance system for the Optical Telescope Assembly is used for the wide field measurements. This fine guidance system is modified to include a high performance reticle. For narrow field astrometry the fine guidance system software is modified to include the necessary multi-target search/measure routines as well as handling the encoded signals as data. The high resolution camera and the area photometer are used for astrometric measurements.

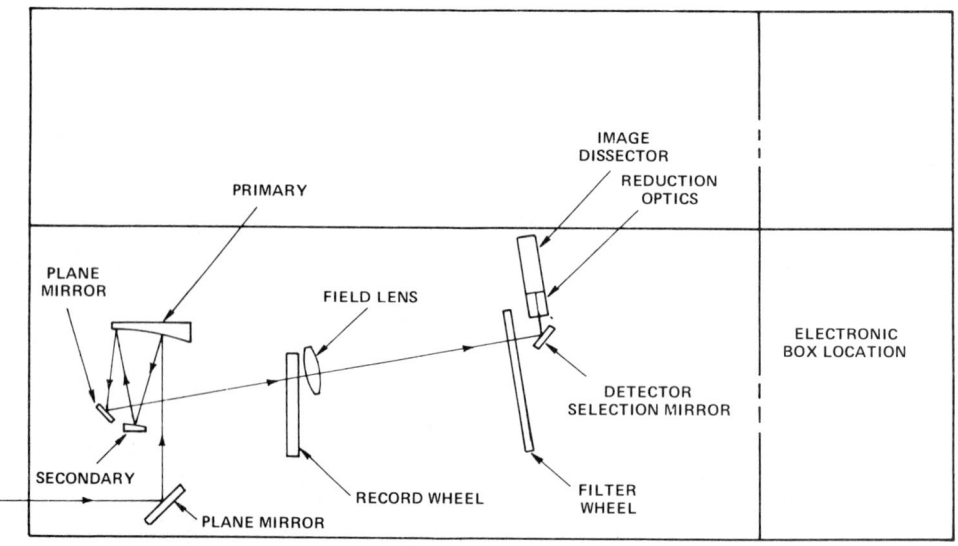

Fig. 9 Astrometer (AMAS) Optical Schematic

MIRROR SUBSTRATE MATERIAL AND MANUFACTURING FOR THE LARGE SPACE TELESCOPE

William C. Lewis*

Corning Code #7971 U.L.E.™ (Ultra Low Expansion) glass has been selected as the material for the primary mirror substrate for the large space telescope.

The basic process for producing this material is the same as that used widely for the manufacture of pure synthetic fused silica, in which vapors of highly purified silicon tetrachloride are introduced into the flames of burners where they react with the water vapor produced by combustion of natural gas to form hot silica soot particles.

The process variation that produces U.L.E.™ material is primarily the introduction of titanium tetrachloride as well as silicon tetrachloride to the flame, thereby producing a binary glass composed of titania and silica.

The titania-silica system has many unique characteristics. For example, titania lowers the coefficient of thermal expansion of silica in direct proportion to its concentration. At 20°C, the instantaneous coefficient of 7.4% titania, 92.6% silica is zero compared to a value of 0.50 ppm/°C for 100% silica. U.L.E.™ material is defined as that part of the titania-silica family in which the coefficient of thermal expansion taken over the 5°C to 35°C temperature interval is in the range of 0.00 ± 0.03 ppm/°C. A second unique characteristic of this system is that a difference in coefficient of thermal expansion due to differing titania levels is a constant that is independent of temperature.

---
*Corning Glass Works

This means that it is not necessary to measure the extent of inhomogeneities for coefficient of thermal expansion at the specific temperature of interest, but rather permits the measurement to be made at a temperature most conducive to accurate data, and the difference found will be true at the temperature of interest also.

A third unique characteristic is that not only is the coefficient of thermal expansion lowered in direct proportion to the concentration of titania, but so also is ultrasonic velocity. This means that by the use of a precision ultrasonic intervalometer and accurate thickness gauging, the coefficient of thermal expansion of various parts of an actual substrate can be determined non-destructively to within a few parts per billion[1], a very important feature as regards forecast of change in image quality as a function of thermal excursions of the substrate.

The fourth unique characteristic of U.L.E.™ material is that it is truly weldable in the full sense of the word. Because of the low expansivity, two parts can be welded together in the open room by heating the edges with gas-oxygen flames until the material is softened on each part. The pieces can then be merged, the flames removed, and the parts allowed to cool. The entire operation can be done without need for controlled heating or cooling. This feature allows for a substrate to be fabricated of parts preselected for quality appropriate to the needs of each part of the substrate for each specific application.

The Corning Glass Works equipment used in the production of U.L.E.™ material consists of a multiplicity of burners, vertically disposed so as to point down toward a rotating table as the target for the accumulation of the soot. The material builds up slowly over a period of weeks until it produces a boule or disc that is about 60" in diameter and normally 3" to 5" thick. All articles to be made of U.L.E.™

material must be made from these discs.

In the manufacture of a mirror substrate we need to convert the material into two plates, an egg-crate like core, and usually both an inner and outer ring. The front plate is the one that is accurately figured and polished, and usually ends up either flat or concave. The piece of glass selected for this is one of selected quality for low inclusions in order to provide for favorable polishability.

If the mirror is over 60 inches in diameter, the material for plates is cut from the boule in extra thickness as necessary so that when the piece is taken to about 1600°C and allowed to flow to the large diameter, the thickness at the new diameter is appropriate. Plates up to 120 inches in diameter and 1 1/2 inches thick have been obtained in this manner.

Manufacture of the core is more complex. First, if a 12-inch high core is wanted the necessary number of boules are stacked vertically to produce a stack that is 60 inches in diameter and 12 plus inches high. This entire stack is placed in a furnace and heated to about 1600°C at which temperature the boules fuse together to form a single monolithic piece.

The stack is then used to saw out individual cell struts and ring sections, which are then ground to exact thickness, i.e.; 0.200 inch. The individual struts are then welded together to form a number of 90° ells, which are the building elements for the core. A number of these are initially welded to a flat plate in an adjacent position and then the square cell accomplished by welding a second row. This procedure is continued row by row until the overall size is adequate, at which point the core is circled and ground to exact height.

The rings are made by sagging sections of straight plates to the proper radius by heating them to about 1600°C on a male mold. The individual sections are then welded together to form complete rings.

The completed parts are then placed in their proper positions to make the complete structure, and the entire assembly is heated to about 1600°C where it fuses to initially form a plano-plano monolithic structure. For the convex-concave configuration, the entire monolith is placed on a male mold, heated and sagged to conform to the mold.

In order to provide optical performance, the two plates must be closely matched for coefficient of thermal expansion. This is commonly done to within 0.01 ppm/°C.

However, differences in expansivity are also purposely employed to enhance the strength of the final substrate. The ends of the struts are geometric stress concentrators, so it is desirable to produce residual compressive stress at these locations. This is done by purposely mismatching the expansivity of the ring and core materials. As the blank cools from the set point of 925°C, the differential expansion of the two is such that the rings pull the plate down against the strut ends, putting desirable residual compression at those points, and yet modest tension in the rings.

Reference

[1] Henry E. Hagy, "High Precision Photoelastic and Ultrasonic Techniques for Determining Absolute and Differential Thermal Expansion of Titania-Silica Glasses". Applied Optics, Vol. 12, No. 7, July, 1973 pages 1440-1446.

Henry E. Hagy and W.D. Shirkey, "Determining Absolute Thermal Expansion of Titania-Silica Glasses: A Refined Ultrasonic Method". To be published in the September, 1975 issue of Applied Optics.

FABRICATION AND TEST OF 1.8-METER-DIAMETER, HIGH QUALITY ULE MIRROR

Richard J. Wollensak*
Clarence A. Rose*

The purpose of this investigation was to determine the feasibility of producing large diameter, high quality optical surfaces similar to those that will be required for some of the planned orbital observing units such as the Large Space Telescope (LST). Current requirements for the Large Space Telescope require that a $\lambda/20$ rms wavefront be produced at the Cassegrain focus. A reasonable distribution of error sources would, therefore, demand that the primary mirror be figured to a tolerance of $\lambda/64$ rms on the surface. NASA has funded the initial step toward demonstrating the ability to meet that requirement. An optical fabrication demonstration program was undertaken using an available 1.8-meter-diameter ULE monolithic mirror blank weighing on the order of 1,200 pounds. The mirror that was used had no central perforation and was approximately 12 inches thick with 1-inch front- and back-plates. In order to accomplish this task, a variety of support and test equipment was designed and fabricated and the actual optical fabrication and test were conducted over a period of approximately 8 months. The results of this effort yielded a mirror with a surface quality of 0.015 wave rms or $\lambda/65.3$. (λ = 632.8 nanometers.)

The demand for large diameter, high quality optical surfaces requires that many engineering disciplines participate in the design and development of a suitable support system for the mirror during fabrication and test, and in a constant upgrade of the test equipment, test optics, and methods of data reduction. One of the key factors in the successful endeavor is providing the support equipment with appropriate sensitivity and a high degree of repeatability in testing the surface. It is critical that the total system used in developing the mirror surface demonstrate confirmed uniformity and repeatability from test to test.

The mirror blank used in this experiment was a lightweight, monolithic ULE blank manufactured by Corning Glass Works. The mirror blank was initially an f/2.7 meniscus-shaped element which was subsequently reslumped to an f/2.2 curvature prior to the start of any optical fabrication. The purpose of the reslumping was to demonstrate fabricability of a faster mirror, which would be compatible with the requirements currently planned for the LST program. The mirror blank was successfully reslumped by Corning Glass Works and the finished weight of the mirror was 1,203 pounds.

*Both authors are with Itek Corporation, Lexington, Massachusetts 02173.

Properly designed support and test equipment is crucial in the development of any high quality surface. We can categorize these requirements in essentially three areas:

1. The mirror support system used to repeatably maintain the surface figure during polishing and testing.

2. The optical test equipment. The most important component of this equipment was a three-element null lens. It was decided in the definition phase of this activity that the most reasonable surface to demonstrate would be an f/2.2 parabola. Consequently, a null lens was required in order to produce a spherical wavefront which would be compatible with the interferometer requirements.

3. The third major consideration would be the optical tooling. A number of different-sized laps with varying degrees of stiffness and flexibility were used during the fabrication of the surface.

The support system chosen for this program was a multi-point, passive liquid mount, consisting of 27 fluid pistons properly spaced and interconnected so as to allow flow from one piston to another. The type of support is shown schematically in Fig. 1 and an actual photograph of the finished mirror mount is shown in Fig. 2. A comprehensive structural analysis was undertaken to determine the number, diameter, and location of the fluid piston supports. In order to minimize any deformation contribution by the mount to the mirror surface, primary considerations in mount design are total amount of sag between any two support points, and also the area of the fluid pistons in order to minimize any highly stressed areas on the back of the mirror blank. Normally, loading on the fluid pistons is kept in the area of 3 to 5 pounds per square inch. A theoretical goal for the total error contribution of the support system was set at $\lambda/150$ rms. As shown in Fig. 2, there were 18 separately adjustable assemblies that were used to laterally restrain the mirror blank from any transfer motion. The restraint assemblies were designed so as to contact the mirror blank on the edges of both the top and bottom plates. These restraints were firmly adjusted to contact the mirror during the actual polishing cycle, and then backed off completely during the testing of the surface so as not to influence the mirror surface.

Fig. 3 schematically shows the null lens configuration. This optical system consists of three elements, each having easy-to-fabricate spherical surfaces. The null lens design also accounts for a 1-inch-thick BK7 window which was the entrance port in the vacuum test chamber. The lens was designed with a certain amount of flexibility in terms of spacing the elements in order to permit compensation for the differences in the refractive index

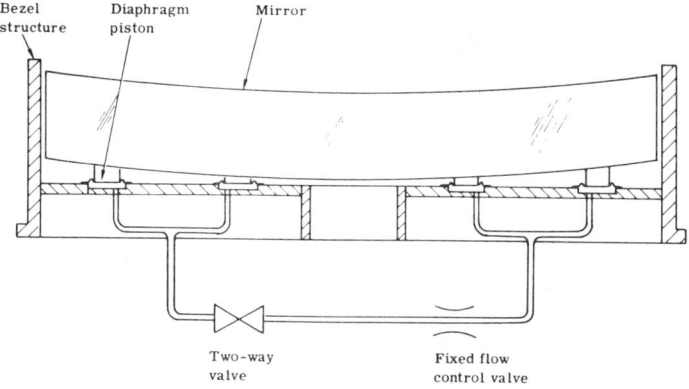

Fig. 1 — Passive Liquid Mount

Fig. 2 — Top View of Bezel

• PRELIMINARY NULL LENS DESIGN TO BE USED IN CONJUNCTION WITH LASER UNEQUAL PATH INTERFEROMETER

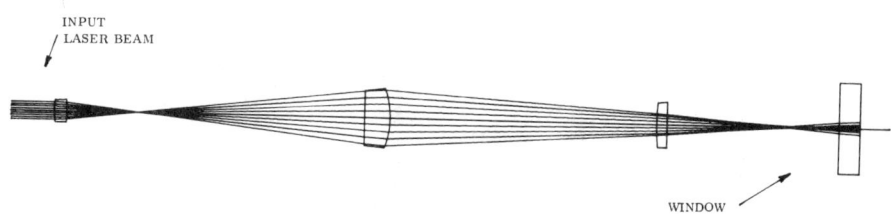

• USES THREE SMALL EASY TO FABRICATE REFRACTING ELEMENTS
• ALL SPHERICAL SURFACES

Fig. 3 — Preliminary Null Lens Design To Be Used in Conjunction with Laser Unequal Path Interferometer

between air and vacuum. In the initial phases of the fabrication, the majority of the tests were conducted in a one atmosphere environment. As the quality of the surface improved, it was necessary to perform the optical testing in a partial vacuum, normally on the order of 150 millimeters of mercury in order to minimize the effects of air turbulence during measurements.*

Critical to the performance of the null design is proper mounting and support of the optical elements. Fig. 4 shows a photograph of the finished null lens cell prior to its attachment to the LUPI (laser unequal path interferometer). As shown in the photograph, the null lens cell was made of several components with highly precisioned internal mounting surfaces. In terms of concentricity and alignment of the optical elements, the design goal for the total mechanical error contribution of the three-element null lens was $\lambda/200$ rms.

As previously stated, a number of variously configured tools and laps were used in the fabrication process, ranging from full size stiff laps for the initial generating work, through to laps with varying degrees of flexibility for the final polishing down to laps in the order of 6 and 8 inches which would be used during the final hand correction. The basic fabrication technique employed in this demonstration was to first generate and partially polish a spherical surface of the proper radius. The spherical surface establishes a baseline and also ensures a surface generally free of astigmatism prior to starting the aspherization process. Conventional techniques were then used to rough grind in the proper aspheric departure from the base sphere. The piece was then polished, tested, and figured using a variety of ring laps and special pitch lap configurations to obtain a smooth surface of $\lambda/50$ rms. From this point on, the major activity was one of hand correction to reduce the local zonal errors until a mirror surface of the required quality was achieved. Fig. 5 shows an overview of the handling equipment which was used in transferring the mirror and its support system from the polishing machine into the test chamber. Fig. 6 shows the workpiece during the initial phases of polishing.

Critical to the production of large diameter, high quality optical surfaces is a well conceived test and data analysis program. Fig. 7 shows a block diagram of the test and data handling sequence. When an optical fabrication cycle has been completed, the piece is then removed from the machine to the test chamber for an interferometric test. Fig. 8 schematically describes the mirror positioned in the vacuum tank for a typical test cycle.

---

*Fischer, R. E., Null Optics for Testing a 1.8-Meter-Diameter Paraboloid, J. Opt. Soc. Am., 64:1,369 (1974) and Null Lens Mapping Errors, J. Opt. Soc. Am., 61:655 (1971).

Fig. 4 — Null Lens

Fig. 5 — Overview of All Handling Equipment

Fig. 6 — Mirror During Polishing

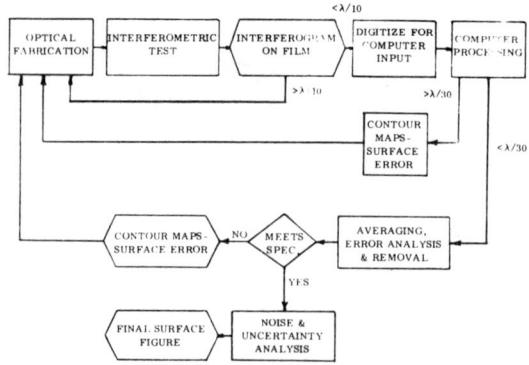

Fig. 7 — Test/Data Analysis Progress

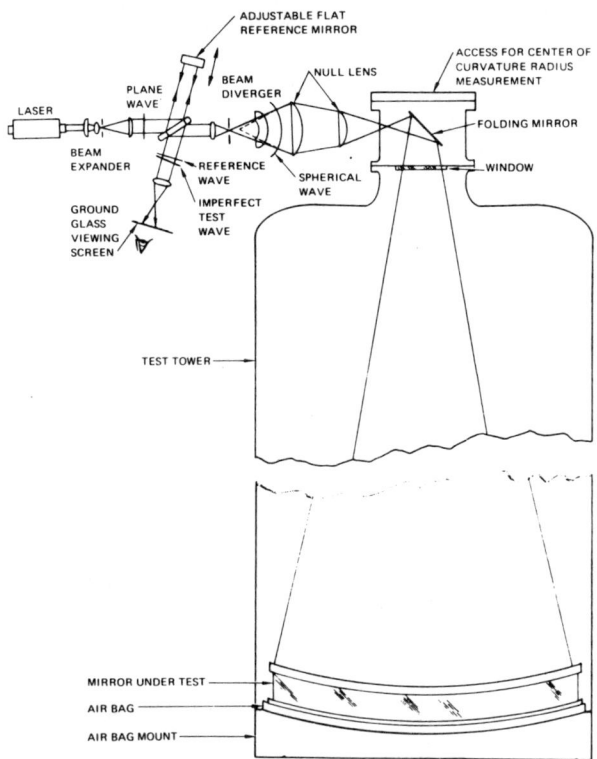

Fig. 8 — Test Configuration

The interferograms produced by the LUPI* are recorded on Polaroid film. The first stage of the fabrication nominally to about the $\lambda/10$ rms level is normally reduced and interpreted by the optician directly. This can be done quite readily using a straight edge or set of parallel rulers and counting the fringes.

---

*Houston, J. B., Jr., Buccini, C. J., and O'Neill, P. K., A Laser Unequal Path Interferometer for the Optical Shop, Appl. Opt., 6(7):1,237 (1967).

Beyond the tenth wave level, the reduction becomes more challenging. At this point a test set of interferograms (e.g., usually 12 photos, six showing horizontal fringe orientations, and six showing vertical fringe orientations) are digitized, sent through the computer, and contour maps of the surface areas are produced. The contour maps are used by the optician in determining the length and location of his next fabrication cycle. This technique is used until the surface quality approaches the order of $\lambda/30$ rms.

From the $\lambda/30$ rms level to completion of the surface, errors in the total system are considered and subtracted from the contour plots that are produced by the computer. These errors are basically those contributed by the null lens, the primary mirror support system, and the test optics. Several tests are run during the course of the fabrication in order to determine the magnitude of these errors. The determination of these errors is accomplished by rotating various components in the test system with respect to each other. In this manner, errors that rotate with the test system components may be separated from the residual errors actually present on the optical surface. This is readily accomplished using a computer program that evaluates the mirror or the optical test component in three rotational orientations. As an example, a set of test data would be taken in the −45-, 0-, and +45-degree rotational orientation. These interferograms would then be digitized and reduced, and by a subtraction routine, the −45-degree position would be subtracted from the 0-degree position and the +45-degree position would be subtracted from the 0-degree position. These errors would then be averaged and subtracted to determine the magnitude of the residual error in the workpiece. Fig. 9 schematically demonstrates this procedure.

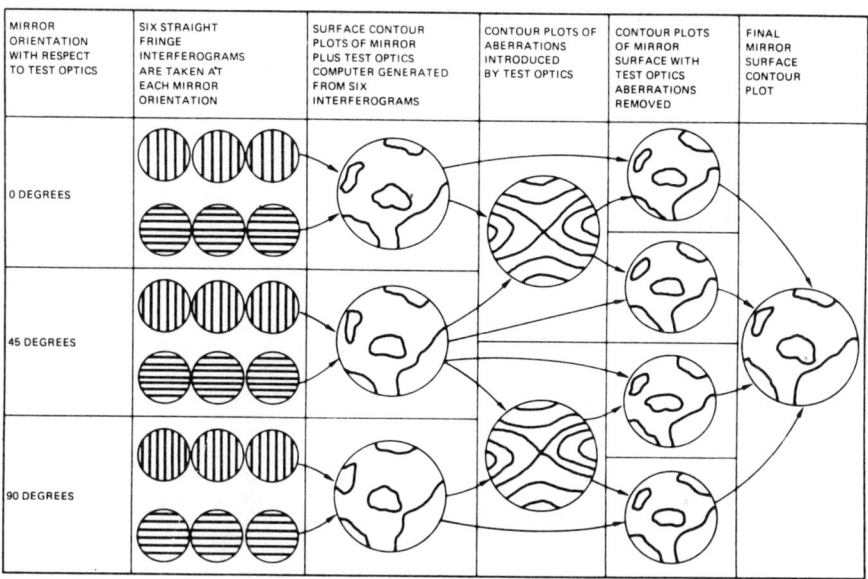

Fig. 9 — Flow Diagram of Data Reduction Process

There are basically three error sources that must be considered in the interferometric error removal process. These are asymmetrical errors, symmetrical errors, and random errors. The asymmetrical errors can readily be defined by the rotation of the workpiece with respect to the test optics. This would hold true for the null optics system with respect to the test piece as well as the primary mirror support system with respect to the workpiece. These rotations can be done individually, and analyzed and accounted for. Symmetrical errors can generally be accounted for in the original error budgets. This is accomplished by precisely measuring the radius, element thickness, and spacing of the null lens assembly. Considerable previous experience has developed a high degree of confidence in accepting these measured values. Random errors in the interferometric testing are generally accounted for by averaging several interferograms. In the initial stages, a fewer number of interferograms are digitized and reduced. As mentioned previously, 12 photos are acceptable up to a level of $\lambda/30$ rms. Beyond this point, depending on the test conditions, primarily influenced by vibration and turbulence effects, it may be necessary to take as many as 20 pictures in any one data set. The main objective is to reduce a sufficient number of interferograms in order to obtain confidence in the repeatability of the test data and minimize the noise contribution.

Fig. 10 shows a plot of actual optician hours plotted against rms surface improvement. The numbers directly above the curve indicate the hours expended between the various functions, such as generating grinding, polishing, and then into the figuring operation. The numbers directly below the curve show the number of test cycles that were used throughout the total fabrication cycle. As shown by the curve, there were approximately 25 test cycles performed in order to achieve a figure $\lambda/2$ rms. In total, the mirror was

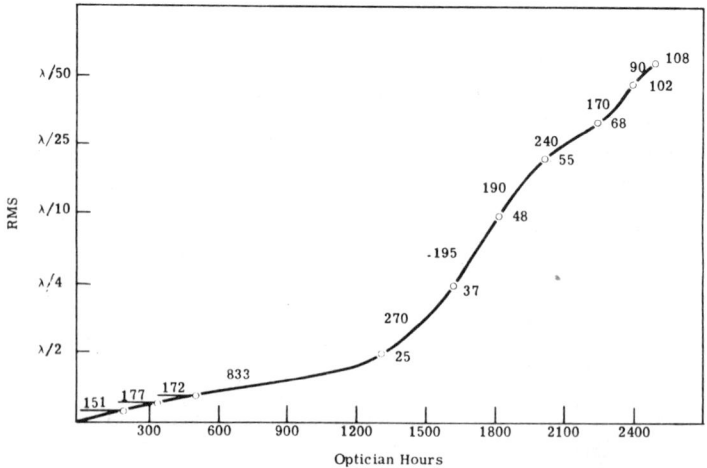

Fig. 10 — Curve Showing RMS Improvement Versus Optician Hours—Phase I
($\lambda/62$ RMS = 0.16; Null Lens Subtracted)

cycled from the polishing machine or through a hand figuring operation into the test chamber, interferograms were taken, the data was reduced, and an evaluation was made, totalling 108 cycles.

Fig. 11 shows a typical contour plot of the finished mirror. The contour increments in this plot are 0.01 wave peak-to-peak.

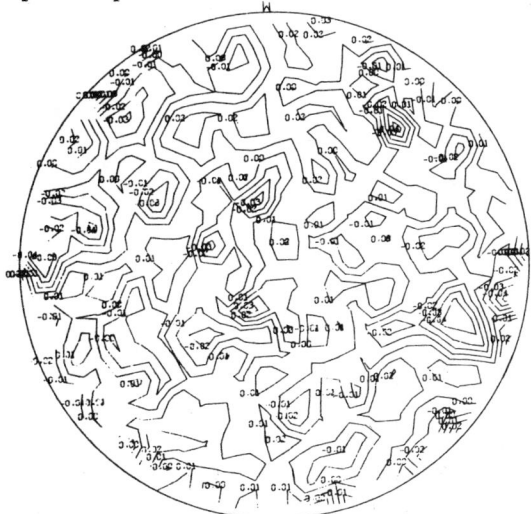

Fig. 11 — Contour Plot (Contour Increment = 0.01λ; Mirror 0.015λ RMS, λ/65.1 Surface)

Fig. 12 is the contour of the mirror surface at .02-wave increments. These contour plots are the result of an average of approximately 24 interferograms, 12 with horizontal fringe orientations, and 12 with vertical fringe orientations. Typical interferograms for data reduction contain approximately 20 to 22 fringes across the surface of the mirror (Fig. 13).

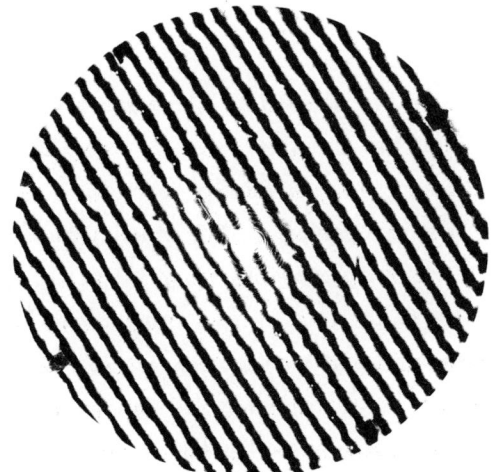

Fig. 12 — Contour Plot (Contour Increment = 0.02λ; Mirror 0.015λ RMS, λ/65.1 Surface)

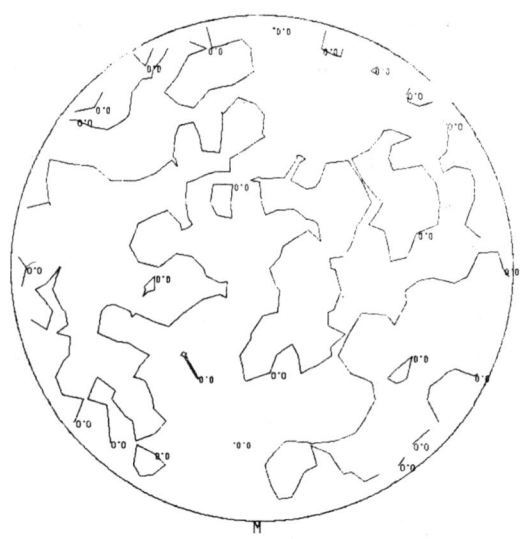

Fig. 13 — Final Interferogram—NASA 72-Inch Mirror ($\lambda/65.3$ RMS)

This gives us a data density on the mirror surface of approximately 1 fringe for every 3 inches of mirror surface. These interferograms are digitized, scanning the interferogram with 21 scans that produce nominally on the order of 367 data points on the mirror surface for each interferogram. These interferograms are then averaged, the known error sources in the system are subtracted, and the result is the contour plot of the actual mirror surface.

The described experiment resulted in a mirror with a surface quality of $\lambda/65.3$ rms over a 70-inch-diameter clear aperture. The average error contribution of the primary mirror support and the test optics resulted in a combined error of $\lambda/114$ rms.

CONCLUSIONS

While this program has resulted in a substantial step forward in developing the technology associated with the production and testing of large diameter, high quality surfaces, it is not at all clear that we have reached an upper limit of fabrication and test capabilities. Clearly, a substantial step forward has been achieved. The production of these surfaces is a painstaking and demanding task requiring much attention to detail and a constant upgrading of all of the various components of the total system. Much progress has been made in the last 5 to 10 years in the production of high quality surfaces. Better test equipment and better analytical tools have helped us to predict and anticipate many of the complex problems. It would, therefore, seem reasonable that higher quality surfaces can be fabricated and that surfaces of comparable quality could certainly be fabricated on large diameter mirrors.

Fig. 14 – Mirror Overview

AAS 75-177

DESIGN AND TESTING WITH A REFLECTIVE NULL SYSTEM

L. Montagnino, A. Offner*

INTRODUCTION AND SUMMARY

To obtain aperture limited performance from a large optical system such as the proposed 2.4 meter aperture LST, the component elements, their spacings, and their relative orientations must be very close to nominal. In particular, to keep the contribution to image degradation of figure errors of the large concave primary mirror consistent with this goal, it is necessary to manufacture this mirror with an rms figure error of about $\lambda/100$ at $\lambda = 6328\text{Å}$. Auxiliary optical systems which can be manufactured and used to measure the contours of a large concave aspheric mirror to this order of accuracy are the subject of this paper. These auxiliary optical systems are retroreflective null correctors. In these systems, the spherical wavefront from a point source is modified by the auxiliary optical system so that its shape is that of the desired aspheric. This wave is then reflected back through the auxiliary optical system which restores its spherical shape. The retroreflected spherical wave can then be compared with an accurately known spherical surface in a high finesse spherical wave interferometer (SWIM).[1]

Three forms of null-correcting optical systems have been developed.[2] Each of these consists of a small field lens and one or two spherical mirrors, so that each of the components can be tested by itself. A form which consists of two spherical mirrors and a field lens arranged in line was designed as a null corrector for a 2.4 meter f/2.3 LST Ritchey-Chretien primary mirror. For null mirror diameters of 15 and 25 centimeters, the nominal residual wave aberration of the retroreflected wave was less than $\lambda/1000$ rms. When practical tolerances are applied to the manufacture of this null corrector optical system, it is shown that when used in conjunction with a high finesse interferometer the contours of the hyperboloid can be measured with an uncertainty which re-

---
*Perkin-Elmer Corporation, Norwalk, Conn.

sults in an rms contribution of less than 0.02λ to the imagery obtained with the Ritchey-Chretien system of which the hyperboloid is the primary mirror.

TOLERANCE ANALYSES

To build and use a null corrector of the form shown in Table 1, it is necessary to know the effects of departures from nominal values of the null corrector elements and of their positions and orientations with respect to each other and with respect to the aspheric mirror being tested. For this purpose, a two-mirror null corrector was designed for use with both a 2.4 meter f/2.3 hyperboloidal mirror to be used for the LST Ritchey-Chretien primary and a 60-inch LST model mirror. This form was chosen as the easiest to align and use, and since its dual application requires only a slight re-spacing of the mirrors and a new field lens. It can be used in conjunction with a SWIM[1] with no additional elements except for the interferometer field lens whose aberrations can be compensated by a slight modification of the spacings of the null corrector.

To obtain accurate values of the tolerance sensitivities, the residual rms wave aberration of the null corrector design was reduced to less than $\lambda/1000$ at $\lambda = 6328\text{Å}$.

The effects of departure from nominal of the various parameters on the rms wavefront departure from the closest reference sphere were determined. At the same time, changes in spacing or orientation which compensate for departures of parameters from nominal were determined. The results indicate that manufacturing tolerances on the radii of the null mirrors and the radii and thickness of the field lens are not prohibitively tight provided that these quantities are measured accurately after manufacture so that their departures from nominal may be compensated for by changes in the spacing. Manufacturing tolerances of ±0.1 mm on the radii of the null mirrors and the thickness of the field lens are not difficult to achieve. The resulting maximum departure of the figure of the aspheric mirror from its nominal value will then be determined by the accuracy of the measurement of the radii and separations of the null corrector. A set of uncertainties due to measurement which we can meet with a high degree

TABLE 1

MEASUREMENT UNCERTAINTIES AND INFLUENCES

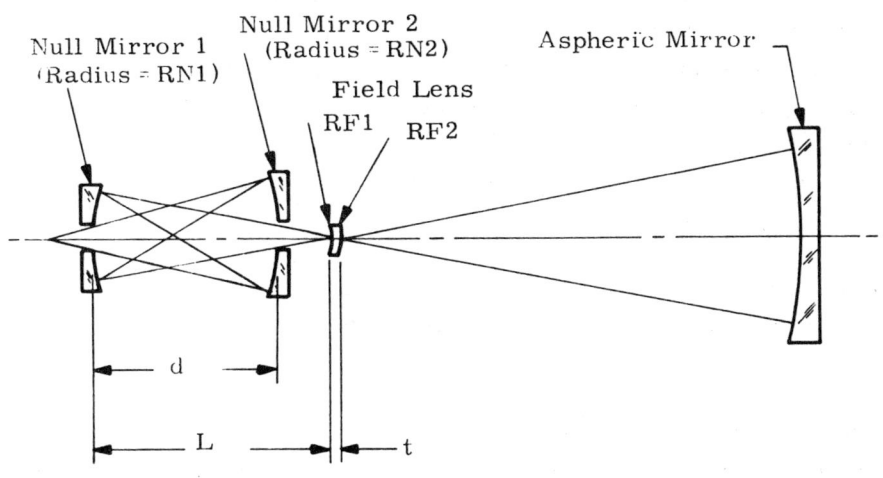

Parameter	Measurement Uncertainty	Figure Uncertainty (λ RMS)
t	±0.010 mm	0.0018
L	±0.010 mm	0.0018
d	±0.010 mm	0.0017
RN1	±0.010 mm	0.0026
RN2	±0.010 mm	0.0016
RF1	±0.100 mm	0.0008
RF2	±0.020 mm	0.0016
Field Lens Decenter	±0.025 mm	
Field Lens Wedge	10 arc-sec	0.0004
Field Lens Tilt	10 arc-sec	
Field Lens Index	±0.00002	
Null and Reference Mirror Figures	0.004λ RMS	0.0086
Longitudinal Position	±0.30 mm	0.0020
Lateral Position	±0.67 mm	0.0020
Tilt Orientation	±0.33 arc-sec	0.0020

Total Uncertainty ~ 0.010λ

of confidence with our present measuring techniques was chosen. These measurement uncertainties and their influences are shown in Table 1. As can be seen, the total null corrector/interferometer measurement uncertainty in terms of mirror figure contour is approximately 0.01λ rms.

CELL DESIGN

To reduce the reflective null lens design to practice, a cell design was formulated which could achieve the objectives of mirror figure quality, and optical alignment and figure stability. Stability is of utmost importance since the null corrector performance must be certified by the evaluation of individual elements and alignments. Once certified, it is essential that the optical performance remain within tolerance throughout the period of performance under all required environmental conditions.

A schematic of the cell design is shown in Figure 1. The null corrector will be mounted to a vibration isolated vacuum tank with its optical axis aligned vertically.

Axial spacing of key elements is controlled by design configuration and selection of materials. Lateral alignment is controlled by symmetry about the optical axis. The mirrors are Cervit. The outer shell is aluminum. Lateral spacing of each is controlled by three steel tangent bars equally spaced at 120 degrees, which connect the outer shell to the mirror.

The effect of gravity on mirror figure is controlled by supporting the weight of each mirror by 52 low rate springs arranged in a grid pattern. The force of each spring is adjusted to a level that is calculated to minimize mirror deflection. The spring forces are trimmed at assembly to provide minimum net reaction at the three position control points to avoid local figure effects. The number of springs supporting each mirror is more than required for adequate figure control. To enable in situ measurements of the mirror figures, ports in the aluminum shell were provided which give access to the mirror centers of curvature.

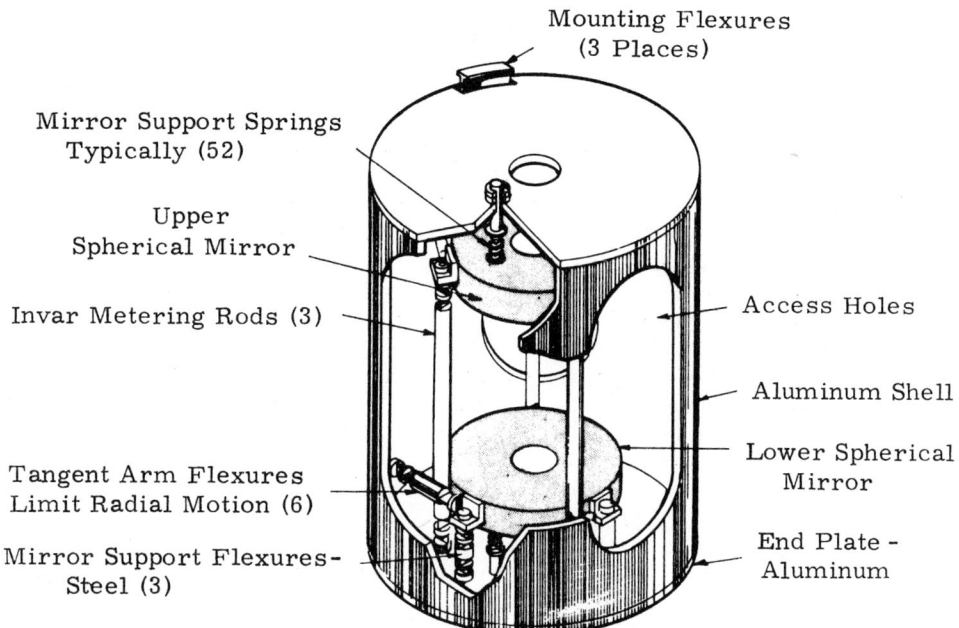

Fig. 1. Null Corrector Cell Design

REFERENCES

1. L.R. Heintze, H.D. Polster, and J. Vrabel, Appl. Opt. <u>6</u>, 1924 (1967).

2. A. Offner, "Reflecting Null Correctors for Conicoid Mirrors", paper presented at Optical Society of America Spring Meeting, April 25, 1975, Washington, D.C.

AAS 75-178

TEST RESULTS ON HOMOGENEITY OF EXPANSION FOR A
1.8-M ULE LIGHTWEIGHT MIRROR

G. Friedman and G. Gasser*

Interferometric tests were performed on a 1.8-m ULE lightweight mirror at different uniform temperature conditions to determine figure changes due to expansion coefficient inhomogeneity. The conclusions of the experiment were: (1) the rms value of surface figure changes due to material inhomogeneities were found to be 0.036λ over a 42°C soak range at 6328 Å. (2) Time variations of surface figure change were found to be insignificant over periods of up to 6 days. (3) No hysteresis remained after returning to figuring temperature (the measurement uncertainty was less than 0.023λ rms at $\lambda = 6328$ Å). (4) Analytical predictions of surface figure change based on α inhomogeneity data were in close agreement with measurements.

INTRODUCTION AND OBJECTIVES

Interferometric measurements of figure changes caused by inhomogeneity in expansion coefficient for various low expansion mirror substrate materials were performed most recently by Paquin and Goggin[1] and in the past several years by Goggin et al.[2] and Bloxsom et al.[3] These measurements were obtained on samples of less than 25 cm in diameter and, therefore, could not account for material variations in a large optical element and the technique used to fabricate lightweight elements. It was the intent of this experiment to measure the over-all optical effects due to inhomogeneity in expansion coefficient in a large lightweight optical element. The element tested was fabricated of Corning code 7971 ULE fused silica.

The specific objectives of the program were to (1) measure the effect of material inhomogeneities on mirror surface figure under uniform temperature changes; (2) determine the hysteresis level remaining after temperature cycling; (3) verify the analytical model used to predict surface figure changes.

Testing was conducted in two parts. To accomplish the first objective during part 1, the mirror temperature was uniformly changed from 24°C to -18°C and interferograms compared at both conditions. During part 2, data were obtained at an intermediate temperature level, -1°C, and apparent figure-time changes observed during part 1 were further investigated. To

---

*The authors are with Itek Corporation, Optical Systems Division, Lexington, Massachusetts 02173

accomplish the second objective during part 1, interferograms were taken on return to ambient and compared with initial interferograms. Data after cycling between 24°C and −18°C were obtained during part 2, in addition to obtaining data after temperature cycling between −12°C and 55°C. To develop understanding of the mechanism of figure change, a structural model of the test mirror incorporating local values of coefficient of expansion was developed. This model is described in Sec. III.

CONCLUSIONS

The conclusions of the experiment were: (1) The rms value of surface figure changes due to material inhomogeneities were found to be equal to 0.036λ over 42°C at 6328 Å. (2) Time variations of surface figure change were found to be insignificant. (3) No hysteresis remained after returning to figuring temperature. (The measurement uncertainty was less than 0.023λ rms at 6328 Å. (4) Predictions of surface figure change were in close agreement with measurements (within ±20%).

ANALYTICAL MODEL

An integrated thermal/structural/optical analytical tool developed at Itek was used to predict surface figure changes due to material inhomogeneities. The major element in this tool is an 880 node EASE[4] structural model of the test mirror. In addition to standard materials properties, local values of the coefficient of expansion for material comprising the faceplate, backplate, core, and ring stack were determined on the basis of edge measured values and estimates of manufacturing variation of expansion coefficient by Corning Glass Works.

Edge measured values of the coefficient of expansion for the faceplate, backplate, core, and ring stack are shown in Table I. A radial variation of coefficient of expansion in any boule was estimated to be 0.035×10^{-6} /°C (lower at the center than at the edge). This was attributed to oven temperature variations during the early production of ULE. Recent improvements in over design have reduced this variation to 0.015×10^{-6} /°C. No circumferential

Table 1. Measured Expansion Coefficients ULE Lightweight Mirror

	Edge measured α, in./in. °C $\times 10^{-6}$
Front faceplate	0.000
Back faceplate	−0.017
Core and ring stack	−0.009
	−0.003
	−0.017
	−0.010
	+0.018

variation in the coefficient was found to exist in a boule. Random distribution of this material property was assumed for the core. Surface deflections from the EASE model were analyzed and restated in terms of residual rms surface errors after removal of phase and tilt coefficients and change of focus. Analytical results that were obtained prior to collecting test data are further discussed in Sec. IX.

TEST CONFIGURATION

The test configuration is shown schematically in Fig. 1. The mirror was positioned on a three point kinematic mount. The mirror (Fig. 2), a monolithic ULE lightweight structure, was 1.83 m in diameter with an overall height equal to 30.0 cm. The core consisted of cells 7.62 cm square. The faceplate and backplate thicknesses were 2.09 cm and 1.20 cm, respectively. The surface figure prior to testing was approximately 0.1λ rms over 90% of the clear aperture.

TYPICAL RESULTS

The values of the surface contours were obtained, as was pointed out, by averaging the surfaces obtained from several interferograms. For each point comprising the mean difference contour, the standard deviation of the mean was computed. By examining the magnitude of the apparent surface changes compared to the uncertainty of measurement, on a point-by-point basis, the significance of the results may be appreciated.

Figure 3 compares the mean profile of the mirror difference surface, across one diameter, for two conditions. The profile for each condition is plotted along with a curve bounding twice the standard deviation determined for each point. The probability that the real surface difference lies between the two boundary curves is 0.95 and only 0.05 that the real surface lies outside the band.

COMPARISON OF MEASUREMENTS WITH PREDICTIONS

Measurements and predictions of the change in surface figure under uniform temperature changes are shown in Fig. 4. Calculations were made both for a radial variation in expansion coefficient within the boules from which the mirror was fabricated equal to $0.035 \times 10^{-6}/°C$ (early material) and $0.015 \times 10^{-6}/°C$ (current material). A random distribution of material within the core was assumed. Due to the uncertainty involved in knowing the actual distribution of expansion coefficient within the mirror, agreement between predictions and measurements within ±50% would have been considered reasonable. The data are in good agreement with predictions and further are grouped about the upper line (within ±20%) which is representative of the material variations typical at the time material for this mirror was produced. Data taken at an intermediate temperature of $-1°C$ were also in reasonable agreement with predictions.

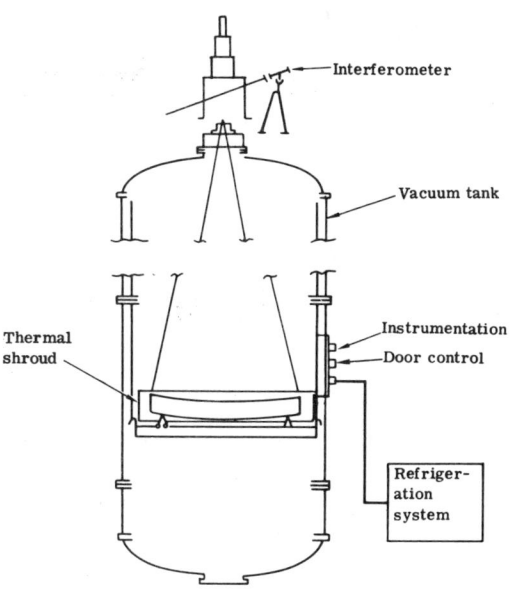

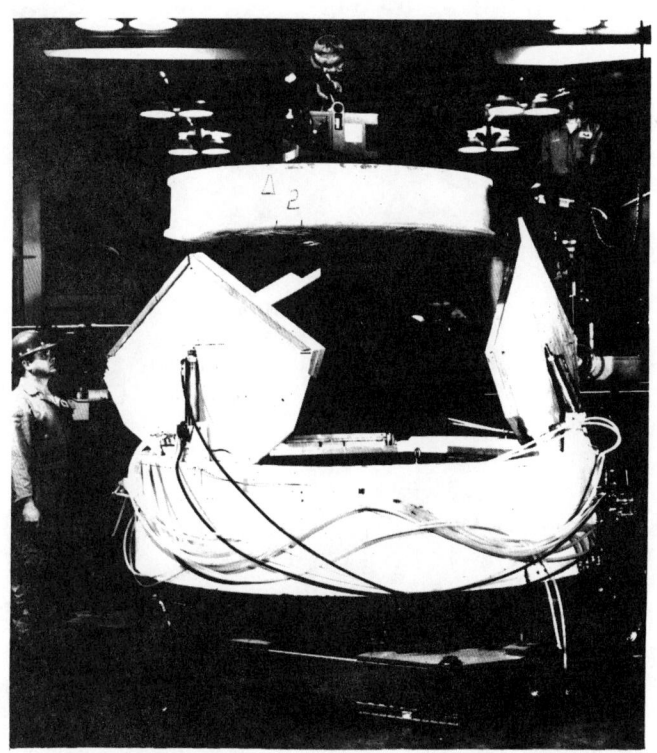

Fig. 2. Test mirror installation into thermal shroud.

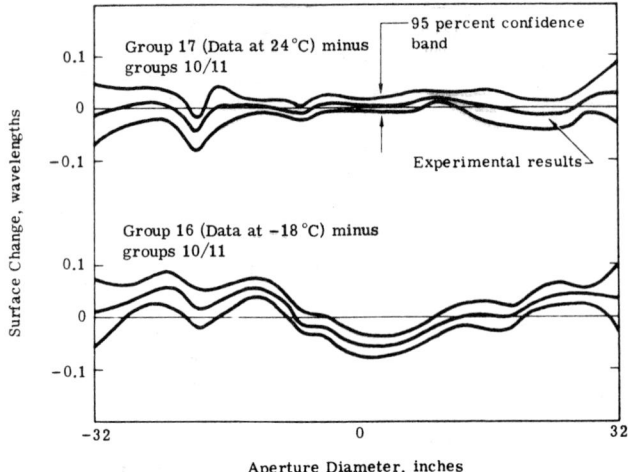

Fig. 3. Comparison of mirror difference surface mean profiles across one diameter for two test conditions.

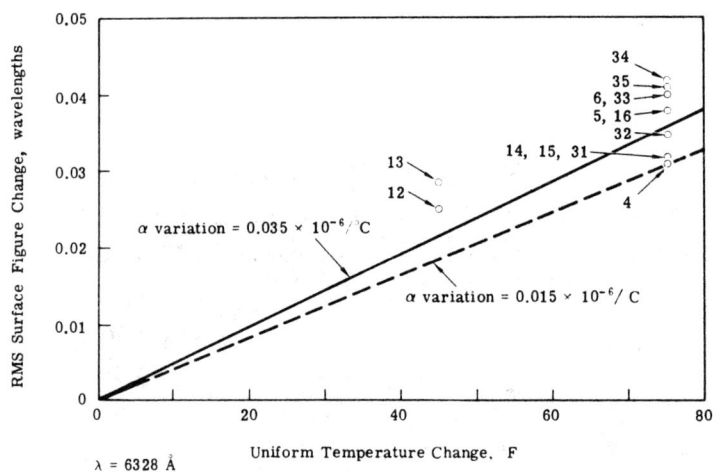

Fig. 4. Uniform temperature change comparison of predictions and measurements.

REFERENCES

1. R. A. Paquin and W. R. Goggin, Perkin-Elmer Report 10657, Contract NAS5-11327 (May 1971).

2. W. R. Goggin et al., Perkin-Elmer Report OOD-41, Contract DAAH-01-68-C-0018 (February 1970).

3. J. T. Bloxsom et al., Perkin-Elmer Report 9383, DDC AD 842-227, Contract DAAH-01-68-C-0018 (September 1968).

4. EASE is a structural analyzer computer program developed by the Engineering Analysis Corporation.

AAS 75-202

SCIENCE PERFORMANCE CONSIDERATIONS
FOR THE DESIGN OF THE LST

Damon D. Ostrander [*]

James C. Tuttle [†]

INTRODUCTION

Any attempt to quantify science performance necessarily hinges on using appropriate and valid performance criteria. For the purposes of this paper a combination of several performance criteria are used to describe the quality and quantity of the data which LST will provide. Computerized mathematical models of the LST optical system (including detectors) are used to evaluate these performance measures. These models are analytical in nature as opposed to the ray tracing approaches used for optical design and assume that distortion and aberrations of the image are negligible. These assumptions are valid for the purposes of preliminary design and greatly reduce computational costs. All of the performance measures are based on a determination of the optical Point Spread Function (PSF) and the Modulation Transfer Function (MTF). These performance criteria are used to evaluate the effect of pointing instability, wavefront error and aperture size on total performance. Three aperture sizes are evaluated each at different levels of performance. Combinations of wavefront error and pointing instability which give performance at each level are determined.

METHOD

Performance Criteria

A summary of the performance measures chosen and their salient characteristics is provided in Table 1. The measures are of two basic

---
[*]Engineer, Mission and Operation Analysis Section, Martin Marietta Corporation, Denver, Colorado

[†]Senior Engineer, Mission and Operation Analysis Section, Martin Marietta Corporation, Denver, Colorado

Table 1

PERFORMANCE MEASURE CHARACTERISTICS

PERFORMANCE MEASURE	DEFINITION	MEASURE TYPE			PARAMETERS					
		LIGHT SENSITIVITY/ EXPOSURE TIME	SPATIAL RESOLUTION	APERTURE DIA.	SECONDARY MIRROR OBSCURATION	DETECTOR CHARACTERISTICS	OPTICAL WAVE FRONT ERROR	LOS POINTING STABILITY		
FAINT OBJECT SENSITIVITY	LIMITING m_v ATTAINABLE TO S/N = 10 IN 4 HRS IN DIFFRACTION DISK CENTRAL MAXIMA	+	–	+	–	–	+	–		
50% MTF RESPONSE	SPATIAL PERIOD IN ARC-SEC WHICH GIVES 50% MTF RESPONSE		+	+	–	+	+	+		
CONTRAST (RAYLEIGH CRITERION)	SEPARATION IN ARC-SEC OF TWO IDENTICAL STARS WHICH GIVES AN IMAGE CENTER INTENSITY 26.5% LESS THAN THE PEAK INTENSITY		+	+	+		–	+		
ENCIRCLED ENERGY	ANGULAR RADIUS IN ARC-SEC CONTAINING 60% OF TOTAL ENERGY ENTERING APERTURE	–	+	+	+		+	–		
FULL WIDTH HALF INTENSITY (FWHI)	PSF DIAMETER AT HALF INTENSITY POINT IN ARC-SEC		+	+	–		+	–		
SPECTRAL SENSITIVITY	SPECTROGRAPH LIMITING m_v ATTAINABLE TO S/N = 2 IN 4 HRS IN CENTRAL MAXIMA OF MONOCHROMATIC LINE IMAGE, SLIT = 0.1 X 1 ARC-SEC	+	–	+	–	–	+	–		

NOTE: + = HIGH SENSITIVITY (MAJOR FACTOR)
– = LOW SENSITIVITY

types: measures related to the light sensitivity or "speed" of the telescope and therefore related to the quantity of data obtainable in a given time, and measures of the spatial resolution of the telescope which relates to the quality of data obtainable. Some parameters of the models used for performance measure evaluation are shown in the table, and their influence on the performance measures is rated on a high vs low sensitivity basis. The large variation in sensitivity to telescope parameters between the various performance measures emphasizes the need to understand and choose suitable measures.

RESULTS

Aperture Size Performance Comparison

The LST science performance measures were used to evaluate the effect of aperture size on mission performance. The data in Table 2 shows a performance comparison of three aperture sizes, each with an rms pointing error of 0.005 arc seconds and an rms wavefront error of $\lambda/20$. For this comparison the performance measure of Faint Object Sensitivity was evaluated using a mathematical model of an f/24 camera with an SEC Orthicon detector. The 50 percent response is determined for an f/96 camera with an SEC Orthicon detector. Spectral sensitivity is determined for a faint object spectrograph using a slit width of 0.1 arc seconds, an SEC Orthicon detector and a spectral resolution $[\frac{\lambda}{\Delta\lambda}]$ of 1600. An equal

Table 2

PERFORMANCE COMPARISON

Performance Measure	Aperture Diameter		
	1.8 Meter	2.4 Meter	3 Meter
Faint Object Sensitivity (Visual Magnitude Limit)	26.8	27.4	27.8
Encircled Energy (Within 0.075 sec Radius)	63%	66%	76%
Contrast Ratio (Separation = 0.1 sec)	53%	88%	96%
Full Width Half Intensity	0.072 sec	0.054 sec	0.044 sec
Spectral Sensitivity	21.2	22.2	23.0

weighted sum of these measures shows that from a science data aspect a 2.4 meter telescope achieves about 77 percent of the total performance achieved by a 3 meter system and a 1.8 meter telescope about 53 percent.

For each aperture, a selection of pointing stability and optical quality requirements can be made based on science related performance data. For example, in Fig. 1 curves are presented which indicate values of wavefront error and pointing error which give performance equivalent to a 3 meter near diffraction limited (NDL) system. The shaded region represents the combinations of design values that would result in a system capable of meeting 3 meter NDL performance requirements for all the indicated performance measures. The measures that drive the design requirements are observed as the boundaries of the shaded region. A cost effective system can be found by selecting the minimum cost point on the boundary.

The same concept can be used to compare different aperture size systems. Fig. 2 shows the boundaries of the acceptable design regions for 3 meter, 2.4 meter, and 2 meter NDL system. The dotted curves indicate the extent that a larger diameter system can be degraded and still reach the NDL performance of a smaller system.

CONCLUSIONS

Improvement of line of sight pointing stability and optical quality (wavefront error) provides limited improvement of telescope performance as compared to improvement obtained by increasing aperture size. A

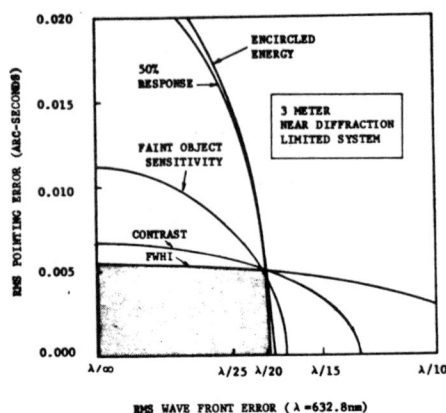 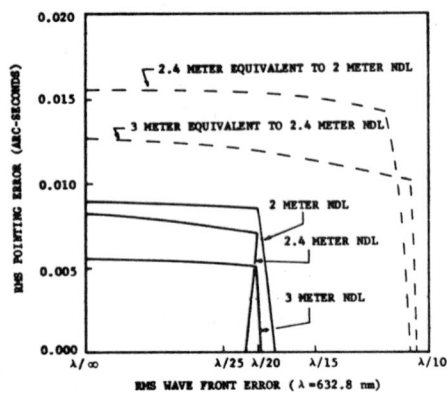

Fig. 1 Equivalent Performance Design Parameters

Fig. 2 Equivalent Performance with Different Aperture Diameter

large aperture system of modest quality can easily outperform a diffraction limited system of smaller aperture size. A large aperture allows design margins and relaxed tolerances which may result in lower total cost to get performance comparable to smaller systems. If Faint Object Sensitivity and Spectral Sensitivity are taken as performance measures of more than average importance, then the sensitivity of performance to aperture diameter is even more pronounced. A cost effective combination of optical quality and pointing stability should be chosen based on data relating the effects of each on total performance as provided in this paper.

AAS 75-203

OPTICAL PERFORMANCE CONTROL

Terence A. Facey *

To achieve maximum scientific utilization of the LST, the optical performance of the system must be essentially equal to the theoretical optimum. To ensure this, LST will be equipped with an optical performance control subsystem that will monitor the optical performance of the telescope periodically and provide corrective data to ground controllers.

The sensing elements of the performance control subsystem are the Focal Plane Wavefront Sensors, three of which are required. A plane wave from a stellar source entering the telescope is aberrated by imperfect optical surfaces and misalignment of optical elements. The function of the wavefront sensor is to quantify the aberrations in this wavefront and telemeter to the ground station a set of pupil optical path difference (OPD) maps for the telescope. On the ground, these pupil OPD maps will be used to generate the error signals that can be used to realign the secondary mirror and, if necessary, correct the figure of the primary mirror.

A set of general control equations has been developed that allows separation of the secondary mirror position errors in the presence of small primary mirror figure errors. The residual errors due only to the primary mirror, are then available for figure correction use by actuators, or for post-exposure image processing.

A perturbation analysis was undertaken for the LST telescope optical design.

Motions of the secondary mirror in each of its five significant degrees of freedom (roll about the optical (Z) axis is of no consequence) are manifest as changes in five aberration coefficients. Thus, secondary mirror position is potentially describable by a set of five simultaneous linear equations.

---
*Perkin-Elmer Corporation, Norwalk, Conn.

Such a set of equations may be expressed, in matrix form as

$$\left[M_{ij}\right]\left[\varepsilon_j\right] = \left[A_i\right] \quad (1)$$

where

M_{ij} are the influence coefficients

ε_j are the alignment perturbations

A_i are changes in the aberration polynomial coefficients

Numerically, then, Equation (1) becomes, for the LST,

$$\begin{bmatrix} -0.24 & -0.24 & -23.34 & 1.67 & 1.67 \\ 0 & 0 & 0 & -0.08 & 0 \\ 0 & 0 & 0 & 0 & -0.08 \\ 0 & -2.57 & 0 & 1.56 & 0 \\ -2.57 & 0 & 0 & 0 & 1.56 \end{bmatrix} \begin{bmatrix} x \\ y \\ z \\ \theta \\ \phi \end{bmatrix} = \begin{bmatrix} A_4 \\ A_5 \\ A_6 \\ A_7 \\ A_8 \end{bmatrix}$$

The determinant of the matrix is 0.9865 and its elements are in units of microns aberration per millimeter perturbation (or micron per milliradian). This equation may be inverted (Equation (2)) to solve for the secondary mirror errors, given the aberration polynomial coefficients deduced from the OPD map of the telescope pupil, as measured at a single point in the field:

$$\begin{bmatrix} 0 & 0 & -7.59 & 0 & -0.39 \\ 0 & -7.59 & 0 & -0.39 & 0 \\ -0.043 & -0.816 & -0.816 & -0.004 & -0.004 \\ 0 & -12.5 & 0 & 0 & 0 \\ 0 & 0 & -12.5 & 0 & 0 \end{bmatrix} \begin{bmatrix} A_4 \\ A_5 \\ A_6 \\ A_7 \\ A_8 \end{bmatrix} = \begin{bmatrix} x \\ y \\ z \\ \theta \\ \phi \end{bmatrix} \quad (2)$$

Thus, if we assume a perfect (or at least statically imperfect) primary mirror, the secondary mirror may be correctly aligned in all of its five degrees of freedom from the information contained in a pupil OPD map measured by a single wavefront sensor located off-axis at a known field position in the telescope focal plane.

The ability to separate tilt and decenter components of the secondary mirror misalignment is dependent on the ability to measure small amounts of astigmatism. However, for misalignments that cause astigmatism of amounts below the measurement limit of the wavefront sensor, it is practical to correct the coma by either tilt or decenter adjustment - whichever is most easily accomplished. This obviously assumes that the sensor is more sensitive to astigmatism than any other instrument in use.

The wavefront sensors are located toward the edge of the scientific data field and therefore 'see' more astigmatism than do scientific instrument entrance apertures.

More generally, Equation (1) is written

$$[A_i(h)] = [M_{ij}(h)] [e_j] + [P_i] \qquad (3)$$

where $M_{ij}(h)$ expresses the field dependence of M_{ij} in Equation (1), and P_i are the aberration polynomial coefficient changes due to primary mirror distortions.

Equation (3) then, suggests that the field independent nature of the primary induced coefficient changes will allow them to be separated from A_i's to yield secondary mirror position errors uncorrupted by primary mirror effects, if measurements are made at more than one field position.

The odd field dependence of misalignment induced aberrations suggests the use of two orthogonal locations in the field. The focus error in the wavefront introduced by axial (Z) misalignment of the secondary is also field independent, and therefore yields a singularity in the 5 x 5 matrix of $[M_{ij}(h_1)] - [M_{ij}(h_2)]$.

Elimination of focus information (A_4) and the axial degree of freedom (Z) from Equation (3) allows elimination of $[P_i]$ to yield secondary position errors in x, y, θ, ∅ uncorrupted by primary mirror effects.

$$[A_i(h_1)] - [A_i(h_2)] = \{[M_{ij}(h_1)] - [M_{ij}(h_2)]\} \{e_j\} \qquad (4)$$

The secondary mirror position control equation becomes:

$$\begin{bmatrix} 3.79 & -3.79 & 1.94 & -1.94 \\ -3.79 & -3.79 & -1.94 & -1.94 \\ -6.25 & -6.25 & 0 & 0 \\ 6.25 & -6.25 & 0 & 0 \end{bmatrix} \begin{bmatrix} A_5' \\ A_6' \\ A_7' \\ A_8' \end{bmatrix} = \begin{bmatrix} x \\ y \\ \theta \\ \phi \end{bmatrix} \quad (5)$$

in which the A_i' are the differences $A_i(h_1) - A_i(h_2)$.

Thus the secondary mirror may be controlled in 4 degrees of freedom (everything except focus) without confusion from primary mirror effects, by two figure sensors located at equal field heights but disposed at right angles to one another in the focal plane.

The focus error signal remains a combined error resulting from secondary despace ΔZ, primary mirror radius of curvature change Δr, and small contributions resulting from decenter and tilt errors.

These effects may be separated by observing changes in the focus coefficient, A_4, and in the 3rd order spherical aberration coefficient, A_{11}, and by addition of the measured coefficient changes from two diametrically opposite field positions.

In this way, small corrupting effects of focus errors due to secondary tilt or decenter are removed due to their odd field dependence, allowing the focus errors due to secondary axial position error (ΔZ) and primary radius of curvature error (Δr) to be separated by a pair of simultaneous equations.

$$\begin{bmatrix} A_4(h) \\ A_{11}(h) \end{bmatrix} + \begin{bmatrix} A_4(-h) \\ A_{11}(-h) \end{bmatrix} = 2 \begin{bmatrix} P_{11} & P_{12} \\ P_{21} & P_{22} \end{bmatrix} \begin{bmatrix} \Delta Z \\ \Delta r \end{bmatrix} \quad (6)$$

In the event, and this is the case for LST, that the system thermal and structural designs are such that no significant pure radius of curvature change of the primary is likely, then this last procedure is unnecessary, and all focus error may be attributed to secondary despace and corrected accordingly.

IMPACT OF FOCAL PLANE DYNAMICS ON IMAGE QUALITY

William J. Pragluski, Peter W. Abbott,
Jack F. Eastman*

INTRODUCTION

The LST has a large aperture telescope which has the potential for (near) diffraction limited performance. This performance potential provides the possibility of very high resolution photography, photometry, and spectroscopy as well as viewing of very faint objects, well beyond the performance of the largest ground based telescopes.

To achieve this potential, the LST's pointing and stability requirements are quite precise. A pointing accuracy of 0.01 arc-second and stability of 0.007 arc-second are design goals, compared to a diffraction limited image size of 0.13 arc-second (i.e., 37 μm at the focal surface). Even the smallest microvibrations of the telescope structure can induce apparent image motions at the focal surface which are of the same order of magnitude as the image size. If true, this would seriously impact the image quality and, hence, the performance of the LST scientific mission.

The parameter used herein to compare the effects of focal plane dynamics on image quality is the size of the "blur circle." This parameter defines the area over which over 400 rays, passing through the telescope assembly, are splayed at the focal surface over and above the normal diffraction image size. Although other criteria, such as rms wavefront error, Strehl ratio, or modulation transfer function were evaluated, the use of the blur circle parameter and its impact on the image quality is the easiest to visualize.

---

* Martin Marietta Corporation, Denver, Colorado

STRUCTURAL DYNAMICS ANALYSIS

The precise pointing requirements of LST require an accurate mathematical model of the structure if the impact of microvibrations at the focal surface are to be reliably predicted. The system disturbances are both known (i.e., reaction wheel static and dynamic imbalance, internal moving parts, etc.) and undefined (i.e., bearing noise, thermal creaking, etc.).

Mathematical Model

Structural mass, stiffness, and damping are critical characteristics in the analysis. The model developed is broken into three basic structural elements, Support Systems Module (SSM), Optical Telescope Assembly (OTA), and appendages. Each of these elements are further broken down into several substructures and are combined using the technique of modal coupling with appropriate boundary conditions. The coupled system includes 177 degrees of freedom and 130 modes below 100 Hz.

OPTICAL PERFORMANCE ANALYSIS

Optics Model

The telescope modeled in this analysis is a standard 2.4m two-mirror Ritchey-Chretien design. Both mirrors are figured as hyperboloids. The system is designed to be free of spherical aberration and coma. The image quality in the field is limited by astigmatism.

Optical Performance Analysis: Sensitivities

The on-axis sensitivity data for defocus is approximately linear with displacement. The slope is nearly equivalent to the first order approximation of (SM magnification minus one) squared. Using an (arbitrary) criteria that the blur circle diameter should be less than 1/10 the Airy disc diameter (37 μm) results in a maximum allowable deflection of 1.0 μm.

The on-axis sensitivity to decenter is somewhat lower than that for defocus (an order of magnitude). The sensitivity, over the range of decenter investigated, is very nearly proportional to (SM magnification) times the displacement.

The basic data at field positions of 80 mm and 135 mm show that the effect of astigmatism masks the impact of structural deflections until the deflections are greater than 2 to 5 µm. The blur circles, in either case, are of the same order in size as the Airy disc diameter. If the effect of this aberration were completely compensated for, the conclusion on the sensitivities would be the same as those described above for the on-axis results. However, physical limitations will play a part in the amount of compensation that can be accommodated. A representative case is shown on the figure, indicating that structural deflections begin impacting the image quality at approximately 0.25 µm.

The sensitivities to tilt of both the primary and secondary mirrors was evaluated over a range of tilts between 10^{-8} to 10^{-9} radians. The range was selected as representative based on a survey of the structural dynamics analysis results. The data shows that deflections of this order of magnitude do not impact the data quality. The effect of primary mirror tilts on the center-line image quality is just beginning to be felt at tilts of 10^{-8} radians.

Optical Performance Analysis: Composite

The structural analysis presented above results in the apparent star motions at the focal surface shown in Figure 1. These data show relatively small responses at low frequencies and excessive responses at around 30 Hz. The low frequency data (1 to 10 Hz) reflect the characteristics after the lower frequency elements (notably antenna booms and solar array structure) had been beefed up, using the weight margin available on Shuttle, as a result of earlier analyses. This frequency regime is critical

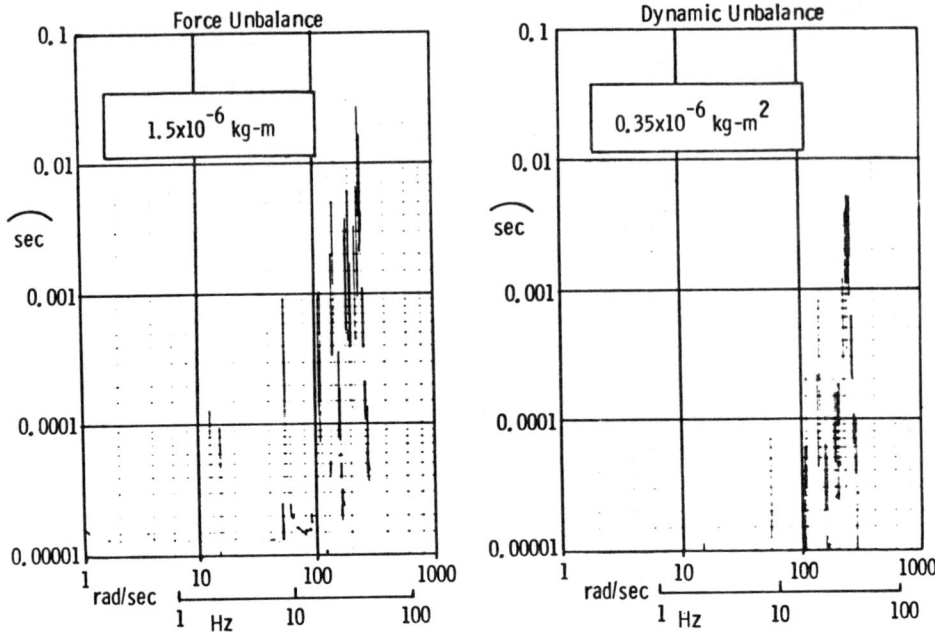

Fig. 1 *Image Motion Induced by Reaction Wheels*

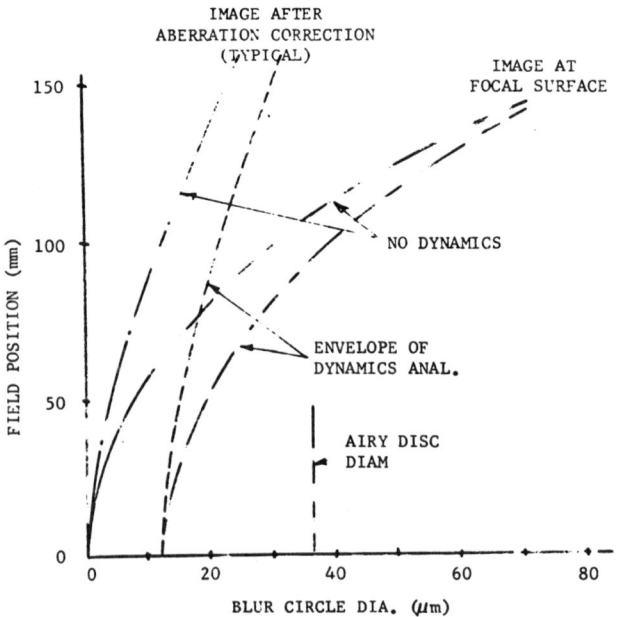

Fig. 2 *Blur Circle Diameter; Composite Dynamics*

from a vehicle control system interaction (stability) point of view and weight margin used there is well spent to minimize potential problems.

The excessive responses (compared to the 0.007 arc-second stability limit) near 30 Hz result primarily from the bending of the secondary mirror support structure from its' attachment near the primary mirror. A more detailed analysis is required to evaluate the possibility of alleviating these adverse responses by stiffening the structure. However, the probability of exciting these modes has been minimized by modifying the LST control law (momentum management) such that the reaction wheel speeds will be well below 30 Hz during science observations. The higher wheel speeds will only be used during large LST slew maneuvers, between targets, when no science data is being taken.

The composite effect of structural deflections on blur circle diameter is shown in Figure 2. These data show both the static (no dynamics) diameter as a function of field position as well as the envelope of the five dynamics cases (below 25 Hz) investigated. Also shown is the basic image quality at the focal surface both with and without typical compensation for aberration. In all cases, the degradation due to dynamics is a small, but significant per cent of the Airy disc diameter. The dynamic blur circle diameter delta, in all cases, represents the dispersion at the maximum structural deflection.

CONCLUSIONS

The analyses summarized above do indicate that the impact of focal place dynamics on image quality can be controlled to an acceptable level. However, this does require that this facet of the LST design be carefully monitored and controlled during the design phase.

AAS 75-205

STRAY LIGHT FROM OUT OF FIELD SOURCES
Robert J. Noll[*]

INTRODUCTION

Typically, there are three classes of stray light problems of concern:

a. Scatter from primary optics

b. Diffraction effects from various telescope edges

c. Specular and diffuse reflections from telescope walls.

In this paper an attempt to describe these effects in terms of a BRDF is presented.

MIRROR SCATTER

Stray light from mirror scatter is typically the result of high spatial frequency random irregularities in the surface profile of the mirror. With a plane wave incident upon the primary optic, the reflected amplitude in terms of the aberration function is well known and can be used to write the BRDF for normal incidence as

$$\text{BRDF}(\theta\varphi \mid 00) = \left(\frac{k}{2\pi}\right)^2 \frac{1}{A_o} \left| \int\int_{A_o} d\vec{r}\, e^{i[k\Phi - \vec{k}\cdot\vec{r}]} \right|^2 \tag{1}$$

where the quadratic Fresnel phase factor, as well as all dependence on the dielectric constant of the mirror, has been neglected.

When the aberration spatial frequencies are large (mirror scatter), it is convenient to consider a statistical representation for the scattered light. The average scattered intensity can be written as

$$<I> = <U>^2 + <|U - <U>|^2> . \tag{2}$$

---

[*] Perkin-Elmer Corporation, Norwalk, Connecticut

EDGE DIFFRACTION

Diffraction as a stray light problem in telescopes was studied many years ago by B. Lyot. He found that diffraction effects could be minimized by introducing a stop at the image plane of the diffracting edge. An analytic evaluation of the efficiency of such a stop has been given by the author. The Lyot stop is effective because diffraction comes only from edges. At an image plane of a circular entrance aperture, the out of field source produces a bright ring image. The energy in this ring is the diffracted energy that reaches the focal plane.

At Perkin-Elmer, extensive analysis of edge diffraction has been carried out, using a general unwanted energy rejection program written that not only traces rays reflected from various telescope surfaces, but also the diffracted rays. The conclusion of this analysis is that typically there are only a few points on an edge that determine the magnitude of the diffracted light reaching a detector. This fact greatly reduces the amount of computer ray tracing required to compute diffraction effects. The edge points act as sources of astigmatic rays which are the diffracted rays. To calculate the diffraction BRDF, the points on the edge which give rise to diffraction must first be determined and then the strength associated with each point.

SPECULAR AND DIFFUSE REFLECTIONS

Baffle fins are placed inside telescopes for two reasons: (1) to make specular rays undergo many more reflections before reaching the detector and (2) to reduced the illuminated area inside the telescope seen by critical surfaces. Generally, baffle fins make specular reflections from walls a negligible contributor of total stray light. In a well baffled system, the primary cause of stray light will result from either diffraction or diffuse reflection from the baffle fin edges. Diffraction tends to dominate for diffraction angles less than 20 degrees and reflection for scatter angles greater than 30 degrees.

The first term in Eq. (2) describes specular reflection and the second, diffuse or scattered light. If the BRDF is taken as the average intensity scattered per unit of incident flux, it can be written as the sum of two terms, a specular and a diffuse term. The effects of non-normal incidence as well as dielectric constant effects are far from negligible. The physical importance of the dielectric constant of the mirror appears to be most significant when the incident light wavelength is in a range where mirror dispersion effects such as surface plasmons are important.

To obtain a qualitative understanding of mirror scatter, only the dependence on the autocorrelation will be considered. For this case the BRDF can be written for a circular mirror as

$$\text{Specular BRDF} = \pi \left(\frac{a}{\lambda}\right)^2 \left[1 - (2k\sigma)^2\right] \left[\frac{2J_o(ka\theta)}{ka\theta}\right]^2 \quad (3)$$

$$\text{Diffuse BRDF} = \frac{1}{2\pi} (2k\sigma)^2 (Tk)^2 F(kT\theta) \quad (4)$$

where

$$F(y) = \int_o^\infty x\, C(x)\, J_o(yx)\, dx. \quad (5)$$

The specular term is simply the aperture diffraction pattern modified by a factor that depends only on the rms height variations of the surface, and sometimes called the "strehl factor". The diffuse or scatter term depends on both the rms height and correlation lengths and is functionally the Fourier transform of the surface autocorrelation function. Effects such as polarization, index of refraction, and arbitrary angles of incidence have been left out, because these effects generally perturb the results only slightly.

Computer calculations at Perkin-Elmer show that sources less than 17 degrees off-axis illuminate the primary mirror, making mirror scatter the dominant stray light mechanism. From 17 to 30 degrees, diffuse reflection from the support struts is important; and from 30 to 90 degrees diffuse reflection from baffle edges followed by diffuse reflection from the primary mirror dominates. Stray light diffusely scattered to the detector from the secondary mirror, inside the secondary baffle, and inside the primary baffle are all negligible. The secondary baffle calculation was done with no baffle fins anywhere in the telescope. The scattered flux reaching the detector for this system from any given source is easily obtained by substituting the BRDF shown in Figure 1 into Eq. (1).

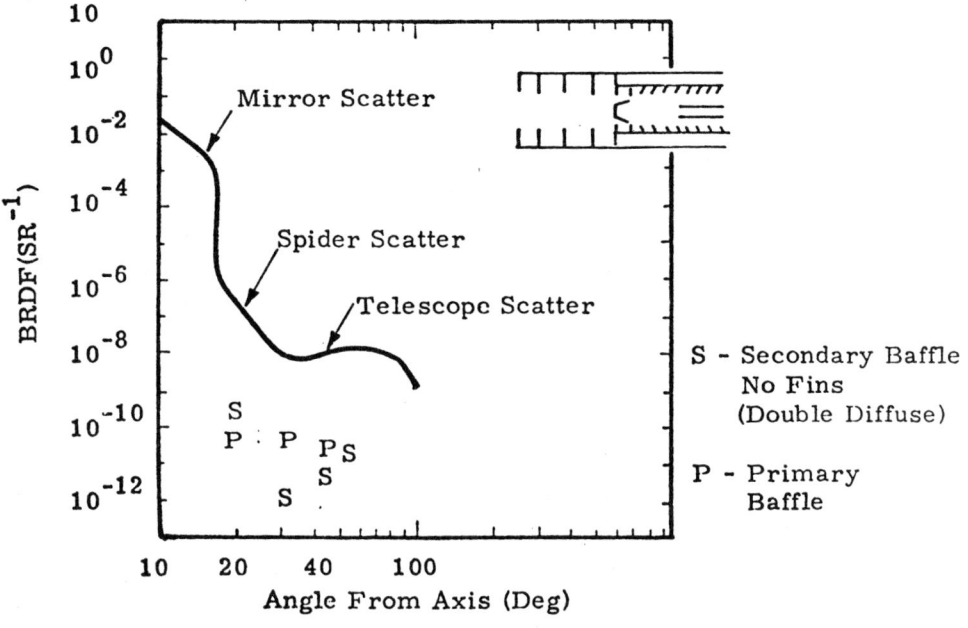

Fig. 1. Telescope BRDF

CONCLUSION

The specific concept of a BRDF as a surface reflectance characteristic has been extended to include general telescope stray light performance. It is shown that the BRDF is proportional to the telescope point spread function in the field, and describes scatter outside the field. It has been determined that the three basic classes of stray light can be described by a BRDF. In particular, the Perkin-Elmer GUERAP program allows BRDF computation for rather complicated but realistic telescope stray light problems. The program has been applied to a prototype LST design, and the results are given in Figure 1.

AAS 75-206

DESIGN OF HIGHLY STABLE
OPTICAL SUPPORT STRUCTURE

Michael H. Krim*

INTRODUCTION

As spaceborne optical systems increase in diameter to achieve improved resolution the stability requirements imposed on structures approach values which were unthinkable only several years ago. To achieve the capabilities of these apertures optical path errors must not exceed a specific fraction of the wavelength of light, and this fraction, typically $\lambda/20$ rms in the focal plane, is independent of system size. The Optical Telescope Assembly (OTA) is installed in the LST spacecraft. The vertex-to-vertex spacing of the primary and secondary mirror is 193 inches. To achieve satisfactory optical performance, this spacing must be maintained constant to a precision of $\pm 1\mu$ for observation periods up to 10 hours. During this time it may be necessary to alter the spacecraft attitude with respect to the sun, which, would change the temperature levels and gradients within the structures. It is believed that by exploiting the use of novel, tunable graphite-epoxy truss elements, the stringent alignment requirements can be satisfied with a nominally passive structure.

THERMAL LOADS

It will be shown that the thermal changes, which occur between factory and orbit, are not critical to system performance. But the thermal changes that can occur subsequent to arrival on station are of consequence. The thermal design envelope was developed from analyses of various vehicle pointing attitudes and heater system control accuracies. The envelope represents the maximum antipated change that might occur during a single target observation.

---

* Perkin-Elmer Corporation, Norwalk, Connecticut

ALIGNMENT REQUIREMENTS

The working depth of focus at the f/24 image plane is ±432μ at 0.63μ wavelength. By applying the simple lens-makers formula to the two mirrors successively, the following relationship is obtained relating defocus to changes in mirror radii and spacing.

$$\Delta f = \left\{ \begin{array}{l} \left[(2\ell_1 - R_1 + R_2) \frac{R_2}{2} + \left(\frac{R_1 R_2}{2} - R_2 \ell_1 \right) \right] \partial R_1 \\ \left[(2\ell_1 - R_1 + R_2) \left(\frac{R_1}{2} - \ell_1 \right) - \left(\frac{R_1 R_2}{2} - R_2 \ell_1 \right) \right] \partial R_2 \\ \left[(2\ell_1 - R_1 + R_2)(-R_2) - 2 \left(\frac{R_1 R_2}{2} - R_2 \ell_1 \right) \right] \partial \ell_1 \end{array} \right\} \div (2\ell_1 - R_1 + R_2)^2$$

With the above equation the tolerance on despace, $\partial \ell_1$, is developed. Based on the system thermal analysis, temperature induced defocus contribution of the optical elements alone is 175μ. Allowing for a focus sensing and correction (initial alignment) error of 100μ and another 50μ for growth of ℓ_2 (7.5) μ/F°) and location uncertainty of the four Science Instruments, 107μ remain for the effect of a change in ℓ_1, as $\partial \ell_1$ equals approximately 1μ.

DESIGN APPROACH

To calculate the truss axial expansion, an equation was developed relating the axial expansion of a truss bay (which can be summed over the total number of bays) to the ring and strut geometry.

$$\Delta S = \frac{1}{S} \left[\ell^2 \alpha_S T_S + R^2 \alpha_R (\cos\theta - 1) \right] \left(T_B + T_T \right)$$

From this equation, the truss member expansivities α_S and α_R can be determined.

This equation shows that α_R / α_S = 10.6 for a nominally balanced design and also that α_S = 0.01 ±0.01 ppm/F°. The fundamental problem facing the structural designer is twofold. One, the variability of α from a design nominal is on the order of ±0.05 ppm/F° for graphite-epoxy laminates and two, the expansivities at the various nominal temperatures along the truss length may be different.

To solve the variability problem, a means of adjusting the expansivity of the elements forming the structure is required. Ideally this adjustment should be performed in the post-cured condition so that process variables can be compensated. Further, the adjustment on tuning should not introduce additional process variables but should be a discrete operation. Finally, the tuning elements should not locally alter the thermal diffusivity of the member, otherwise transient behavior would be compromised.

The Dual-α Tunable Graphite-Epoxy Strut accomplishes these objectives. Here a single strut is constructed from a single graphite-epoxy system but with different laminate (i.e., layup) geometry in the left- and right-hand sections. On the left, for example, the laminate would be designed so that its expansivity, considering the process variables is $0 \pm 0.1 \times 10^{-6}$. To the right of the transition zone it would be $0 \, {}^{+0.1 \times 10^{-6}}_{-0}$. Obviously a net thermal expansion of zero could be achieved if the left- and right-hand sections could be proportional such that $\alpha_L \ell_L + \alpha_R \ell_R = 0$.

It is initially constructed oversize and trimmed after curing to satisfy the equation, $0 = \alpha_0 = \dfrac{\alpha_1 \ell_1 + \alpha_2 \ell_2}{\ell_0}$. Expansion measurements are made, using coupons cut from either end, to determine the as-cured values of α_1 and α_2 such that $\alpha_1 = -\dfrac{\alpha_2}{\alpha_1 - \alpha_2} \ell_0$. These coupon measurements are made at the nominal temperature the strut will operate at, thus solving the $d\alpha/dT$ problem.

The sensitivity of the overall truss assembly of off-nominal (non-zero) values of α may be assessed by use of the ΔS equation. This equation was also employed as a check on a finite element model of the truss which was then used to determine the deformations caused by more complex thermal conditions, such as axially varying side-to-side temperature gradients.

From the closed form solution, the change in length of each truss bay is given by

$$\Delta S = 74.75 \, \alpha_S T_S - 3.53 \, L_R (T_B + T_T) \text{ inches}$$

Note that if α_R is 10.6 α_S, ΔS is identical to zero when $T_S = T_B = T_T$ or the temperature change of the system is uniform or when $T_S = \frac{T_B + T_T}{2}$. This form of temperature desensitization of structures has previously been used in telescope structures before the use of composites. In those instances the geometry of the sturcture was configured to satisfy α ratio constraints for titanium/aluminum combinations. However, since this form of athermalization depends on the small difference of relatively large numbers (α in the case of metallic pairs) it is impractical except in those situations of precisely uniform soak or static gradient situations.

For the LST truss design under discussion here, the preceding equation can be used to examine the effect of laminate α tolerance on performance.

For a $\pm 0.05 \times 10^{-6}$ in/in/F° uncertainty or spread in the nominal strut (α_S) and ring (α_R) tolerances, the RSS despace error is 99×10^{-6} inches or 2.5μ. This is in excess of the 1μ budget.

If tuned struts are employed where the nominal strut uncertainty is ± 0.005 ppm/F° and the ring tolerance remains ± 0.05, the RSS uncertainty is reduced to 0.33λ. The wider ring tolerance is retained because

    a. The system is less sensitive to ring α uncertainties.

    b. The ring is more difficult to tune.

To accomplish uniform soak athermalization, as opposed to a zero-α approach

$$\alpha_R/\alpha_S = \frac{74.75}{2 \times 3.53} = 10.588$$

If, in this instance, $\alpha_S = 0.05 \times 10^{-6}$ and therefore $\alpha_R = 10.588 \times 0.05 \times 10^{-6} = 0.5294 \times 10^{-6}$, the change in length caused by the gradient change is nominally zero as expected. But, by applying a ±0.05 ppm/F° tolerance on α_R and α_S, a 2.1μ error in the first bay alone is possible. If the temperature change over a bay is not linear then the $T_S = \frac{1}{2}(T_T + T_B)$ relationship is invalidated and the α_R/α_S balance defeated. In this instance, low absolute values of α are demanded to ensure satisfactory performance.

Based on the results of temperature uncertainty and α tolerance and despace, it was concluded that the struts need to be constructed with a nominal α no greater than 0.012 ppm/F° and a dispersion not in excess of 0.01 ppm/F°. This implies individual parts measurements and as a logical extension, tuning. The Dual-α strut will accomplish this.

AAS 75-207

LARGE FORMAT SEC ORTHICON INTEGRATING
TELEVISION SENSOR FOR LST

John L. Lowrance*

The Large Space Telescope will employ a 70 mm format SEC type television camera tube as the data sensor for one or more scientific instruments. The characteristics of this tube are described. The current performance is characterized by an MTF of 50% at 20 line pairs per mm and a signal-to-noise ratio that is about one-half the theoretical based on the square root of the number of photoelectrons. The paper presents some astronomical results and the status of the environmental testing of the tube. Other topics such as radiation sensitivity and thermal design are also discussed.

Introduction

In the mid 60's NASA recognized that television type sensors would be very useful in space astronomy if they could be made to integrate for long periods of time with high photometric accuracy. A study completed in 1965 concluded that a magnetically focused SEC tube made by Westinghouse had the best chance of meeting these requirements. This choice was based on the almost indefinite storage capability of the SEC's potassium chloride target which also exhibited a gain of approximately 100 to overcome readout noise.

Through a series of SR&T grants and contracts a magnetically focused SEC tube with a 70 mm (51 x 56 mm) format has been developed for scientific photometric applications, and in particular, for the Large Space Telescope f/24 camera and High Resolution Spectrograph. The basic image sensor requirements for these two instruments and summarized in the following table:

---

*Dept. of Astrophysical Sciences, Princeton University

TABLE I

Instrument	Format	Spectral Response
f/24	2000 x 2000 pixels 50 x 50 mm format	S-20 on MgF_2
High Resolution Spectrograph	2000 x 2000 pixels 50 x 50 mm format	C_sI on MgF_2 Bi Alkali on MgF_2

SEC Orthicon

The SEC type television camera tube has been described in detail elsewhere and will only be summarized here[1,2]. The Westinghouse WX-32193, 70 mm magnetically focused SEC tube is shown in Figure 1. The image section is made of ceramic rings to make the tube more rugged. This also makes it possible to process various photocathodes in the tube using metal tabulations that can be pinched off after the photocathode is evaporated. Considerable design effort has gone into eliminating most of the magnetic Kovar parts normally used in making these type tubes.

The u.v. transmitting Magnesium Fluoride window is sealed to the tube's metal flange using a heavy gold foil that allows for the differential expansion of the MgF_2 and the metal flange.

Resolution

We are currently working on optimizing the image quality over the 70 mm format. The video signal corresponding to an image of a 20 cycles/mm on the 70 mm SEC tubes' photocathode is shown in Figure 2. Sixteen scan lines have been averaged together in a digital computer to reduce the noise and thereby allow more accurate measurements at high spatial frequencies where the statistical fluxuation in the number of photoelectrons per half cycle is a correspondingly greater percentage of the signal. And, of course, the spatial frequency response is lower. The MTF curve obtained in this way is shown in Figure 3.

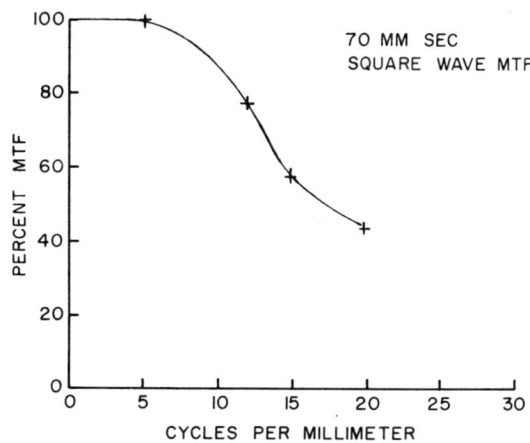

Fig. 1 Westinghouse WX-32193, 70 mm SEC Tube

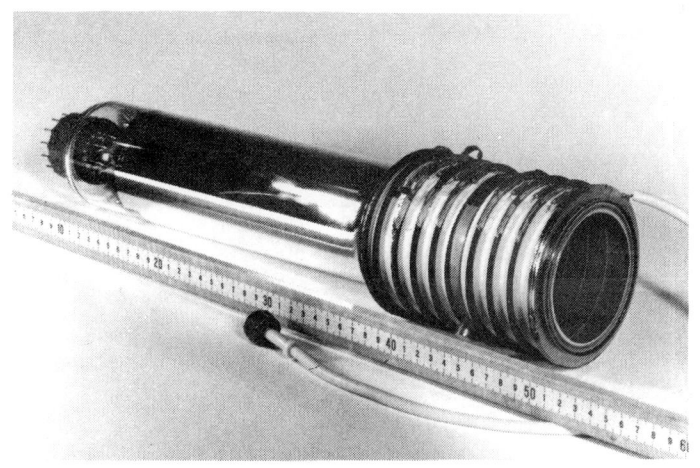

Fig. 2 Average of Video Signal from 16 Scans Across Resolution Test Pattern

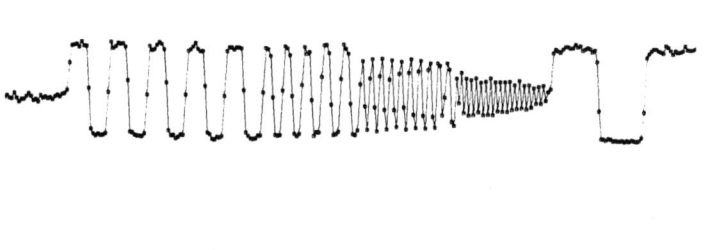

Fig. 3 70 mm SEC Square Wave MTF

Photometric Performance

The photometric performance of the SEC tube is of considerable importance in the LST mission.

Accurate photometric measurements are difficult to make and have been found to be very time consuming in dealing with a detector with such a large number of picture elements. Work has been going on for the past year to measure the stability of the photometric transfer function and to measure the signal-to-noise characteristics[3].

The following data has been obtained with the 35 mm format SEC tube but is expected to be the same for the 70 mm tube since the target characteristics are the same.

In an ideal tube, the signal S, defined as the total number of photoelectrons measured in a single pixel, would be related to the corresponding noise, N, by the question

$$N = S^{\frac{1}{2}} \qquad (1)$$

The tests show that within a square pixel, 50 microns on a side the noise level is increased above equation (1). Representative results are shown in Table II from a tube cooled to -50°C, but are representative of operating the tube at about -10°C. The dominant thermal effect is the photocathode dark current.

TABLE II

SIGNAL AND NOISE IN 50 PIXELS

S	79	156	312	625	1250	2500	5000	photoelectron/pixel	
N	20	26	35	50	70	122	265	"	"
S/N	4	6	8.8	12.5	17	21	19	"	"
$\frac{S^{\frac{1}{2}}}{S/N}$	2.2	2.1	2.0	2.0	2.1	2.4	3.7	"	"

As shown in the last row of Table II, the S/N ratio is evidently less than the ideal value predicted from equation (1). At low exposure the signal-to-noise is dominated by the readout noise, which is primarily the preamplifier noise. At intermediate exposures the signal-to-noise ratio S/N is about half the ideal value, presumably because of statistical fluxuations in the electron multiplication in the KCl target and some loss in photoelectrons due to reflection and absorption by the

Al_2O_3 target substrate and Aluminum signal plate. At high exposure the target begins to saturate the electrical signal per photoelectron decreases, the tube becomes noisier resulting in a decline in S/N.

The net result is a dynamic range from about 50 to 2500 photoelectrons per 50 micron pixel, with a S/N ratio over this range that is about half of its value for an ideal detector.

Astronomical Observations

The 35 mm format SEC tube WX-31718 has been used by Morton and Crane at Princeton for a number of ground based observations. There has been one Aerobee rocket flight carrying an u.v. echelle spectrograph with the 35 mm SEC as the data sensor. Unfortunately the optics became contaminated during launch but the SEC Camera worked during the flight and after the parachute landing.

Figure 4 shows the spectrum of a Seifert Galaxy NGC 1068 taken by Morton with the SEC camera on the Hall 200" telescope's Coude spectrograph. The spectrum is from 3800 to 4120 Å and is 27 arc seconds wide. This galaxy has strong emission lines as well as absorption. It is interesting to note the strong Calcium II absorption in the Seyfert galaxy due to the rotational velocity of the stars about the nucleus of the galaxy.

Environmental Tests

The LST Shuttle launch environment has been interpreted in terms of quantification tests levels for the subsystems as shown in Table IV.

TABLE IV

	Level	Range
Acceleration	33 g	
Acoustic	152 db	8 to 8000 Hz
Shock	57 g to 225 g	200 Hz to 400 Hz
	225 g	400 Hz to 1250 Hz
	225 g to 300 g	1250 Hz to 1600 Hz
	300 g	1600 Hz to 4000 Hz

Radiation Background

Trapped energetic charged particles and cosmic rays are of particular concern in sensitive electro-optical detectors because of the background

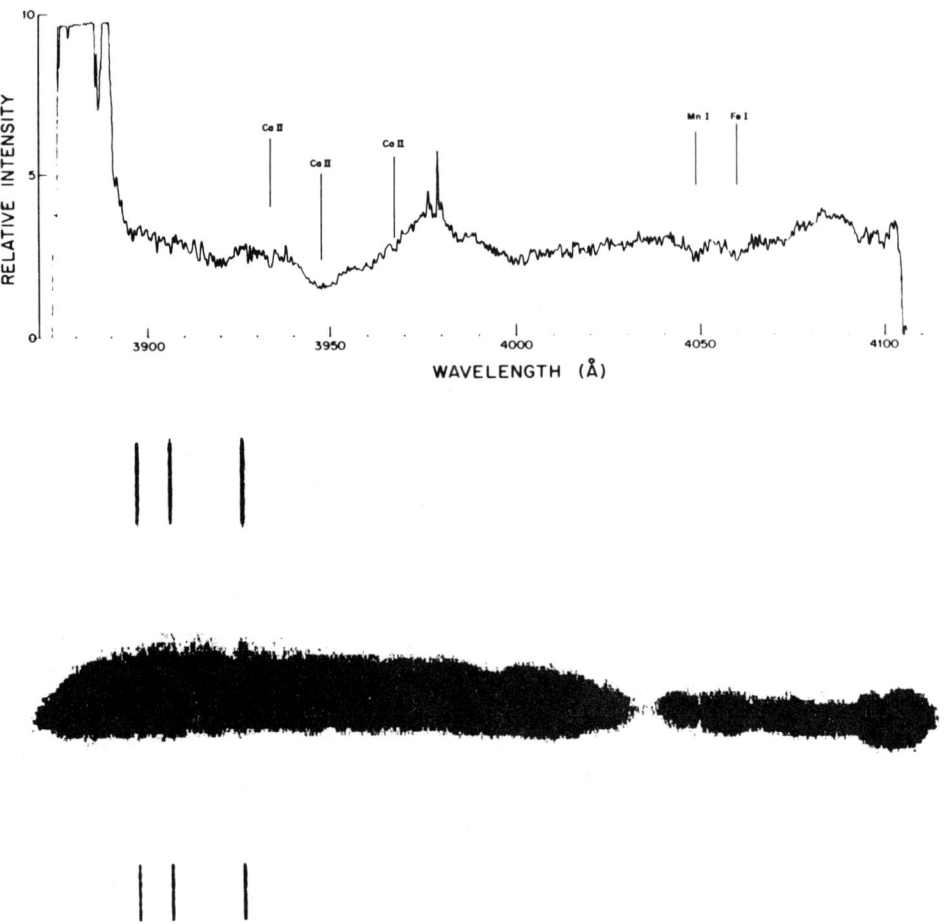

Fig. 4 Spectrum of NGC-1068 Madd with SEC Television Camera on Hale 200" Telescope's Coude Spectrograph

signal generated by these particles striking and/or passing through the windows and dynodes of the detector. Recent measurements by Viehmann, et. al.,[4] of the fluorescence of MgF_2 indicate that in the LST orbit one can expect about 100 photoelectrons cm^{-1} sec^{-1} from an S-20 photocathode.

This is comparable to the night sky surface brightness of 23 magnitude/arc sec^2 when imaged at f/24 through an optical filter 30 Å wide. Therefore, the radiation background will not be significant for most of the f/24 camera observations. The radiation background from the bi-alkali photocathodes planned for the High Resolution Spectrograph is expected to be less than 20% of the S-20 photocathode's radiation induced background due to the long wavelength insensitivity of these photocathodes. However, the radiation background will probably set the lower limit on the system sensitivity since, in this case, the spectral filtering on the night sky background is much greater.

Radiation background measurements on the SEC tubes over a wide energy range are planned for later this year to confirm these estimates.

Acknowledgements

Most of the results reported in this paper were obtained by Paul Zucchino and Don Long. Dr. D. Klinglesmith, Goddard Space Flight Center has been helpful in the video image processing. The astronomical observations of NGC 1068 were made by Dr. D. C. Morton of Princeton. The tubes were made under the supervision of Dr. J. Pietrzyk at Westinghouse. This work is supported by NASA Contract NAS-5-20833.

References

1. Goetz, G.W. In "Ad. Electronics & Electron Physics", Vol. 22A, pp. 219-227, Academic Press, New York (1966).
2. Boerio, A.H., Beyer, R.R., and Goetz, G.W. In "Ad. Electronics & Electron Physics", Vol. 22A, pp. 229-239, Academic Press, New York (1966).
3. Lowrance, J.L. and Zucchino, P. In "Methods of Experimental Physics", Vol. 12, Part A, Chapter 6.3, pp. 277-313, Academic Press, New York (1974).
4. Viehmann, W., Eubanks, A.G., Pieper, G.F., and Bredekamp, J.H. "Fluorescence and Phosphorescence of Photomultiplier Window Materials Under Electron Irradiation. GSFC X-Document X-755-74-210, July 1974; revised April 1975.

AAS 75-190

THERMO/STRUCTURAL DESIGN CONSIDERATIONS TO ACHIEVE THE LARGE SPACE TELESCOPE LINE-OF-SIGHT REQUIREMENTS*

Domenick J. Tenerelli[†]

The Large Space Telescope (LST) (Fig. 1), which is scheduled for launch in 1982, is a long-life precision-pointing, earth-orbiting satellite requiring structures that provide high dimensional stability, minimum thermal distortion and minimum response to on-board dynamic environments (e.g., reaction wheels). The LST must be developed, tested, and manufactured at the lowest system cost. This vehicle will be placed in orbit by the Space Shuttle, which will also be utilized for orbital servicing and return of the LST.

Some of the design, analysis, and test studies that have been conducted to meet the requirement goals that have been stipulated for the structure are described in this paper.

Fig. 1 Large Space Telescope

---

*This work was done under Contract NAS8-31313.
[†]Staff Engineer, Lockheed Missiles & Space Co., Inc., Sunnyvale, California

The LST is to be an astronomical facility developed by the National Aeronautics and Space Administration (NASA) under the direction of the George C. Marshall Space Flight Center (MSFC), Huntsville, Alabama. It will be designed as a national observatory, capable of utilizing a wide range of scientific instruments.

As currently configured, the LST will be an assemblage of three major modules — an optical telescope assembly (OTA), scientific instruments (SIs), and a support system module (SSM). The OTA will be a Ritchey-Chretien optical system, consisting of a 2.4-meter-diameter, clear aperture reflecting telescope with an f/2.3 (approximately) primary and a secondary mirror combination. The SI package will include field cameras, low- and high-dispersion spectrometers, and ultraviolet and infrared sensors to give the LST a broad capability for spectral analysis.

Located within the SSM will be the LST's very precise stabilization system for attitude sensing and control. NASA's development goal of 0.007 arc-sec guidance stability for periods of up to 10 hr is akin to keeping the view within the area of a dime nearly 400 miles away, or from Washington, D.C., to Boston. Such fine pointing accuracy is necessary to obtain full utilization of the OTA's optical capabilities. The SSM will also contain power and communications systems and provide environmental control and data management for the SIs. Electrical power for the system will be supplied with the aid of solar panels.

The orbiting telescope will measure approximately 13.13 meters (43 ft) in length and from 3.03 to 4.24 meters (10 to 14 ft) in diameter. This vehicle will circle the earth at an altitude of approximately 500 kilometers (270 nm) and at an inclination to the equator of 28.8 deg.

The LST will contribute significantly to studies relevant to the origin and structure of the universe, the study of energy processes that occur in galactic nuclei, the study of early stages of stellar and solar systems, and observation of such objects as supernova remnants and white dwarfs. The LST will be capable of viewing galaxies 100 times fainter than those seen by the most powerful ground-based optical telescopes.

The LST configuration was modeled for analysis by the SNAP (Structural Network Analysis Program) finite element computer code. The purpose of the analysis was to determine the effect of SSM thermal deformations on the alignment of the optical system. The early designs of the LST had the OTA/SI and SSM meeting at the main ring. It was assumed that if the in-orbit thermal deflections of the OTA/SI structure could be maintained to the listed requirements the performance of the LST would be guaranteed. A representative finite element computer model was developed depicting a design configuration that had the OTA/SI and the SSM joined at the main ring. To ensure the ability to differentiate thermal deformation effects of the SSM structure on the optical system from those of the OTA/SI structure, the OTA/SI structure was maintained at ambient conditions. The results of this detailed thermostructural computer analysis of the total LST structural system showed that an accumulated tilt and decenter misalignment of over 100 arc-sec will exist between the optical elements and the focal plane during certain orbit conditions. This amount of misalignment results in unacceptable performance output of the telescope. A solution to this situation is to design the OTA/SI structural system such that it is essentially isolated from the external structure; i.e., the SSM. Isolation of the optical system can be achieved by a three-point support with either flexure joints or spherical bearings. This type of design was also modeled for finite element computer analysis. Results of a thermal/structural deformation analysis showed that the optical system is unaffected by SSM thermal deformations when the LST structural system is designed with a thermal isolation joint between the OTA/SI and the SSM.

A strong candidate for the type of configuration that should be used as the main element of the metering structure that maintains the alignment between the primary and secondary mirrors is a ring-stiffened, thin-walled circular cylinder made from graphite epoxy. An analysis was carried out to determine the effect on the despace between the primary and secondary mirrors because of the incompatibility of displacements occurring at the ring-shell junctures of a ring-stiffened, thin-walled cylinder as the result of a uniform increase in temperature (5 deg F) in the rings and shell. The results indicate that the shell axial deformation is minimized as the CTE of the ring in the hoop direction is increased in the positive direction. This result is logical; i.e., when the hoop expansion of the ring approaches the hoop expansion of the shell, the normal forces between two components are minimized. This trend in turn reduces the Poisson's ratio effect. The results also showed that the CTE of the ring base should be close to the shell CTE in the axial direction, although a complete match at CTE = -0.04125 in./in.-deg F still causes an increase in shell length. This is due to the Poisson's ratio effect caused by the mismatch of the hoop deformations between the ring and the shell. The variation of the shell length with the adhesive thickness indicates that the axial deformation of the shell is significantly affected by the thickness of the adhesive layer. Large adhesive thicknesses result in large increases in the length of the shell as a direct result of the adhesive extensional stiffness increase.

Material and vibration studies conducted with respect to the LST are also discussed.

AAS 75-191

DESIGN OF LOW-THERMAL-DISTORTION LST METERING STRUCTURE

John R. Lager*

High-precision performance of the Large Space Telescope (LST) proposed for future astronomical observations from Earth orbit requires that the lens support structure remain dimensionally stable under variable environmental conditions. A fibrous structural material--continuous graphite filament--offers potential for design of structures with near-zero thermal distortion characteristics. This is possible because, unlike other structural materials, individual graphite fibers have a negative coefficient of thermal expansion (i.e., contract when heated) in their axial direction. Composite laminates using these fibers and epoxy matrix material can be used to design truss structures that remain dimensionally stable when subjected to temperature variations.

The LST metering structure shown in Fig. 1 was designed, and a representative section was fabricated and tested to demonstrate the potential of a near-zero thermal-distortion structure. The selected overall structural concept shown in Fig. 2 is a truss with individual members fabricated of continuous graphite fiber, epoxy resin, and fiberglass cloth. A zero-thermal-distortion truss structure is defined as one for which the node points at truss member joints do not move when the overall truss is subjected to gross thermal gradients. However, other points on the individual truss members can, and in general will, move. For example, the requirement for angle plied materials in the end regions of truss members (e.g., for attachment purposes) causes these regions to

---

*Martin Marietta Aerospace, Denver Division, Denver, Colorado

have a slightly positive axial thermal coefficient that is offset in the center portion of the member by a laminate configuration that provides a slightly negative coefficient. Therefore, the overall gross axial movement of the truss member is a net zero even though there is relative movement of individual points along the truss member. The truss members also exhibit relatively large radial movement, which is not detrimental because relative movement of truss node points remains near zero.

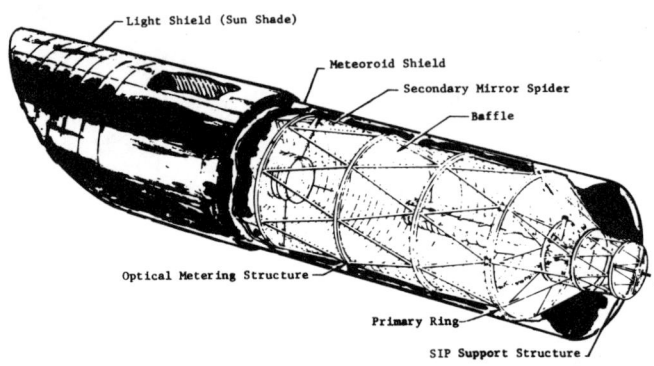

Fig. 1 Large Space Telescope Structural Support Assembly

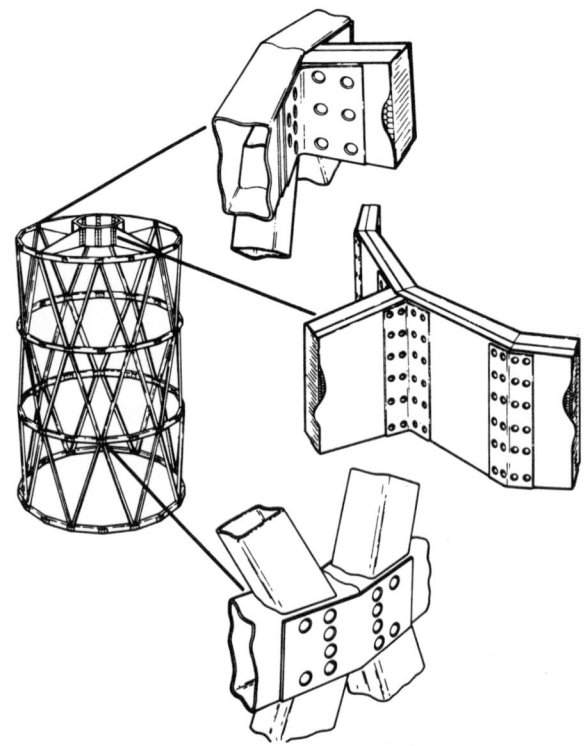

Fig. 2 Low-Thermal-Distortion Graphite/Epoxy Metering-Truss Concept

Finite-element analysis of overall truss performance and application of basic strength-of-materials-type theory to truss components have been shown to be entirely appropriate and satisfactory for predicting structural response. Fortunately, the type of structure and basic composite laminates required to provide a zero thermal expansion coefficient lend themselves to the use of uncomplicated theoretical analysis methods. A unique feature of LST truss design is that a specific value of overall thermal expansion (i.e., zero) is required, with very little tolerance for error on either side of the desired value. This is in contrast to typical aerospace structural characteristics, such as stiffness and strength, in which minimum values are established, with a relatively large margin for error on the overdesign side. Therefore, the analytical expression used to predict the axial thermal expansion coefficient of individual truss members was subjected to a Monte Carlo probability analysis. Each of the fourteen influencing variables was assigned a mean value and assumed to have a normal distribution of variation with an assigned standard deviation. The mean value and standard deviation of the strut axial thermal expansion coefficient determined from the Monte Carlo analysis indicated that design allowables of the LST metering truss could be satisfied without imposing undue restrictions on the normal variation of the influencing parameters.

A triangular section (Fig. 3) of the full-scale metering truss was fabricated to demonstrate feasibility and refine proposed techniques. Fabrication of a single truss strut member used a dissolvable plaster mandrel that was grooved along two opposite sides to provide potential for fabrication of high-quality wrinkle-free components. The truss strut was debulked several times before final cure. Each debulk cycle involved applying a resin bleeder system and vacuum bag and then heating the bagged system to 160°F for 2 hours. After debulking, a considerable amount of epoxy resin had bled off; however, the component remained uncured. The debulking cycles provided high fiber content of the final component, which is required for a low axial thermal expansion coefficient. Intermediate debulking also minimized resin removal during final cure, resulting in good surface appearance.

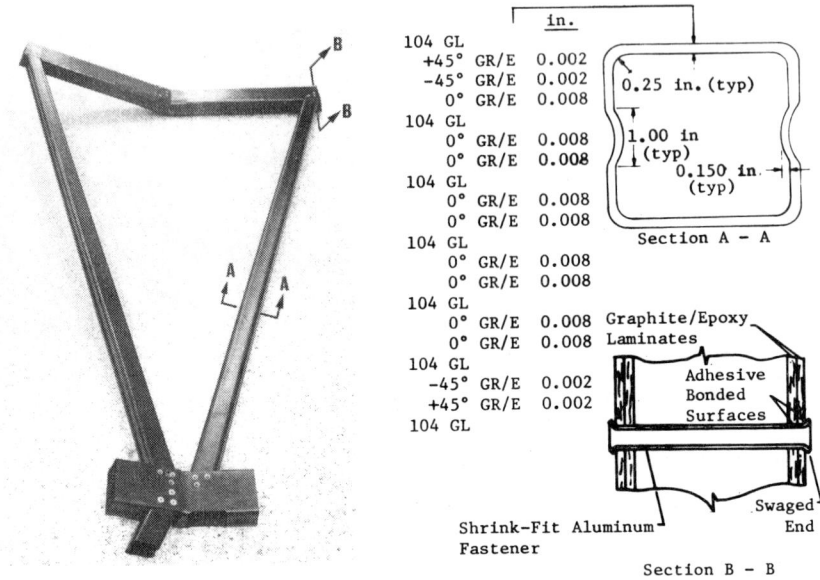

Fig. 3 Full-Scale A-Frame Section

Final cure used a vacuum bag system and autoclave with applied pressure of 100 psi and maximum cure temperature of 375°F. After final cure, the plaster mandrel was washed out by passing hot water through the center aluminum tube. The ends of the struts were cut to finished dimensions using a diamond cut-off wheel on a standard table saw.

Struts, frames, and splice plates were assembled using the attachment concept shown in Fig. 3. In general, fibrous composite laminates have low interlaminar shear and bearing strength. The effect of these characteristics on the design of mechanical attachments is that fasteners in single shear should be avoided and fasteners with a large bearing area are desirable. The attachment concept shown has these desirable features, with the added advantage that it is lightweight, inexpensive, and very rigid. The fastener consists of a hollow aluminum tube that is shrunk fit, bonded, and swaged in place.

Successful fabrication of the metering-truss A-frame fully demonstrates the potential for fabricating high-quality low-thermal-distortion truss structures. Fabricability and fabrication cost were given full consideration throughout design concept development. The result is that the final truss concept uses component parts that are very easy to fabricate; the attachment concept is lightweight, rigid, and inexpensive; and the process techniques developed result in a very appealing appearance of the finished structure.

Predicted strength and stiffness of the end attachment method were verified through successful structural tests of a series of test specimens. The ultimate failure load of approximately 2500 lb of a single-pin unbonded specimen provides for design of joint attachments in which the pins resist all of the design ultimate load, and the epoxy adhesive, although significantly strengthening and stiffening the joint, need not be relied on for structural integrity.

One of the full-scale 73-in. 2x2-in. truss struts was tested to verify the predicted low value of the thermal expansion coefficient. The optical extensometer used has the ability to detect relative motion of approximately 0.0002 in. and, therefore, a relatively large change in temperature is required so that small movements associated with the expected low thermal expansion coefficient can be properly monitored. This method of measurement was used because it was relatively inexpensive and satisfied the test objectives to determine feasibility rather than fully and precisely characterize the thermal distortion characteristics of the strut. The measured coefficient of thermal expansion was -0.05×10^{-6} in./in./°F over the temperature range of 72 to 200°F.

AAS 75-192

3 AXIS SIMULATION OF THE LST POINTING CONTROL
SUBSYSTEM - A MULTI-DISCIPLINE ACTIVITY*

W. W. Emsley ‡
T. D. Fehr ‡
D. C. Fosth ‡
D. L. Knobbs ‡

The pointing control subsystem for the LST vehicle represents one of the most complex and comprehensive space vehicle control system simulations BAC has attempted. This simulation is complicated by a large number of interfaces with other subsystems and systems and a requirement for state of art performance from most of the hardware. A three Axis air bearing simulator has been used to define interface and sub-tier requirements. Areas covered include: Three axis slew law (including flexible appendage effects), detailed on board computer requirements including computer word-length and input/output quantization, detailed software design including adaption for a failed actuator - known and unknown, and detailed system hardware characteristics - noise levels, resolution, and interface characteristics. Results from completed tests are presented along with descriptions of current work.

INTRODUCTION AND SUMMARY

The Boeing Aerospace Company started research related to orbiting telescopes in 1965. The initial effort was a laboratory investigation of precision pointing capability where a "rigid" vehicle was three axis stabilized with performance indicated at better than 0.05 arc-sec peak-to-peak. This early demonstration of the feasibility of precision pointing led to detailed studies of the actual spacecraft implementation of pointing control systems.

---

*This work was accomplished on Boeing Aerospace Company IR&D
‡Member of LST PCS studies group
‡Member of Control System Research Group

The analog control systems were judged inflexible, and costly relative to long-life and diverse operating modes. The follow-on research emphasized digital control technology to overcome the limitations. The initial efforts were analytically oriented to develop the background and tools for advanced digital control concepts. An inter-disciplinary approach resulted in the development of a succession of analytic tools followed by hardware prototypes and software development to validate the analyses. This tool evolution resulted in two key computer simulations for the LST application: The Vibration Analysis Program and the Digital Three Axis Attitude Control Simulation. These programs allow analysis of the LST Pointing Control Subsystem (PCS), and interfaces including structures and optical paths, in the frequency and time domains respectively.

With the establishment of the analytic base, the emphasis shifted to investigations of the multi-disciplinary aspects of synthesizing a PCS. The proof-of-performance testing was being actively pursued by MSFC. The BAC decision was to compliment this effort with a three-axis air bearing simulation emphasizing the total range of PCS functions and interfaces. (This choice was prompted in large part by the difficulties experienced in this area by NASA and DOD contractors in space telescope related applications.) The three-axis air bearing simulation (facility and simulated vehicle) is shown in Figure 1. The sensory inputs for the

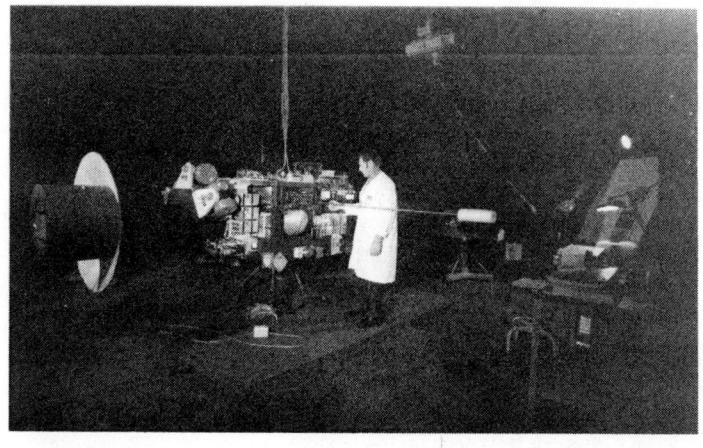

FIGURE 1 Three Axis Air Bearing Simulator

simulated PCS are provided by an array of state-of-art, strapdown, rate-integrating gyros, two cadmium sulfide star trackers with a 1 Milliwatt Helium laser and collimating mirrors for a source, and a modified Canopus tracker with a collimated star source. The actuators of the simulated PCS are six prototype control moment gyros mounted in a skewed array. The signal processing is provided by a prototype aerospace computer (24-bit, 8K memory, 0.8 μsec add, 3.4 μsec multiply) interfacing with 16 channels each of 15 bit A/D and D/A conversion and three channels of V/F converter - updown counters. Other interfaces for the PCS included the onboard power system, simulated structural modes and articulated appendages including an antenna gimbal system, (See Figure 2). The antenna gimbal system base motion simulator duplicates the motion of the air bearing simulated vehicle. Software for the system was developed at the assembly level and tailored to the simulated LST application.

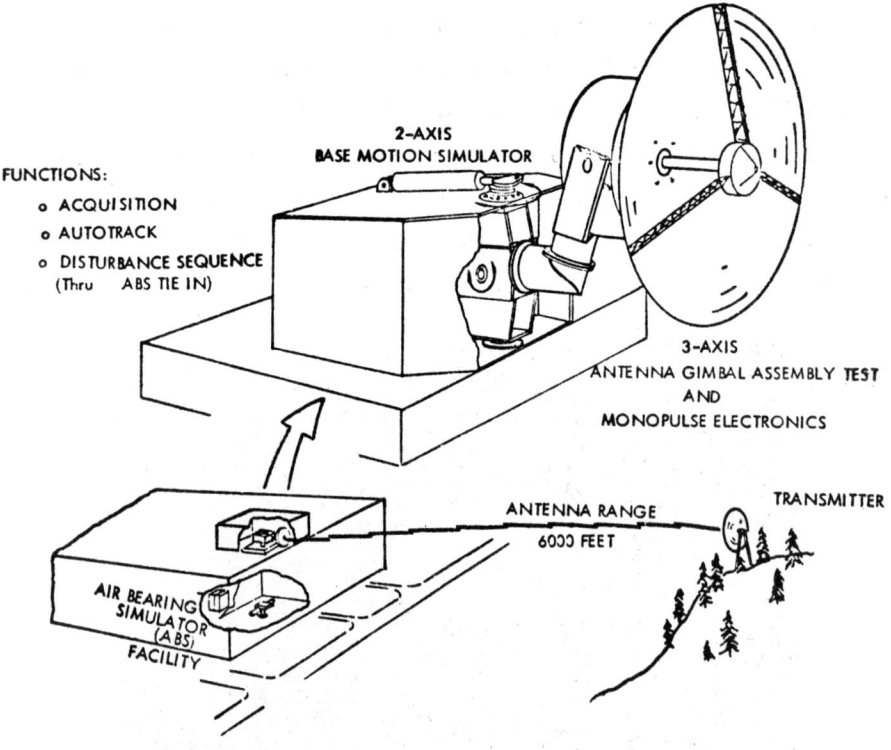

FIGURE 2 Communication System Test Facility

The test program is progressing successfully. The integration of the various interfaces (and disciplines) has been demonstrated. The actuator control law (pseudo-inverse) has been verified and a low gain caging loop added to retain a nominal null configuration permitting maximum momentum utility. The pointing demonstrations were followed by an assessment of the slew mode performance. The slew law was developed and validated which applied jerk (derivative of acceleration), acceleration, and rate limits to a single axis (Euler) rotation for any maneuver. Appendage motion was monitored in these tests to assess slew excitation and has been judged minimal. Reacquisition was assessed and implemented effectively using the slew law for even the smallest rotations. The antenna gimbal system performance was assessed and pointing performance approaching 5% of the beamwidth was demonstrated using monopulse tracking data. Signal lock was maintained during simulator vehicle slewing (with base motion following).

A significant result of this test program is a detailed preliminary design and validation of the PCS software. The pointing performance of the simulated vehicle, on the order of 0.05 arc-sec peak-to-peak is not representative of the LST application. However, the correlation of the measured performance with that predicted using the analytic simulations serves to validate the digital computer program models. This validation lends credibility to the analytic predictions for LST pointing performance. Furthermore, the air bearing simulation has resulted in further detailed definition of the interface and sub-tier requirements.

AAS 75-193

LST DATA MANAGEMENT AND MISSION OPERATIONS CONCEPT[*]
R. Walker, F. Hudson, L. Murphy[+]

LST studies in progress have accentuated the need for a clear understanding of the interaction between spacecraft and ground facilities. This paper describes a candidate design concept for an LST ground facility. The design objective was strongly influenced by a desire to use NASA institutional (existing or planned) hardware, software and facilities whenever practical to reduce development cost. A second objective was to maximize efficiency of telescope usage.

The LST is a space-based observatory with a complement of Scientific Instruments that may be changed upon retrieval of the vehicle.

The telescope is oriented so as to place the target to be viewed on the focal plane of the viewing instrument and within the field of view of the selected electronic detector. Fundamental to the operation are the pointing accuracies and the stability of the telescope, once acquisition of the object has been achieved. Table 1 shows the performance requirements of the LST Pointing Control Subsystem (PCS) which is being designed to maintain a stability of 0.007 arc sec for all viewing states where guide star pairs are available.

The nominal sequence for experiment pointing is as follows. A slew maneuver to the commanded celestial coordinates is accomplished by use of navigation instrumentation. Fine guidance sensors are used to acquire two guide stars within +30 arc sec field-of-view. Closed-loop commands through the PCS, position the telescope to the accuracies stated for Modes II and III, shown in Table 1. However, knowledge of the position

---

[*] Work performed in part on NASA, Marshall Space Flight Center, Alabama, Contract NAS8-31312

[+] IBM Federal Systems Division, Huntsville, Alabama

Table 1

POINTING CONTROL PERFORMANCE REQUIREMENTS

	Line-of-Sight (LOS) Accuracy (Diameter)	Stability RMS	Maximum Exposure Period	Target Acquisition Mode
Photometer (IR)	0.1 $\widehat{s}$	0.03 $\widehat{s}$	1 hr	III
Faint Object Spectrograph (FOS)	0.03 $\widehat{s}$	0.03 $\widehat{s}$	10 hrs	I[1]
High Resolution Spectrograph (HRS)	0.01 $\widehat{s}$	0.01 $\widehat{s}$	10 hrs	II
Astrometry (ASTR) Wide	1.0 $\widehat{s}$	0.007 $\widehat{s}$	10 min	III
Astrometry (ASTR) Narrow	0.03 $\widehat{s}$	0.007 $\widehat{s}$	10 min	III
High-Speed Area Photometer (HSAP)	0.03 $\widehat{s}$	0.05 $\widehat{s}$	5 hrs	I[2]
High Resolution Camera (HRC) Field Camera	1 $\widehat{s}$	0.03 $\widehat{s}$	1 hr	III
High Resolution Camera (HRC) Planetary Camera	1 $\widehat{s}$	0.007 $\widehat{s}$	5 min	III

Note: Mode I[1] Ground Control, Real-Time Operation Slit Jaw Camera, 160 x 100 Pixel Frame
Mode I[2] Ground Control, Real-Time Operation 160 x 100 Pixel Frame
Mode II On-board, Image Dissector Detection, Closed-Loop PCS
Mode III On-board, Program Coordinates Used for Pointing.

of the target with respect to the guide start without direct target viewing is expected to be ±0.3 to 1.5 arc sec. This level of accuracy is a natural goal since this is the type of accuracy one can expect to routinely derive from all-sky ground based photographic surveys.

Further refinement of pointing in Mode II is accomplished by on-board direct object viewing with loop closure through the PCS. In Mode I correction commands are derived from direct target display on the ground referenced to LST body coordinator.

On an average day it is expected that 30×10^8 bits of Science data including calibration, engineering and housekeeping data will be generated by the LST. Approximately 10 percent of the Science data will be processed in real time for quick look or for target acquisition verification.

Figure 1 shows the major data interfaces of the candidate ground system. All data from LST is communicated to the Mission Operation Center (MOC) in real-time and to TELOPS as well. MOC and Science Operations (SO) real-time operations are performed with the Science Institute (SCI) acting in an advisory and monitoring role. In the non-real-time mode engineering data is processed through TELOPS and the MOC but the Science data goes only to TELOPS and subsequently to the SCI. This does not preclude SO from requesting a real-time operation on demand. From the time the science data is received by TELOPS the data processing may be viewed as a "batch operation". The SCI uses the Master Data Processor and Image Data Processor facilities to perform specific image processing with LST unique algorithms.

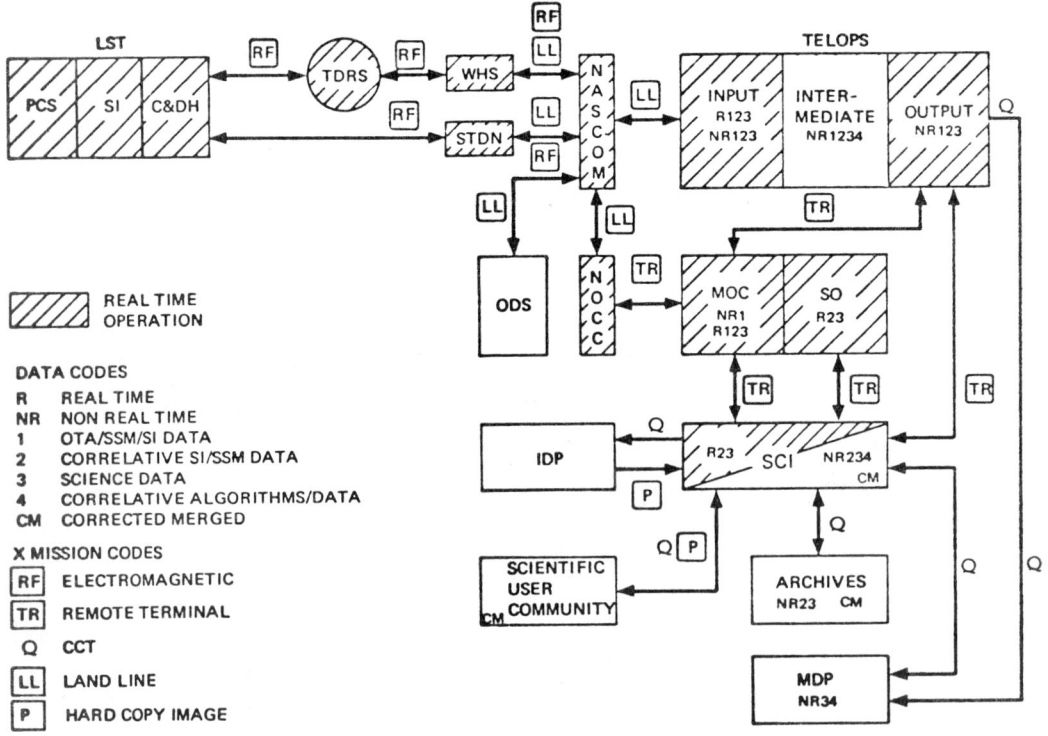

Fig. 1 Major Data Interfaces

Of considerable interest for display and computer manipulation of data by the SCI and the SO in the LST program is the Earth Resources Interactive Processing System (ERIPS) which is designed to analyze data from the Earth Resources Technology Sattelite (LANDSAT). Primary analysis software presents 'menus' of interactive questions for call-up of data manipulation programs, imaging of digital data, pattern recognition and statistical handling of multi-spectral images.

CONCLUSION

The LST, as a vehicle, has highly autonomous guidance, navigation and control functions. This uncouples ground mission operations from serious time constraints and facilities workaround and interactive science procedural investigations. The criteria of high efficiency in data taking is best achieved by exact modeling, mission planning and execution of the mission plan in such a way as to minimize lost time due to vehicle slewing to new targets and to occulations. Once the science data crosses the NASCOM interface, the problem switches to one of management of high data rate information and data processing. The thesis of this paper is to encourage in these formative years, the formulation of LST requirements for NASA multi-user institutional facilities. Development cost is a prime concern, and it must be realized that image data processing is extremely demanding of data processing capacity and that the facilities now being procured for work in the allied field of earth resources digital image processing must be shared by LST.

AAS 75-194

DATA MANAGEMENT FOR LARGE SPACE TELESCOPE[*]

by G. R. Hope, Jr.[†] and T. J. Rasser[††]

Data management for LST must encompass the identification of the type, rate, quantities, and source of all data within the system and the flow of the data including sink and intermediate way-stations. The processing and/or computing required, and the presentation requirements of data users, in both mission and science roles must also be determined.

Key requirements relate to quantity, quality, and flow rates in the system. These requirements include up to 1.6×10^8 bits per 95 minute orbit and up to 3×10^9 bits per day, with a bit error rate less than 10^{-5} for data, 1 false command in 15 years of operation, and rates of up to 1 MBPS. Services available to LST include the post-1979 space Tracking and Data Network, including as a primary service the Tracking and Data Relay Satellite System, and including the point-to-point services of NASCOM.

The data management system must provide for functional control of the spacecraft, acquisition and processing of vehicle subsystem health/status data, and provide both mission (or vehicle) and science (or experiment) support. The vehicle must have some degree of autonomy, since full-time communications to earth are precluded.

The vehicle contains instrumentation, communications and data management subsystems, including a general purpose computer that controls the vehicle as a stable observation platform and the complement of Science Instruments (SI) used in collecting science data. The ground-based system is composed of a Mission Operations Center, which provies command/control data to LST and monitors its health, a Science Institute, where instrument data is processed to final form, and a communications service for the space and point-to-point data flow.

---

[*]This work is supported by Contract NAS8-31313 with the NASA-Marshall Space Flight Center, Al.
[†]Lockheed Missiles & Space Company, Inc., Sunnyvale, Ca. 94088
[††]Guidance Systems Division, Bendix Corporation, Denver, Co. 80236

The **Large Space Telescope** (Fig. 1) is an orbiting national astronomical observatory which will provide useful data for the scientific community from a variety of science instruments. The present complement of scientific instruments includes an F/24 field camera, F/48 and F/96 planetary cameras, faint object and high resolution spectrographs, astrometer, and infrared spectroradiometer and a high speed area/point photometer. Major detectors envisioned for use in the SI are a 70 mm SEC orthicon (SECO) with a format of 2000 x 2000 picture elements (pixels) and two types of semiconductor devices, a 400 x 400 element charge coupled device (CCD) and a 100 x 160 element intensified charge coupled device (ICCD). In addition, auxiliary sensors, ICCDs and image dissectors, are used for target centering in instrument and guidance sensor apertures.

Because the Large Space Telescope (LST) will be a useful observatory for many years, provisions are being made in the design to allow instrument replacement as technology or observational needs warrant. Instrument replacement may occur on-orbit using the Shuttle and crew EVA, or ground return, refurbishment, and subsequent relaunch. Flexibility is therefore the keystone upon which data management for the LST must be founded. There must be sufficient

Fig. 1 Large Space Telescope

flexibility to accommodate different instruments, different detectors, and different observational requirements in response to changing needs of science. The challenge to the system designer is to provide this flexibility while also providing a system at minimum cost.

There are two significant differences between the LST and previous NASA programs. The first difference is, of course, the advent of the Space Transportation System (also referred to as Space Shuttle or Orbiter). The capabilities and flexibility of Shuttle have a profound impact on system design. For instance, we are no longer facing the severe weight constraints of the past; in fact, we can trade addition of weight in structure to provide larger factors of safety against costs of both static and dynamic structural tests. Probably the largest impact, though, is the capability Shuttle provides to perform on-orbit maintenance of components or modules, which are called Orbit Replaceable Units (ORU). Thus, a failed transmitter can be replaced or an SI updated on-orbit or, if need be, LST can be brought back to earth for more extensive refurbishment.

The other difference between LST and current systems is the support provided by the Tracking and Data Relay Satellite System (TDRSS), a new addition to the Space Tracking and Data Network (STDN). There is no longer the constraint of relatively short and infrequent ground station contact opportunities for command, telemetry, and payload data dumps. TDRS provides both forward (command) and return (telemetry) narrow band services over about 85 percent of the orbital period at the present altitude (500 kilometers). The operational groundrules are that science data transmission (at 1.1 MBPS exclusive of the 1/2 rate convolutional coding required by TDRSS) can be scheduled for up to 30 minutes per orbit when in view of either or both relay satellites. The capabilities of TDRS exceed present needs, so any constraint due to high-rate data communications services does not exist. This nearly full-time command and engineering data coverage contributes to the ability to provide flexibility. In addition, as a contingency mode of operation in the event that the TDRS link is unavailable, the ground stations of the post-1980 STDN can be used to receive science and engineering data, and transmit command loads, although at greatly reduced coverage.

Both on-board and ground-based elements (Fig. 2) must be considered because LST data management (Fig. 3) addresses the end-to-end flow of data from SI to ultimate user, the astronomer/scientist, and also the command/control and health monitoring of the vehicle. The onboard portion consists of the instrumentation, communications, and data management subsystems. The instrumentation subsystem provides the transducers and sensor that monitor the vehicle and payload health and status. The instrumentation system isolates, by telemetry processing, failures on the vehicle and payload to the ORU. The communications subsystem provides the facilities to transmit science instrument and engineering data to the TDRS or STDN ground station, and receive commands from the relays or ground stations.

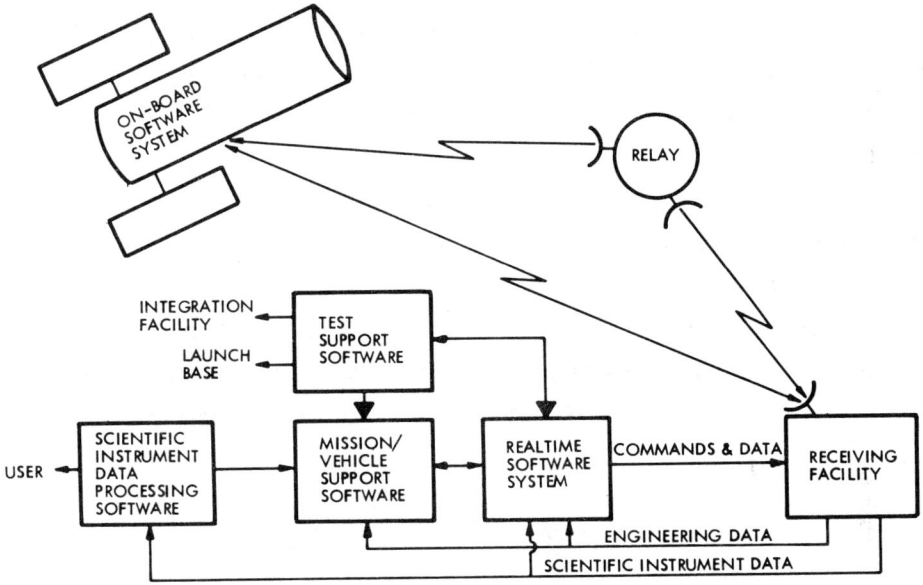

Fig. 2 Function Allocation Overview

- ACQUIRE, FORMAT, TRANSMIT AND PROCESS SCIENCE DATA QUANTITIES OF FROM 1 FRAME (1.6×10^8 BITS) PER REV TO 3×10^9 BITS PER DAY, AT RATES THROUGH SYSTEM OF UP TO 1 MBPS.

- USE TDRSS, BOTH SSA AND MA CHANNELS, AS PRIMARY SPACE LINKS, WITH POST-80 STDN AS BACKUP, FOR FORWARD AND RETURN LINKS. USE NASCOM FOR POINT-TO-POINT DATA TRANSMISSION.

- GENERATE ON-BOARD SOFTWARE AND HARDWARE REQUIREMENTS FOR VEHICLE COMMAND/CONTROL, HEALTH AND STATUS DATA ACQUISITION, FORMATTING, SCIENCE DATA ACQUISITION AND ROUTING, TELEMETRY TRANSMISSION, COMMAND RECEIVING.

- GENERATE GROUND SOFTWARE REQUIREMENTS FOR VEHICLE CONTROL, MISSION PLANNING, SCIENCE SCHEDULING, SCIENCE PROCESSING AND DISPLAY, TEST AND CHECKOUT, SOFTWARE MAINTENANCE/UPDATE, HEALTH/STATUS PROCESSING AND DISPLAY.

Fig. 3 Key Requirements

The data management subsystem functions are: (a) acquisition, formatting, storage (if required), and routing of engineering and science data to the communications subsystem; (b) decoding, verification, storage and distribution of commands to the vehicle subsystems and instruments; (c) data processing to support the pointing control subsystem in maneuvering, stabilizing, and fine pointing the vehicle to allow scientific observations. Stabilization and target reacquisition requirements necessitate extreme accuracy: 0.007 arc-seconds. Observations can last from several milliseconds to several hours.

The ground-based portion of the data system has two major facilities, the Mission Operations Center (MOC) and the Science Institute (ScI), supported by the TDRSS ground terminal (TDRST) located in the continental United States, and NASCOM, which provides point-to-point communications services between NASA facilities.

Determination of the functions to be performed and allocation of functions to and among the system elements are the primary tasks of the system designer. He must then configure hardware and software to implement the functions. Figure 4 describes the functions that must be performed by the data management system, and Fig. 5 shows the allocation of functions to spacecraft and ground elements.

VEHICLE SUPPORT
- VEHICLE COMMANDING
- VEHICLE HEALTH AND STATUS
- TELEMETRY PROCESSING
- SUBSYSTEM CONTROL
- CALIBRATION UPDATE
- TELEMETRY FORMATTING

- SOFTWARE SUPPORT
  - DATA BASE MANAGEMENT
  - EXECUTIVE CONTROL
  - ASSEMBLER/LOADER
  - DISPLAY HANDLER
  - MAINTENANCE & UPDATE

MISSION SUPPORT
- POINTING CONTROL
- MISSION PLANNING
- COMMAND GENERATION
- EPHEMERIS TABLE CONSTRUCTION
- SCIENTIFIC INSTRUMENT CONTROL
- SCIENTIFIC INSTRUMENT DATA PROCESSING

- TEST SYSTEM SUPPORT
  - COMMAND GENERATION
  - TELEMETRY PROCESSING
  - VEHICLE MODELING
  - CALIBRATION PROCESSING
  - REALTIME TEST SUPPORT/CONTROL
  - VEHICLE STATUS

Fig. 4 Basic Software Functions

ON-LINE VEHICLE SUPPORT	AIRBORNE SOFTWARE	OFF-LINE (BATCH) PROCESSOR	TEST SOFTWARE
• REAL-TIME EXECUTIVE	• EXECUTIVE PROCESSOR	• CALIBRATION PROCESSOR	• CALIBRATION PROCESSOR
• COMMAND PROCESSOR	• COMMAND PROCESSOR	• MISSION PLANNING	• VEHICLE LOGIC MODEL
• TELEMETRY PROCESSOR	• TELEMETRY FORMATTING & SYSTEM STATUS	• COMMAND GENERATION	• COMMAND GENERATION
• DISPLAY HANDLER	• POINTING CONTROL & UPDATE	• ORBIT DETERMINATION & SPACE POSITIONING	• EXECUTIVE PROCESSOR
• POINTING CONTROL (SELECTIVE)		• DATA BASE MANAGEMENT	• COMMAND TRANSMISSION
• SCIENTIFIC INSTRUMENT DATA PROCESSING (SELECTIVE)		• TELEMETRY PROCESSOR	• TELEMETRY PROCESSOR
		• ON-BOARD COMPUTER ASSEMBLER/LOADER	• POST-TEST ANALYSIS
			• ON-BOARD COMPUTER ASSEMBLER/LOADER

SCIENTIFIC DATA PROCESSING

- EXECUTIVE PROCESSOR
- DATA CONTENT INTERROGATOR
- ANCILLARY DATA PROCESSOR
- TELEMETRY PROCESSOR
- CALIBRATION PROCESSOR
- SI DATA PROCESSOR

Fig. 5 Software Allocation

As mentioned previously, one of the drivers in the system design is minimum cost, the other is flexibility, and a careful balance must be struck between the two. To that end, each function must be examined to determine where it can best be performed. The philosophy is that functions that can only be performed onboard will be performed onboard, and all others will be performed on the ground. Experience has shown that space hardware and software is many times more expensive than ground hardware and software.

It has also been shown that software to perform real-time processing is several times more expensive than batch processing software. Therefore, only those functions which must be controlled by real-time data feedback or which require near-real-time analysis in the MOC or ScI will use real-time software.

Consider, for example, spacecraft health and status processing. This function could certainly be performed onboard, and, in other systems, is. However, with TDRSS, there is near full-time orbital coverage of telemetry and command. Review of each spacecraft and payload subsystem for time-criticality of response to an anomaly revealed few areas which could not stand the time delay of transmission of telemetry to the ground and formulation and transmission of a command to the LST in response to the anomaly.

Review of the relative costs of software onboard and on-the-ground to process the health and status data led to choosing to perform the function on the ground. A similar decision was reached concerning pointing the high gain antennas (HGA) at TDRS. Ephemerides of both spacecraft are known and the necessary software and computing capacity could be made available onboard, however, cost of software led to a decision to do the computation on the ground and uplink stepping commands to control the HGA.

Availability of only near full-time TDRS coverage and the delay caused by the space-ground-space communication forced us to incorporate processing capability for fine pointing of the spacecraft onboard. But slewing between target objects, and the computation required to formulate the steering commands is done on the ground and uplinked to LST. These examples serve to show the interplay between the elements of the total data system and the trades which must be made to achieve a minimum cost but flexible data management system.

Mission planning and spacecraft control must be addressed early in the conceptual design of the overall system for they can have a profound impact on data management.

Spacecraft control is the responsibility of the MOC (Fig. 6), while mission planning is a cooperative effort between the MOC and the ScI (Fig. 7). A typical sequence of events would be as follows. Months before an experiment or observation is performed by LST, (typically 6 months to as much as a year) the astronomer/scientist submits his proposal for observation/experiment, which is screened and put in a roughly outlined observational sequence. About 3 months prior to observation, the detailed planning begins, with an ordering of observations/experiments, and this is transmitted to the ScI for more detailed work. Software processed in the MOC computer sorts the observations using criteria such as viewing constraints, state of health of the request instrument, etc.

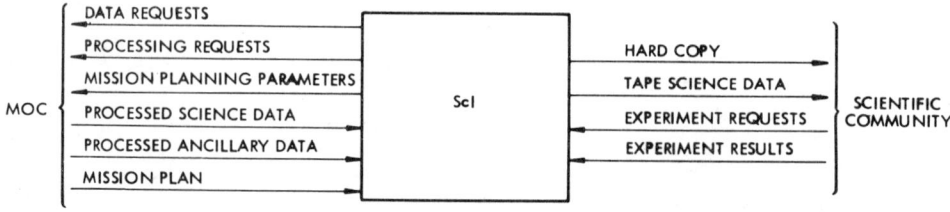

Fig. 6 Science Institute Interfaces

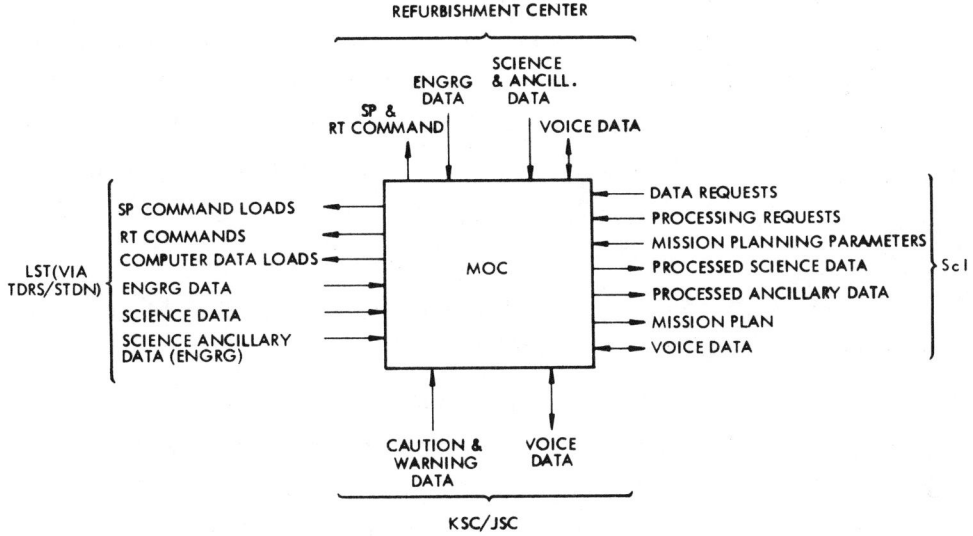

Fig. 7 MOC Interfaces

About 1 month prior to observation, details of the observation are produced; for example, slewing constraints are added, a preliminary time-line of events constructed and a final integrated sequence of observations is produced. About a week prior to observation, the MOC produces and the ScI reviews the detailed observational sequences required. Based on these sequences a spacecraft command load required to command/control the spacecraft and instrument is produced. This command load is verified in a software logic model of the spacecraft. Twelve to 24 hours prior to the observation, the command load is uplinked via the TDRS to LST. This command load provides the command/control of the instrument, the support systems, slewing commands to the pointing control subsystem, tracking commands for the high gain antennas, schedules for the on-board storage devices if required, and the other functions which support the observation. In the meantime scheduling of TDRS, NASCOM, and other support functions and services has taken place.

The observation begins with the on-board data management subsystem issuing the slewing commands to the pointing control subsystem. The spacecraft settles, and the required guide stars are acquired. If quick-look data is required to verify the validity of an observation or ensure target acquisition, the first frame of data from a given observation will be processed under control of the ScI in near realtime and displayed immediately in the ScI and at the astronomer position in the MOC. Based on the quick-look analysis any necessary commands to implement pointing corrections or modifications to the SI function will be generated and transmitted as real-time commands (RTC). After quick-look processing and analysis and after necessary adjustment of pointing and SI parameters, observation continues. Instrument data is acquired and routed by the on-board data system to the relay thence to the MOC/ScI. There it is formatted, corrected and enhanced if necessary, calibration applied and presented to the observer in the form requested.

The foregoing sequence shows the heavy involvement of the data management system (Fig. 8). The MOC with extensive computer support prepares the rough observation list, sorts the observation for optimum spacecraft utilization, subject of course to manual override, prepares and checks the command load, and, under control of the ScI, processes the resultant science data. Concurrently controlled by the real-time software, the MOC determines the state of health of the spacecraft and maintains the interactive operation data base and the terminal system which distribute data throughout the MOC and ScI.

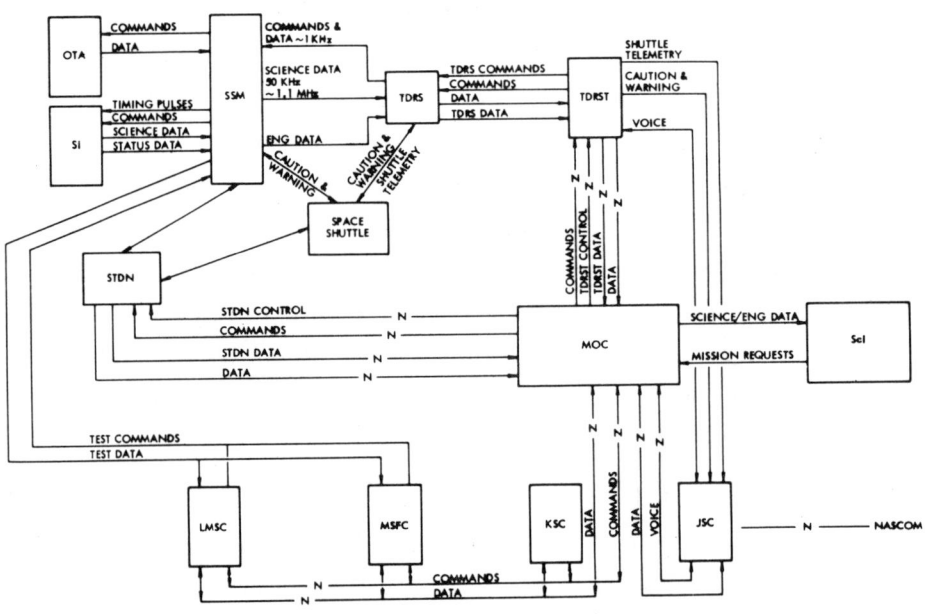

Fig. 8 LST System Data Interfaces

Data management does not stop with design of the on-board system, but runs the gamut of vehicle, control/command, ground data handling and processing, and user interface. For a system such as LST, the planning functions are as important as the actual handling and processing of the data. Efficient use of all the capabilities of the vehicle, its dedicated command/control and processing facilities and the supporting systems, such as Shuttle, STDN/TDRS, and NASCOM provides the system designer with the means to configure a flexible and minimum cost system.

AAS 75-195

A COST EFFECTIVE DATA MANAGEMENT SUBSYSTEM FOR THE LST[*]

John A. Dougherty, Thomas D. Patterson, and Albert E. Cole[+]

INTRODUCTION

The LST DMS concept development and selection challenge lies in the need to provide a low cost, simple and reliable approach to meeting the data management requirements of all LST modules. That approach must be fitted to the initial science payload and subsystem requirements and provide the operations flexibility that is responsive to near real time changes by the astronomer in the observing requirements. In addition, reasonable hardware flexibility and growth margins must be provided that minimize the impact of the science payload and subsystem changes that can be anticipated over the long mission life. Finally, the DMS approach must be responsive to the overall Project requirements[1] and to the use of presently and potentially available NASA standard components and subsystems[2].

The Large Space Telescope (LST) is currently projected for launch in the last quarter of CY 1982. As an unmanned Space Shuttle Orbiter (SSO) payload, it will be inserted into a near-Earth orbit to perform a science mission of many years in accumulated on-orbit life. The LST will comprise an Optical Telescope Assembly (OTA), Science Instruments (SI) and a Support Systems Module (SSM). The overall objective will be to acquire spectral, imaging and astrometric data from celestial objects and planetary bodies, with a significant improvement over data obtainable from Earth observation.

DISCUSSION

The approach to solving the problem required accomplishing the activities identified in Figure 1. These included rigorous definition of immediate

---

[*] Work sponsored by the NASA Marshall Space Flight Center under Contract NAS8-31312.

[+] Martin Marietta Aerospace, Denver, Colorado 80201

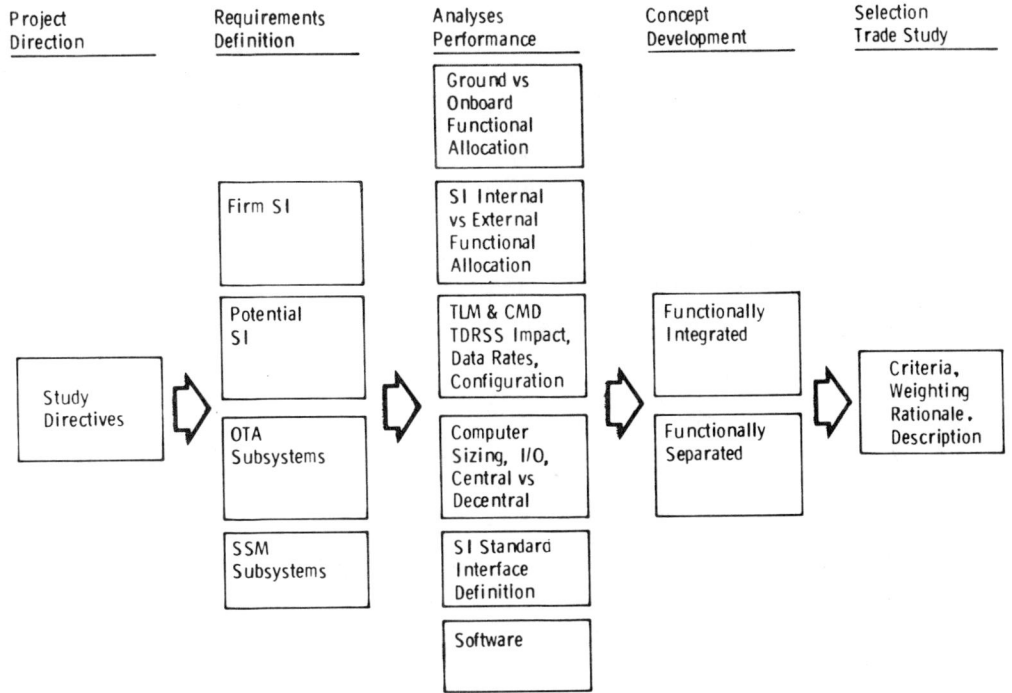

Figure 1 LST DMS Concept Development

and potential DMS requirements, and the identification and performance of key analyses. Results of the analyses supported the development of candidate DMS concepts for trade study and subsequently resulted in selection of the most cost effective DMS approach for the LST.

Seven candidate SIs are now being considered for accomplishing the initial LST scientific objectives. The NASA provided the primary definition of SI DMS requirements[3]. Using these and other selected NASA, LST SI documentation, firm requirements were derived for each of the SIs. Potential SI requirements were also examined and this assessment factored into the DMS concept evaluation.

OTA and SSM DMS requirements were derived. These were telemetry, command, timing and control. The OTA optical performance control and thermal control subsystems do not require on-board computer support. The LST Pointing Control Subsystem (PCS) drives the on-board computer sizing. Computer growth margins for throughput speed and memory were added based on an assessment of the maturity of individual requirement definition and the potential growth, after initial definition, as experienced for similar requirements on Gemini, Viking, Skylab, etc programs.

Key analyses performed were (1) ground vs on-board functional allocation, (2) SI internal/external functional analysis, (3) distributed vs centralized telemetry, command and computer systems and (4) definition of a standard SI DMS interface. In addition, software analysis was performed to determine a low cost approach.

To examine flexible cost effective approaches to the LST DMS, two basic concepts were developed using results obtained from the analyses previously discussed. One provides integration of the DMS functions while the other implements maximum separation of the functions. In the integrated concept, the computer I/O is a data bus. For the separated concept, a digital hybrid computer I/O is employed. The functionally integrated concept is shown in Figure 2. This approach utilizes a powerful central SSM computer and would require a throughput speed and main memory of 40 kops and 8 k words, respectively, larger than that of the SSM computer of the functionally separated concept.

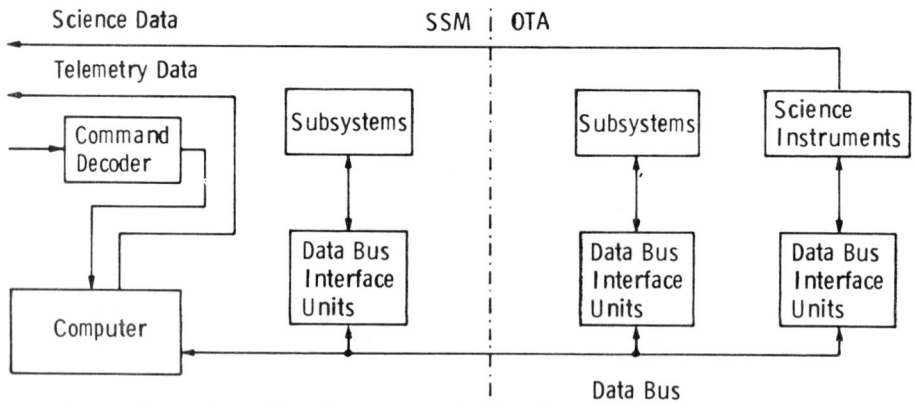

Figure 2. Functionally Integrated DMS Concept

The functionally separated concept is shown in Figure 3.

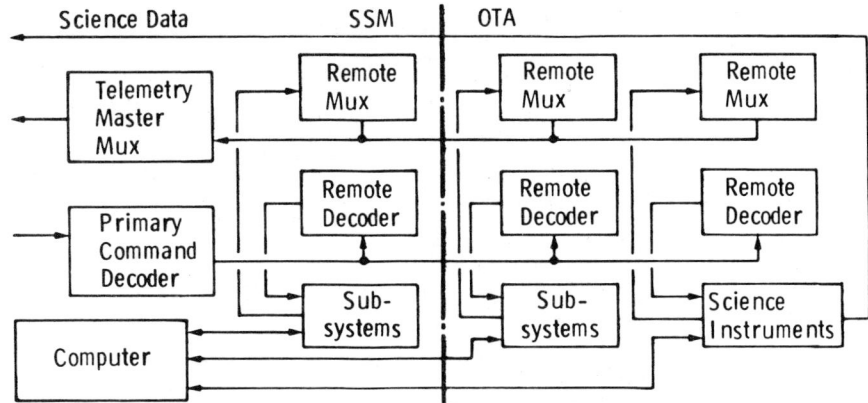

Figure 3 Functionally Separated DMS Concept

Separate command and telemetry hardware elements are used. The central SSM computer is limited primarily to implementation of on-board, closed loop pointing of the LST.

Computer decentralization is an intended option for each concept, should future SI requirements dictate its feasibility.

RECOMMENDATION

A functionally separated concept having a central SSM computer with hardwired digital I/O, separate telemetry and command systems and individual SI RAM/ROM subsequencers is the recommended LST DMS. This concept has the following advantages.

1. Most cost effective
2. Most potential for use of NASA Standard Components
3. Least interface complexity
4. Offers good approach for selective use of redundancy
5. Separation of functions increases operations flexibility
6. Hardware flexibility/growth permits addition of telemetry and command remotes and potential for an SI computer at a later date.
7. Low schedule/development risk
8. Software is minimized by limiting computer function primarily to onboard pointing of the LST.

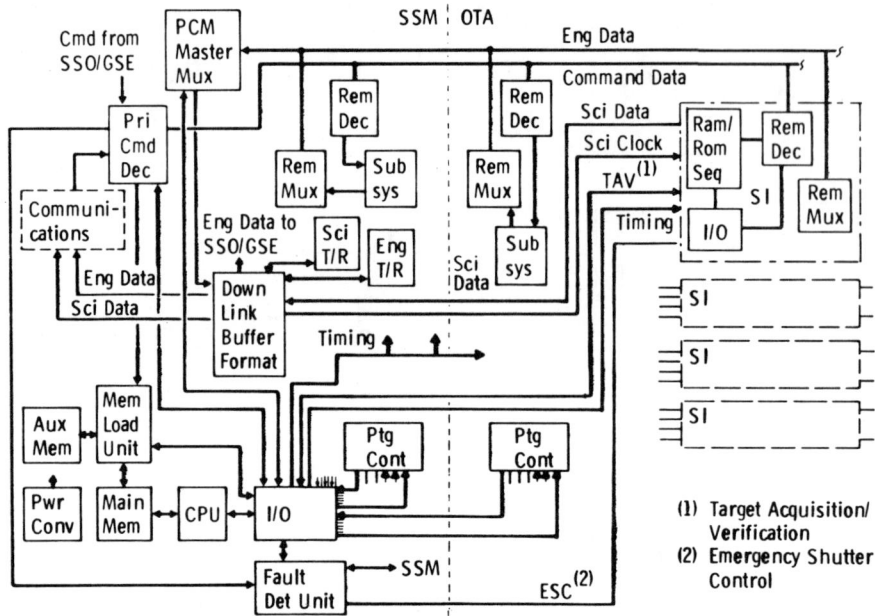

Figure 4 Recommended LST DMS Concept

The selected LST DMS is cost effective in satisfying the LST Project design goals. Its development and selection is based on a thorough understanding of the firm and potential DMS requirements and on the results of analyses of the critical considerations that drive the design.

REFERENCES

1. LST Project Requirements and Guidelines Document - Revision 2, NASA/MSFC, 25 June 1975.
2. NASA Standard Modular Spacecraft Specification, Section 5.0, May 1975, NASA/GSFC.
3. Data Systems Laboratory, Branch Technical Notes, EF23-75-1A through -7A, NASA/MSFC, 4 April 1975.

AAS 75-197

SYSTEM CONSIDERATION, DESIGN APPROACH AND TEST OF A LOW GAIN
SPHERICAL COVERAGE ANTENNA FOR LARGE SPACE VEHICLES*

Richard E. Ferguson, Thomas D. Patterson, Manuel R. Moreno[+]

INTRODUCTION

The objectives of this paper are to assess the system requirements on an antenna with spherical coverage using the Large Space Telescope (LST) as an application example, to develop a design approach and to describe the antenna and LST modeling measurement techniques which were used to obtain the early empirical antenna pattern data essential to the overall command and data system definition.

Large Space Telescope Application

The prime communications links will be through the Tracking and Data Relay Satellite System (TDRSS)[1]. The Spaceflight Tracking and Data Network (STDN)[2] ground stations will be used normally for backup and during the deployment and retrieval of spacecraft by the Shuttle.

In summary, the needed communications operational capabilities with the omni-directional antenna system are as follows:

1. During orbital deployment and retrieval operations by the Shuttle: receive commands from and tramsit engineering data to the STDN remote sites under conditions of both retraction and deployment of the solar arrays.
2. After deployment from the Shuttle: receive commands and transmit engineering data to both the TDRSS subnet and the STDN remote site subnet. Also, transmit science data to the STDN remote site.

Antenna Placement Considerations

The antenna elements should be placed so as to provide an overall spherical gain pattern and to avoid occlusions and distorting multipath reflections by the vehicle structure. If the elements are to be arrayed to provide a fixed omni-directional pattern, the placements must also minimize the gain reductions due to the interferometry effect. To fulfill

---
* Work partially sponsored by the NASA Marshall Spaceflight Center under Contract NAS8-31312.
+ Martin Marietta Aerospace, Denver, Colorado

these criteria, the mounting of two turnstile-over-cone RCP antenna elements, each having hemispherical response patterns were tentatively selected to be on the ends of the solar panel booms, as shown in Fig. 1.

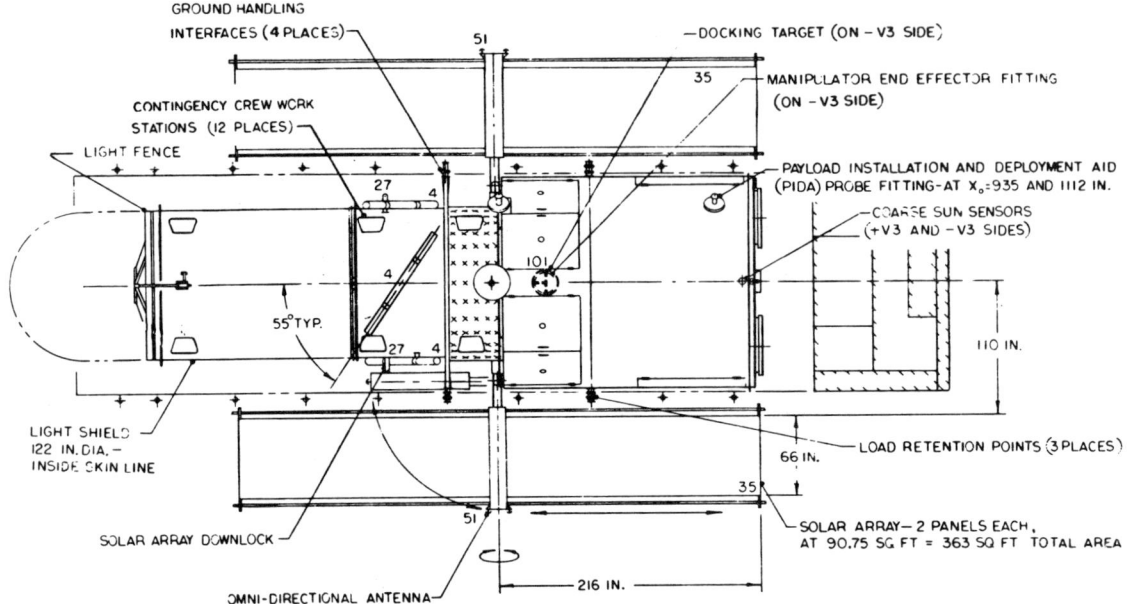

Figure 1   LST With Solar Panels and Omni-Directional Antennas Deployed

<u>Trial RF Link Performance Analysis</u>: A trial RF link performance analysis was accomplished to determine the required antenna gain versus bit rate for command reception, and for engineering and science transmission links appropriately to the TDRSS and to the STDN considering the services available. Tables were constructed with the known factors included and summarized to the quantities available for dividing between antenna gain and bit rate for the command link, and, additionally, the LST RF transmitter power for the data return link. For the STDN and TDRSS services, the command bit rate is plotted versus the LST antenna gain in Fig. 2, and return data rate is plotted versus antenna gain and LST transmitter RF power in Fig. 3. Note that in Fig. 2 a 3.5 dB loss is included in the known factors for passively splitting the received signal power to the redundant receiver to make it unnecessary to command a switchover to the redundant receiver in the event of prime receiver failure.

<u>Omni-Directional Array Scale Model Pattern Measurements</u>

Gain pattern measurements were conducted on a 1/5 scale model of the LST

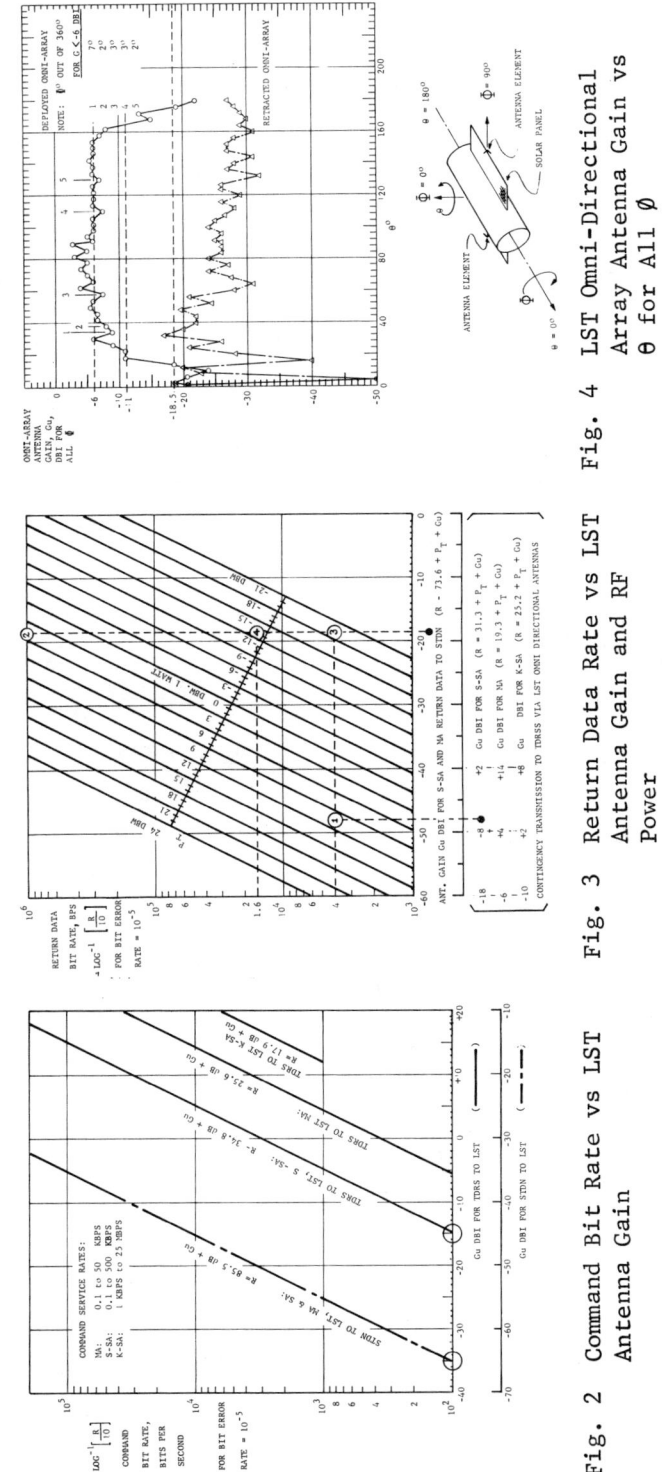

Fig. 4 LST Omni-Directional Array Antenna Gain vs θ for All ϕ

Fig. 3 Return Data Rate vs LST Antenna Gain and RF Power

Fig. 2 Command Bit Rate vs LST Antenna Gain

with the antenna elements mounted at the ends of the booms of simulated solar panels. The model was constructed to set up the conditions of solar panel deployment, retraction and angular positioning of the deployed panels about the axis of the boom. Outputs of the antenna elements were connected together in phase by a coaxial T-junction to the test receiver. The minimum pattern response in a $360°$ rotation of the LST about its longitudinal axes ($\emptyset$) at each azimuth look angle (θ) increment is plotted versus θ in Fig. 4. Since the antenna gain minima reduce nearly by the same amount versus θ as θ approaches the longitudinal axis at either end of the LST, a model for spherical coverage is suggested. The angle, θ, to a selected response minimum approximates the half angle of cones at both ends of the LST within which the antenna response is less than the minimum. From this geometry, the percent continuous spherical coverage is equal to at least 100% cos θ.

Selection of the Command and Return Data Service, Rates, and Antenna

Analysis of observational sequences and command history shows that a 100 bps command bit rate is satisfactory. The S-SA service is selected. Fig. 2 shows that it requires a minimum antenna response of (-)14.8 dBi at 100 bits per second for communications with the TDRSS. Fig. 4 shows that the deployed arrayed antenna response is at least (-) 11 dBi minimum to within $18°$ half cone angles from the longitudinal axis at either end of the LST. Thus, 95 percent (100% cos $18°$) continuous coverage is achieved with 3.8 dB of margin. For commands from the STDN ground station subnet, Fig. 2 indicates that a minimum antenna gain of -65 dBi is needed; Fig. 4 shows that all gain minima are no less than -50 dBi which results in 100 percent continuous spherical coverage with 15 dB of margin.

Based on the numbers of items needed to be monitored, the return data rates, of 4 and 16 kbps for engineering were chosen. A high rate channel of 1 mbps for science data is a requirement. RF power allocated to the above rates is -7 dBw, -0.97 dBW and +12.8 dBW respectively. The minimum antenna response needed is obtained from Fig. 3. Add to the minimum antenna response for the engineering rates the factors, 3.9 dB noise loss due to multiplexing with the science data channel and one dB channel loss due to a multiplex with a PN range and tracking code. The percent continuous spherical coverage may be obtained from Fig. 4 as found for the command channel.

Table 1 contains the summary of the LST communications interface with the TDRSS and STDN using the proposed fixed beam omni-directional array for the LST. Included also for completeness is the interface for the high gain antenna.

Note	Ant. Resp. dBi	% Coverage
1	-11	96
2	- 6	80
3	-50	100
4	-18.5	97
5	-40	99.9
6	-24	100

Table 1
LST COMMUNICATIONS INTERFACE WITH THE TDRSS/STDN

	LST Rad. EIRP, dBW	Transmit Channel	Modulation	LST Antenna	Command and Data Rates, kbps				Deployment, Contingency Return Engr, kbps	Tracking & Range Rate	RF Link Margin from Design Case, dB	Average Access Time per Orbit, minutes
					Forward Commands	Return Data						
						Engr. kbps	Science. kbps					
TDRS												Per Satellite
S-SA Receiver	--	--	SQPN	Omni 1	0.1						+3.8	30
Transmit												
DG 1	+8.9	I	SQPN	HGA		4				X	+3.9	30
DG 1	+14.9	Q	SQPN	HGA		16				X	+3.9	30
DG 2	+28.7	I&Q	PSK	HGA			1000				+4.7	30
DG 2	+2.1	I&Q	PSK	Omni 2				4			+2.1	30
STDN				Omni 3 Deployed/								Per Station
S-SA Receive	--	--	SQPN	Retracted	0.1						+15	1L 5
Transmit												
DG 1	-30.2	I	SQPN	Omni 4		4				X	+7.2	1L 5
DG 1	-24.1	Q	SQPN	Omni 4		16				X	+7.2	1L 5
DG 2	-10.4	I&Q	PSK	Omni 4			1000				+7.9	1L 5
DG 2	-33.6	I&Q	PSK	Omni 5 Retracted				4			+8.7	1L 5
DG 2	-17.6	I&Q	PSK	Omni 6				4			+24.7	1L 5

DG = Data Group

REFERENCES:

1. Tracking and Data Relay Satellite System (TDRSS) User's Guide, Revision 2, STDN No. 101.2, Goddard Space Flight Center, Greenbelt, Maryland, May 1975.

2. STDN User's Guide Baseline Document. Revision 2, STDN No. 101.1, Goddard Space Flight Center, Greenbelt, Maryland, May 1974.

AAS 75-296

SYSTEM APPLICATIONS OF THE FAULT TOLERANT MEMORY[*]

L. J. Murphy[+]

CONDENSATION

INTRODUCTION

In recent years there has been a marked increase in the long life and high reliability required of computers. Attaining these goals is complicated by an attendant increase in required equipment capability. Computer memories with large, fast arrays and memory sizes in the range of 24,000 to 64,000 words are not uncommon. These larger memories present some system design challenges in meeting overall program reliability goals.

IBM has been involved in recent studies that required long term, highly reliable performance of an on-board computer system. Two of the more stringent sets of reliability requirements are imposed by the Solar Electric Propulsion Stage (SEPS) and the Large Space Telescope (LST).

The reliability goal for the SEPS Command Computer System (CCS) is 0.95 for a three year, unmaintained mission. The system includes CPU, memory, input/output (I/O), power supply and tape recorder.

The initial reliability goal for the LST computer system is 0.975 for a one year, unmaintained mission. The most current LST Computer System concept includes CPU, memory, I/O and digital interface units (DIU). Three redundant power supplies (an optimization of the basic design) are used to increase system reliability slightly. Both the reliability goal

---

[*] The complete paper appears in Vol. 25, AAS Microfiche Series
[+] IBM Federal Systems Division, Huntsville, Alabama

and the configuration may be modified in the current phase B studies, but the trade study described in this paper will remain valid.

Reliability analyses were conducted as part of the computer system selection trade studies which were performed. The results led to the selection of a fault tolerant memory over standard memory modules for both SEPS and LST applications because of savings in cost, weight, and volume.

RELIABILITY AND TRADE STUDY ANALYSIS

To determine the computer system configuration, a reliability analysis of candidate configurations was performed to determine the number of redundant elements required in each candidate configuration to meet the established reliability goal. The candidate configurations chosen for analysis encompass both single-string and cross-strapped concepts and systems with conventional as well as fault tolerant memories. After the candidates were sized, a trade study was performed to determine the most cost effective solution. Trade study parameters were cost, weight, power, and volume.

The candidate configurations evaluated for SEPS included redundant simplex systems with conventional memories, cross-strapped systems with conventional memories, simplex systems with a shared fault tolerant memory, and cross-strapped systems with a shared fault tolerant memory. Cross-strapped systems have higher reliability but suffer the disadvantage of having more complex interfaces. When the best candidate was selected, the design was optimized to minimize the hardware required to meet the reliability goal. Based on the reliability analysis, the design selected was the three cross-strapped system with a five fault tolerant memory. This design was further optimized to a cross-strapped system with three CPU/translators, three power supplies, two I/O's, two tape recorders and a 16K 16-bit five fault tolerant memory. The reliability of the optimized system was 0.95039 for a three year unmaintained mission.

The system selected for the LST was two single strings with a five fault tolerant memory. This was optimized to include three cross-strapped power supplies leading to a system reliability of 0.977.

To demonstrate the advantages of the fault tolerant memory system over those using conventional memories for these particular applications, an analysis of relative costs, weight, power consumed, and volume of the competing configurations was performed. All costs were recurring and all parameters were normalized to the selected configuration in each case (SEPS and LST). The lowest value for each selection criteria is given the value one (example: lowest cost earns a one), and designated the winner. In both cases a fault tolerant memory configuration was lower in cost, weight, and volume, with a penalty of 20-30% in power.

FAULT TOLERANT MEMORY

This section will discuss the fault tolerant memory advantages in more detail. A detailed technical description of memory operations is given in a recent technical paper titled "A Memory System Design Which Can Tolerant Multiple Storage Array Faults". The authors are McCarthy, C.E., of IBM Federal Systems Division, Carter, W.C., of the IBM Thomas J. Watson Research Center, and White, J.B., of the NASA Marshall Space Flight Center.

The fault tolerant concept is not totally new. Error correcting codes have been used in some models of the IBM System/360 and System/370 for some years. Field experience showed that significant increases (10:1) in reliability could be achieved by using an error detecting, error correcting scheme. This is based on the fact that many memory errors are transient in nature, and if detected and corrected the system could continue to operate. The concept that is new in the fault tolerant memory discussed here is the memory plane replacement, which represents the limiting case modular memory replacement.

Strategies for replacement of memory planes will depend upon program needs. If desired, a plane could be replaced when a single bit error occurs. However, since bit errors are frequently random and since they are corrected by the error correcting logic, a better scheme may be to replace a plane after either a hard failure or multiple single errors in the same plane. The selected strategies can be implemented in software.

SUMMARY

The fault tolerant memory has much to offer when the system requirements include high reliability, long life, and large memory size. The reliability of the memory approaches unity for missions as long as two years and memory arrays as large as 32K 16-bit words. This improved memory reliability allows a reduction in the redundancy required as compared to designs using conventional memory techniques, which results in reduced cost, weight, and volume. Power penalties for the system using the fault tolerant memory approach 30%.

The fault tolerant memory concept has been proven in non-aerospace applications. Field experience has shown a 10:1 improvement in reliability. It appears that based on performance experience to date and anticipated requirements in the future, the fault tolerant memory deserves the attention of system designers.

AAS 75-198

A SCIENTIFIC OPERATIONS PLAN FOR
THE LARGE SPACE TELESCOPE

Donald K. West*

INTRODUCTION

The LST Science Institute (LSI) has been proposed for the LST. If this science institute is adopted by NASA it will be responsible for proposal selection, telescope-time schedules, science planning, and the implementation of investigator's programs. Guest Investigators as well as Principal Investigators will normally come to the Institute to perform their astronomical observations. During an observing run, the investigator will participate in the verification of the observing plan, real-time target acquisition, the analysis of quick-look data, changes to the observing plan, and the specification of data reduction procedures. Data processing is performed to the user's specification to the point where scientific judgment is required. At this point, the data is given to the user in the form and format of his choice for scientific analysis.

The purpose of this paper is to describe an LST ground system which is compatible with the operational requirements of the LST. The goal of this approach to the ground system is to minimize the cost of post launch operations without seriously compromising the quality and total throughput of LST science. The resulting system is able to accomplish this goal through optimum use of existing and planned resources and institutional facilities.

REQUIREMENTS

Mission and data operations conceptual studies for the LST have been completed at Goddard. The results of these studies showed that all of the mission operations requirements could be met through the slightly modified use of hardware and software resources developed for OAO. Scientific operational requirements, however, call for the development of new resources in three areas.

The first of these new requirements involves the development and operation

---
*Goddard Space Flight Center, National Aeronautics and Space Administration

of a scheduling and science planning system for the LST Guest Investigators. The support requires sophisticated hardware and software systems to accomplish long-range observatory scheduling and science planning.

The second new requirement concerns the real-time operation of the scientific instruments (SI) and spacecraft pointing control system to accomplish closed-loop-to-ground target acquisition and positioning in the SI apertures. Extended real-time contact periods and ground target acquisition systems are needed to meet this new requirement.

The third and most demanding new requirement is in science data management. The LST will transmit several images per orbit which will require preprocessing and image processing of millions of data points per day. Preprocessing of raw telemetry data is needed to sort out the science data and output it in a format suitable for data reduction tasks, such as image processing. Processing of 10^9 bits/day is an LST requirement which identified the need for a significant new resource.

COST CONSTRAINTS AND GUIDELINES

Ultimately, an LST Science Institute will desire autonomous dedicated computer facilities to maximize the scientific flexibility and data throughput in the planning of observations and the final science data processing and analysis.

The Goddard study has identified a scientific operations solution which will meet the LST requirements within the limits of current cost ceilings. The plan is centered on the use of existing and planned data facilities at Goddard which can be shared by LST. The hardware and software facilities identified for LST have broad data handling and data processing capability extending well beyond the LST requirements.

In order to develop a completely viable LST ground system which will effectively interface the Science Institute with the MOC and the GSFC shared facilities, it is necessary to define the following set of system guidelines:

1. The Mission Operations Center will be located at Goddard Space Flight Center.
2. The LST Science Institute will be located at Goddard or a part of its operational staff will be co-located at Goddard.
3. Raw data will be acquired, stored, and preprocessed by TELOPS.
4. Routine image processing hardware and system software is to be provided by the GSFC Image Processing Facility (IPF).
5. The Science Institute will be responsible for providing

scientific applications programs and calibration data for image processing to the IPF.
6. Custom data processing is to be done by the Science Institute.

IDENTIFICATION OF NEW RESOURCES

Following the above cost constraints and ground system design guidelines, a study of existing and planned resources was performed. The results of this study revealed that most of the LST's requirements could be met by observational and data processing software and hardware resources currently in development at GSFC. These facilities will be fully operational several years prior to LST launch, and can be made available to the LST Program if their effective use by LST can be assured. The effective use of these shared facilities is strongly dependent on the proximity of the user. Three distinct resources have been identified.

1. The International Ultraviolet Explorer (IUE) is an international space observatory which will host at least 50 guest observers per year. Software and hardware for long-range guest observer scheduling and real-time target acquisition systems are currently being developed for a 1977 launch. These IUE systems are ideally matched with the LST requirements for guest observer support and real-time closed-loop-to-ground target acquisition.

2. The LST requirements for the preprocessing of 1×10^9 bits per day of science data can be satisfied by using the Telemetry Operations Processing Systems (TELOPS). This system which is currently being built at Goddard is sized to handle data volume and preprocessing requirements several orders of magnitude greater than those of the LST.

3. The Image Processing Facility (IPF) under development at Goddard is a large scale computer system which will process up to 2×10^{11} image bits per day. In addition, IPF is able to accommodate the scientists' need for making changes in calibration data sets, scientific algorithms and applications programs.

TELEMETRY OPERATIONS PROCESSING SYSTEMS (TELOPS)

TELOPS serves as an on-line digital interface between the network and the LST ground system. It provides spacecraft and user support in near real-time with a storage and processing capacity to handle large volumes of data. The features of the system include: data capture, data staging, data editing, rapid access on-line storage, off-line data archival, 24 hour-seven day/week operation, and a fail-safe protect system. System capacity characteristics are: on-line data input from network on 22 lines simultaneously, 20 lines up to 108K bit rate each, combined peak rate of up to 2.6M bits, processing capacity of 450M data points per day, and on-line storage of 169 billion bytes.

IMAGE PROCESSING FACILITY (IPF)

The general function of IPF is to provide standard image processing support to users. Image Processing functions include: preprocessing, annotation and guiding, radiometric calibration, geometric corrections, editing, data merging, reformatting, product generation, near real-time transmission, enhancement and differencing. The characteristics of the IPF are: digital radiometric and geometric corrections, geometric corrections are performed on the basis of image correction data and reference control points (relative or absolute); primary purpose of geometric corrections is temporal registration of data; resampling algorithms using nearest neighbor and six x/x approximation (at least cubic); processing volume of 10^{11} bits/16 hours; and input-output in the form of high density tapes or general computer compatible tapes.

GROUND SYSTEM PLANNING AND DATA FLOW

The major functional elements which will make up the LST ground system are shown in Figure 1. Those elements which provide institutional support services to LST are shaded. The Mission Operations Center (MOC) and the proposed LST Science Institute (LSI) will contain all of those elements dedicated to LST.

An observation request will originate with the principal or guest investigators in the form of target coordinates, time and duration of observations, measurement accuracy, and instrument configuration. The planning computer receives observing plans from the LSI via terminals, punched card or tape input. Mission control inputs current spacecraft constraints to the planning computer which translates the observing plan into 24 hours of LST commands. This command message will be approved by the MOC for transmission to the LST via STDN (TDRSS). Real-time operation, such as target acquisition, will be accomplished by direct commanding of the LST by the MOC computer in response to requests from terminals in the MOC and LSI.

The LST transmits science and engineering data which will be relayed by STDN (TDRSS) to the MOC and TELOPS at GSFC. Engineering data is available in the MOC for spacecraft monitoring and analysis. Science data in quick-look form will be available to the users in the LSI by the MOC computer or TELOPS. TELOPS receives, stores, and preprocesses large quantities of science data for image processing by the Image Processing Facility (IPF) which will then output the reduced data to the LSI. Final data reduction and products will be processed to the point where scientific analysis is required. Reduced data is given to the user for his analysis and to the NSSDC for timely distribution to the general scientific community.

SCIENTIFIC INTERFACES

The question of how guest astronomers interface with the LST is a central one. Staff astronomers, night assistants, and technicians will assist guest astronomers in the daily functions: specification of target sequences, target acquisition, identification, and slit monitoring; scientific instruments (SI) command requirements, SI operation in real-time and stored command memory mode; the specification and evaluation of all science and SI engineering data processing, storage and distribution, the specifications and requirements for all science operations ground support hardware and software systems and changes thereof; and the maintenance of SI calibration and image processing data sets, programs and algorithms.

The interface between astronomers and the LST in orbit is the planning computer. Standard computer terminals will be used to initiate programs on the planning computer. Similar interactive display/control computer terminals will interface with the operations (command-control) computer in the MOC for real-time operation.

The interface between astronomers and the data processing facility will be standardized and pre-defined in terms of the number and type of data reduction processes available and in the types of output and formats available to the user for his analysis. The user will have the capability of reviewing scientific data in condensed (quick-look) form prior to image processing. This interface will also allow for the timely update of scientific algorithms, applications programs, and calibration data.

FUTURE DEVELOPMENT

This scientific operations plan was developed to meet the immediate requirements of the LST within the pre-launch budgetary constraints. The plan described here not only accomplishes this goal but in addition, contains modular design concepts which offer considerable flexibility for future development of facilities dedicated to the LST.

AAS 75-199

LST OPERATIONS, A TYPICAL DAY

William J. Pragluski* and Robert H. Brown*

INTRODUCTION

The development of operational timelines provides visibility between the degree to which science objectives of the LST mission can be achieved and realistic design requirements on the flight and operations systems. The mission timeline presented here encompasses approximately three weeks of mission time corresponding to a point in the mission after the initial on-orbit checkout of the LST. The over 300 orbits of timeline using several target types allows using a statistical approach to evaluating the LST design requirements and their impact on science mission design flexibility.

The results of this analysis demonstrate the significant contribution the LST can make in the science of astronomy. It can not only provide new, fundamental data on the nature of the universe, it will also provide more accurate data on the yardsticks used in virtually all astronomical analyses.

LST SCIENCE OBJECTIVES

Cosmology is high on the list of priorities. The LST's light gathering potential allows observing targets over five magnitudes dimmer than possible from the surface of earth. This allows taking direct measurements at a distance where the many theories on the nature of the universe (i.e., flat and infinite, curved space, closed, origin?) result in theoretically different results.

---

*Martin Marietta Aerospace, Denver Division, Denver, Colorado

Precise measurements of interstellar medium, binary starts, and stellar parallaxes will provide far more accurate data on the abundance of mass in the universe and the relationship between the target's stellar characteristics and the relative distances in space. These yardsticks are fundamental in properly interpreting the data gathered by all telescopes; both orbiting and on the earth's surface.

The LST can also make significant contributions to solar system astronomy. It cannot compete with highly sophisticated spacecraft of the Viking type which can take in-situ measurements, but it will provide considerable data on the outer planets and comets which equals the quality and quantity of other precursors of the Pioneer type.

MISSION OPERATIONS PROFILE

A timeline for mission operations and typical instrument usage sequences is simulated to determine the data quantity performance of the LST. The event times and standardized presentation plots for the mission operations profile are generated by a timeline computer program. The timeline program includes an ephemeris generator and a set of routines to calculate the event times shown for orbital day/night cycles, sidereal revolution numbers, viewing of representative science targets (constrained by sun, moon, earth, and stray light), South Atlantic Anomaly encounters, target slewing maneuvers, and ground station/tracking and data relay satellite contacts.

COMMAND AND DATA OPERATIONS SUMMARY

Several operations profile segments have been developed and are representative of integrated science sequences that could be used during the LST operational mission phase. Analysis of the information presented by these profiles furnishes summary statistics related to uplink command storage volumes and science data quantity.

Command Storage

LST is being designed as an autonomous, on-orbit operations system for periods of up to 24 hours. The onboard data management subsystem has the responsibility for receipt, storage, and dissemination of commands for the subsequent automatic execution. These time tagged commands are generated in the Mission Control Center and uplinked periodically to LST via the TDRSS. Requirements are therefore established for ground generation and onboard storage systems to accommodate the command volume required for the scientific operations of LST.

The scientific instrument and supporting subsystem command requirements for a 24 hour period depend upon the particular observation sequence performed. The previously mentioned operations profile segments have been evaluated for determination of command requirements. The data of Figure 1 presents the results of these evaluations.

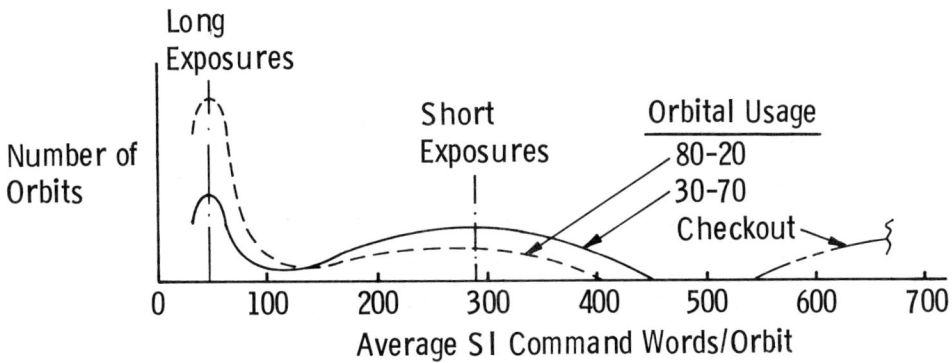

Figure 1 Typical Command Storage Distribution

The command storage volumes were determined on an orbit-by-orbit basis for the various profile segments. The number of orbits requiring particular volumes of commands are plotted versus the volume of commands (Figure 1). The distribution indicates three distinct groupings. The first grouping is the result of long exposures where the commands required per orbit are low. The second grouping represents orbits in which the exposure times are short, several frames of science data are taken, and many commands are required. The third grouping is representative of the initial operational checkout phase where an extremely large number of data frames are taken per orbit and even larger volumes of commands are required. The volume of commands required on a daily basis is a result of the number of orbits requiring long exposures versus those requiring short exposures for data acquisition. This relationship is referred to as the orbital usage percentage. The standard planning cycle for LST would suggest equal probabilities of the orbital usage being anywhere between a 10-90 relationship (extensive calibration segments) to a 90-10 relationship (extensive observations on faint objects). The scientific instrument checkout usage is a special case that approaches a 0-100 relationship.

The data of this figure indicates mean values for long, short, and checkout exposure distributions. The long exposure commands have minimum impact on daily command storage requirements. The short exposure mean is the predominate factor in the derivation of a 24 hour requirement for the operational mission. This factor, when combined with orbital usage considerations, has resulted in a recommended daily storage capability of 1500 words. The checkout usage mean places a two-orbit requirement on command storage. The data of this figure shows that the recommended 1500 word storage capability is satisfactory for accomplishment of two orbits of science instrument checkout exposures.

Science Data Quantity

Scientific data quantities, in terms of data bits and data frames, are also dependent upon the particular observation sequence being performed. The various profile segments were again evaluated on an orbit-by-orbit basis to derive these quantities for the orbital usage relationships. The data of Figure 2 shows bit volume ranges per orbit for long exposure and short exposure data frames. The trends of these data curves in the short exposure range show only minor variations with orbital usage. The apparently large deviation in the long exposure range is due to a difference in downlinking philosophies. Onboard, instrument detector integration over multi-orbits was assumed on the 80-20 orbital usage profile segment, whereas, ground integration on an orbit-by-orbit basis was assumed on the 30-70 orbital usage segment. Both procedures are viable candidates for LST operations. An accumulation of scientific data bits from the various profile segments shows overall average for recorded and downlinked data of 2.8×10^9 bits per day and a maximum of 3.7×10^9 bits per day.

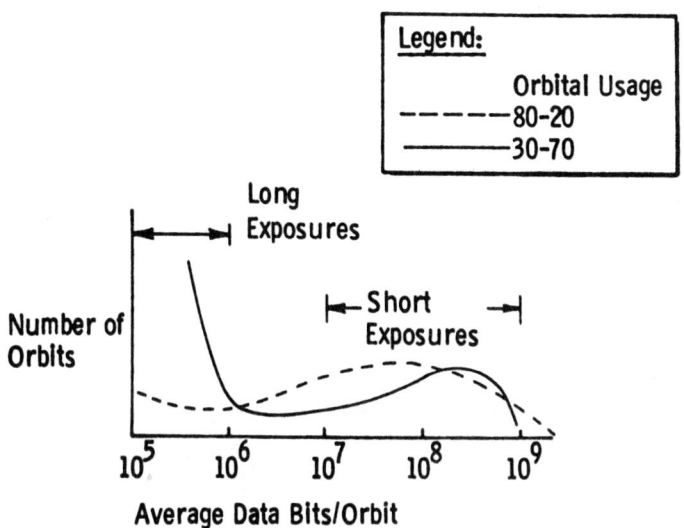

Figure 2   Typical Science Data Bit Distribution

The distribution of individual science instrument frames, based on the accumulation of data from the various profiles, is presented by the data of Table 1. This distribution shows that the instrument usage being planned for LST will furnish data for significant contribution.

Table 1
LST/SCIENCE OBJECTIVES SUMMARY

Scientific Instruments	% Frames	Contribution to Science Objectives
f/24 Field Camera	36	Cosmology, Extragalactic, Young & Binary Starts, Stellar Parallaxes, Solar System
FOS	24	Cosmology, Extragalactic, Quasars, Interstellar Medium, Binary & Old Stars
HSAP	22	Cosmology, Quasars, Stellar Parallaxes, Old Stars
IR Photometer	16	Extragalactic, Young Stars, Solar System
HRS	2	Quasars, Interstellar Medium, Young Stars, Solar System
Astrometer	--	Binary Stars, Stellar Parallaxes
f/48-96 Planetary Camera	--	Quasars, Solar System

CONCLUSIONS

The timeline developed herein demonstrates that the LST, with a cost effective payload of SI's, can perform a wide variety of astronomy experiments. With careful mission planning, the LST utilization (i.e., useful science time per orbit) is high, allowing it to support a large segment of the astronomical community in a timely fashion. In addition, it can perform this mission without requiring overly expensive requirements on either the flight hardware design or operations system.

AAS 75-200

LARGE SPACE TELESCOPE MISSION ANALYSIS[*]

Frank M. Friedlaender[†]

INTRODUCTION

The Large Space Telescope (LST) will be one of the first large scientific payloads to be launched and supported by the Space Shuttle. This payload will have a 2.4-meter-diameter, near-diffraction-limited mirror in conjunction with four interchangeable instruments. The current family of instruments from which these four instruments will be selected for the initial launch in 1982 consists of an f24 wide field camera, faint object spectrograph, high-speed point/area photometer, infrared photometer, f48/f96 planetary camera, astrometer, and high resolution spectrograph. The LST is designed for long-duration operation with periodic maintenance and refurbishment cycles after extended observational periods.

The instrument complement can be changed for any specific observation period, and allows measurements from 1200 Å in the ultraviolet to 1 mm in the infrared. The interchangeability of instruments as well as of other spacecraft components provides the long-term capability to this observatory. Maintenance can be performed on orbit with support of the Space Shuttle or, using the Space Shuttle as the transporter, major refurbishment can be accomplished on the ground.

This earth-orbiting observatory will open a new era of astronomy because the LST can see 7 times farther and 350 times as much volume as the best ground-based telescope. It also has 10 times better resolution and 10 times the frequency spectrum of ground-based systems. It is this greatly expanded astronomical capability in conjunction with the environmental constraints imposed by an orbital observatory that provides the mission planner with a unique challenge.

SCIENTIFIC OPERATIONS

The planning for the scientific observation is shown in Fig. 1 and contains both a long-term and short-term sequence. The long-term sequence is performed several months prior to the observations, and the short-term sequence is performed just days prior to the observation. The interaction of the targets, instruments, observation time frame, and environment provide the ingredients that

---

[*]This work was done under Contract NAS8-31313.
[†]Frank M. Friedlaender is responsible for LST Mission Analysis at Lockheed Missiles & Space Co., Inc., Sunnyvale, California.

make up the fabric of each experiment. Each experiment, in turn, consists of a target and one or more instruments operating in a number of modes. As part of the long-term planning sequence, the optimum time must be chosen to conduct the experiment, with consideration given to both the time of year and the environment. The environment includes background phenomena such as zodiacal light and integrated starlight radiance, proximity to bright objects, passage through the South Atlantic Anomaly, and whether the observation is in sunlight or in the earth's shadow.

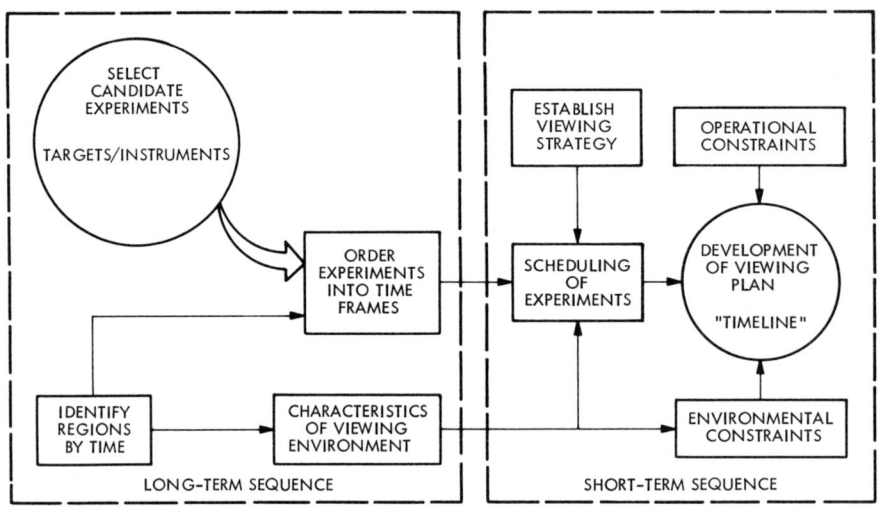

Fig. 1 Science Observation Planning Sequence

SUPPORT OPERATIONS

Although the scientific mission operation phase is the key mission phase in satisfying the scientific objectives of the LST, it is only one of several flight phases.

To gain some perspective on the overall operation, Fig. 2 shows the flight phases in context of the total mission scenario. This figure shows all of the events from the integration of the spacecraft until its final return to Earth. The optical system and the instruments are integrated into a payload element, which is then mated to the spacecraft bus. This is now the LST which goes to systems test, to launch site, and into orbit. Several on-orbit maintenance actions are then repeated until the mission is completed. On-orbit maintenance actions are expected to have a frequency of 1 to 2 years, with earth return 4 to 7 years. Planning for all of the mission phases is considerably broader in scope than planning for the science portion only.

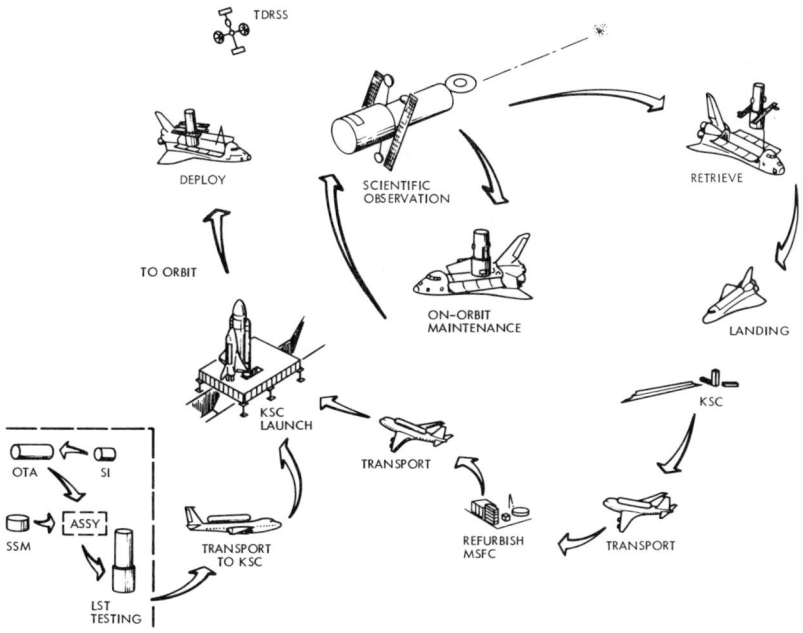

Fig. 2 LST Operational Phases

The Large Space Telescope deployment, shown in Fig. 3, is one of the first significant events in the flight operations sequences. The Orbiter remote manipulator system is used to erect the telescope on the sill of the Orbiter bay, the solar arrays are deployed, the antennas are deployed, a direct communication link is established, and the telescope is raised from the sill and released as the Orbiter backs away. The LST is then completely checked out while the Orbiter flies in escort mode for several days.

Another area of particular interest to the mission planner is the maintenance and refurbishment of the telescope throughout its life-cycle. Both on-orbit maintenance and earth refurbishment have been included. On-orbit maintenance consists of the replacement of those elements called orbit-replaceable-units which can be readily replaced on-orbit. The replacement consists of routine replacement of limited-life components such as batteries, components that have failed during the operational period and thereby cause reduced system redundancy, and components that are not expected to last until the next maintenance action. Ground refurbishment consists of major actions such as recoating the mirror, changes that affect vehicle interfaces, and updating of instruments and equipment.

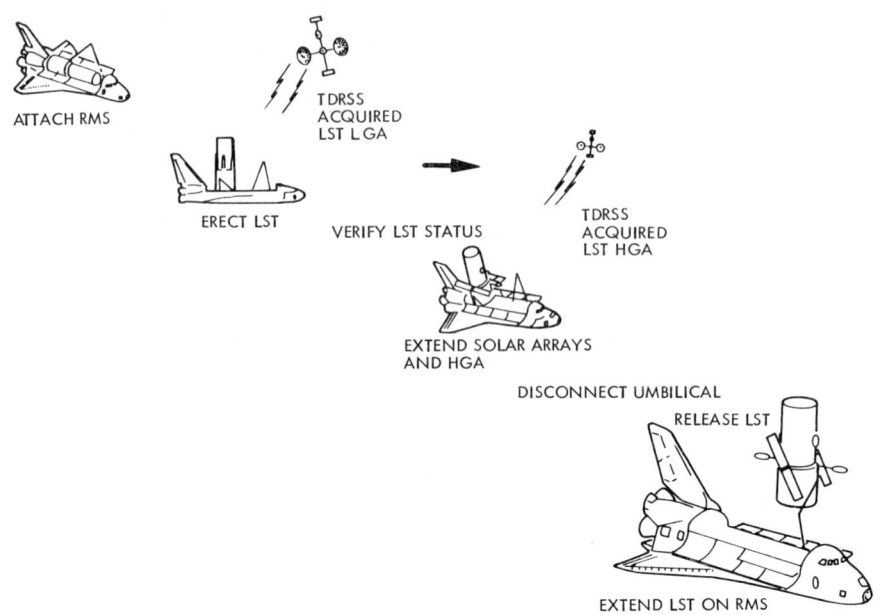

Fig. 3 Deployment of LST

The periodic maintenance action also permits the planned replacement of selected electronic components that are potentially subject to radiation damage. Radiation damage is caused primarily by repeated passages through the South Atlantic Anomaly. Electronic components have different levels of susceptibility to radiation damage, depending upon their composition. Although the minimum lifetime is approximately 15 years, it is advisable to plan for the replacement of these components on a regular schedule. The periodic replacement of components can provide both the lifetime and flexibility for an extended mission.

CONCLUSIONS

The inherent characteristics of the periodic Space Shuttle revisits and earth return capability provide this astronomical observatory with a capability for growth that should be incorporated into every phase of its development. Not only must the design allow for a change in technology, such as allowing higher data rates, more computer storage, and greater instrument capability, but the integration and testing of these systems with new capabilities must be accomplished without having to rebuild the spacecraft whenever an improvement occurs. It is, however, this basic growth process in the development that will permit the LST to become an outstanding astronomical facility initially and to grow with the constantly improving technology that will allow the observatory to continue probing the frontiers of science.

AAS 75-201

AUTOMATION OF THE LST

William W. Warnock*
C. William Case*

The Large Space Telescope (LST), to be launched into Earth orbit in the early 1980's by Space Shuttle, will have an effective aperture of 2.4 meters, observe wavelengths from about 115 nm to approximately 1 mm, and be capable of operation for 15 years. The LST will possess many observational advantages compared to ground-based telescopes: diffraction-limited performance, faint object sensivity, no atmospheric absorption or refraction, no mechanical perturbations from gravitational flexure, full celestial sphere viewing, and continuous observing capability. A variety of scientific instruments will provide the capabilities for imagery, spectrography, photometry, astrometry, and polarimetry.

Automation of the LST is obviously required, to some level. This space observatory should be automated optimally for maximum science return within cost constraints. The automation of the LST can be separated into two categories: on-board and ground. The on-board subsystems must have the capabilities for (1) a variety of processing functions related to the operation of the scientific instruments and (2) spacecraft control functions. The Data Management Subsystem (DMS) is the major contributor for providing these capabilities, and its design is consequently crucial. Automation of ground systems, with high data volumes, is also important. Science operations and mission operations must interact efficiently for proper design, control, and execution of a mission plan. These requirements call for a judicious design of automated ground systems.

---

* Dr. Warnock and Mr. Case are affiliated with Martin Marietta Aerospace, Denver Division, Denver, Colorado

INTRODUCTION

Phase B studies are being conducted by several contractors for the three sections of the LST: the Optical Telescope Assembly (OTA), the Scientific Instruments (SI), and the Support Systems Module (SSM). In addition, a wide variety of supporting disciplines are being researched under Advanced Technological Development contracts. Candidate scientific instruments for LST are listed in Table 1, along with one important characteristic for each.

Table 1
CANDIDATE LST SCIENTIFIC INSTRUMENTS

Instrument	Characteristic
Astrometer	Positional Accuracy = $\pm$ 0.002 arc-second
Faint Object Spectrograph	Spectral Resolution = 10^3
High Resolution Camera(s)	Imagery with High Spatial Resolution
High Resolution Spectrograph	Spectral Resolution = 10^5
Infrared Photometer	Wavelength Range = 2 - 1000 μm
UV - Visible Photometer	Time Resolution = 100 μsec

AUTOMATION CONSIDERATIONS

Maximum science return from LST requires maximum utilization of the observatory, which will produce large data volumes (3 x 10^9 bits per 24^h). The automation of the LST can be separated into two categories: on-board and ground. Trade studies are being conducted by Martin Marietta and the other two SSM contractors for determining the best designs for accomplishing the automation functions.

On-board Automation

The on-board subsystems must have the capabilities for (1) a variety of processing functions related to the operation of the scientific instruments and (2) spacecraft control functions. General instrument-related processing functions include: observation sequencing; target acquisition and slit centering; data encoding; merging of science data and engineering data; fault detection and safing; and bulk data storage management. Some specialized on-board functions are Doppler shift correction for the High Resolution Spectrograph, data buffering for the high-speed UV-Visible Photometer, and data storage for photon-counting detectors. On-board spacecraft control functions include: pointing control, SSM subsystem fault detection and safing, and engineering data formatting and command handling.

The Data Management Subsystem (DMS) will provide most of these functional capabilities. Six basic DMS concepts have been considered, with emphases being placed upon cost-effectiveness and SI accommodation flexibility. The difference among the DMS concepts lies in the number of computers employed and the separation/integration of command and telemetry functions with the computers. The need for an SSM computer to perform pointing control functions has been firmly established. The currently preferred concept has separate command and telemetry systems, an SSM processor, and a subsequencer/microprocessor within each scientific instrument. This approach enables standardization and yet gives great flexibility and growth potential for later instrument designs.

Ground Automation

Automation of the ground systems required for flight operations is necessary for efficient use of the LST. A brief description of the functional requirements for these systems is best provided by the following NASA guidelines[1], subject to change as a result of trade study conclusions:

1. All LST scientific operations shall be managed by an LST Science Institute and the responsibilities of the Institute shall include control of science viewing requirements, science mission planning, pointing verification, quick-look science evaluation and attendant mission changes, final calibration determination, guest observer selection and interface, and science data management, including preprocessing, processing and analysis.

2. The Science Institute will establish observation requirements and perform the science viewing mission planning and an LST Mission Operations Center (MOC) will perform the integrated mission planning including the science requirements, spacecraft constraints, support requirements, and Shuttle support as required. Execution of the mission plan shall be the responsibility of the MOC. Responsibilities shall include such items as spacecraft command and control, spacecraft status monitoring and contingency control.
3. The MOC will perform the data processing and evaluation of the engineering and selected science data in support of overall evaluation of the spacecraft.

CONCLUSIONS AND SUMMARY

The LST observatory will function very similarly to a ground-based, automated, general-purpose, optical observatory, with some exceptions which include the following: (1) an RF communications link will be required by LST for commands and telemetry; (2) on-board computer(s) will be required by LST for several types of processing and control; (3) operating constraints will be different; (4) observatory status monitoring will include different sets of variables; and (5) fault detection, isolation, and recovery will require different techniques.

Automation of astronomical observatories can be used to maximize science return within cost constraints. While automation of ground-based optical observatories is not required for many types of observations, it must be used for LST. Trade studies are being conducted to determine the best methods for satisfying the large variety of LST automation requirements.

REFERENCE

1. LST Project Requirements and Guidelines Document - Revision 2, NASA/MSFC, 25 June 1975.

AAS 75-179

LARGE SPACE TELESCOPE EXTERNAL INTERFACES*

Richard E. Collart[†]

INTRODUCTION

The Large Space Telescope is the first of the permanent large space observatories made possible by the development of the Space Shuttle. The LST is designed for on-orbit maintenance from the Shuttle as well as retrieval for ground-based refurbishment if required. This capability will permit high-confidence cost estimating and lower total costs. The LST will be an international facility similar in operations to ground-based observatories such as at Kitt Peak.

The success of the LST as an outstanding science achievement in the latter part of this century will be directly affected by the definition of its interfaces. The requirements of the LST external interfaces will determine the design and operation of the LST. Also, the LST requirements will impact those areas it interfaces with. The earlier the interfaces can be defined, the lower the total program costs.

Interfaces have been established for expendable payloads of the past, but the LST presents unique new interfaces, for it is one of the first large satellites that will use the Shuttle as both a launch vehicle and a service vehicle. Because the LST will be an international facility operating for years to come, long-term operational and scientific interfaces are of paramount importance. An important new interface will be with the Tracking and Data Relay Satellite System (TDRSS), which will be developed in the same time frame as the LST and will augment the Spaceflight Tracking and Data Network (STDN). Figure 1 shows the key external interfaces for the LST project.

KEY INTERFACES

Key interfaces can be directly related to the mission phases, which include ground test and integration, prelaunch and launch operations, ascent, deployment, operations, on-orbit maintenance, retrieval, de-orbit and landing, and refurbishment. As the LST spacecraft progresses through these mission phases, the following key interfaces evolve.

---

*This work was done under Contract NAS8-31313.
[†]Systems Engineering and Integration Manager and Chief Systems Engineer for the LST Program at Lockheed Missiles & Space Company, Inc., Sunnyvale, California (Department 61-51, Bldg. 562, (408) 743-0914.

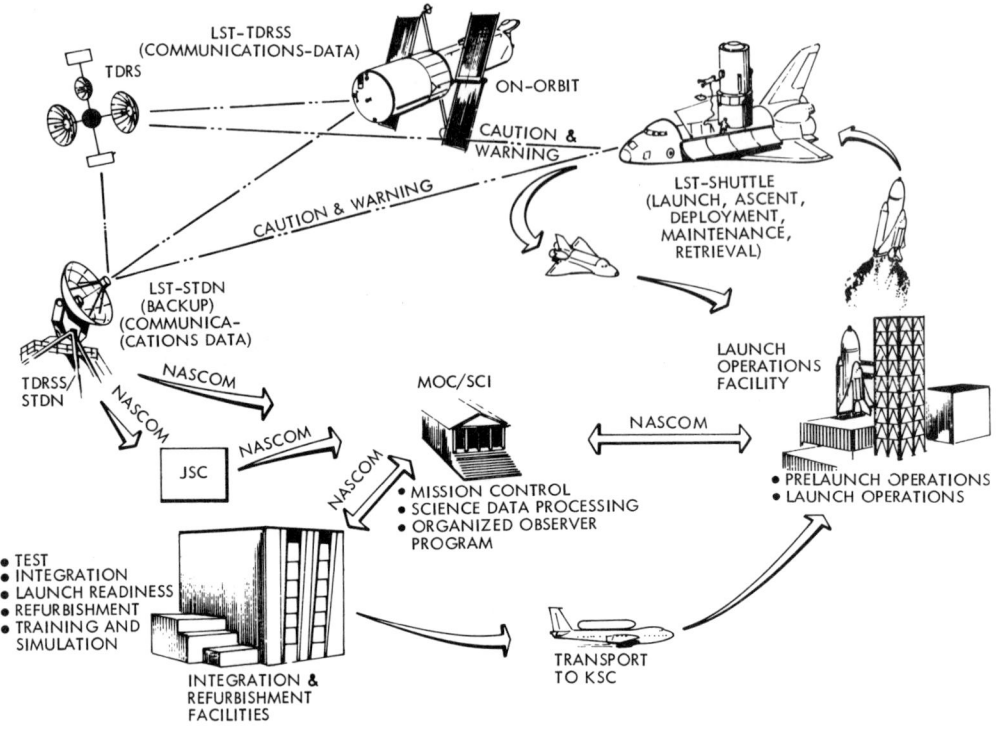

Fig. 1 External Interfaces

Test and Integration Interfaces

The LST-to-test-and-integration interfaces are primarily in two areas — facilities and ground support equipment (GSE). A typical LST integration facility such as that located at LMSC provides the interfaces required to integrate and test the LST spacecraft prior to launch operations. Floor area, crane capacities, hook heights, door sizes, and test chamber capabilities at LMSC are adapted to LST requirements and configurations. The close grouping of the pertinent areas will keep handling/mobility operations and risk of contamination to a minimum and will require little relocation of supporting GSE throughout LST operations prior to shipment to the launch base.

LMSC is adjacent to the common boundary of the LMSC facility and Moffett Field Naval Air Station, thus precluding the need to transfer the LST over any public roads and permitting virtually door-to-door air shipment of the LST from the integration facility to the launch site by means of the 747/payload pod.

Launch Operations Interfaces

The LST-to-launch-operations interfaces are similar to these required for test and integration; again these interfaces are primarily facilities and GSE.

Shuttle Interfaces

The interfaces between the LST and the Shuttle are unique new interfaces that must be developed because the Shuttle will be used both as a launch and as a service vehicle. These interfaces are caution and warning, electrical, mechanical, and on-orbit maintenance.

The Shuttle places requirements on users to ensure safety and provides facilities for monitoring potentially hazardous elements and initiating safing commands upon detection of a warning. All caution and warning (C&W) signals from both the Shuttle and the LST are routed to the Integrated On-Orbit Crew Station on the aft flight deck. The Aft Crew Station has the capability to provide checkout and diagnostic services to payloads and orbiter subsystems; it also has the C&W Status Panel for displaying parameters. A C&W Annunciator Panel in the Forward Crew Station relays alarm conditions from the Aft Crew Station.

To minimize interface and LST program costs, the electrical interfaces between the LST and the Shuttle are simple and consistent with the factory-to-pad test concept. Several Shuttle to ground umbilicals are available to the payload users, but at this time only the T-0 umbilical with the digital forward (command) and return (health and status telemetry) data are planned to be used. The caution and warning capability through the Shuttle will provide sufficient command and monitoring of the LST for the last few hours the Shuttle is on the ground. Other electrical interfaces include battery charging, deployment, instrumentation and monitoring devices, and status monitoring.

The Shuttle payload bay available for the LST mission is 180 inches in diameter and 720 inches in length, minus the OMS tank kit length. These net available payload bay dimensions (15 feet diameter by approximately 50 feet long) must accommodate the LST spacecraft and associated support equipment. The LST spacecraft envelope must allow 48 inches clearance of the forward payload bay bulkhead for hatch opening clearance. Mechanical connection of the LST to the Shuttle payload bay occurs at four points: three on a plane at Shuttle station 1010.0 and a fourth at Shuttle Station 951.0.

The key interfaces for on-orbit maintenance of the LST are the Shuttle and its crew. The tilted payload system is an example of a system used to allow servicing of the LST while it is attached to the Shuttle. The LST is swung out of the payload bay on pivot points located on both sides of the LST; firm sill bridge attachment of the payload is maintained throughout the vertical erection and locking process. The tilted payload position permits full 360-degree access to the LST. The astronauts are fully constrained with built-in crew aids (mostly GFE equipment from Skylab and other NASA programs).

Data (TDRSS/STDN/NASCOM) Interfaces

The LST will be supported by an extensive network of existing and planned facilities; these will be an extension of the current STDN and NASA communications (NASCOM) network operated by GSFC. Other major facilities completing the total support network include the MOC and the ScI.

Mission Operations Interfaces

The principal operational interfaces are between LST and the MOC, ScI, and science community. MOC personnel will control all real-time and near real-time activity of the LST by the transmission and execution of real-time command and stored program commands. MOC mission controllers will use displays to monitor the onboard activities, health, and status in near real-time. MOC will also provide near real-time processing of a limited quantity of quick-look science data to the ScI.

The ScI will develop experiment requirements that are based on requests from astronomers and transmit the data to the MOC. The ScI will use a remote terminal to control the processing of science data, the preparation of output topics, the listing of items for analysis by the ScI, and the transmittal of experiment requests. The ScI will manage all current and archival science data libraries.

CONCLUSIONS

The LST external interfaces are presently being defined in detail as part of the Phase B Definition Studies. These early definitions and tentative agreements will lay the initial ground work for the C/D phase of the program. NASA and its contractors understand the importance of this task, because LST program costs will be directly proportional to the time used to reach the formal external interface agreements: the earlier in the program this task is accomplished, the lower the total program cost.

AAS 75-180

LST REFURBISHMENT AND SUPPORT

John Henschke*

The Large Space Telescope (LST) is a Space Shuttle
payload with relatively high reliability, but it will
require maintenance support in order to meet the operational life requirement of 15 years. While planning
the LST support program, Martin Marietta considered
having only ground refurbishment. This ground activity
would include optics recoating and realignment. thermal
coat renewal, bench check of all line replaceable
units (LRUs) and replacement of failed or degraded
items. After studying this subject and comparing
costs, it was determined that it would be more efficient and economical to support LST by having four (4)
limited IFM missions and three (3) ground refurbishments. The IFM missions would include replacement of
the four (4) items with a limited operating life. If
another critical component failed prior to the scheduled IFM flight, its replacement on-orbit would be
considered but this would be a real time decision
based on the projected degree of operational degradation before the next ground refurbishment. The LST
program would provide for a ground refurbishment every
four years at a central payload depot. It was recommended that the depot should be located at MSFC and be
manned by a permanent party which would be augmented
on a temporary basis during LST refurbishments by
engineering personnel from the LST contractors. Required spare parts would be acquired incrementally
for the depot based on maximum refurbishment of
repairable items. In order to keep support costs
low, maximum use of existing equipment, parts, transportation, data, and support systems should be planned.

INTRODUCTION

In conjunction with the LST Phase B Definition Study Contract NAS8-31312, the Denver Division of the Martin Marietta Corporation investigated the maintenance/refurbishment and support requirements of the LST equipment. The results of the investigation, discussed below, have been presented to MSFC, but to date have not been approved by NASA.

MAINTENANCE REQUIREMENTS

Since maintenance requirements are based on predicted failures, a detailed reliability study of the LST design, as proposed by Martin

---
*Martin Marietta Aerospace, Denver Division, Denver, Colorado

Marietta, had to be completed. This was accomplished using available LST design data plus that generated from previous space programs, in a Monte Carlo maintenance simulation computer model. The LST items that are most likely to fail in a relatively short time are bearings, image tubes and batteries. A preliminary contamination analysis performed by Martin Marietta indicated that the LST optics may be degraded as much as 20 percent in the 1000 Angstrom Wave Length after four years of operation.[1] The proposed silver-coated teflon thermal coating/cover has a predicted life time of ten years, but mechanical and/or thermal degradation will not occur uniformly on all sections. It can be assumed that after four or five years, certain portions of the thermal coat will be degraded below the level that is acceptable. Other LST functional components may fail, but the predicted failures will occur at much greater intervals.

INFLIGHT MAINTENANCE (IFM)

During the study it was confirmed that contingency IFM must be included in all LST maintenance programs. Contingency IFM in this case covers those on-orbit crew actions necessary to prepare the LST for deployment or return to earth when the external equipment such as arrays or antennas do not operate as planned.

Limited IFM

A limited type of IFM was considered for LST as a compromise between no IFM capability and complete IFM capability. Only the batteries, reference gyros, two SI modules and the fine guidance sensors would be replaced on every mission. During orbital operations, if another critical component fails prior to a scheduled IFM mission, its replacement would be considered as part of the next IFM mission. For these added tasks, the required tools and spares would be acquired, the additional procedures prepared, and the necessary training planned and conducted, as special tasks.

IFM Vs Ground Refurbishment

After a detailed analysis,[2] it was determined that having all ground return or refurbishment would be the least expensive and having complete

IFM with minimum ground return would be the most expensive. A combination of minimum ground return plus limited IFM would cost approximately ten (10) percent more than all ground return. Since the probability of requiring maintenance between one and two years is vastly increased without limited IFM, it was decided that limited IFM should be included in the LST maintenance program.

GROUND REFURBISHMENT

After analyzing the maintenance requirements and determining the extent of the planned IFM, a complementary ground refurbishment program was established.

General Concept

After the LST has been removed from the Orbiter at KSC, it will be shipped in the assembled configuration to the refurbishment site. The LST will then be disassembled into the major modules, and each functional component will be removed and bench checked. The optics assembly will then be shipped to the optics contractor for the recoating and realignment. Each failed or degraded and limited life item will be replaced, required in-place repairs accomplished, and all items including the optics assembly installed. The thermal coating will also be renewed. The controlling factor concerning the total ground time is the refurbishment of the optics assembly. Providing spare SI modules, a spare SSM structure as well as a spare optics assembly including the metering truss would reduce ground time to the practical minimum, but the cost would be excessive. A spare set of mirrors would cost about 20 percent of one refurbishment, and reduces the ground time from 45 to 35 weeks, so this was proposed.

Refurbishment Site

It was determined that the average cost of one refurbishment at a payload depot located at MSFC would be about 75% of the estimated cost for a refurbishment at a LST contractor's facility. Furthermore, if a depot was not established and the refurbishment was accomplished at MSFC using personnel on temporary duty, the average cost would be increased by about 35%. Since cost was the primary evaluation factor, it was proposed that a Payload Depot be established at MSFC and that it should be the LST refurbishment site.

Payload Depot Operation

For efficiency and cost effectiveness, it appears that other payloads for which MSFC is responsible should be refurbished in the depot. This would require a great deal of detailed planning, especially with regard to the scheduling of work in order to prevent unacceptable peaks and valleys in manning requirements. The depot could be organized similarly to one of the laboratories at MSFC. The proposed method for manning the depot would be to have a permanently assigned group of aerospace-experienced personnel which would be augmented by personnel on a temporary basis from the payload contractors.

REFURBISHMENT SUPPORT

Transportation

The 747 (Pod) is baselined, but it was determined that the Super Guppy should be considered as the primary backup mode for shipment of the assembled LST. The availability of the Super Guppy in the required time period is uncertain so the river barge could be the secondary back-up. The C-5A could be a primary backup for the optics assembly.

Spares/Supplies

The proposed provisioning method is to procure or build the necessary new spares and supplies after the first launch and prior to the first LST refurbishment, and to refurbish all repairable failed items. It is estimated that refurbishment of repairables would reduce spares costs for the LST approximately 40%.

Ground Support Equipment

The ground equipment necessary to support transportation, checkout, disassembly, repair, assembly, servicing, integration and test will be required at the depot. If the same GSE is required at other locations, it will normally be shipped rather than duplicated. GSE not required elsewhere will be stored at the depot.

Technical Documentation

Necessary technical documentation to support the refurbishment must be provided, but available data should be utilized to the maximum. It is proposed that operation and maintenance (O&M) procedures for LST ground support equipment be prepared. It is not believed that written detail step-by-step procedures will be necessary to cover the remove and replace tasks for the LST since the depot personnel will be experienced with aerospace hardware. For the same reason, it is believed that detailed formal procedures for shop repair of failed items removed from the LST flight article are not necessary.

Personnel and Training

It is assumed that the present skills training at MSFC can be expanded to cover LST requirements. In addition, courses will be conducted at the depot on the LST subsystems. Before any person will be allowed to operate, test, or maintain any LST equipment, he must be certified. Proper recertification of personnel will be especially important in case of LST because of the relatively long intervals between ground refurbishments.

Facilities

Building 4755 at MSFC is scheduled to be the Central Integration Site (CIS) and meets the LST requirements for refurbishment. Repair shops equipped with normal bench equipment are required to perform the bench check and repair of items removed during the flight article refurbishment and these are also existing at MSFC.

CONCLUSIONS

1. Based on available reliability data, four (4) IFM missions and three (3) ground refurbishments scheduled alternately after every two years of orbital operations, is the optimum maintenance program for LST.

2. From the overall program cost standpoint, the ground refurbishment of the LST with exception of the optics, should be accomplished at a central payload depot located at MSFC.

3. Maximum use of existing equipment, data, facilities and logistics systems should be emphasized in order to keep refurbishment support costs to the lowest practical level.

ACKNOWLEDGEMENT

The data, and helpful suggestions received during the preparation of this paper from Mr. A. Howard Kent, Jr., of the Martin Marietta Corporation is gratefully acknowledged and appreciated.

REFERENCES

1. L. Bareiss, <u>Preliminary LST Contamination Analysis (2.4m Option)</u>, MMC LST 75 203671-508, March 1975.

2. H. Kent, <u>On-Orbit and Ground Return Maintenance vs All Ground Return</u>, MMC LST TSR 11.1V, July 1975.

AAS 75-208

THE INTENSIFIED CHARGE COUPLED DEVICE AS A
PHOTON COUNTING IMAGER

Jack T. Williams[*]

SUMMARY

Introduction

The sensor is the limiting factor in any imaging system. This is especially true in astronomy where imaging low light level requirements are extreme. Sensors currently in use such as film, PMT's and TV cameras have severe limitations in astronomical applications which we feel can largely be overcome using a solid state intensified charge coupled device as a photon counting sensor.

ICCD

The ICCD is an extremely simple device consisting of three basic parts:
1. The photocathode for wavelength discrimination and generation of electrons,
2. Focussing method which can be proximity, electrostatic or magnetic, as required,
3. CCD for signal storage and readout.

The CCD itself is a near perfect analog shift register in which signal charge is transported over large distances with a silicon chip to a low noise amplifier. The arrays being used are TI 100 x 160 arrays that have been thinned for backside illumination to eliminate problems with the front side gate structure. These devices are three phase, buried channel arrays with a resolution element size of 0.9 x 0.9 mil. The CCD itself is a near perfect analog shift register in which photon or photoelectron generated signal charge is transported over long distances

---
[*] Goddard Space Flight Center, National Aeronautics and Space Administration

within a silicon chip to a low noise charge sensitive amplifier. The use of a thinned backside-illuminated device allows us to overcome the degradation in performance that occurs in frontside illuminated devices due to the necessity of acquiring signal either through or between the electrode gate structure with resultant absorption and interference effects. Buried channel devices were selected because of their higher charge transfer efficiency (CTE) and low noise characteristics at room temperature. When operating, a CCD charge is stored within the pixel elements and then by proper sequencing of the gate electrodes, the signal is clocked in a parallel manner, one line at a time, into a serial shift register where it is clocked out through a charge sensitive amplifier one pixel at a time.

Performance

In any photon counting device the goal is to be photon noise limited. This means that in CCD application, the inherent CCD noise must be less than the noise associated with the photoelectron generation process at the photocathode. To accomplish this a number of considerations have to be taken into account. These include:

1. Gain which is accomplished by secondary generation of electrons within the CCD at the rate of 1 for each 3.5 eV of accelerating potential.
2. Losses within the CCD due to surface dead layers and charge transfer efficiency (CTE).
3. Possibility that the signal from a photon event may be divided evenly among 4 adjacent pixels.

Once all these are taken into account one finds that the required performance characteristics are as shown in the following table.

Parameter	Required	Measured
Gain	6000	<6000
CTE	0.9992	0.9996
RMS Noise Electrons	100	80

This table also shows the measured values of currently available CCD arrays, indicating that current CCD technology can produce arrays of sufficient quality for photon counting.

Results

To date actual laboratory results show that CCD will operate in the intensified mode and can survive the processing required to place the array in a tube bottle with a photocathode. This was accomplished in a cooperative program with the Army Night Vision Laboratory which resulted in the successful fabrication of a proximity focussed ICCD which is currently undergoing testing at GSFC.

Future

The proximity tube cannot, nor was it intended to, do photon counting, however, a magnetically focussed device has been designed and is being fabricated for delivery late in 1975, which should be able to do single photoelectron detection.

AAS 75-209

THE INFRARED CAPABILITIES OF THE LARGE SPACE TELESCOPE

D. E. Kleinmann*

The conceptual design of the IR photometer proposed for the LST is described. The performance to be expected from the LST and its IR photometer is evaluated, and some possible applications to far-infrared astronomy are indicated.

INTRODUCTION

One of the seven instruments that have been proposed for the four instrument bays aft of the primary mirror of the LST is the infrared photometer whose design requirements have been specified by the LST Infrared Instrument Definition Team, consisting of G. Neugebauer[†], R. I. Hall[‡], T. Kelsall[#], and the author. This instrument in concept will use at least two detector channels to cover the 2 - 1000 micron wavelength range; present plans call for either a Si:As or a Si:P photoconductor for use from 2 μm out to 24 μm or ~30 μm respectively, and a bolometer for the longer wavelenghts. The photometer will feature cooled interchangeable field stops and filters for each channel. The passband of the filters will nominally be 10% of the effective wavelength, although some may be as wide as 50%. The field stops will range from the diffraction limited beam size for the shortest wavelength in a channel up to 10 times the diffraction limited beam size for the longest wavelength in that channel. The detectors, filters, and apertures will be

---

*Center for Astrophysics, Cambridge, Massachusetts, (617)495-7453.
[†]California Institute of Technology, Pasadena, California,(213)795-6841.
[‡]Aerospace Corporation, El Sequndo, California, (213)648-7012.
[#]Goddard Space Flight Center, Greenbelt, Maryland, (301)982-6045.

placed in a dewar and maintained at a temperature less than 2 °K for an on-orbit lifetime of at least one year. Chopping will be accomplished by wobbling a mirror on which the primary mirror is imaged. This chopping mirror is located in the warm relay optics forward of the dewar. The chopping frequency will be adjustable over 5 - 35 Hz and the amplitude will be adjustable from 0.4 arcsec to 210 arcsec on the sky.

DETERMINANTS OF SYSTEM PERFORMANCE

Absence of the Atmosphere

For the LST there will be no atmosphere to attenuate the signal, or to produce a position-dependent, time-varying background, or to degrade the angular resolution because of "seeing". In giving the angular resolution of several large telescopes, Figure 1 shows that unless the seeing disc is smaller than 1 arcsecond, the LST will provide higher angular resolution than any ground-based telescope out to ~5 μm. It also shows that the larger ground-based telescopes do not enjoy the full advantage of their size until 20 or 30 μm, where atmospheric attenuation begins to preclude all ground-based observation out to 1 mm, except for a rather occasional window at 350 μm.

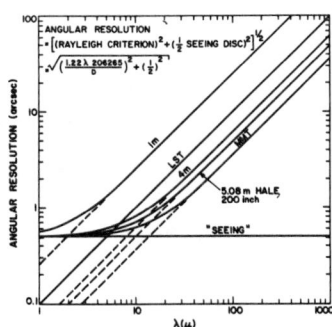

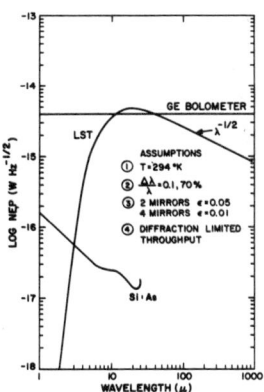

Figure 1. The Effect of Atmospheric Seeing on Angular Resolution. Diffraction limited optics and a seeing disc 1 arcsec in diameter are assumed.

Figure 2. Individual Contributions to the System Noise. Specific imputs to the calculation of the background radiation noise power conform to the properties of the LST and of the IR photometer described in the text.

Noise Power

Figure 2 shows the individual noise powers of each of the most significant contributors to the system noise power, which is the square root of the sum of the squares of the noise power from each element. The curve labeled LST is the fluctuation in the background power that is incident on the detector through a diffraction limited beam. It is the NEP that could be obtained if a perfectly efficient detector were available. The background NEP is a fundamental limit that can be relaxed only by reducing the background on the detector --- e.g. by lowering the temperature or emissivity of the background, or by reducing the optical throughput ($A\Omega$), or by reducing the bandpass of the cold filter.

Figure 2 also shows the intrinsic noise for two detectors of possible application on the LST. The performances indicated for these detectors in Figure 2 have not been achieved in fact; nevertheless, the projections are modest, and do not require significant development in the state-of-the-art.

SYSTEM PERFORMANCE

The NEP measures the system performance at the detector. From an astronomical point of view, it is more useful to know the system performance at the entrance aperture, indicated by the noise equivalent flux density (NEFD) given in Figures 3 and 4. The NEFD gives the flux into the telescope that would produce an rms signal-to-noise ratio of 1 in a 1 Hz bandpass, or equivalently, in 1/2 second of integration time.

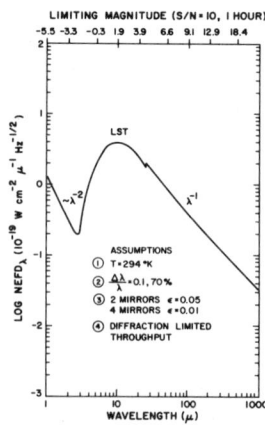

Figure 3. NEFD and Limiting Magnitude as Functions of Wavelength.

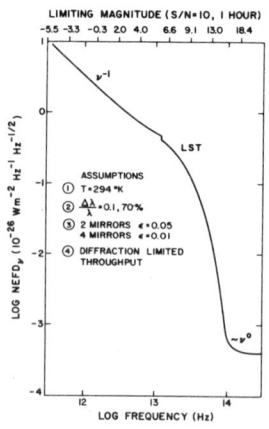

Figure 4. NEFD and Limiting Magnitude as Functions of Frequency.

Limiting magnitudes are also shown at the top of Figures 3 and 4. For this paper these are defined to be the visual magnitude of the A0 star (T = 9200 °K) that could be measured with an rms signal-to-noise ratio of 10 in a one hour integration.

SOME POSSIBLE APPLICATIONS IN FAR INFRARED ASTRONOMY

Figure 5 puts the performance of the LST into astronomical perspective by addressing the question, "How far could some of the bright, well-known far-infrared sources be removed before the LST instrument would require 1 hour to make 10 sigma measurement?" Galactic objects

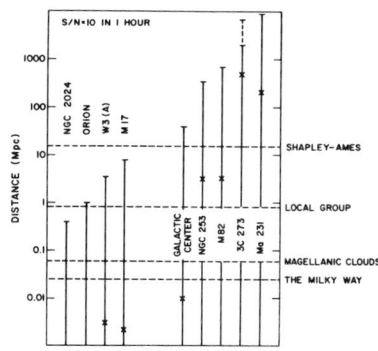

Figure 5. The Far Infrared Observability of Galactic and Extragalactic Sources.

are shown on the left in Figure 5 and extragalactic objects are shown on the right. Several distance milestones are indicated by the dashed horizontal lines, each of which is labeled on the right. The actual distance of each source is indicated by the X on its line, except for NGC 2024 and Orion which are too nearby to be accomodated within the boundaries of this figure. Figure 5 was prepared by comparing Figures 3 and 4 with far infrared flux measurements (preferably those with effective wavelengths between 65 µm and 100 µm) compiled from the literature[2,3,4,5,6,7]. Note however, that neither 3C 273 or Markarian 231 have been measured at ~100 µm. An estimate of the 100 µm flux from 3C 273 was obtained by interpolating between the ~20 Jy seen at 3 mm[8] and the 0.4 Jy seen at 10 μm[9]. The far infrared flux from Markarian 231 was estimated by assuming that that resemblance between M 82 and Markarian 231 from 10 - 20μ (Rieke, private communication) continues out to ~100 μm. From the observed ratio of their 10 μm fluxes[10] one therefore would expect that the 100 μm flux for Markarian 231 would be about 1/20 as bright as that observed for M 82.

Figure 5 suggests several intriguing observational programs. It would be possible for example to determine the number and location of of all the sources similar to the Orion nebula, W3(A), or M 17 in each of the galaxies in the local group. It should be possible to monitor the quasi-stellar source, 3C 273, to determine whether there is variability at 100 m, and if so, how it compares with that reported at 3 mm[8]. It would be possible to make a far-infrared survey of all 1149 galaxies in the Shapley-Ames catalogue to determine which ones have phenomena like that found in the Galactic Center.

REFERENCES

1. Smith, R.A., Jones, F.E., and Chasmar, R.P. 1968, The Detection and measurement of Infra-red Radiation, 2nd edition, Oxford Univ. press. see p. 214f and p. 288f.
2. Hoffmann, W.F., Frederick, C.L., and Emery, R.J. 1971, Ap. J. (Letters) 170, L89.
3. Harper, D.A., and Low, F.J. 1971, Ap. J. (Letters) 165, L9-L13.
4. Harper, D.A., and Low, F.J. 1973, Ap. J. (Letters) 182, L89-L93.
5. Emerson, J.P., Jennings, R.E., and Moorwood, A.F.M. 1973, Ap. J. 184, 401-414.
6. Fazio, G.G., Kleimann, D.E., Noyes, R.W., Wright, E.L., Zeilik II, M., and Low, F.J. Proc. Eighth ESLAB Symposium, Frascatti, Italy, pp. 79-85.
7. Olthoff, Henk 1975, PhD. Dissertation.
8. Fogarty, W.G., Epstein, E.E., Montgomery, J.W., Dworetsky, M.M., 1971, A.J. 76, 537-543.
9. Kleinmann, D.E., and Low, F.J. 1970, Ap. J. (Letters) 161, L203-L206.
10. Rieke, G.H., and Low, F.J. 1972, Ap. J. (Letters) 176, L95-L100.

AAS 75-210

DEVELOPMENT OF AN INFRARED SPECTRORADIOMETER

W. H. Alff and J. G. Thunen*

An infrared spectroradiometer for the Large Space Telescope is being planned that will allow astronomers to carry out observations in the IR with sensitivities and angular resolutions impossible to achieve using earth-based, balloon-borne, or aircraft-mounted instruments. To perform the required measurements, the basic instrument has been designed to operate diffraction limited over the spectral range of 2 to 1000 microns with approximately 10 percent resolution. The dynamic range capability (target brightness) will be at least 10^5 and special emphasis will be given to automated operation and data processing. The detectors will provide essentially background limited detection over the spectral range. To approach this sensitivity, it will be necessary to cool the detectors to approximately 2 deg K with supercritical liquid helium. The cooler system is thus integral to the performance of the radiometer and must have a lifetime of at least one year.

BACKGROUND LIMITED PERFORMANCE

The background noise results from statistical fluctuations in the radiation incident on the detector from thermal background sources. For example the primary and secondary mirror of the Large Space Telescope (LST) and the optical elements in the relay optics radiate according to Planck's Law. The magnitude of the statistical variations in the power radiated from the various background sources is defined as the noise equivalent power (NEP). The NEP is a function of electrical bandwidth detector area and wavelength and defines the systems ability to detect low level signals. To achieve background limited performance, the detectors must be cooled to eliminate generation and recombination noise which is internally generated in the case of photodetectors, and thermal noise (phonon noise) in bolometers. This requires the short wavelength detectors to be cooled to 4 deg K and the long wavelength detectors to be cooled to 1.8 deg K.

―――――――
*Lockheed Missiles & Space, Co., Inc., Sunnyvale, Calif.

CRYOGENIC SYSTEM

The IR detectors are maintained at 2 deg K or lower by a superfluid liquid cryostat. The $He^4 II$ is maintained nominally at 1.8 deg K at its saturation pressure of 12.466 mm Hg. These conditions are controlled by a vent line orifice whose diameter determines the flow rate for a given dewar heat leak. The heat load governs the evaporation rate. An increase in heat load, due to use of the electronics for example, causes an increase of dewar pressure. This increases the vent rate and pushes the system back toward the design point. A similar reaction occurs for a decrease in heat leak. The vent gas first passes through a porous plug which prevents the expulsion of liquid in zero gravity. The vapor then circulates through two vapor cooled shields which are used to intercept the heat leak through the multilayer insulation, fiberglass struts, fill line, wiring, and detector cavity aperture. The warm vapor then passes through the vent orifice and is dumped overboard through a pair of opposed nozzles directed tangentially to the LST skin.

MODULATION NOISE

It is necessary to have the detector alternately view the source and then the background alone to remove the background radiation during measurement of a source signal. This can be accomplished using many different techniques. The method used in this design is to place a tilt in a pupil plane to generate the image translation. The telescope pupil (primary mirror) is imaged onto a flat mirror in the relay optics which is oscillated angularly with variable frequency (3 to 30 Hz) and variable amplitude (0.4 sec to 210 sec field angle). No more than ten percent of the cycle time is required to move the mirror frame from one position to the other.

In practice, the most difficult source of thermal radiation to predict and control is that which can cause a difference in the background between the signal and reference channels. The most common sources of this noise are emissivity and temperature gradients in the telescope baffles and optics and in the relay optics prior to the scan mirror. Scanning of these gradients develops a modulation noise. In the long wavelengths, diffraction effects further complicate the analysis.

Several steps have been taken in the design of this radiometer to reduce the level of modulation noise. Initial analysis indicates these steps will minimize thermal radiation differences seen by the detector to a level which can be successfully handled in the data processing.

DIFFRACTION EFFECTS

It can be assumed that all optical aberrations throughout the radiometer are negligible for wavelengths greater than a few microns. Diffraction, on the other hand, is not negligible. At wavelengths over 100 microns the finite size of several critical instrument apertures diffract energy which can significantly degrade the expected performance. Some of the results are as follows:

- A loss in signal intensity from a point source

- Incomplete modulation due to diffracted energy from the object into the adjacent field

- Increased background emission (higher NEP) reaching the detector from diffracted baffle sources just outside the pupil

- Increased modulation noise due to variations in diffracted background radiation.

Predicting the distribution of diffracted energy at the detector is accomplished by initially considering coherent imaging which treats diffraction from a point source in object space. The other half of the problem considers diffracted energy from a source originating in the plane of the primary mirror. Coherent imaging describes the effects of diffraction on predicted signal levels, while incoherent imaging is used to describe the background noise level.

For the specific design considered here, this analysis predicts a signal-to-noise degradation over that predicted geometrically by a factor of four at 1000 microns. A revised instrument design is not expected to improve on this loss. These effects are relatively minor, however, when compared to the major increases in signal level offered by the large aperture optics and space environment of LST.

AAS 75-211

FAINT OBJECT SPECTROGRAPH

William P. Devereux*

To solve the secrets of quasars and distant galaxies, a spectrograph capable of analyzing faint objects is needed. Such an instrument can be built for the LST using photon counting detectors and a highly efficient optical system. A magnitude threshold of 26 is not beyond achievement.

INTRODUCTION

A faint object spectrograph is an instrument specifically designed to record the distribution, as a function of wavelength, of the energy emanating from a faint celestial object. Such an instrument is desired by the astronomical community in order to provide more data, hopefully leading to a better understanding of the nature of quasi stellar objects and the validity of Hubble's law when applied to very distant galaxies.

NATURE OF A FAINT OBJECT SPECTROGRAPH

Several factors combine to make a faint object spectrograph. They include a telescope with a large collecting area, a detector which conserves the information content of every photoelectron generated, a design that is highly efficient in terms of the number and quality of surfaces used, and a spectrograph concept that makes efficient use of the available detector format and does not provide resolution higher than is necessary. The first two items are the LST and the ICCD,

---

*Senior Member Technical Staff, Ball Brothers Research Corporation, Boulder, Colorado

both of which are discussed in separate papers. The design features between the telescope and the detector are the proper subject of this paper.

DESIGN FEATURES

To assure high efficiency in the design, three separate detectors are used in the spectrograph, each peaked for a different region, namely the visible, the near ultraviolet, and the vacuum ultraviolet. In addition, each of the detectors is used with two gratings so that operation is always near the blaze. The astronomical community agreed on a resolution of 1000 as meeting its needs, so that is the design value. The number of reflections was kept low by using a Wadsworth mount, and all mirror surfaces are coated to provide highest reflectance. A further increase in observing efficiency is achieved by the use of an image slicer which effectively triples the length of the detector format and allows the recording of more spectrum for each observing hour.

RESOLUTION CHANGING OPTIONS

To permit the observation of even fainter objects, a collapsed dispersion is made available. In this mode the mean resolution is decreased by a factor of ten, permitting more photon to fall on each picture element. In the visible and near ultraviolet ranges, prisms are used as the dispersing elements, and a grating is used in the vacuum ultraviolet where a suitable prism material does not exist.

For occasional observation of the fine detail of spectral features there is also a high dispersion mode with a resolution capability ten times greater. This is achieved at the expense of sensitivity and simultaneous spectral coverage, and can be used either with or without the image slicer.

A fourth mode of operation removes dispersion altogether by positioning an imaging mirror in the grating station and providing a near diffraction limited, photon counting camera for viewing the area in the immediate vicinity of the target object.

ACCESSORY FEATURES

Other features of the instrument include a slit jaw camera, which is used as an aid in pointing the telescope at the object of interest and a wavelength calibration lamp which makes possible absolute wavelength determination of spectral features.

CONCLUSION

With the spectrograph it is possible to record spectra of 21 magnitude stars at a resolution of 1000 with three percent photometric accuracy in a ten-hour exposure. By reducing the accuracy to ±10 percent, and using the collapsed dispersion, this threshold can be dropped to 26 or 27 magnitude. A faint object spectrograph worthy of the LST is within our group.

ACKNOWLEDGEMENT

The encouragement provided by LST program personnel at NASA, by the scientific community represented by the Faint Object Spectrograph Instrument Definition Team, and my colleagues at the Perkin-Elmer Company and Ball Brothers Research Corporation is gratefully appreciated.

AAS 75-212

HIGH SPEED AREA PHOTOMETER
CONCEPTUAL DESIGN AND INTEGRATION

William Bloomquist[*]

Fred Steputis[*]

INTRODUCTION

The LST Photometer will have the highest sensitivity and radiometric precision of all ultraviolet and visible instruments used on the LST and will provide a vast improvement over ground based photometry. Advantages of orbital operation, exploited by the Photometer, are accessibility of UV wavelengths absorbed by atmosphere, reduced veiling effects of air-glow and auroras, and freedom from spurious intensity and position modulation by inconstancies in the atmosphere.

Orbital operation makes the instrument inaccessible and places it in a harsh environment, requiring a highly reliable and tough instrument. The use of the shuttle makes possible the recovery of a malfunctioning or obsolete instrument but does not relieve the photometer design from the necessity of meeting environmental challenges and performing well over a great length of time without maintenance.

The design presented here has both point and area photometry capability with provision for inserting filters to provide spectral discrimination. The electronics provide photon counting mode for the point detectors and both photon counting and analog modes for the area detector which produces small field images with high resolution, precise photometric accuracy and wide dynamic range. The analog mode allows the area detector to be used as a target locating device for the point detectors.

The design originally had provision to analyze the state of polarization. The elements to provide this have been removed from the optical train. However, the design is such that these elements could easily be included.

---

[*] Martin Marietta Corporation

The design and development approach will reflect the capability of the shuttle to retrieve an LST that is not performing, either initially or one that suffers failures during operation. The retrieval capability justifies the assumption of greater risk and the resultant savings through reduction of expensive design verification, development, qualification and acceptance testing.

INSTRUMENT INTEGRATION

All of the LST scientific instruments are packaged into common axial modules that are identical in exterior dimensions and in mechanical, structural and electrical connection interfaces. The HSAP will be assembled as two submodules, opto-mechanical and electronics, and they are integrated into the larger common axial module.

INSTRUMENT GENERAL DESIGN

The instrument design concept introduces light from the OTA past a contamination shutter and dark shutter to focus on the entrance aperture wheel where the desired portion of the image is selected. The light then diverges to two filter wheels in tandem, and after spectral filtering the light is reflected by one of four mirrors selected by a carousel. Two of these mirrors are used to image the telescope's exit pupil on point detectors. Two mirrors are used as relay mirrors with secondary mirrors to reimage the light at either f/24 or f/96 and simultaneously to correct the OTA astigmatism.

All the optical elements and detectors are attached to graphite epoxy optical bench rods. With an almost negligible coefficient of thermal expansion, the rods maintain the alignment. The rods are supported at mid point and aft end on flexures to minimize stresses induced from deformations in the shell.

An aluminum radiator rejects heat from the thermoelectric device which cools the area detector to $230^\circ K$.

Mechanisms Design and Mounting

Typical of the mechanisms which operate the optical elements is the aperture wheel assembly containing a variety of apertures used to limit the field of view. The smallest is two Airy disks in diameter or 72 μm.

The need to center a point source within this opening is a controlling requirement on alignment tolerances. Initially calibrated using the area detector, it is required to have precise repeatability in the positioning of the apertures.

The wheel is driven by a stepper motor and positioned by a spring loaded cam which seats into detents on the rim of the aperture wheel to assure repeatability of positioning. An optic encoder controls the positioning and indicates position. The entire assembly is mounted to the graphite epoxy bench rods and is adjustable to slide along the bench rod for initial adjustments with clamps to lock the assembly in place.

Instrument Electronics Support

Functionally the electronics consist of several elements. Area detection consists of the detector, its intensifier and grid, high voltages, drive clock, signal conditioning and a 16k x 16 bit memory. In photon counting mode the 160 x 100 element array is read out when information reaches 4 bit accuracy, then, through use of an adder, added into the memory.

Point detection is accomplished by either of two channel electron multiplier phototubes. There are two discriminators and photopulse counters for redundancy. Output from the point detectors can be stored in the area detector's memory.

A microprocessor handles the timing and control functions and stores sequences of instrument operations. A small buffer memory stores commands from outside the Photometer.

OPTICS DESIGN AND PERFORMANCE

The Photometer consists optically of two conventional or point photometers and two imaging photometers which resemble cameras.

Fabry mirrors are used for the point modes and relay primary mirrors are used for area detection magnifications.

The first optical element encountered by the light in the Photometer is a set of entrance apertures, located at the telescope cassegrain focus which serve as field stops to select the portion of the sky whose brightness is to be measured or which is to be imaged. There are circular, slit, and rectangular apertures.

Spectral discrimination is accomplished in either point or imaging modes through use of two filter wheels each having eight positions including an open position thus allowing each filter to be used separately or in combinations.

Area photometry, performed by the camera type optical system, accepts the astigmatic off-axis image from the telescope and reimages the light onto an Intensified Charge Coupled Device (ICCD) at either f/24 or f/96. The reimaging adjusts the plate scale of the image and removes virtually all of the telescope's image aberration.

The f/24 system image resolution is limited almost equally by optical image quality and by inherent detector resolution depending on the wavelength. At f/96 the system image resolution is limited by the optical image quality which in turn is limited by diffraction. The f/96 pictures are expected to contain nearly all of the image information possible from the 2.4m telescope. The high resolution of f/96 is obtained at a price, of course. The image irradiance is less than that at the f/24 input image by a factor of at least sixteen. For many astronomical objects the obtainable resolution will be limited by photon statistical fluctuations.

System Performance

The Photometer will have the highest light sensitivity of all LST instruments: to magnitude 26.5 or better. It will have the highest time resolution: $100 \mu s$ or better. And it is designed to have the highest photometric precision: 0.1% is the goal. The Photometer will also have spatial resolution limited only by diffraction to about 0.1 arc seconds over a relatively small field-of-view. With minor modification the Photometer can also do highly precise polarimetry.

The LST Photometer provides a useful scientific instrument which fully utilizes and complements the unique astronomical capabilities of the Large Space Telescope.

AAS 75-213

HIGH RESOLUTION SPECTROGRAPH

Keith Peacock[*]

INTRODUCTION

The High Resolution Spectrograph (HRS) is one of four instruments sharing the focal plane of the Large Space Telescope. It occupies a 90° segment centered on the LST optical axis. The telescope is an F/24, 57.6 meter focal length Ritchey-Chretien whose primary aberrations are astigmatism and focal plane curvature.

The design requirements have been established by a team of Astronomers under the Chairmanship of Dr. A. Boggess of the Goddard Space Flight Center. It is required to cover the spectral range 115 to 410 nm with resolutions of 10^3, 3×10^4 and 10^5. The detector is an SEC Orthicon with a 51 x 56 mm target and available with Bi-alkali/SiO_2 and CsI/MgF_2 photocathode/window combinations.

The design presented here meets the requirements, supplies all the desired modes of operation, has minimum aberrations and yet contains a minimum of moving parts.

OPTICAL DESIGN

The optical layout of the HRS is shown in Figure 1.

A slit mechanism permits the selection of any one of eight fields-of-view between 0.1 and 10 arc second square. After collimation by an off-axis parabolic collimator with a focal length of 800 mm the beam falls on an

---
[*]Dr. K. Peacock is a Staff Physicist with the Bendix Aerospace Systems Division in Ann Arbor, Michigan.

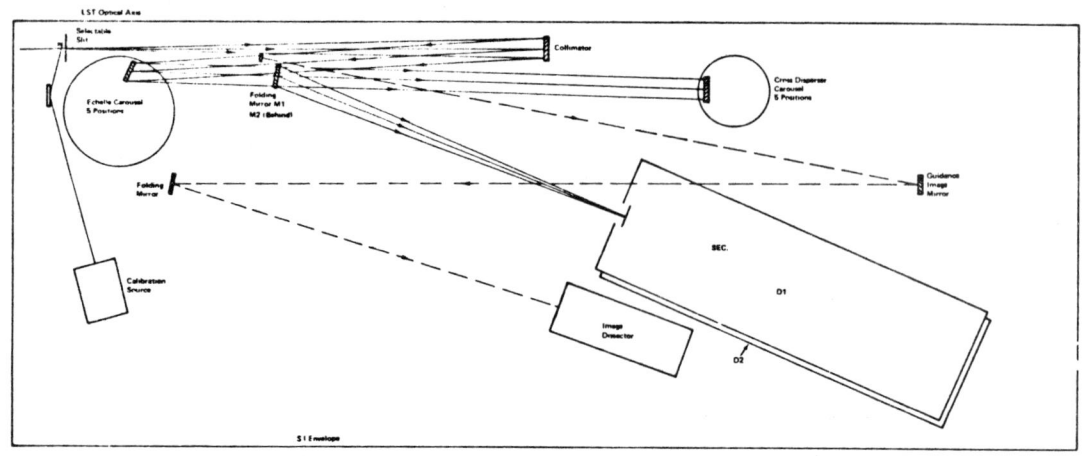

Figure 1. Optical Schematic

echelle carousel which supports 4 selectable echelle gratings and a plane mirror. The carousel has a sinusoidal rotation with ±15 arc sec amplitude for correction of the orbital Doppler shift. The diffracted beam passes to a second carousel which holds three selectable, spherical, cross-disperser gratings and two off-axis parabolic mirrors. One cross disperser is blazed to send the diffracted beam to folding mirror M1 which reflects the spectrum to short wavelength detector D1. The CsI/MgF$_2$ photocathode/window combination of this SEC orthicon limits its range from 115 to 170 nm. Any of four images can be displayed on this detector; (1) a cross dispersed spectrum at a resolution of 3×10^4 using one of the echelle gratings, (2) a cross dispersed spectrum at a resolution of 1×10^5 using another echelle, (3) a single low dispersion spectrum at 10^3 resolution using cross dispersion only with the plane mirror in place of the echelles, and (4) a direct image using a figured mirror in place of the cross-disperser.

Two other spherical gratings image either of the wavelength bands 170-265 or 265-410 nm across the bi-alkali photocathode of detector D2. With the additional two echelle gratings the total operational modes are the eleven listed in Table 1.

Table 1

MODES OF OPERATION

Mode	Echelle Carousel	X-Disperser Carousel	Spectral Range (n.m.)	Spectral Resolution	Detector
1	E1	G1	115-170	3×10^4	1
2	E2	G2	170-265	3×10^4	2
3	E2	G3	265-410	3×10^4	2
4	E3	G1	115-170	1×10^5	1
5	E4	G2	170-265	1×10^5	2
6	E4	G3	265-410	1×10^5	2
7	M1	G1	115-170	1×10^3	1
8	M1	G2	170-265	1×10^3	2
9	M1	G3	265-410	1×10^3	2
10	M1	P1	115-170	------	1
11	M1	P2	170-600	------	2

A small folding mirror picks off a few percent of the collimated beam to provide an image at the image dissector. The dissector is used for target identification, acquisition and reacquisition after an exposure interruption. It operates in a closed loop mode through the Systems Support Module to the telescope guidance.

The HRS contains a hollow cathode lamp to act as a spectral calibration source. Its energy is reflected through a pinhole close to the target slit.

The system optical efficiency, including detector quantum efficiency, varies between 0.2 and 1.5% in the required wavelength range.

The abberations of the instrument are very small and do not significantly degrade the image. Diffraction, wavefront error, detector resolution, and image motion are the main limitations and they produce a resolution of 3.5 lp/arc sec at 50% MTF and 9 lp/arc sec at 10% MTF for a wavelength of 1714Å.

Mechanical, Thermal and Systems Design

The optical components and mechanisms are mounted to an "optical bench" which consists of 2 longitudinal beams attached to a manufactured piece of each end. These end pieces attach to the three instrument package support points; one on either side near the focal plane and one three-quarters of the way along the package. The optical bench must be kept at a constant temperature to maintain alignment, focus and to allow precise reacquisition of the spectrum after an exposure interruption by earth occulation. A thermal control subsystem adjusts the power to various heaters so that the temperature at strategic points is constant.

The weight of the HRS including the package is estimated to be 203 kg.

The electronic system is autonomous requiring inputs of power, reference clock and gross operational commands only from the SSM. A continuously powered Digital Interface Unit (DIU) links the instrument with the rest of the spacecraft. After recognition of the HRS command a microprocessor assumes control of the HRS and using stored, or transmitted functions, it controls power, heaters, detectors and exposures. The main mode of exposure control is electronic shuttering of the detector high voltage power supply using one of the system clocks for timing. The SECO target is read out as straight video at 10^6 bps with frame sync supplied by the microprocessor through the output selector gate and the DIU line driver.

The average operating power drawn by the HRS is 120 watts and the peak power is 160 watts.

CONCLUSION

This design study has revealed no major technical problems which could prevent manufacture of an HRS capable of accomplishing the scientific objectives of the investigators.

This work was supported by ITEK subcontract 8237-A-0001 under prime contract NAS 8-29949.

AAS 75-294

THE EUROPEAN SPACE AGENCY STUDY OF PHOTON COUNTING IMAGING FOR LST

R. J. Laurance[*]

The European Space Agency is presently discussing with NASA, possible cooperation on the LST programme. ESA would supply one of the focal plane instruments - the Faint Object Camera. This camera uses an imaging detector working in the Photon Counting mode. This paper presents the results of studies undertaken by ESA on the definition of the detector and its electronics. The incorporation of the detector into the Faint Object Camera is also demonstrated.

INTRODUCTION

In early 74 an agreement was reached between the then European Space Research Organisation and NASA on possible cooperation on the LST programme. It was agreed that ESRO would consider the possibility of providing one of the focal plane instruments for the telescope. During 1974 preliminary studies were conducted and two Focal Plane Instruments, the Faint Object Spectrograph and the f/96 High Resolution Camera were selected as possible candidates for European development. One of the important features of these instruments was that they were proposed to use imaging detectors working in the photon counting mode. This requirement had been established by the NASA Instrument definition teams of these instruments, if the full potential of the LST performance was to be obtained.

During 1975 the instrument choice has been narrowed down to the camera, and a change of name to the Faint Object Camera (FOC) has been made. This is to emphasise the requirement for imaging faint Objects which is so important for the cosmological studies of LST. The European Space Agency is now actively engaged in studies on the Faint Object Camera and its detector system.

---

[*]ESTEC, Amsterdam, Netherlands

OPERATIONAL CONCEPT OF THE DETECTOR

The operational concept of the Image Photon Counting System IPCS is shown in Fig. I. Incoming photons are first converted to photo electrons by photoemission from a photocathode. By means of a suitable image intensifier each photoelectron is amplified to a sufficient degree so that unambiguous detection can take place. The detectors in the systems we are considering are image tubes, but other detectors like CCD's could also be considered. The coupling between intensifier and image tube may be fibre optic coupling or a lens relay, if sufficient gain is available. A number of solutions are possible for the Intensifier/Image tube combination, ranging with solutions with high gain in the intensifier and low sensibility in the image tube to those with low gain intensifiers and high sensitivity image tubes.

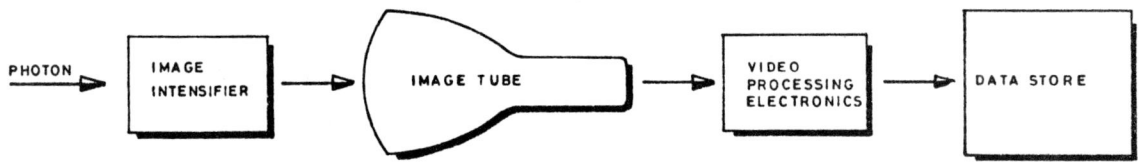

FIG. I. CONCEPT OF IMAGE PHOTON COUNTING SYSTEM

The image tube continuously scans the output of the intensifier. The photo electron events appear as light scintillations and will be detected on several lines of the television scan. It is therefore necessary to identify the center of each photon event. Failure to do this would give different weight to each detected photon resulting in a loss in image signal to noise ratio. This is carried out by a set of pattern recognition logic in the video processing Electronics. The output of the processing electronics is the X and Y coordinates of each detected photon. This information is passed to the data store and the work location corresponding to these X, Y coordinates is incremented by one. Prior to an exposure all locations in the store are set to zero. While the exposure is in progress the image may be examined and the exposure terminated when the image signal to noise ratio is adequate.

One advantage gained by the inclusion of centering logic is the regaining of some of the resolution lost in amplifying the image. Most systems will require more than one stage of amplification as shown in Fig. 2. A photoelectron leaving a point on the primary photocathode A deviates slightly from the axis according to the magnitude of its initial velocity components and lands on the first dynode B at a position governed by a two-dimensional probability function, the shape of which is identical to the large signal Point Spread Function (PSF) of the first stage (i.e. the integrated electron density profile at the first dynode B resulting from a large number of input electrons). Due to multiplication at the dynode, a large number of input electrons is emitted, and acceleration through subsequent stages of the system results in a recorded scintillation (with a finite spatial profile determined by the imaging characteristics of the system after the first stage) on the final target. C. The video processing Electronics locates the center of the scintillation and records its position in a memory location D which is conjugate with the position in which the primary photoelectron struck the first dynode B.

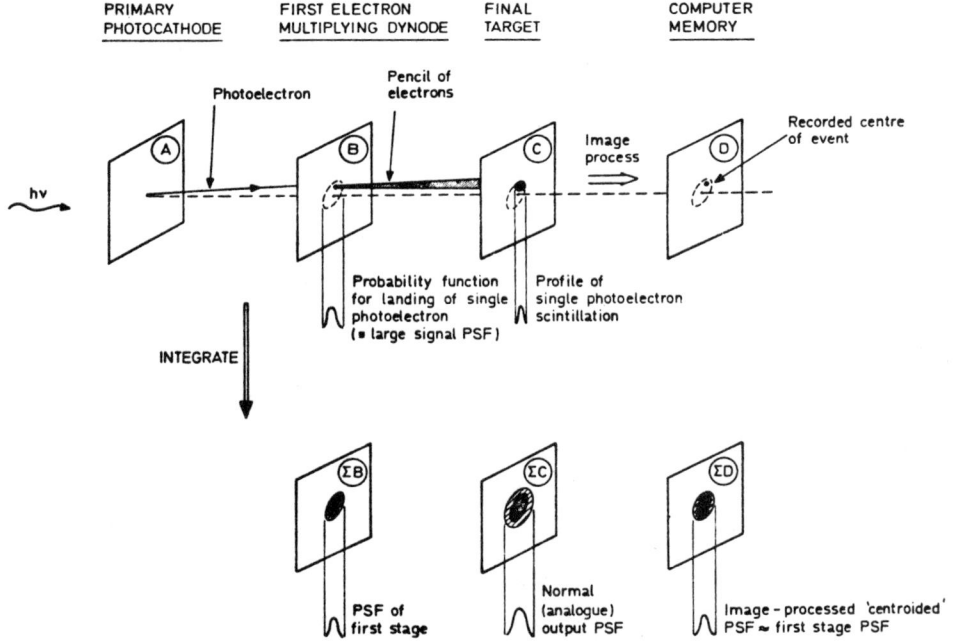

FIG. 2. DIAGRAMMATIC REPRESENTATION OF THE EFFECT OF CENTROIDING ON SYSTEM POINT - SPREAD FUNCTION

For a large number of electrons originating from the same point on A, integration in the external memory Σ D of the image-processed events results in an "image-processed PSF" closely similar to the first stage PSF Σ B, and considerably sharper than the PSF normally obtained by analogue integration on the final target Σ C(= convolution of first stage PSF and scintillation profile).

It is also possible to eliminate the effect of ion scintillations in the video processing electronics. Ion scintillations occur in nearly all image intensifiers. As they consist of a group of some 15 electrons they are given disproportionately higher recording weight in analogue systems. However because of their large amplitude compared with a single photoelectron scintillation, they are easily discriminated against in the logic of the Image Photon Counting System.

CONFIGURATIONS UNDER STUDY

At the begining of this year two studies were initiated by ESA with the task of defining the preferred configuration of components for an Image Photon Counting System. These studies identified three configurations worthy of further consideration.

The first configuration identified is based on a system developed at University College London, and which is now in extensive use for ground based astronomical observations. It consists of a 4-stage magnetically focussed cascade intensifier having a blue light gain in excess of 10^7, which is optically coupled via a lens relay to a plumbicon image tube. This system has been well described in the literature see ref. 1 & 2. For use in the LST application a number of changes must be made. The image intensifier must first be made sensitive to the vacuum Ultra Violet, and for this it is necessary to change the input window to Mg F2.
In the ground based system the focus solenoid of the 4-stage intensifier dissipates $\sim$ 350 watts to obtain the double loop focus field of 280 gauss. This power is far in excess of the power available to the instrument in LST.

It may be possible by careful optimization of the coil design and by operating the intensifier at single loop rather than double loop focus to reduce this figure to 70 watts. A scheme using this concept is shown in Figure 3. Even this power level is likely to be excessive in the LST application and so an alternative system using a permanent magnet focus assembly has also been studied. A scheme using this concept is shown in fig. 4.

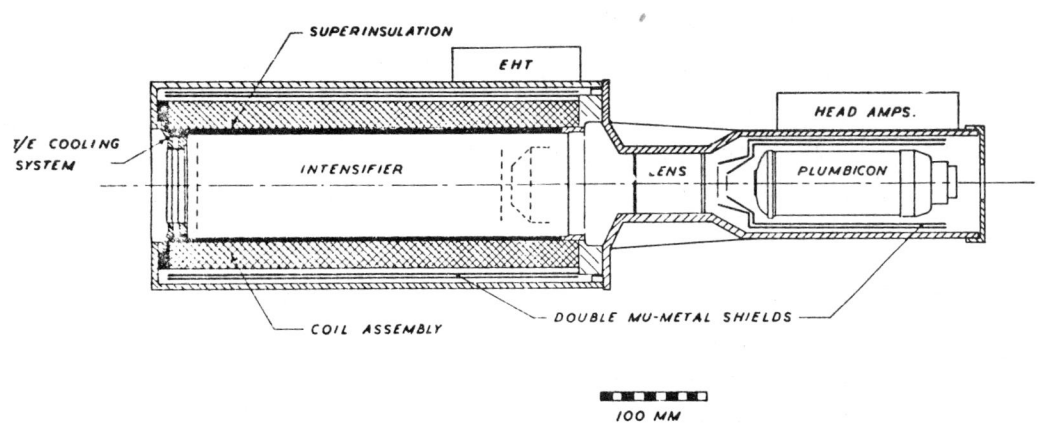

FIG. 3   IPCS SYSTEM USING 4-STAGE INTENSIFIER WITH MAGNETIC FOCUS SOLENOID

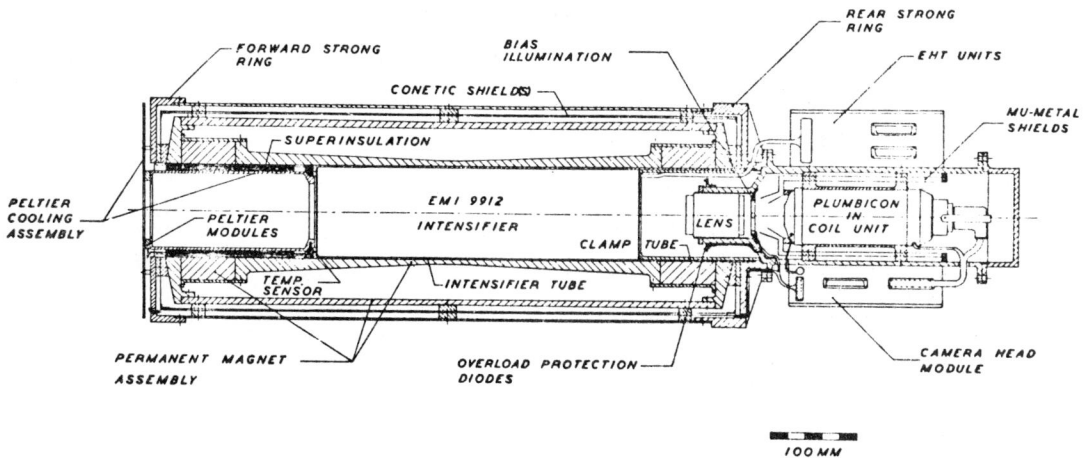

FIG. 4   IPCS SYSTEM USING 4-STAGE INTENSIFIER WITH PERMANENT MAGNET ASSEMBLY

Because of the good resolution performance of the intensifier magnetically focussed first stage the system is capable of achieving a high number of independant pixels although the resolution when measured in an analogue mode will be poorer. For a usable image diagonal of 35 mm on the input face plate it is possible to obtain 1230^2 pixels, which is in excess of our requirements. The pixel size is 20 μm and therefore only modest cooling will be required to keep the dark current to an acceptable level.

The disadvantage of this system for the LST application is its large size and mass compared with the other detector systems being considered. The attraction is however its good performance and its already proven operation in ground based systems. The only problem foreseen is in the modification of the 4-stage intensifier to meet the launch and landing environments. A programme of work is currently under way to procure modified intensifiers and a breadboard system will be evaluated next year.

The second configuration being considered is an intensified EBS system. The EBS image tube (because of its higher sensitivity) allows a lower gain in the image intensifier. If the EBS is operated with an electrostatic intensifier, it is also possible to fibre optically couple them together. This is not possible in the first solution because the magnetic field of the intensifier would disturb the operation of the image tube. This leads to a further lowering of the required intensifier gain because of the absence of the loss in lens coupling (usual efficiency $\simeq$ 1°). It is not possible however to operate the system with a single stage of electrostatic intensification because the input photocathode is curved. When working in the visible it is customary to use a fibre optical input window to correct from the flat input to the curved photocathode. This is not practical in the Ultra Violet because of absorption of fibre optics in this range. It is therefore necessary to include a proximity focussed image converter tube fitted with a flat Mg F2 input window at the input of the system.

Such a system is shown in Fig. 5. It uses a 40 mm proximity focussed converter (EEV P 8103 or ITT F 4122), a single stage intensifier (VARO 8605) and an EBS Camera Tube with DEMC target (Westinhouse WX - 32432/ 32719). This system will provide 465^2 pixels when operating with the 32 mm target EBS (WX 32432) which will increase to 575^2 pixels with the introduction of the 40 mm EBS (WX 32719).

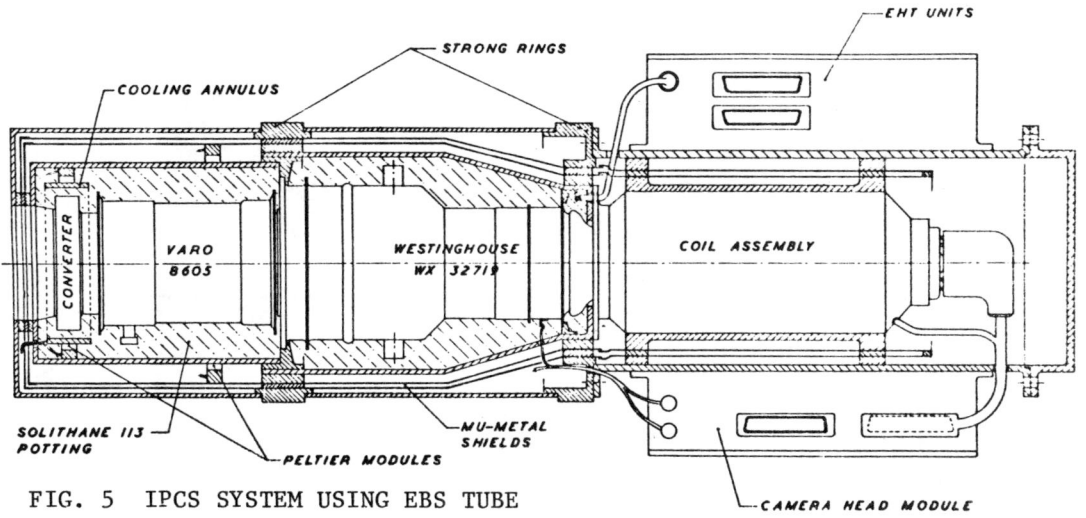

FIG. 5   IPCS SYSTEM USING EBS TUBE

It should be noted that the long term integration is being carried out in the data store and the EBS tube is only integrating and storing photon events for a maximum of a frame time ($\leqslant$100 milliseconds). It is therefore not necessary to cool the EBS target as is usually the case when working in an analogue mode. Cooling of the converter photo-cathode to 0°C may be required to keep the dark current to an acceptable level.

Although an EBS IPCS system has never been practically demonstrated it uses commercially available components which should not be too difficult to space qualify. Theoretical analysis indicates it has a good chance of successful operation, with a well peaked pulse amplitude distribution. The only uncertainties are whether the effects of structure in the fibre optical coupling and EBS target could produce practical problems. A breadboard of the EBS detector system will be assembled and evaluated next year as part of the Programme of LST Detector development by ESA.

The third detector system identified uses a microchannel Plate image intensifier (MCP). Considerable interest has been raised in the MCP because of its high gain and compact size. The negative exponential pulse amplitude distribution characteristic of the MCP with straight channels is however unsatisfactory for an IPCS system. The development of the MCP with curved channels allows operation in the saturated mode

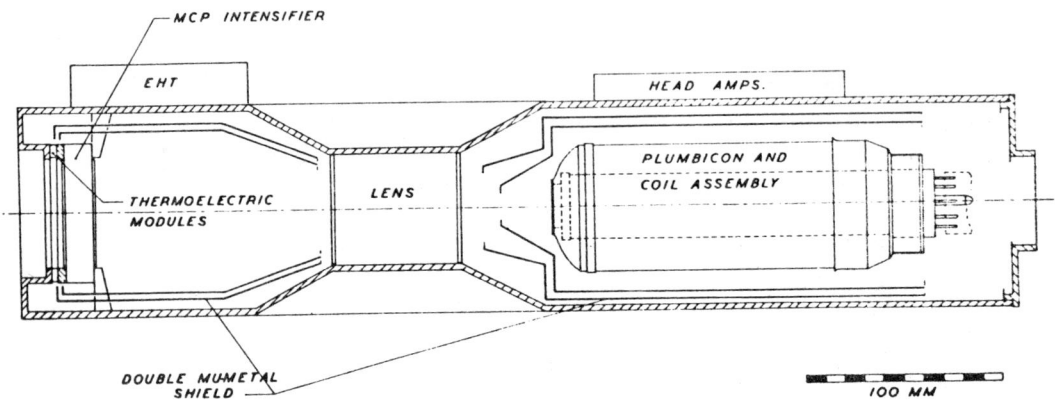

FIG. 6   IPCS SYSTEM WITH MCP USING CURVED CHANNELS

and gives a well peaked pulse amplitude distribution which is ideal for this application. The configuration proposed for the LST detector system is shown in Fig. 6. The MCP with curved channels is incorporated into a tube with proximity focussed input and output. This allows the use of a flat faced Mg F2 input window and results in zero image distortion. With a MCP with a length to diameter ratio of 70-80 a gain of 10^6 should be possible. The tube can either be fibre optically coupled to a 2 in.Vidicon or as shown here, lens coupled to a Plumbicon. It is expected that a 40 mm MCP with 12 or 15 μm channels would provide a picture format of 500 x 500 pixels. The system is likely to have a higher detective quantum efficiency than either of the other two solutions because of the increased detection efficiency of the wafer structure. Because of the uncertain development problems of the MCP intensifier with curved channels this system is not presently being proposed as a candidate detector for the LST. It is thought to be a worthwhile area of development for future IPCS systems and plans are being prepared for its further development on a longer timescale.

DATA STORE REQUIREMENTS

All the detector systems discussed are dependant on the provision of a data store for carrying out the image integration. This store forms part of the IPCS and its performance may well limit the capability of the total system. The baseline design being considered has 500 x 500 pixels. The data store must therefore have a quarter of a million words, each word consisting of 16 bit. This will allow the accumulation of approximately 65×10^3 photons per pixel. One of the most important parameters of the data store is its access time. The image tube will be capable of reading at a rate of 10 M pixels/sec which dictates an access time of 100 nano seconds. However the counting rate of photons will not be greater than 1 M photons/sec as dead time corrections and counting losses will be excessive for rates greater than this. Two approaches to the data store have been considered. The first consists (see fig.7(a) of a main store buffered by a fast shift register which temporarily stores the X Y coordinates of each photon event recorded. The shift register has a speed capable of accepting the highest instantaneous rates expected from the video processing logic. The main store can thus work at a slower speed corresponding to the average photon counting rate.

The second approach is to have a multiplexed store as shown in figure 7 (b) Each section of the store need only have n x 100 nano seconds access time where n is the number of sections used. The sequencing of the store is synchronized to the scanning of the image tube and when a photon is detected the data work addressed at that instant is incremented by one. Hybrids of these two schemes are also possible. The second approach is presently favoured.

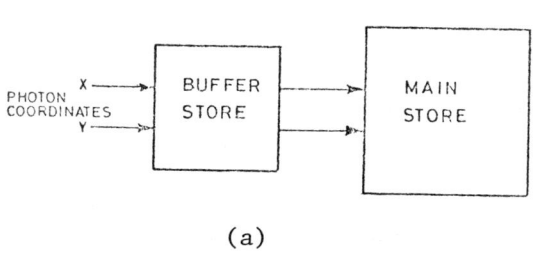
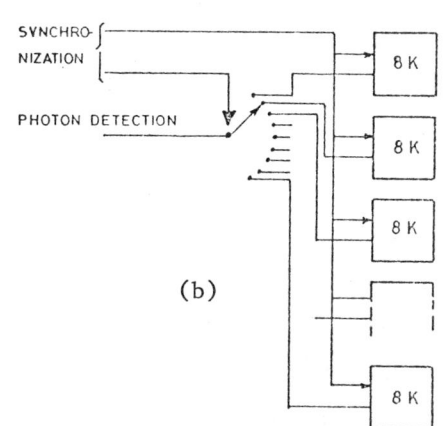

(a) (b)

Fig. 7 CONFIGURATIONS OF DATA STORE

A number of technologies are possible for the data store. Along the most promising are C-mos, magnetic bubbles and core. The baseline design presently considered by ESA is a core store based on technology developed for Helios. It consists of 32 sections of 8 K words, each with an access time of 2.5 μ sec. It will consume $\leq$ 50 watts when working at 10 % duty cycle (corresponding to a maximum counting rate of 1M photons/sec). The required reliability will be obtained by including spare sections. The estimated mass of the data store is $\simeq$ 40 kg and the volume $\simeq$ 50 litres. ESA will be undertaking a detailed examination of this store and the other technologies during the next 6 months.

DETECTOR ACCOMODATION IN FOC

Preliminary studies have been undertaken by ESA on the LST Faint Object Camera and these demonstrate the satisfactory accomodation of the IPCS detector system. The analysis has been restricted to the IPCS system using the 4-stage intensifier fitted with a permanent magnet focus assembly, and using a core data store. This configuration is considered to be the most difficult to accomodate. The configuration of this design is shown in Fig. 7. The camera includes a set of relay optics to allow the focal ratio of the camera to be chosen, so that the resolution of the camera will be determined only by the OTA. The IPCS system with magnetic focussed intensifier has an MTF of 50 % at 25 lp/mm after centroiding, thus a slight magnification of the OTA image will be sufficient to make the influence of the detector on the OTA resolution, negligible. This leads to a focal ratio of F/36. A switchable element allows the focal ratio to be changed to $\simeq$ F/200 allowing a speckle mode of operation to be used. In this mode each pixel corresponds to 0.01 arc seconds. Other modes of operation are also possible including a coronographic mask mode for the selection of faint companions to bright stars, and a dispersive mode.

Two detector heads are included in the FOC for redundancy, a plane mirror being used to switch from one to the other. The electronics of the FOC are isolated thermally from the optics because of the large and variable dissipation of the data store. A general purpose computer may be included to control the various modes of operation of the camera.

The overall design of the Faint Object Camera has been shown to be feasible using the worst case IPCS detector system presently considered. The design presented includes redundancy where required, and provides a flexible instrument including a number of different modes of operation. This instument and its detector system should make a significant contribution to the operational performance of the LST programme.

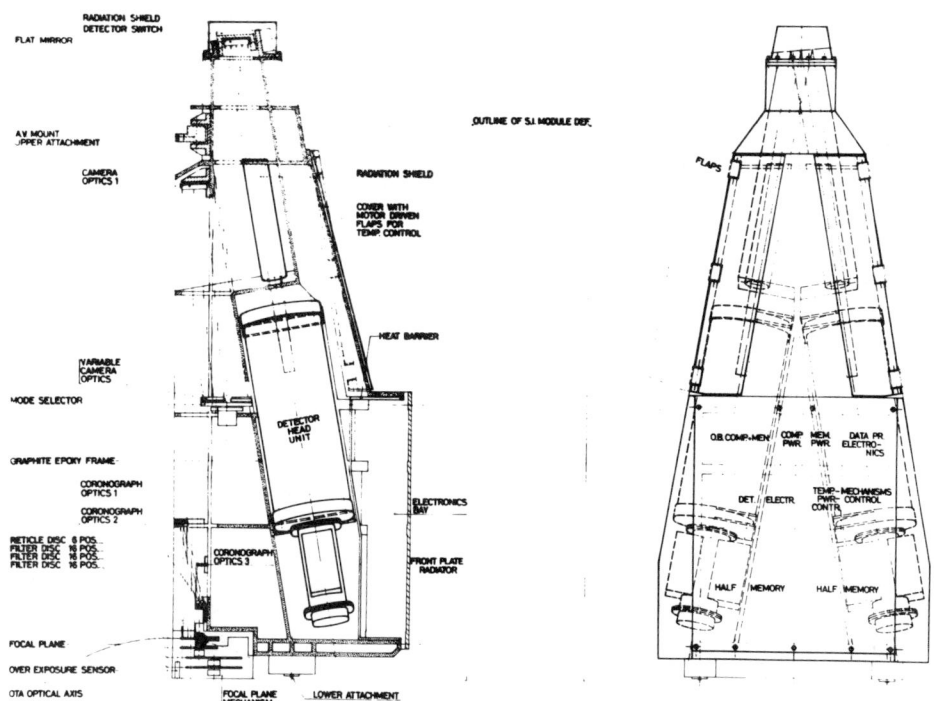

FAINT OBJECT CAMERA FOR LST
FIG. 8

References

1. Image Photon Counting - A. Boksenberg/Applications Spatiales des Tubes de prises de vues - CNES - November 71

2. University College London - Image Counting System
   A. Boksenberg, D.E. Burgess - Astronomical Observations with Television Type sensors - University of British Columbia - Vancouver, May 73

AAS 75-184

AN ANALYTICAL AND EXPERIMENTAL EVALUATION OF ACTUATOR VIBRATION ON LST IMAGE DISTORTION

A. D. Houston, L. W. Hodge, Jr., and T. J. Kertesz*

INTRODUCTION

Described in this paper is a program conducted at LMSC that addressed the overall question of pointing control in the presence of actuator vibrations. The intent of the project was to reduce the technical uncertainties associated with programs in which pointing control is critical.

The program was conducted in three phases. Phase 1 consisted of exploratory shaker testing to evaluate measurement techniques and to develop experimental transfer functions. Phase 2 testing, in which operating Bendix MA500 CMGs were used, employed the same analog data acquisition and processing system used in the Phase 1 tests. Phase 3 of the experimental work was conducted with an advanced analog-digital data acquisition and processing system (MODALAB) and included additional transfer function studies and preliminary evaluation of a Sperry reaction wheel.

DISCUSSION

Test Specimen. The LST SDTV used in the experimental program consists of three basic sections: the optical telescope assembly (OTA), the support system module (SSM), and an 18-in.-deep conical monocoque adapter, which interconnects the OTA and the SSM. The OTA section consists of a cylindrical semi-monocoque ring-stiffened aluminum shell and contains the basic elements of the LST optical system, including primary and secondary optical surface and focal-plane simulators. The specimen accelerometer and shaker locations are shown in Fig. 1.

Instrumentation. Ultrasensitive vibration transducers (1000 mV/g sensitivity) were installed on the SDTV, to monitor motions of the primary and secondary optical surface simulators, and at the focal plane. Both Kistler QA116 and Unholz-Dickie 1000 PA transducers have been used successfully in the program.

---

*Aeromechanics Department Research Specialists for LMSC's Space Systems Division

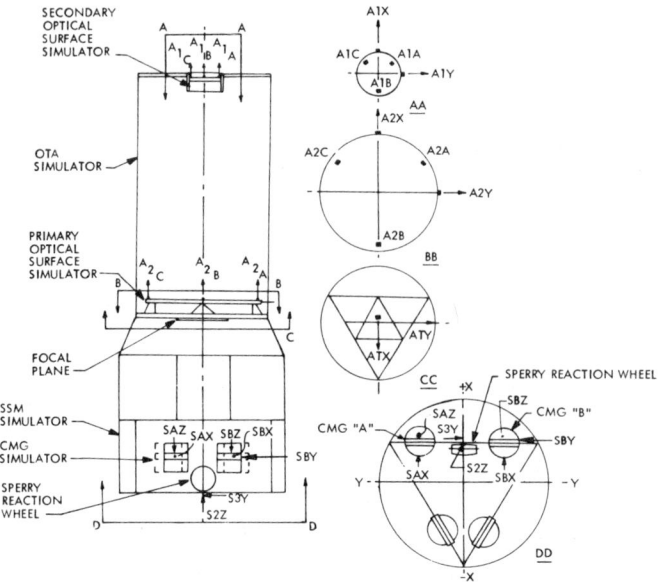

Fig. 1 SDTV Test Instrumentation and Actuator Locations

The location and orientation of this instrumentation (Fig. 1) was such that translational and rotational motions of the primary and secondary optical surfaces relative to the X and Y axes could be determined. Instruments were installed to monitor translation of the focal plane along the X and Y axes.

Data Acquisition. Phase 2 testing on the specimen employed an analog data acquisition and processing system. Signals proportional to shaker force, which were held constant, provided absolute value ($|A|$) inputs to each of two co-quad analyzers. Image distortion signals ($\ddot{\epsilon}_x$ and $\ddot{\epsilon}_y$) were passed through 2-Hz tracking filters and provided absolute values ($|B|$) inputs to the co-quad analyzers.

Phase 3 testing utilized the bulk analog-to-digital capability of MODALAB as the primary data acquisition, storage, and processing system.

Analytic Simulation

Mathematical Model. The conical adaptor SDTV mathematical model was constructed using a half model approach and the LMSC finite element code SNAP. The final model consisted of 2600 equivalent full model degrees-of-freedom (DOF). The model used 417 plate and 366 beam elements.

Major LOS error contributions will result from modes (see Fig. 2) involving significant relative optical element and focal-plane motions with respect to local support structure.

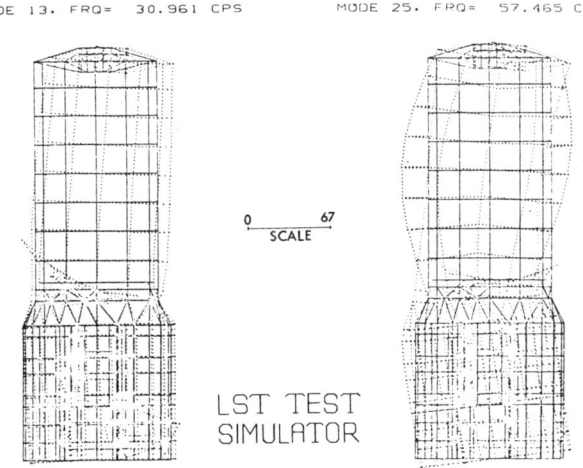

Fig. 2 Significant LOS Error Analytic Model Modes

Transfer Functions

Using computed optical eigenvectors, transfer functions are computed directly for force and moment inputs by using an LMSC steady-state response code STEDY9. All transfer functions are computed assuming uniform damping of 0.5 percent, based on previous SDTV test experience.

EXPERIMENTAL AND ANALYTIC RESULTS

A typical comparison of analytic and empirical results is shown in Fig. 3. Note that the significant analytical modes 13 and 25 have excellent correlation with empirical results.

CONCLUSIONS

Transfer function tests conducted with a force amplitude of 0.05 lb, which is typical for allowable disturbance on spacecraft such as the LST, showed that:

1. State-of-the-art instrumentation and data processing can reliably measure dynamic response to such a small force in a 13500-lb structure
2. About 30 − 40 Hz, pointing errors in different directions are about the same regardless of the direction of excitation
3. Analytical simulations based on detailed finite element models can predict in a conservative manner the frequency and amplitude characteristics of the response in the plane of excitation; however, they tend to underestimate the response in directions normal to the plane of excitation
4. Taking the vector sum of analytically obtained response and assuming it to be applicable in any direction proves to be a good estimator

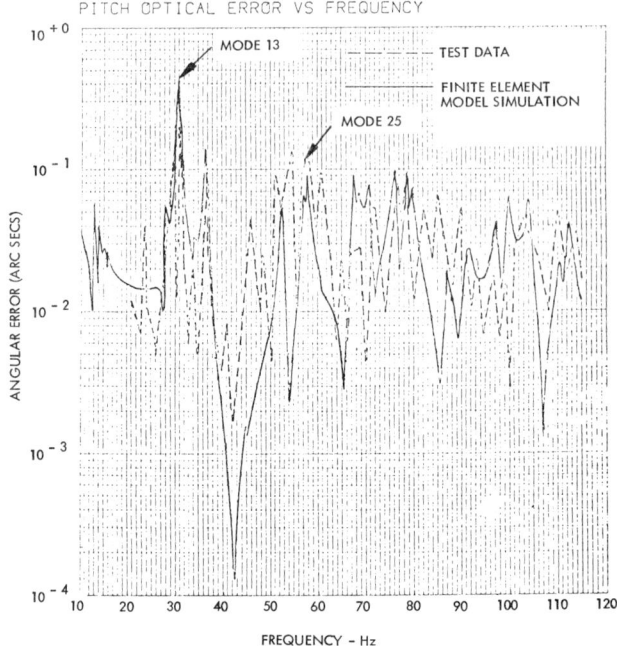

Fig. 3 Test and Analyses Simulation Results

5. Large amplitude regions of the response, particularly at high frequencies, are clustered in frequency bands associated with resonances of secondary mirror supports

6. The key to ensuring high pointing accuracy is to either have CMGs whose speed can be changed by 10 – 15 percent, thus moving them from these frequency clusters, or to design secondary mirror supports that have no resonances within 10 – 15 Hz of the CMG's operating speed

7. Three-axis isolators can attenuate disturbance caused by very low level forces; however, the effectiveness of attenuation varies by as much as 10 dB, depending on the orientation of the CMG spin vector

8. Variation of isolation effectiveness with direction can be minimized by

    a. Ensuring that the CG of the CMG is in the plane of the four isolators

    b. Ensuring that the CMG support structure has resonances 3 – 4 times greater than the isolation frequency

9. Nonlinear "beating" could not be induced in the experimental structure within a 45-dB amplitude range with reference to the response at the CMG's spin frequency

10. Pointing stability requirements budgeted for high-frequency vibration, such as 0.0035 sec for the LST, is attainable with state-of-the-art CMG designs and isolation systems; furthermore, the performance of structures at typical spin speeds can be predicted with a high degree of confidence, thus lowering development risks.

AAS 75-185

DEVELOPMENT OF A LARGE-INERTIA FINE-POINTING
AND DIMENSIONAL STABILITY SIMULATOR

R. L. Gates, D. H. Wine, R. W. Seiferth, and N. A. Osborne*

Martin Marietta has developed a fine-pointing
attitude control physical simulator incorporating
important advances. The effective inertia is
approximately that of the LST (large space telescope).
The long term pointing stability has been improved
by use of a graphite/epoxy instrument truss structure.
The short-term jitter stability is 0.0001 arc-s RMS.
The long-term stability was designed to test state
of the art gyros.

The large moment of inertia was achieved by using a
unique mass augmentation system employing a cylinder-
piston arrangement filled with mercury. The effective
inertia is a function of the piston area, the piston
to cylinder gap area and of the mass of the mercury
within the total gap.

The overall structure consists of a 22-foot length
aluminum outer truss connected to the mass augmentation system and an inner graphite/epoxy truss for
instrumentation mounting. The structure has a minimum system frequency of 12 Hz and a bending mode of
the instrument truss of 25 Hz.

A control console provides a flexible means to provide
gain, compensation, filtering, and power to the
attitude control system, including the sensors and
actuators. Four independent sensors are used for
attitude reference, all capable of less than
0.001 arc-s resolution.

---

*Martin Marietta Aerospace, Denver Division, Denver, Colorado

INTRODUCTION

The LST attitude control requirements have placed an increased emphasis on simulation techniques to prove the validity of the requirements. The 0.007-arc-s stability requirement and the long-term pointing requirements require that the hardware component models be more complete for scientific computer simulations. Proper experiment design requires that the errors in physical simulation equipment, particularly the instrumentation, be about an order of magnitude less than the spacecraft operational performance requirement.

In 1972 Martin Marietta built a fine-pointing simulator that obtained a pointing stability below 1 milliarc-s.[1] This physical simulator, which demonstrated excellent pointing stability, was considerably below the desired LST spacecraft inertia. It has a maximum inertia of 2000 kg-m^2 instead of the 6×10^4 kg-m^2 inertia of the 2.4 meter LST. This meant that the data had to be scaled by a factor ranging from 30 to 90. When nonlinearities are involved, scaling becomes questionable. With a full inertia simulator, scaling becomes unnecessary.

Other limitations of the first generation simulator is that the rotation axis is vertical and the main beam was made of aluminum. Long-term testing of gas bearing rate integration gyros was difficult because vertical earth rate required compensation and the aluminum truss was susceptible to dimensional changes due to temperature variations.

A new physical simulator referred to as the fine-pointing and dimensional stability (FPDS) simulator was completed early this year. The unique features of this simulator are a mercury mass augmentation system to provide 1×10^5 kg-m^2 inertia and a graphite/epoxy instrument truss to insure long-term stability to gyro testing (Fig. 1).

The FPDS simulator is a single-degree of freedom simulator with the pivot axis oriented east-west to minimize the earth rate effects on gyro testing. The pivot freedom is provided by specially designed flex pivots that provide a rotation angle of two degrees. The pivot support and the inner truss, which is center connected with the flex

Fig. 1 FPDS Simulator

pivot, are made of graphite/epoxy. A larger outer aluminum truss is then connected to the graphite truss by flex hinges, which are stiff in rotation but allow the aluminum truss to change dimension with temperature while keeping the graphite stress to a minimum. Four mercury mass augmentation systems are connected to the aluminum truss. These systems not only provide proper equivalent mass but also give approximately 4000 pounds of buoyant force to float the larger aluminum truss so the pivots are nearly unloaded.

The total simulator is enclosed in plexiglass to isolate it from the room air currents. The system is quite sensitive to acoustical coupling.

The pivot support, mass augmentation system, and reference piers all are mounted to a very large inertia block that rests on bedrock. This particular location has a very low seismic background noise and provides a very precise testing environment. The floor area is spring-isolated from the testing piers (Fig. 2).

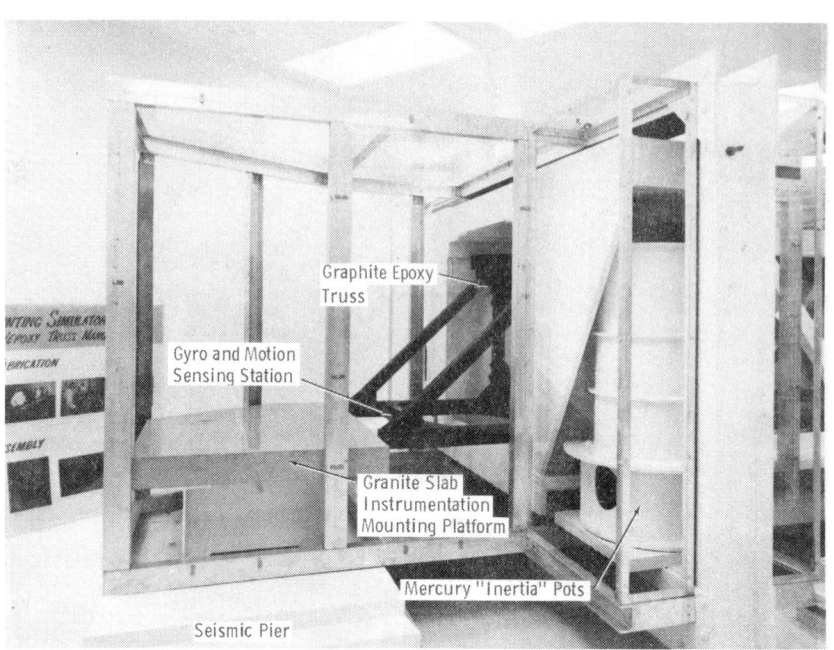

Fig. 2 FPDS Simulator Instrumentation Station

AAS 75-186

EVALUATION OF COMMUNICATION ANTENNA DRIVE SYSTEM DESIGN
REQUIREMENTS TO ALLOW TDRS TRACKING
DURING LST FINE POINTING

A.J. Besonis[*] and C.J. Chang[**]

The NASA Large Space Telescope to be operational in the early 1980's will have to provide very precise pointing over image integration times ranging up to ten hours. To insure that such tight performance requirements are met the Pointing Control System (PCS) design has to reflect all potential sources of attitude disturbances. Design trade-off have to be performed for the total LST system to insure that the system design is optimized to be compatible with the PCS without imposing major cost penalties through equipment operational constraints or on the equipment design. Presented is an evaluation of the design requirements for a high gain antenna that could be used for TDRS (relay satellite) tracking during fine pointing of the LST. A precision computer simulation model is used to establish the design requirements/modifications of an existing design. It is shown that with moderate mass balancing of the antenna dish and momentum compensation of the antenna drive system, TDRS tracking is feasible during fine pointing of the LST.

---

[*] Lead Engineer LMSC for the Pointing Control System for the LST, Member AIAA.

[**] Research Specialist, Lockheed Missile & Space Company, Sunnyvale, California

INTRODUCTION

Orbiting telescope systems such as the Large Space Telescope (LST) have to provide a pointing stability accuracy of ± 0.007 arcsec (RMS). This accuracy has to be maintained over a frequency spectrum from essentially 0 to frequencies ranging beyond several hundred Hz.

Effect of all potential disturbance sources has to be evaluated to insure such extreme pointing accuracies can be maintained in the operational environment of the telescope. Major attitude disturbances to the system include external torques such as gravity gradient, aerodynamic, solar torques, and reaction torques produced by operating equipment internal to the LST. The latter category of disturbances include those that would be produced by communications antenna motion if required to track the TDRS.

A major trade-off in terms of cost savings through the use of existing antenna and its drive system design has been identified. This would require antenna tracking during certain periods of LST observations. An additional important pay-off would result with this design since the required transmitter power would be limited to levels compatible with state-of-the-art. This again can have a significant impact on cost and will certainly minimize potential design risks.

The key trade-off consideration between fixed and tracking antennas is summarized in Table 1.

An existing communication antenna design developed by TRW[*] has been identified for LST application. The system is depicted in Figure 1.

The major objective of this study was to determine by computer simulation if such a design with a HPBW of 8 deg would be adapted to the LST.

---

[*] Support by TRW for this study in particular by Mr. Roy Acker for providing design data on the antenna system is acknowledged.

The selected approach was to simulate the complete dynamic system consisting of the LST, the antenna with its drive system and the antenna structural support system. Required antenna motion was then introduced while the LST was inertially stabilized with a control system bandwidth compatible with the fine pointing requirements dictated by considerations other than antenna tracking motions.

Table 1

FIXED POINTING VS TRACKING HIGH GAIN ANTENNA

o Line of Sight from LST to TDRS Changes A Maximum of $11°$ during an Orbit

o Baseline Configuration has Open Loop HGA Tracking to Maintain a Boresight Gain of 25.7 dB $\pm$ 0.2 dB

o Use of Optimized Fixed Pointing (One Slew per Orbit Pass) Causes Antenna Boresight to be a Maximum of $\pm 5.5°$ off Angle to TDRS Giving a Gain Loss of 4.9 dB

o Transmitter Power is Increased by 3.1 Times over Baseline (from 10 Watts to 31 Watts)

FIXED POINTING

Advantages	Disadvantages
- Antenna Inertial Compensation not needed	- Needs 3 times more power
- Computer use for one slew vs continuous pointing	- Amplifier is a SOA development
	- Possible Interference with other MA users because of 4.9 dB Signal Strength variation over a single pass (3 dB max. per Rev. 1, TDRSS User's Guide)

SIMULATION RESULTS

The simulation evaluated antenna/drive system design for the existing 1.1 m antenna with the dish offset from the elevation drive axis. Evaluated were the existing configuration, the effect of mass balancing of the antenna and the effect of adding momentum compensation to each axis of the antenna drive. Two types of antenna mounting support (booms) were also considered. The first type of support consisted of aluminum alloy booms configured in an A-frame in the LST y-z plane stabilized in the LST x-z plane by guy wires of stainless steel. The aft guy wires imparted a pre-load to the forward guy to achieve a suitable stiffness. The A-frame had a natural frequency in the y-z plane of 16.4 Hz and the guy wire support had a frequency of 2.25 Hz.

The alternate type of support considered was a rigid single boom of graphite apoxy. The reason for an antenna support that will shift about either axis is that control actuators such as reaction wheels would more readily excite a soft boom producing large dynamic errors. Boom stiffness of 8.0 and 11.0 Hz was considered.

The level of compensation considered was 95% for static balance and 90% and 80% of momentum compensation.

Table 2 summarizes simulation results for the different antenna drive systems considered for the soft antenna support. Results show that with mass balance alone LST pointing errors of the order of 10^{-3} arcsec can be produced. This level is considered excessive based on error budget allocated for the total system. Results for two levels of momentum compensation are also shown in Table 2. As shown, momentum compensation reduces the attitude error to below an order of magnitude of the allowed total pointing stability error for the LST. Momentum compensation to within 80% is quite easily attainable in the drive system considered without incurring a significant design cost.

Simulation for the rigid single boom supports showed little difference for the same level of momentum compensation.

Table 2

COMPARISON OF MAXIMUM LST POINTING ERRORS PRODUCED BY
ANTENNA TRACKING (STEPPING) IN AZIMUTH & EVALUATION
(Step Size 0.0375 deg/step)

Antenna Position	Condition	Attitude Error (arcsec)		
		Roll	Pitch	Yaw
Antenna Vertical (Elevation 90°deg)	No mass balance No momentum compensation	.0112	0.00128	0.00563
	With mass balance (95%)	.00309	0.00116	0.00137
	Mass balance (95%) Momentum compensation (80%)	.00062	0.00023	0.000275
	Mass balance (95%) Momentum compensation (90%)	0.00029	0.000115	0.000138
Antenna Horizontal (Elevation 0°deg)	No mass balance No momentum compensation	0.0094	0.00127	0.00653
	Mass balance only (95%)	0.00287	0.00113	0.0015
	Mass balance (95%) Momentum compensation (80%)	0.00055	0.00023	0.00030
	Mass balance (95%) Momentum compensation (90%)	0.000285	0.000113	0.000148

CONCLUSION

Antenna tracking of the TDRS can produce significant attitude disturbances on the LST if proper design procedures are not adopted to compensate the antenna system. With moderate mass balance and momentum compensation the pointing errors can be held to below one order of magnitude of the pointing stability error specified for the LST. Considerable lattitude exists in the design of the antenna structural support system from the standpoint of stiffness, provided of course it is sufficiently stiff so as not to interact with the dynamics of the Pointing Control System and will not induce large errors due to excitation from momentum actuator unbalance.

A SMALL INSTRUMENT POINTING SYSTEM FOR SHUTTLE SORTIE MISSIONS*

Carl W. Henrikson+
Ewald E. Schmidt++

A small instrument pointing system (SIPS) has been conceptually designed as a means of interfacing smaller astronomical instruments with the ESA spacelab system on shuttle sortie missions. Small instruments are defined as being up to 0.9 meter in diameter, three meters long, and 340 kilograms in mass. The resulting system is shown in Figure A. Two independently supported instruments are deployed on a common pedestal to give a hemispherical field of view. The instruments are each supported at their centers-of-mass in bi-axial gimbal systems for fine pointing. The central pedestal provides coarse azimuth rotation of both instruments simultaneously. Telescoping of the pedestal raises both instruments from their launch locks and returns them to the same locks for landing. A separation plane in the pedestal allows the entire SIPS to be jettisoned if for some reason it could not be restowed and the bay doors could not be closed.

The instruments are each enclosed in a canister (Fig. B) which provide a regulated thermal environment as well as protection from contamination and acoustical damage. The

---

*Based on work performed for Goddard Space Flight Center under Contract NAS5-24048
+Manager, Advanced Control Systems, Ball Brothers Research Corporation
++Mission Technology Division, NASA/Goddard Space Flight Center

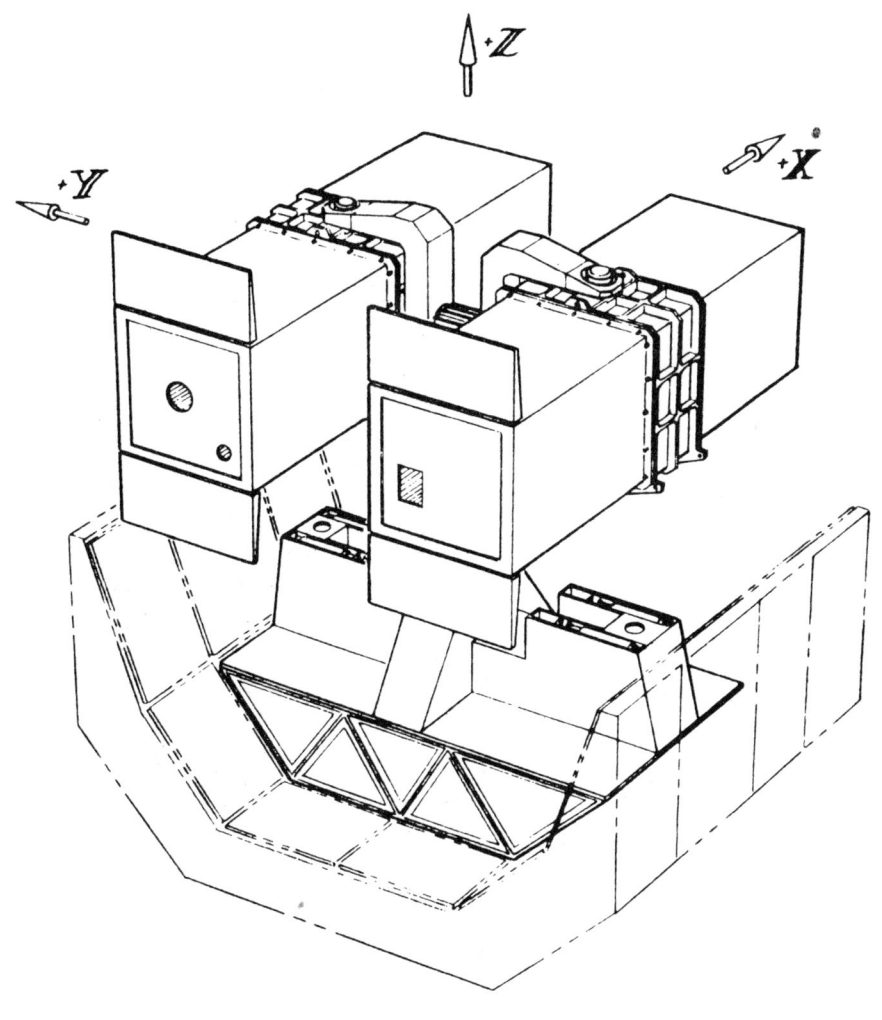

Fig. A Small Instrument Pointing System (SIPS)

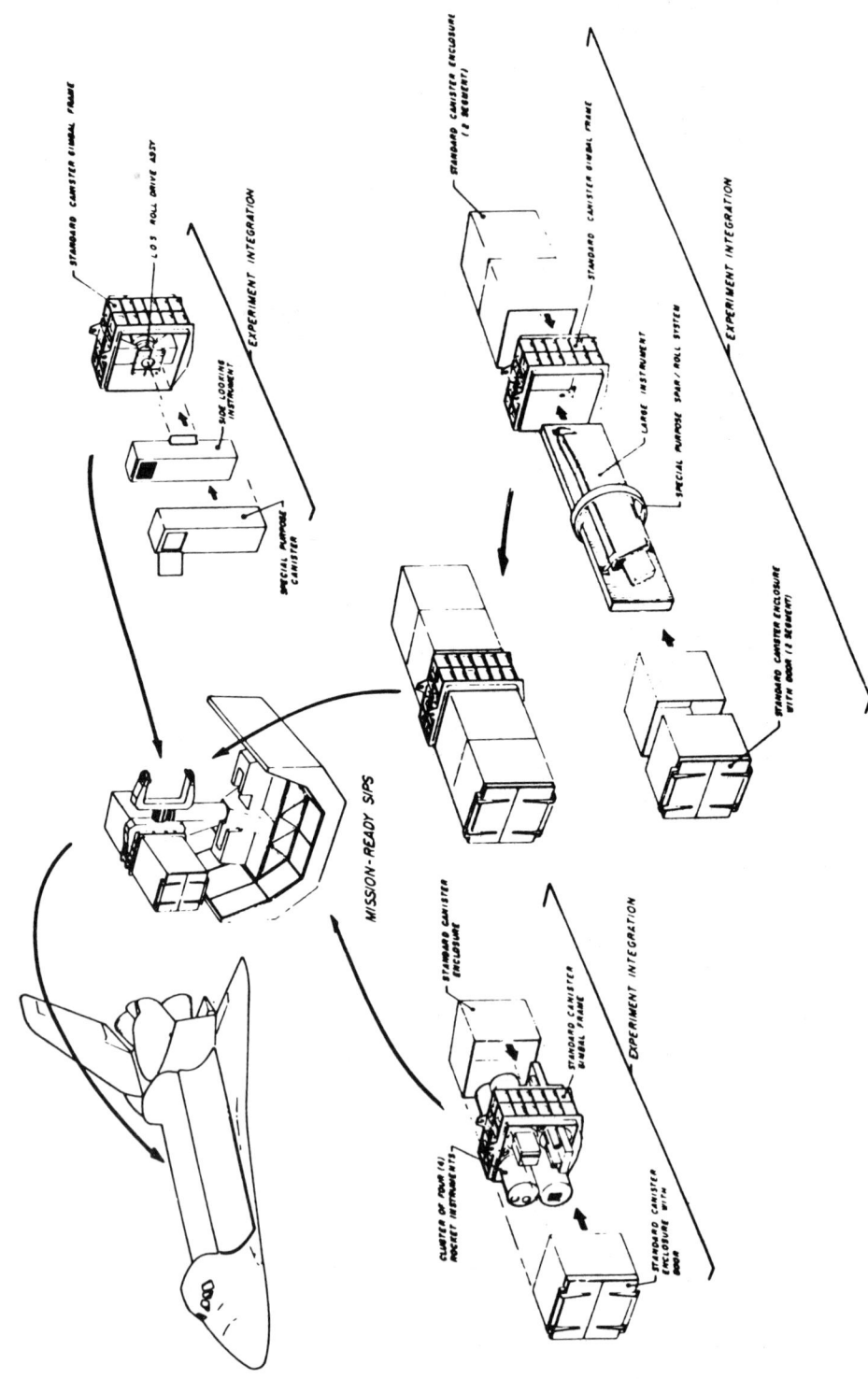

Fig. B SIPS Canister Gives Flexibility and Speeds Instrument Integration

canisters are modular, have a universal aperture door, and are easily removable for access to the instruments. An optional roll gimbal can be used inside the canister to provide control about the line of sight. Spacelab data and power interfaces are also provided within the canister. Temperature control to ±1°C can be provided by a system of heaters, insulation, and surface finishes or by a system of heat pipes. The canister allows instrument integration to take place before attachment to the gimbal system which reduces costs and shortens schedules.

Fine pointing is achieved by the right/left drive in the yoke and the up/down drive in the pedestal top. These drives use precision ball bearings, limited rotation brushless torque motors, and noncontacting encoders. The combination of low friction drives and a balanced payload allows pointing stabilities of 0.1 arc-second to be achieved even in the presence of reaction jet firings on the orbiter and crew member wall push-offs in the cockpit. The sensors used for control include fine pointing sensors (solar or stellar) located on or in the instruments themselves, two rate integrating gyros on the canister gimbal frame, and auxiliary acquisition and roll sensors as required.

The SIPS concept is a practical and economical means of pointing small solar and stellar instruments from the shuttle orbiter payload bay. By providing pointing, environmental protection, and spacelab electrical interfaces, it is a total instrument support system.

AAS 75-276

A FINE-POINTING FACILITY FOR SPACELAB
EXPERIMENTS IN THE 1980'S - THE
INSTRUMENT POINTING SUBSYSTEM*

H. Heusmann and J. Collin Jones[+]

INTRODUCTION

Many of the scientific experiments to be performed on Spacelab missions require accurate and stable pointing significantly better than that provided by the Orbiter vehicle. It is the purpose of the Spacelab Instrument Pointing Subsystem (IPS) to satisfy these requirements by providing a fine-pointing facility capable of adaptation to the diverse dimensions, weights, operating modes, and interfaces of potential Spacelab instruments.

DESIGN CONCEPT

The pointing system design which has evolved during the study phase to meet these diverse requirements is an unusual one. A three-axis gimbal system is mounted between one end of a payload and the pallet as shown in Fig. 1. The gimbal axes are arranged in the opposite order than normal in that the payload is not supported about its CG within the inner gimbal with subsequent gimbals arranged outside this one; the gimbals are in an "inside-out" configuration with the payload attached to the outer, largest gimbal, with the other gimbals arranged inside this gimbal. This technique offers a very compact system with capability to accommodate the diversity of payload physical dimensions, but, in principle, is sensitive to disturbances of the Orbiter because of the large offset distance between the payload CG and gimbal center of rotation. To reduce this effect, the entire gimbal system is mounted on a "softmount", consisting of springs and dampers which act as a mechanical low-pass filter for Orbiter disturbances in all three directions.

---

* The complete paper appears in Vol. 25, **AAS Microfiche Series**
+ Spacelab Project Office, ESA/ESTEC, Noordwijk, Netherlands

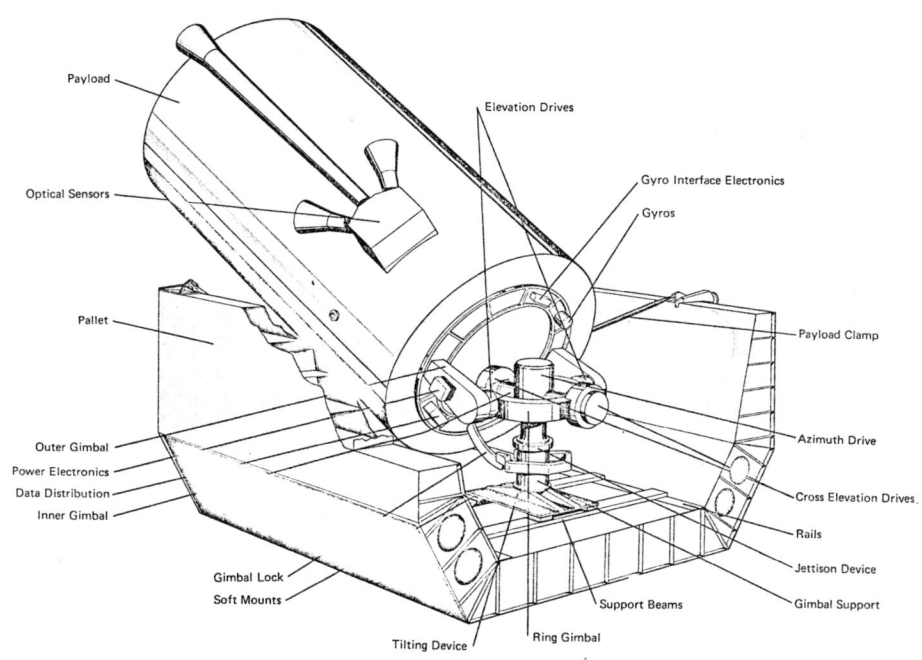

Fig. 1 IPS Gimbal Configuration

Overall attitude control of the payload is based on rate integrating gyros (RIG) error signals processed within the Spacelab computer (CDMS) to generate command signals to DC torquers in each gimbal axis as shown in Fig. 2. The RIG package is located on the outer gimbal and hence, aside from distortions or flexures occurring within the payload, maintains the payload as an inertially stabilized platform.

To correct for gyro drift and to provide an absolute reference attitude, a package of optical sensors is also included. In a stellar mission this would comprise three fixed-head star trackers, and in a solar mission one star tracker would be replaced by a solar sensor. The package will normally be located on the payload, and alignment errors between the experiment line-of-sight and the reference guide star are minimized by an on-orbit alignment measurement technique such that the best possible pointing accuracy within the inherent capabilities of the star-tracker and experiment optics is achieved.

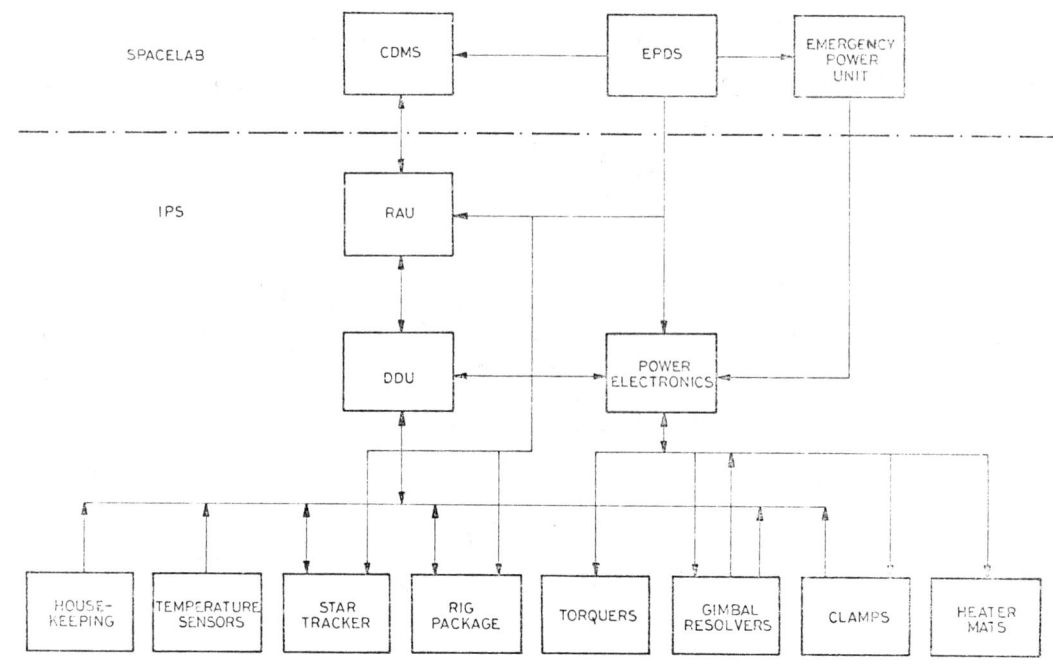

Fig. 2 IPS Functional Block Diagram

PREDICTED PERFORMANCE

Full 3-axis simulations of the IPS have been performed of the motion of the Shuttle Orbiter, the softmount, and the payload under the influence of man-motion and RCS thruster disturbances. In a simplified model of the motions, one can safely assume that the payload does not move instantaneously when the Orbiter is being displaced, but that in the first instance the softmount tilts by whatever angle is required (typically 100 to 300 arc-sec) and the pointing of the payload is essentially undisturbed.

The springs of the softmount then tend to return to the undeflected, pushing gently at the pivot point of the gimbal system. This motion has a characteristic frequency of a few hundredths of a Hertz, based on the rotational spring constant of the softmount and the combined payload/gimbal mass. In this way the g-level of the disturbance is attenuated by several orders of magnitude by the softmount. The attitude control loop of the IPS, the bandwidth of which is a few Hertz, then responds to this disturbance and maintains the pointing error below 1 arc-sec. The soft-mount is so soft (as far as rotation is concerned) that the center of rotation due to the servo loop control is essentially the payload CG and not the pivot point, and therefore the payload behaves almost like a free flyer.

The resulting stability error is sensitive to a number of parameters and conditions, e.g., location in the Orbiter cargo bay, LOS angle relative to the zero position, control loop bandwidth, payload mass and inertia, CG offset distance, gimbal pivot point height, etc. The computer simulations performed have confirmed the early analysis of the body motions involved, and have shown that a peak stability error of less than 1 arc-sec is potentially achievable, even taking into account all the performance-degrading conditions mentioned above.

CONCLUSION

Throughout the design phase every effort has been made to retain as much flexibility to accommodate as large a variety of instruments as possible in the IPS design without introducing excessive complexity. This has been achieved; however, some adaptation on the part of instrument designers and experimenters will be necessary to utilize the IPS to its full potential.

Simulations of the predicted performance of the IPS indicate that if adequate analyses of the optimum IPS/payload interfaces, viewing configurations, operating modes, etc. are made for each payload individually, then the inherent capability exists in the IPS to provide arc second absolute pointing accuracy and pointing stability for a very large percentage of potential Spacelab experiments in the 1980's.

The system is currently under final evaluation by ESA and a decision for phase C/D implementation is expected in the near future. The objective is to have the IPS available in time for the second Spacelab flight in late 1980.

AAS 75-188

LST INTERFEROMETRIC FINE GUIDANCE SENSOR

A.B. Wissinger, R.H. Carricato[*]

INTRODUCTION

This paper describes the LST Fine Guidance Sensor (FGS) preliminary design by answering three questions: What does the fine guidance sensor have to do? How does the sensor function? What is the sensor's predicted performance?

WHAT DOES THE FINE GUIDANCE SENSOR DO?

The LST fine guidance sensor (FGS) has several functions. Its primary function is to detect mispointing of the telescope by measuring the positions of auxiliary or guide-star images in the telescope focal plane. This measurement is converted to an electrical error signal for the Support System Module's (SSM) use in correcting the pointing direction to 0.007 arc seconds (RMS).

Since the guide stars will occur at arbitrary (but known) positions relative to the star or object of scientific interest, the fine guidance sensor must be capable of performing its primary function for any accessible star location. Since the accuracy of some of the spectrographic instruments depends on the location of the star image in the entrance slits, and it is the null point of the fine guidance sensor that locates the star image, the FGS must have the capability of accurately positioning the guide star relative to the science instrument. This means that the null point of the fine guidance sensor must be adjustable and repeatable with an accuracy of 0.01 arc-second to meet the needs of the spectrographs.

The fine guidance sensor must also have sufficient sensitivity to measure faint guide stars. It is required that the FGS shall have sufficient field area and sensitivity to ensure an 85% probability of acquisition in the area of the sky where stars are the least dense (at the galactic poles).

---

[*] Perkin-Elmer Corporation, Norwalk, Connecticut

Given a field area for the FGS sized to achieve the 85% probability of acquisition, the system must then have the capability of searching for the guide star. Once a star is detected, the FGS must check it with the brightness of the selected guide star. When the detected star is the correct star, pointing error signals are sent to the spacecraft so that it can reduce the pointing error to less than 0.007 arc-sec. The simulation described below includes the spacecraft dynamics.

One final function that is included in the FGS is the ability to continuously offset the tracking point (sensor null) in order to make a time exposure of a moving planet and to compensate for the effects of the differential velocity aberration. Differential aberration is the apparent change in the radial position of stars in the telescope field caused by the spacecraft velocity vector. The FGS will correct for this gross direction change, but stars in different parts of the telescope will have slightly different shifts. The effect is as large as three times the LST stability requirement (0.007 arc-second) and must be compensated.

HOW DOES OUR SENSOR DO IT?

A single sensor is capable of detecting errors simultaneously and continuously in pitch and yaw. A second sensor is used to measure the motions of a second guide star caused by roll motions around the axis defined by the line of sight to the star used to control pitch and yaw pointing.

The Perkin-Elmer concept uses an interferometer to detect the tilt of the wavefront that results from a change in the telescope pointing direction relative to the guide stars. The wavefront tilt is best measured at a pupil (which is an image of the primary mirror) because all the rays pass through the pupil for any guide-star positions.

A small gimbaled mirror is also placed at a pupil formed by the relay mirror in the FGS. Its purpose is to direct the star light to a fixed reference location regardless of the guide-star position. At this location, there is a mirror with a small aperture at its center and a beamsplitter in the aperture area.

The gimbaled mirror is preset precisely to an angular position corresponding to the guidestar location. The signal for setting the mirror is a single ground command. However, the mirror can also be used for planet tracking by storing a series of commands representing the planet's trajectory.

The action of the mirror and beamsplitter is to reflect the star image to the image-dissector when the image is outside the field stop and to evenly split the light between the image-dissector and interferometer when the image is within the field stop.

The interferometer output is split into four quadrants and directed to four photomultiplier tubes. The output of the photomultipliers are amplified by photon-counting type amplifier/discriminators. The pulse trains from each amplifier output are used to drive up-down counters, which produce the error signal between opposing pairs of PMT/amplifiers.

THE LST POINTING SYSTEM SIMULATION

In order to test the performance of the fine guidance sensor, a digital computer simulation program was written.

The emphasis of the simulation is placed on the fine guidance sensor. A simple single-axis model of the LST vehicle and its pointing control system is included. The only nonlinearity included in the vehicle control system is a torque limiter.

The fine guidance sensor consists of two error detectors: the image dissector, which is used to sense errors from 2.5 arc-minutes down to 0.1 arc-second. The outputs from these two error detectors are called the coarse error signal and the fine error signal, respectively.

The input to the simulation is a star direction with respect to the vehicle pointing angle; the difference angle is the pointing error. Initially, the fine guidance sensor is in the coarse mode, and it remains in the coarse mode until the coarse error (ϵ_c) indicates less than 0.1 arc-second.

As the star image traverses the beamsplitter field stop, an apparent centroid shift takes place. This shift is computed and added to any system offsets that might occur within the coarse sensor to form the error signal generated by the image dissector star tracker. A 0.1-second time delay is inserted to account for aperture scanning and dwell time required by the image dissector. It is this processed and delayed signal that is used to determine when the error is within the fine error detector capture range.

Once the coarse error signal shows an error less than 0.1 arc-second, control is passed to the fine error sensor. The interferometer transfer function computation simulates the polychromatic transfer function of the fine error detector. Gaussian noise is generated using conventional techniques and is added to the error signal.

This error signal is then fed to the pointing control system. After the fine guidance system has been in the fine mode for 10 seconds, an rms pointing error is computed within the program. In general, the value of the rms pointing error varies between 0.0011 and 0.0014 arc-second. It must be remembered that this is only the error caused by fine guidance internal noise and quantization.

CONCLUSIONS

The fine guidance sensor described has ample performance margins. The effects of nonlinearities, noise, quantization, and mode-switching have been included in the simulation of both the FGS and the SSM. In spite of these effects, performance requirements are met for a variety of simulated operational conditions.

The wide-field image dissector, with its five arc-minute acquisition field, relieves the initial pointing accuracy requirements for the SSM. When compared with earlier requirements, the requirement for the SSVM is reduced by a factor of five.

The combination of photon-counting electronics and the interferometer for fine guidance sensing permits operation on fainter guide stars. The margin of probability for acquiring guide stars is thereby appreciably increased over the 85 percent requirement.

AAS 75-189

PRISMATIC GRATING STAR TRACKER

Allen H. Greenleaf*

A unique star tracking sensor developed at Itek Corporation is eminently suited for LST because of its inherent stability and its ability to obtain diffraction-limited performance at the edge of the LST field of view without a complex set of field correctors to eliminate the telescope astigmatism. Using a 13th magnitude guide star, it can provide a tracking signal to the telescope with noise equivalent to 0.002 arc-second at a 5-Hz noise equivalent bandwidth. It is also suitable for astrometric measurements to determine star parallaxes and proper motions. A breadboard built at Itek fully demonstrates the performance capability of the sensor.

LINE-OF-SIGHT STABILIZATION FOR LST

The Large Space Telescope (LST) will observe celestial objects much too dim to provide line-of-sight stabilization error signals by some kind of target lock-on sensor. The stabilization error signals must be provided therefore by a device that locks on to a brighter star image somewhere else within the telescope field of view. In the Itek version of the LST, there are three fine guidance sensors situated radially around the focal plane area so that star images appearing within three areas at the edge of the field of view can be used for stabilization. The fine guidance sensors must be capable of searching for and acquiring the guide star image within a 1 arc-minute diameter initial pointing error and then providing pointing commands to bring the telescope to the correct orientation.

OPERATION OF PRISMATIC GRATING SENSOR

The basic concept of the Fine Guidance Sensor is to use a fixed, coarse grating at the focal plane upon which the guide star is imaged. This grating breaks the beam from the guide star into different optical channels, and the precise location of the guide star image on the grating determines the relative balance of power in the different channels. The only element in the sensor whose stability is critical for stabilization of the telescope line of sight is the grating itself.

---

*The author is with Itek Corporation in Lexington, Massachusetts.

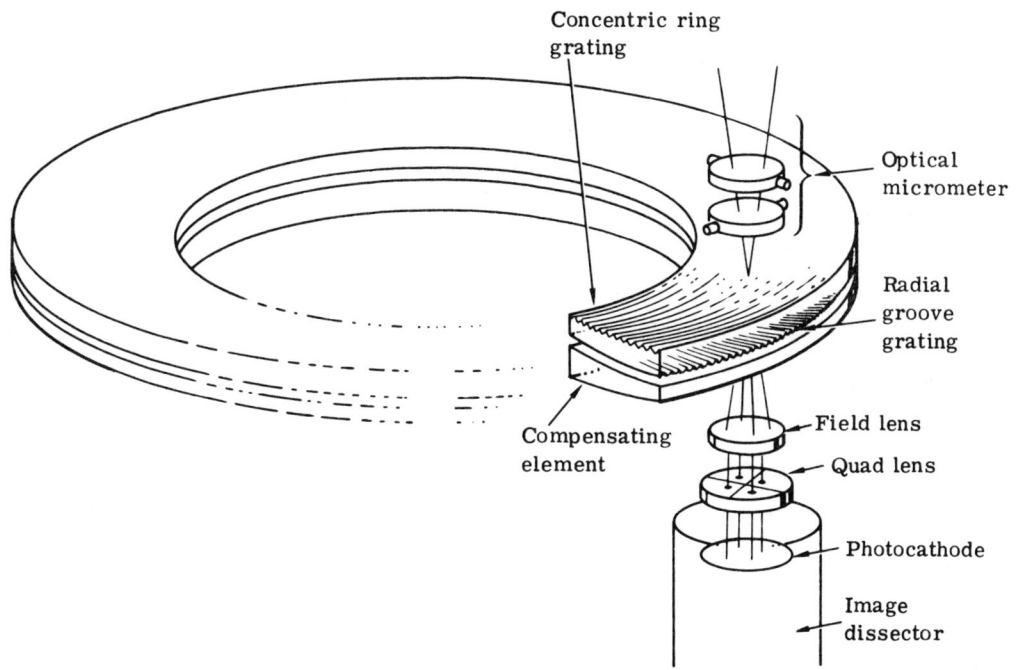

Fig. 1 — Fine Guidance Sensor schematic

Fig. 1 is a schematic of the Fine Guidance Sensor as it is used in LST. The converging beam from the guide star is first passed through an optical micrometer which consists of two gimbaled plane parallel glass plates. The optical micrometer provides accurately controlled displacements of the guide star image over a range equal to the 4 arc-second groove spacing of the grating plate. This provides the means to guide on any guide star image within the field of the grating plate while keeping the guide star image centered upon a groove intersection.

Next, the guide star beam converges to focus at the grating plate. The grating plate is a curved element whose front and rear surfaces conform, respectively, to the tangential and sagittal focal surfaces of the telescope. The grooves on the front surface form concentric rings, since the star image at that surface is blurred tangentially. The grooves at the rear surface run radially since the star image at that surface is blurred in the radial direction. The grating on the front surface divides the light into two beams whose relative power depends upon the radial location of the star image. The grating on the rear surface divides each of the two light beams at that point into two more beams, the relative power of which is dependent upon the tangential position of the star image. This configuration automatically compensates for the astigmatism in the telescope image without the need for additional correctors. The error signal from the sensor is essentially that of a diffraction-limited optical system.

Behind the grating plate is the sensor head, which contains the optical relay elements and the image dissector. The four beams from the guide star image formed by the grating plate are projected onto different points on the image dissector. During fine tracking, the image dissector samples these four beams in sequence, and develops tracking error signals proportional to their differences in power. During initial acquisition, the image dissector also measures the positions of the four beams on the image dissector, and uses this information to provide coarse tracking information. This controls telescope pointing until the guide star image has been brought to the appropriate groove intersection on the grating plate, when fine tracking commences.

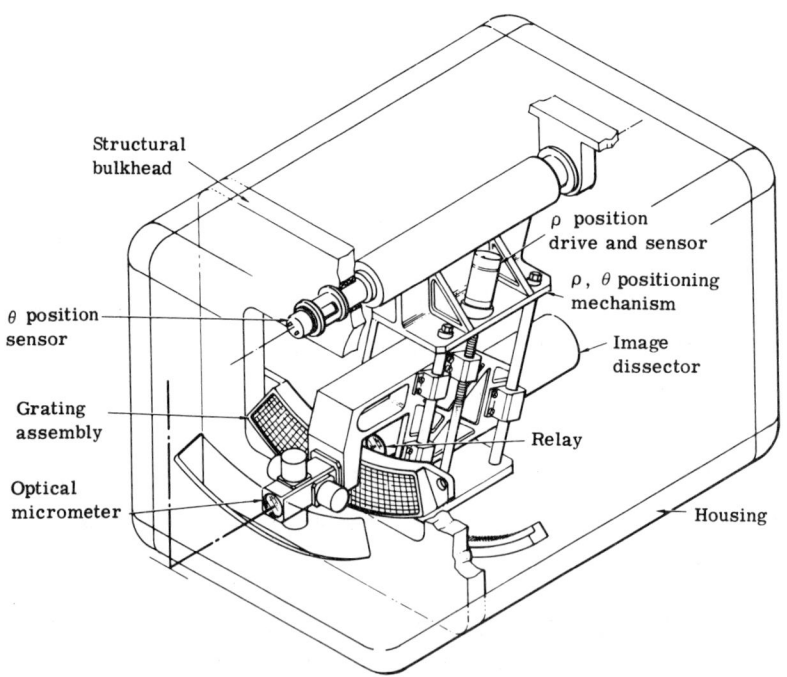

Fig. 2 – Fine Guidance Sensor

Fig. 2 shows an isometric view of one of the Fine Guidance Sensors. The optical micrometer, relay, and image dissector are assembled together to constitute the sensor head and can be positioned at the point in the grating at which the guide star is to appear. The grating is mounted to an Invar structural bulkhead to which the mounting feet for the instrument are attached.

By selecting the photon counting mode of operation for the nulling loops within the sensor, sources of noise other than shot noise in the signal current may be made negligible. At the same time, the background counts in the signal are insignificant compared to the counts received from the guide star, so that the system is virtually limited only by the photon noise of the guide star. The noise equivalent angle as a function of the visual stellar magnitude of the guide star and the effective integration time of the control system is shown in Fig. 3. A control system with a 5-Hz noise equivalent bandwidth has an effective integration time of about 0.1 second, so that for a magnitude 13 star, a stabilization error contributed by the sensor noise of about 0.002 arc-second rms can be expected.

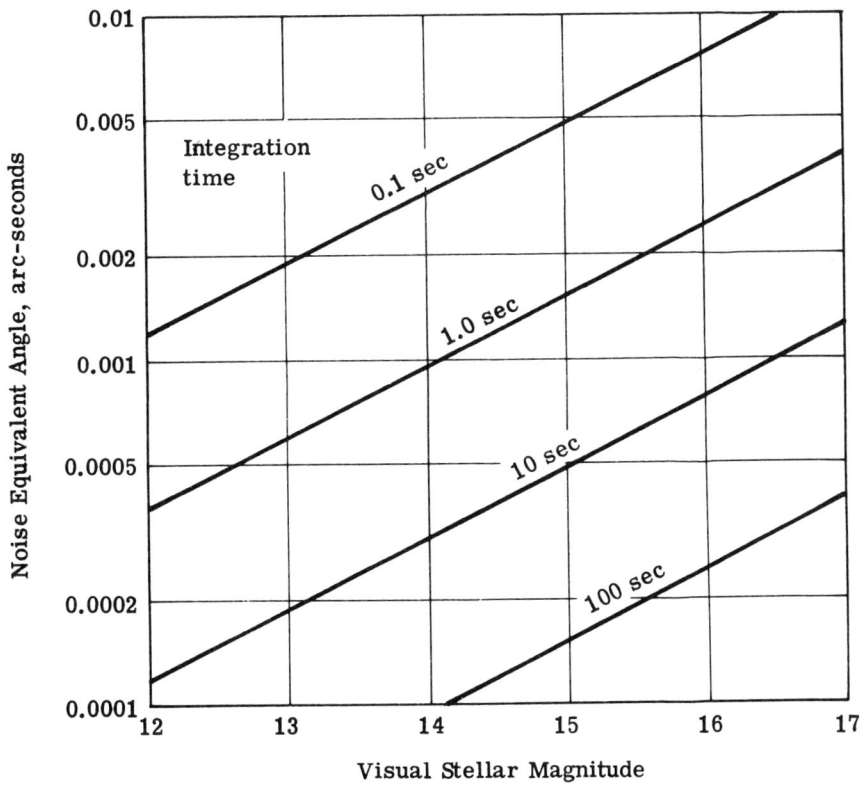

Fig. 3 — Photon noise of Fine Guidance Sensor

The sensor, by virtue of its basic concept, is remarkably free of stabilization errors other than the photon noise error. The bandwidth-independent errors add up to about 0.001 arc-second rms. This also represents the sensor capability in reacquiring a guide star following occultation by the Earth.

ASTROMETRIC MEASUREMENTS

The Fine Guidance Sensor is suitable for making astrometric measurements for star parallax and proper motion determinations, since the basic astrometric measurement consists of measuring changes in the angular separation between stars over an extended period of time.

BREADBOARD VERIFICATION OF THE CONCEPT

A breadboard built at Itek, with astigmatism accommodating gratings, relay optics, and image dissector, fully demonstrates the performance capability of the tracker in searching for, acquiring, and locking on to a guide star. Performance limited by diffraction and photon noise is achieved. The gratings were produced by Bausch and Lomb in the circumferential and radial groove form on curved surfaces, as would be required for LST. Fig. 4 shows the breadboard and electronics.

Fig. 4 — Star Tracker Breadboard

AAS 75-181

SIMULATION OF THE ON-ORBIT MAINTENANCE CYCLE FOR THE LST*

J. A. Donnelly[†]

INTRODUCTION

Projected long system life for the LST program spacecraft is achieved through inherent reliability of the selected hardware, service revisit capability of the Space Shuttle, and ground support activity. Reliability of equipment determines the frequency of corrective maintenance, amount of spares required, and need for preventive maintenance. Therefore, an important input to the system maintenance model is reliability information, which is prepared to define the complete LST system process mode. The initial step in the description of the complete LST system is to establish basic and redundant elements and the functional relationship of each. The final configuration selection is made to achieve the best reliability within the constraints of system weight, volume, margin of safety, and life-cycle cost. The addition of a unit in redundancy to achieve stated reliability goals and a fail-safe capability is an iterative decision process. Level-of-repair optimization concepts must be designed into the system to take advantage of on-orbit as well as ground refurbishment. Optimum maintenance is achieved over the system's total life cycle if the repair level alternatives selected minimize total support cost for a specified level of system effectiveness, as compared with other possible maintenance concepts, procedures, or design alternatives.

The maintenance analyses study flow encompasses the interaction of reliability, criticality assessment, maintenance threshold, program cost, and Shuttle revisit schedule.

---

*This work was done under Contract NAS8-31313.
[†]J. A. Donnelly is a Senior Staff Engineer in the System Effectiveness Organization of Lockheed Missiles & Space Co., Inc., Sunnyvale, Calif.

DESIGN AND OPERATIONAL MODEL REQUIREMENTS

System Definition

The LST consists of three major groups: (1) the optical telescope assembly (OTA), which includes the primary and secondary mirrors plus associated electronics and thermal protection; (2) scientific instruments (SI), which includes a high-resolution camera (f/24), a faint object spectrograph, an IR photometer, and a photopolarimeter; and (3) the support system module (SSM).

An essential part of the assessment of the LST configuration option is the development of the subsystem configuration with its required redundant units. The minimum equipment required to perform the LST mission constitutes the so-called single-thread reliability model. Redundant OTA and SSM equipment is added to the single thread in a logical sequence as a function of the minimum weight, cost, and volume to provide the best incremental improvement in reliability. The so-called state of the system at a given interval is defined by the redundant equipment complement that exists at that time. Simply stated, equipment is added in the most logical order by the computer program until the desired reliability is achieved or the cost, volume, or weight constraints are exceeded. Reliability, therefore, becomes a key driver in the search for the best operational system. Variations in reliability and associated redundancy are directly related to the frequency of Shuttle visits and hence changes in total maintenance program cost.

Critical Equipment

Equipment criticality is determined through the identification of those functions that must occur to assure safe retrieval of the LST. One analytical method to quantify the system probability of loss is to use fault-tree computer graphics. The results of this analysis determine redundant fail-safe requirements over and above those defined for cost or reliability.

Maintenance Threshold and Call-Up Criteria

The need for Space Shuttle visits to either retrieve or service the LST depends on the following factors:

- System redundancy depletion
- Performance of scientific instruments below an acceptable level
- Degradation of the optics solar arrays or other limited operational life equipment

Removal and replacement of failed equipment may be accomplished either on-demand or scheduled. The on-demand approach suggests a threshold of acceptability below which maintenance action is necessary. The number and type of spares carried by the Space Shuttle to maintain the LST above this threshold level is determined through fault isolation/detection or by analysis. The system is inoperative when no more redundant units are available. The degradation of SI performance (loss of channels) is identified as an output of the simulation. Once the threshold is reached, a call for Shuttle and maintenance exits. In effect, the utility of the LST has reached some level where it is not cost effective to continue without maintenance. Suppression of this threshold line in the model tends to reduce program cost at the expense of data acquisition. The simulation run identifies both on-orbit maintenance action and ground maintenance action such as refurbishment of the optics.

Programming the Simulation

The General Purpose System Simulator (GPSS) Univac version was selected to be used in the evaluation of the maintenance model. GPSS is ideal for the model since the language is adaptable to event flow and requires a minimum of program statements.

As a computer programming language, GPSS has the characteristic of being versatile but not very flexible. This is to say that a relatively short program written in GPSS can represent a complex problem, but that operations (such as data matrix manipulation) which are not specifically designed into the language are exceedingly cumbersome to program - if not impossible. As a result, GPSS has the capability of interfacing with a more general programming language, FORTRAN, to accomplish these otherwise cumbersome operations.

The GPSS/FORTRAN interface is a subroutine entitled "HELP." The purpose of this subroutine in the LST maintenance simulation program is to produce plots and system parameters over time. In effect, the HELP subroutine is divided into three parts: an initializer, a recorder, and a plotter. The initializer is called at the start of the program to set up the data matrix. The recorder is called from the main program during the simulation each time there is a change in system data. The change and the time of the charge are then noted in the data matrix. After the simulation is completed, the plotter portion of the HELP subroutine is called to draw a graph of the data matrix created during the simulation.

CONCLUSIONS

The maintenance model became a useful tool in the system evaluation process. A distinct advantage to this program is the quick-look capability with plots available for analysis of the results. Frequency distributions of the time to first maintenance action and the interval between maintenance shuttle visits give an insight into system behavior

A comparison of the case studies shows clearly that an on-demand maintenance policy is the most cost effective from a total program cost standpoint.

Spares provisioning becomes an integral part of the cost sensitivity studies since it is directly linked to equipment selection, redundancy, and life-cycle failures.

ACKNOWLEDGMENT

Reliability and cost models developed by E. A. Polgar were the key elements in his systems approach method in the search for the most cost-effective design alternative. C. J. Sheehan provided the GPSS model and developed the unique subroutine plot capability to display the simulation results.

AAS 75-182

LST POWER SYSTEM
LONG LIFE DESIGN TECHNIQUES*

Owen B. Smith, Richard L. Donovan, and James L. Oberg +

INTRODUCTION

The challenge of the Large Space Telescope (LST) power system design is to design a solar array/battery system with an operating life in excess of two years on orbit. The Space Shuttle provides the capability to replace the batteries and solar array in orbit but the LST program must pay the cost of the Shuttle flight. Even if the cost of the maintenance mission is shared with another space program, the cost to LST will be several million dollars; therefore, it is imperative that the life of the components be the maximum possible within the constraints of the LST program.

SYSTEMS DESCRIPTION

The solar array selected is the Flexible, Roll-up Solar Array (FRUSA) designed and flown by Hughes on the STP71-2 mission. The selection of the FRUSA as the LST solar array permits additional area of active solar cells at an acceptable weight and cost penalty. To increase the FRUSA two year design configuration to a configuration to support a five year life results in a weight penalty of only 20 pounds and a cost increase of $240,000. Although this value represents a significant percentage increase in the cost of the EPS initial design, it decreases the spares requirement and is cost effective for the LST program.

The initial design requirement for the LST EPS specified a two-year mission which requires a battery design with a cycle life of 12,000 cycles. The remainder of this paper will address the design techniques of the nickel-cadmium battery system required to meet or exceed the 12,000 cycle requirement.

---

*Work sponsored by the NASA Marshall Space Flight Center under Contract NAS8-31312.
+Martin Marietta Aerospace, Denver Division, Denver, Colorado

LST CHARGE CONTROL SYSTEM

The long life design techniques which are addressed in the LST research project include the capabilities to provide individual cell level control, operate at a reduced state of charge, provide spare cells, and recondition on orbit. To accomplish the cell control function with reasonable size and weight, a relay interconnection network under microprocessor control is used.

FLEXIBLE CHARGE DISCHARGE CONTROLLER (FCDC)

The 36-cell FCDC LST system shown in Fig. 1 consists of the microcomputer, command decoder, relay drivers, relay interconnection network, multiplexer, current switch network, clock, and ancillary signal conditioning circuits[1]. One relay is required for each cell. Magnetic latching relays, which are energized by a 50-msec pulse, are used to reduce power consumption. Relay contact position is under control of the microcomputer. If the microcomputer detects a cell out of limits (voltage, high or low) a command is issued to a relay that removes the cell from the battery. The LST battery will normally have six cells not connected (spare cells). These cells will be used later in the mission for extending the battery life and when reconditioning other cells.

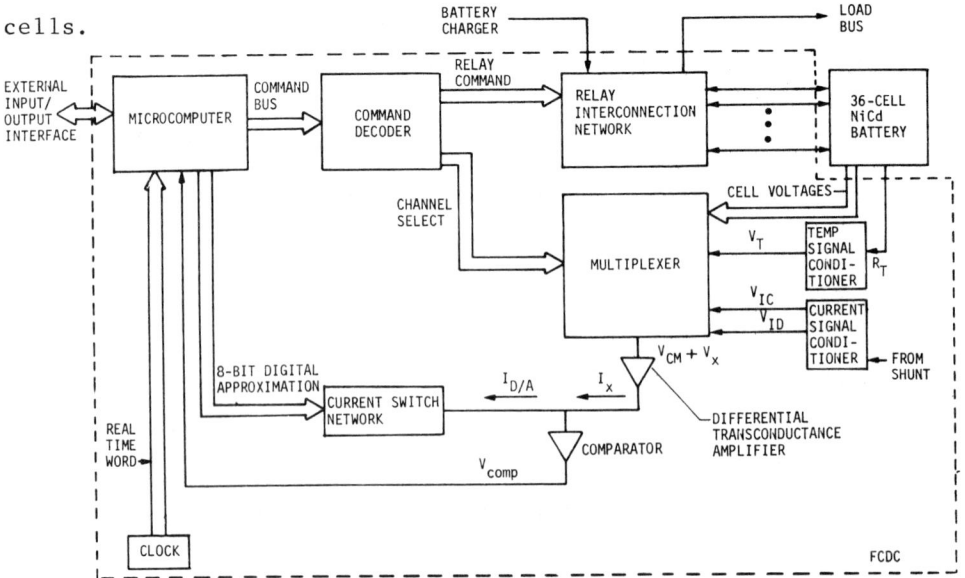

Figure 1 FCDC Block Diagram

The FCDC system also contains a clock that generates a real-time word. The provision for time indication permits numerical integration by the microcomputer of the battery charge and discharge ampere-hour integrals. The ampere-hour integrals can be used to implement charge control techniques that use recharge fraction in addition to cell voltage for terminating battery charging. Recharge fraction control will be the primary battery control with cell voltage as a backup.

A unique feature of the FCDC is reconditioning the cells to regain useful ampere-hour capacity. The microprocessor controls the cell discharge until it detects a cell voltage equal to or less than the low voltage limit.

The discharged cell is then switched back into the battery and is recharged until it reaches 100 percent state of charge as determined by the recharge fraction. This process is repeated until all cells have been reconditioned. In addition to the cell control function, the FCDC also performs a monitoring function. The system parameters monitored are individual cell voltages, battery current, battery voltage, and battery temperature.

FCDC BREADBOARD

The feasibility of the FCDC design as presented herein has been successfully demonstrated by a breadboard system at Martin Marietta in Denver, Colorado. A 30-cell breadboard was developed to demonstrate the operation of a 24-cell battery with six spare cells using the individual cell monitoring and charge/discharge control. The basic software for this system has been developed and the system is currently operational.

FCDC PACKAGING

Figure 2 shows the FCDC mounted to a battery housing. A significant reduction in weight of the battery housing is obtained by using a honeycomb in a dual role of a heat conductor and load-carrying structure. Martin Marietta has studied the use of the honeycomb structure and has established the feasibility of its use in space applications[2]. By using the honeycomb structure, the total weight of the 36-cell battery with FCDC would be about 124 pounds. By comparison, 36-cell Skylab housing weight would be about 137 pounds without the FCDC.

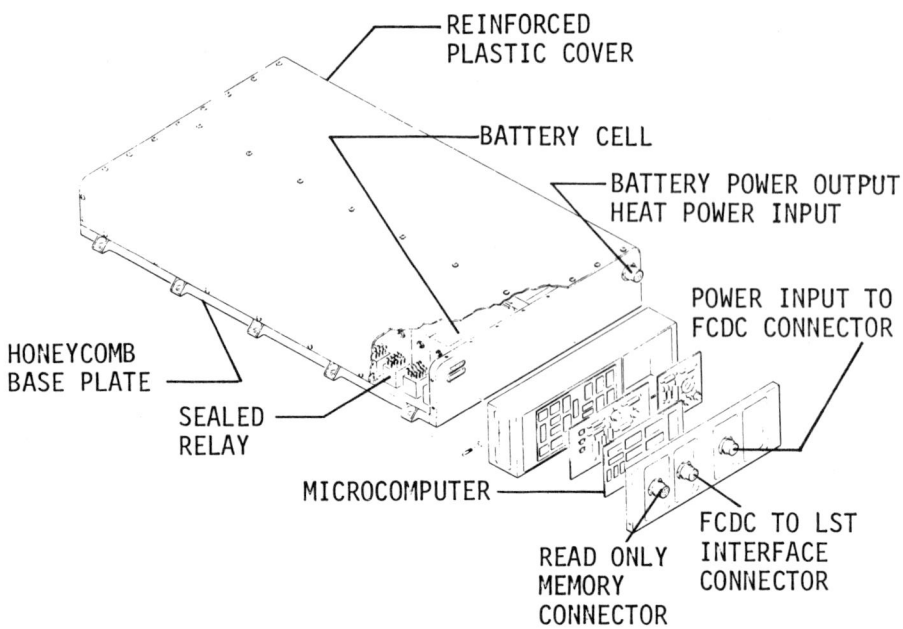

Figure 2   FCDC Mounted to a 36-Cell Battery

SUMMARY

The individual cell charge and discharge control of the LST nickel-cadmium cells offers a potential design improvement that could result in batteries with cycle life well in excess of the 12,000 cycles. The exact value of the FCDC concept must be proved by battery testing or inflight operation. If indeed the battery tests prove that the FCDC significantly increased the battery cycle life, the battery life increase can be weighed against the cost of the FCDC to design the most cost effective system for the LST.

REFERENCES

1. M. S. Imamura, R. L. Donovan, J. L. Oberg, L. A. Skelly, and D. H. Julseth: "Microprocessor-Controlled Battery Protection System", to be presented at the 1975 Intersociety Energy Conversion and Engineering Conference.

2. W. T. Perreault, "Lightweight Housing Study for Batteries and Electronic Packaging, Final Report, MCR-74-71, Martin Marietta Corporation.

AAS 75-183

LARGE SPACE TELESCOPE — ORBITAL CREW
EV MAINTENANCE OPERATIONS*

H. T. Fisher[†]

INTRODUCTION

Initial Human Factors Engineering analyses directly applicable to LST were undertaken in support of the NASA-Industry 1965/1967 Optical Technology Experiment System (OTES) studies. Then, again, in mid-1972, Lockheed undertook a major and since continuous Crew Systems activity including development of full-scale hard mockups and conduct of extensive man-in-the-loop simulations. During this time and in concert with NASA, program objectives were developed, interfaces identified, and ground and on-orbit crew system requirements established.

The primary objectives are to develop an on-orbit EV maintenance capability to minimize program cost, reduce design/operations complexity, improve operational reliability, and reduce development uncertainties. Additional program objectives are: develop flexibility in candidate maintenance approaches, accommodate (where practical) off-shelf equipment items, provide for maximum safety of maintenance and flight crew, and evolve a transitionally flexible and smoothly integrated crew subsystem.

HUMAN FACTORS PROGRAM

In analyzing the LST system and attendant maintenance modes, certain key issues predominate and therefore become the catalyst of crew systems operations and design development. These key issues are listed in Table 1.

---

*Funding for this effort was provided by Lockheed Missiles & Space Company with support from NASA Contract NAS8-31313.
[†]H. T. Fisher is Head, Crew Systems at Lockheed Missiles & Space Company, Inc., Sunnyvale, California

Table 1

KEY LST CREW SYSTEM ISSUES

- SAFETY TO CREW, SHUTTLE & LST (IN THAT ORDER)
- COMPATIBILITY WITH PERKIN ELMER, ITEK, & NASA SI LAYOUT
- LST TO SHUTTLE DOCKING AND/OR BERTHING
- PHILOSOPHY FOR MANUAL DETRACT/EXPAND – DEPLOYABLE DEVICES
- AXIAL VERSUS RADIAL SI REMOVAL/REPLACEMENT
- MAJOR SCIENTIFIC INSTRUMENT MODULE SIZE & CONFIGURATION
- SPARES QUANTITIES, STOWAGE & VOLUME (IF IN P/L BAY)
- CREW EV TIME – INTEGRATED SUIT ECLSS VS STRAP-ON ECLSS
- MANIPULATOR VS CREW VS COMBINATION OF CAPABILITIES
- ON-ORBIT CHECKOUT MODES AND VERIFICATION

As the program has evolved, several fundamental crew requirements have been generated. Each has been based on extensive analyses, mockup design verification, simulation results (both neutral buoyancy and 1-g), and data extracted from the highly successful Apollo and Skylab EV activities (both planned and emergency). These basic requirements are presented in Table 2.

Table 2

BASIC LST CREW SYSTEM REQUIREMENTS

- OPERATIONS BY ONE EV CREWMAN SHALL BE THE DESIGN GOAL
- CREW WORK, TRANSLATION AIDS SHALL BE DESIGNED INTO LST
- COMPONENT/MODULE CHANGEOUT SHALL BE DESIGNED FOR ONE GLOVED HAND OR HAND HELD TOOL
- ALL COMPONENTS/MODULES SHALL BE ORBIT REPLACEABLE
- ALL HARDWARE & SPACEFRAME SHALL BE DESIGNED FOR CREW SAFETY
- SSM/SI ACCESS DOORS SHALL BE SIZED TO PERMIT MODULE, SI, & COMPONENT WITHDRAWAL/INSERTION – GROUND OR ORBIT
- ALL COMPONENTS/MODULES SHALL BE SIZED NOT TO EXCEED TWO MAN HANDLEABILITY ON THE GROUND
- ADEQUATE INTERNAL VOL. FOR GROUND OR EV CREW INTERNAL ACCESS TO & MANEUVERING WITH COMPONENTS, MODULES, & SI's
- LEAST CONTAMINABLE CREW GROUND/EV TRANSLATION ROUTES
- SHORTEST ROUTE TO SPARES PALLET & WORK PLATFORM
- MINIMIZE NEED FOR SPECIAL CREW SSE
- DIRECT UNCLUTTERED ESCAPE ROUTE & RESCUE ACCESS
- OPTION FOR RMS USE TO TRANSFER EV CREW &/OR EQUIPMENT
- MAXIMUM UTILIZATION OF SHUTTLE CREW/SYSTEM CAPABILITIES

EV Maintenance Concept

The preferred maintenance approach, incorporating results of 4 years of human factors studies, is for unaided manned EV suited operations. The maintenance crewman translates up rails to the Equipment Section for external maintenance of support system hardware. Figure 1 illustrates the maintenance mode configuration and the associated spares rack and work platform.

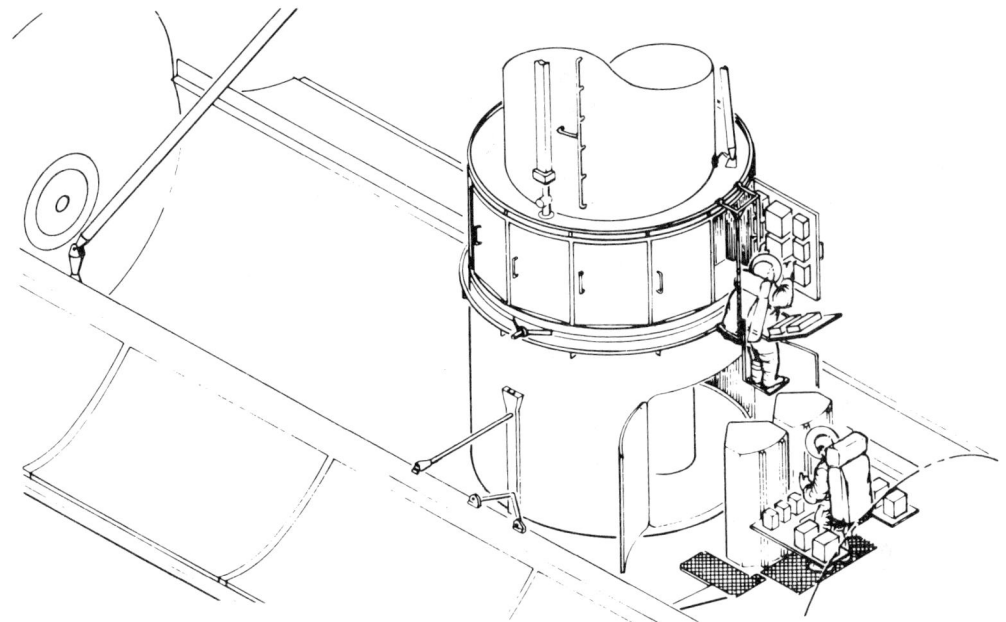

Fig. 1 LST Berthed to Orbiter, Illustrating Spares Pallet/Work Platform

The primary maintenance mode is internal for the SIs (Scientific Instruments) and therefore conducted radially. SI access is quite feasible and can be accomplished from the work platform by translating through either compartment "door," as seen in Fig. 1. Maintenance of the Equipment Section is conducted externally in a mobile work platform, also shown in Fig. 1. A series of external "doors" has been provided to permit radial access to the Equipment Section, and also to provide for ease of ground maintenance access. Figure 2 illustrates the techniques for SI and equipment removal.

Berthing or Docking

A key element in the overall LST maintenance mode is the successful capture and berthing or docking of LST to the Orbiter. On the basis of extensive trade studies conducted to date, it appears that the most simple and cost effective docking technique is to soft berth the LST to the P/L Bay Sill (Fig. 1). Close proximity of the Crew to the work platform (with spares) and the work platform to the LST is one of the driving factors in the selection of this approach.

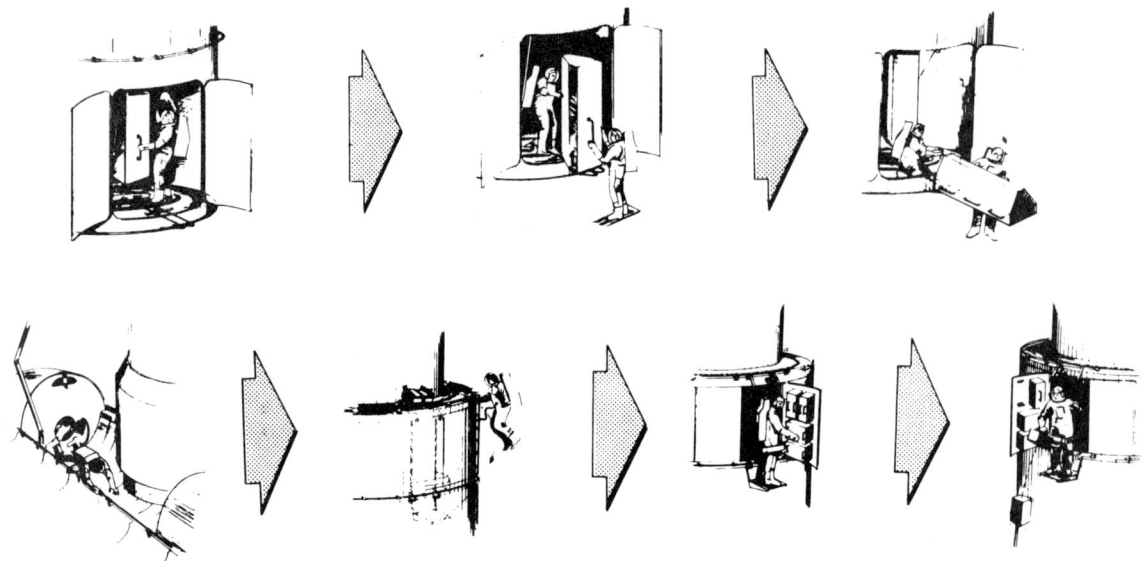

Fig. 2 SI and Equipment Removal Sequences

Mockup Development and Simulation

In mid-1972 a program plan was prepared in concert with NASA-MSFC for the Lockheed independent development of a full-scale high fidelity mockup to permit water immersion for potential neutral buoyancy man-in-the-loop simulation. The mockup was developed as a design and operations verification tool to be used by each of the major engineering and system disciplines. Results of mockup studies have indicated that installation and removal of Equipment Section hardware is rather straightforward.

CONCLUSIONS

All studies to date indicates that on-orbit EV manned maintenance of the LST is not only feasible but can be designed to be easily within the capability of the EV functioning astronaut. Both 1-g and neutral buoyancy man-in-the-loop simulations further verify this point. It is important to recall the success of the Skylab crews in accomplishing the "impossible" mission saving tasks with few, if any, built-in EV mobility and work aids. Therefore, this opportunistic success strongly suggests consideration of the following: (1) What can future crews accomplish if payloads are designed for orbital maintenance? and (2) What potential time and cost savings can be realized by program developers and, ultimately, the principal users?

AMPS PROGRAM
(Atmospheres, Magnetospheres, and Plasmas in Space)

AMPS PROGRAM
(Atmospheres, Magnetospheres, and Plasmas in Space)

Program Chairman		C. A. Lundquist, NASA/MSFC
Assistant		S. R. Schrock, Martin Marietta Corporation
Science		
	Chairman	Alan Rosen, TRW
	Co-Chairman	Frank Bartko, Martin Marietta Corporation
Engineering Challenges		
I	Chairman	Sherman DeForrest, University of Alabama
	Co-Chairman	Robert W. Ellison, Martin Marietta Corporation
II	Chairman	William Roberts, NASA/MSFC
	Co-Chairman	Owen Taylor, Bendix Corporation

AAS 75-132

CHALLENGES
IN SPACE PHYSICS

Billy M. McCormac[*]

Scientific objectives in space physics are discussed
for solar wind-magnetospheric interactions, magnetospheric phenomena, magnetospheric-ionospheric interactions, and using the magnetosphere as a plasma
laboratory. Particular attention is given to the
constraints on AMPS to satisfy these scientific
objectives.

INTRODUCTION

There are many challenges in space physics for the 1980's and for many decades. Since space physics is less than 20 years old, we are not about to run out of important problems. You will hear in the next presentation how atmospheric physics still has many exciting problems after more than a century of research. Astronomy, including solar physics, is not discussed in this presentation.

There is insufficient time to try to explain and justify the space physics problems; therefore, I will cover the spectrum of challenges. I will emphasize those constraints on meeting the challenges wherein AMPS should try to accommodate the physics. I will naively assume that the whole justification for AMPS is to support the atmospheric and the space physics investigations to be discussed in this and the next talk.

For the purposes of this presentation, space physics is defined as including those physical phenomena in the earth's magnetosphere, but not those phenomena clearly falling in the domain of atmospheric

---

[*] Dr. Billy M. McCormac is Manager of the Electro-Optics Laboratory, Dept. 52-54, Bldg. 202, Lockheed Palo Alto Research Laboratory, 3251 Hanover St., Palo Alto, California 94022, (415) 493-4411 Ext. 45314

chemistry and physics. We will cover the physical processes from the solar wind through the magnetosphere to the atmosphere. The earth's magnetic field plays a very dominating role in this region. The major problems fall into four categories.

Solar Wind-Magnetospheric Interactions

Although we believe that the energy to power magnetospheric phenomena comes from the sun, little is known about the details of energy transport and forms.

Magnetospheric Phenomena

It is necessary to determine how energy flows through the magnetosphere.

Magnetospheric-Ionospheric Interactions

The upper atmosphere and the magnetosphere are closely coupled by means of earth magnetic field lines.

Magnetosphere as a Plasma Laboratory

The magnetosphere can be thought of as a gigantic plasma laboratory. There are many plasma experiments which are impossible in the laboratory because of scale lengths, wall effects, etc.

What kinds of problems will remain by the 1980's? The view is often expressed that by then the IMS, ISEE, GEOS, etc. will have discovered the major dynamical processes. There is a semantics problem, but I do not agree. By 1980 the exploratory phase will have been completed and there should be no major surprises in composition and physical characteristics. We will tend to have determined the general configuration and composition of the magnetosphere with the general or average energy, particle, temperature, bulk flow, etc. content. We will know very little about the detailed physical processes, correlation between different observables, transport of energy and momentum, etc. The exploratory phase platforms have only been able to carry a small number of instruments, resulting in observations of limited composition and physical characteristics. Thus, simultaneous measurement of particles, fields, waves, composition, etc. has not been possible. A single instrument platform cannot determine whether an observed change is temporal or spatial. This results in considerable confusion in characterizing the magnetosphere. The space physics problems all involve dynamical processes requiring separation of time and space. By 1980, only ISEE will have had a Mother-Daughter experiment, and it will only have a limited number of experiments. It is absolutely essential to have two or more identical sensor packages separated in space. If the main vehicle of AMPS is too polluted for the observation, then there must be at least two subsatellites.

It is necessary to perform observations as a function of UT (quiet, disturbed solar conditions, etc.), LT, latitude, longitude, magnetic coordinates, seasons, etc. Thus synoptic data over a large array of conditions is mandatory.

Do not get bogged down with models, boundaries, and regions in space. Models are to help explain a concept. It is wrong to argue about models -- rather discuss the physics involved. Boundaries and regions are established on the basis of observables and perhaps on an inapplicable observable. Many important processes may not recognize the boundary or region. Overemphasis of a region or boundary on the basis of a poor observable often has serious consequences on theory and future measurements.

The shuttle can provide a facility that can be used many times to perform space physics experiments. It is hoped by the scientific community that the ratio of dollars for scientific instruments and analyses to the total dollars for shuttle will be increased over that for earlier spacecraft. So far, there is not much indication that the curve is proceeding in the right direction.

One of the most important capabilities of AMPS over all existing space vehicles is the large payload which allows a multi-instrument attack on many problems. Adequate power, data handling, mechanical apparati, telemetry, etc. can be provided. Prior to AMPS we have tended to measure particles, waves, and optical emissions from different platforms. Even when some of these instruments were continued, weight and space tended to limit observations. Prior to the 1980's very few active experiments will have been conducted. AMPS has a mission to provide the best possible facility, within the time and fiscal restraints, for space and atmospheric physics. To meet the challenges, the AMPS facility must accommodate the physics; however, it sometimes seems like the tail is wagging the dog.

STRUCTURE AND CHARACTERISTICS OF THE MAGNETOSPHERE

The magnetospheric structure is complicated and there are so many dynamical processes occurring simultaneously such that it is difficult to separate the various processes. The regions of the magnetosphere are identified in Figure 1.

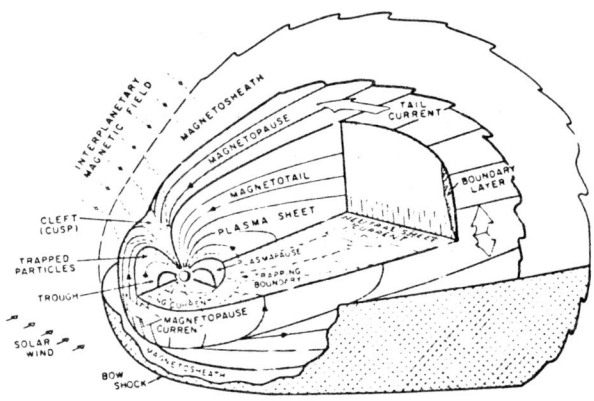

Figure 1. Regions of the Magnetosphere

Solar Wind

The energy to power the various magnetospheric phenomena is believed to come from the solar wind; however, there are problems in correlating observations with the solar cycle. There are no clear solar cycle variations in solar wind properties.

Open Magnetosphere

It is now completely accepted that the magnetosphere is open, i.e., magnetic field lines from the earth connect with interplanetary magnetic field lines. Magnetospheric phenomena are affected by the direction of the interplanetary magnetic field. Much work is yet to be done on the magnetospheric configuration when the interplanetary field is northward vs. when it is southward.

The electric field of the solar wind maps into the polar cap. However, the mathematical techniques for relating the polar cap electric field to the field imposed across the magnetospheric tail by the solar wind are relatively crude.

For many years the evolution of magnetospheric models has been based on the fundamental concept of a closed magnetosphere. Therefore, to a considerable extent the development of open magnetospheric models now follows from a modification of closed magnetospheric models.

An important process in an open magnetosphere is reconnection of the earth's magnetic field lines which were for some part of the earth's rotation connected to interplanetary field lines, then reconnect, trapping some plasma which entered from interplanetary space. The reconnection process has been investigated for a long time; however, details of the reconnection process are poorly understood; for example, we do not know if the electric field across the magnetotail is the cause or effect of reconnection.

Plasmasphere

The plasmasphere is observed to have a bulge on the dusk side and its boundary, the plasmapause, shrinks to lower L shells as K_p increases. The plasmapause is the boundary where hot plasma becomes dominant over cold plasma. The plasma seems to be accelerated to high energies during a substorm to form the hot plasma sheet. The plasmapause is thus the boundary beyond which all of the cool plasma is accelerated to become hot plasma. After a substorm the plasmapause expands to higher L values as the plasmasphere is refurnished with plasma from the ionosphere.

Current Systems

The electric field imposed on the polar cap will extend into the ionosphere and form part of the magnetospheric convention system. More information is needed on the electric fields, magnetic fields, and conductivities if one is to use the available techniques for computing the magnetospheric current systems.

The electrical coupling between the ionosphere and magnetosphere is caused by the high conductivity along the magnetic field lines. The ionosphere is the region of maximum electrical conductivity transverse to the geomagnetic field lines caused by ion-neutral collisions. The magnetosphere is characterized by collision-free MHD processes. The coupling occurs through Birkeland currents and is important for many phenomena such as substorms, plasma convection, auroral particle behavior, instabilities producing double layers, etc. It cannot be assumed that the potential difference between the ionosphere and magnetosphere can be explained by parallel electric fields, those parallel to the local magnetic field lines.

WAVE PARTICLE INTERACTIONS

The research effort in wave-particle interactions has greatly increased in the last few years. In the magnetosphere enough data are now available to clearly demonstrate that non-linear processes are involved and that linear theory is not applicable. On the other hand, there is insufficient data to uniquely determine the specific types of wave-particle interactions for the various parts of the magnetosphere.

Which comes first, the waves or the particle anisotrophy, i.e., do waves cause pitch angle diffusion of particles or vice-versa?

A number of forms of longitudinal waves have been found (Figure 2) which appear to have strong interaction with local plasma particles.

The magnetosphere is divided into two regions separated by the plasmapause. The flux tubes close to the earth which corotate with the earth

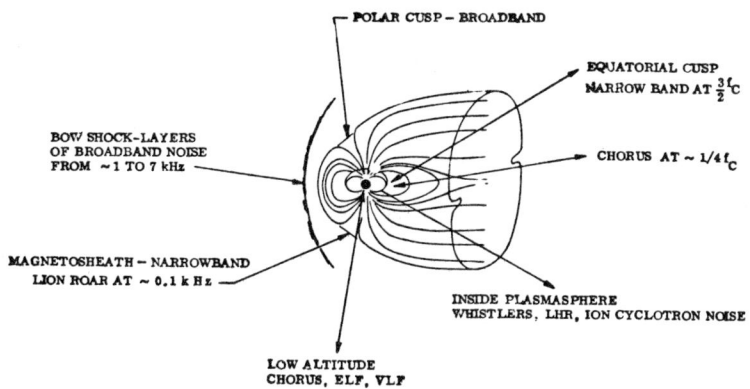

Figure 2. Location in the Magnetosphere of Some of the More Common Intense Waves

never become open and contain high densities of ionospheric cold plasma. At large distances from the earth flux tubes flow from the magnetospheric tail to the dayside carrying hot plasma sheet particles. Wave-particle interactions can play an important role in the various processes by: affecting the structure of the inner edge of the plasma sheet through strong pitch angle scattering; decay of the ring current; and radial diffusion of the hot plasma in the plasmasphere.

SUBSTORM PHENOMENA

The study of magnetospheric substorms is receiving much attention. Details about the morphology are better known than in the past; however, it is difficult to recognize the primary features of a substorm since they are highly variable in time and space. There is much interest in the initiation of the expansion phase to determine the triggering mechanism for substorms. Observations of substorm phenomena must become more sophisticated. More observing stations are required. Multidisciplinary observations are needed, such as optical, plasma, and magnetic field observations. Figure 3 shows the different types of substorms comprising a magnetospheric substorm.

MAGNETOSPHERIC SUBSTORM

POLAR SUBSTORM

1. AURORAL
2. POLAR MAGNETIC
3. IONOSPHERIC
4. X-RAY
5. PROTON AURORAL
6. VLF EMISSION
7. MICROPULSATION

{
a. ABNORMAL IONIZATION
b. LAYER DEFORMATION
c. ATMOSPHERE WAVE SUBSTORM
d. EXPANSION AND CIRCULATION BY HEATING
}

{
1. TRAVELING WAVE DISTURBANCE
2. INFRASONIC WAVE
}

Figure 3. The Various Phenomena Comprising a Magnetospheric Substorm

Morphology

A schematic representation of a substorm is shown in Figure 4. The interplanetary magnetic field turns southward after a long interval of northward field. The substorm begins when the discontinuity reaches the magnetopause. After some delay the magnetopause moves earthward due to the erosion of magnetic flux which is being transported by the solar wind to the lobe of the tail. This increases the tail field. The plasma sheet thins and the tail current moves earthward.

The expansion phase begins when the near-earth plasma sheet becomes extremely thin. Magnetic flux previously stored in the tail lobe is annihilated, decreasing the magnetic field in the tail. Plasma energized at the neutral point is injected both earthward and tailward increasing the partial ring current and expanding the plasma sheet.

The recovery phase begins when the additional energy in the lobe of the tail is exhausted through annihilation. Field aligned currents, westward electrojet, injection into the partial ring current, and the eastward electrojet begin to decay.

It is difficult to determine the onset time of a substorm. A study based on the onset of midlatitude positive bays has been conducted.

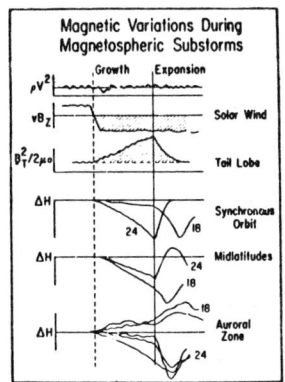

Figure 4. Phenomenological Model of Magnetic Variations During Magnetospheric Substorms

There is a wide variety of substorms apparent at midlatitude stations; there are sharp onsets, slow onsets, multiple onsets, eastward sweeping onsets and westward sweeping onsets. Also, the central meridian of maximum development of substorms may be centered anywhere in the night hemisphere.

Aurora

Remote sensing of optical emissions provides a picture of the magnetosphere as it projects into the polar atmosphere. Various optical features in the atmosphere can be correlated to regions and processes in the magnetosphere. The aurora consists of arcs and regions of diffuse aurora. The plasma sheet maps into the diffuse aurora which are the most persistent and prominent auroral features.

It is difficult with this data to determine if there is really a single auroral oval or if there are overlapping horseshoe structures. Does the auroral oval have any physical significance beyond a statistical one, or is it a composite of several distinct but overlapping regions of diffuse and discrete aurora? Discontinuities occur frequently, especially near midnight.

MAGNETOSPHERE AS A PLASMA LABORATORY

In addition to the laboratory plasma physics experiments associated with investigating the magnetosphere, similar type experiments can be conducted in the magnetosphere.

Energetic Particle Trapping

When we have a magnetic field immersed in a cold or hot plasma we still do not know how many energetic particles can be trapped on the magnetic field lines. Does it depend on the energy density at the mirror points, at the equator, or what? The ratio of the magnetic field energy density to that in energetic particles varies from 10^{-4} at the inner trapping boundary to 1 in the outer magnetosphere.

Wake and Sheath Studies

Measure the characteristics and mechanisms for production.

Antenna Studies

The efficiency for long wave length antennas for receiving and transmitting is important for interpreting data and planning experiments. The estimated efficiencies for small antennas make their use very marginal.

Wave-Particle Interactions

Instabilities involving waves and particles play a very key role in the form of energy, its propagation, and deposit. These interactions are especially important in establishing the phase space distribution of charged particles and in the transfer of ring current energy into ionospheric heating. Controlled experiments are needed to investigate non-linear effects.

Sources of Waves

The mechanisms for producing different types of waves under various plasma conditions, etc. can be investigated by controlled experiments. The characteristics and propagation of these waves are of interest.

Momentum Transfer

How is momentum transferred in the magnetosphere if a plasma cloud is detached or injected (such as Ba)? Inside the magnetosphere the Alfvén velocity is much larger than the ambient velocity. The cloud loses its momentum relative to the ambient plasma by propagation of its electric polarization field along the flux tube with the Alfvén velocity. Thus, a steady increasing amount of matter shares the initial momentum of the cloud.

AMPS FACILITY

The physics problems that I have briefly discussed place certain requirements on AMPS. In considering the challenges in space physics, I have tried to be realistic about the shuttle support.

The scientific community will insist on performing better measurements on AMPS than have been made in the 1970's. They are not going back to geiger counters and brownie cameras.

I want to emphasize three problem areas, whose solution must be integrated into experimental concepts and design: AMPS pollution, data processing, and the lack of sufficient interaction between the scientific and the vehicle communities.

AMPS Pollution

A few passive instruments in a very well designed small satellite may not significantly pollute the space environment. However, if you build a city in place of a horse barn, you will lose the "country air". You cannot have a large shuttle vehicle with many active instruments without seriously affecting the space environment. The environmental problems are divided into two categories: 1) EMI and 2) dust and gas.

Many of you are aware that many instruments will not operate in the near field of a radar, in a capacitor bank room, or in a dusty environment.

The EMI will be more severe for shuttle than for any other vehicle. Currents will flow between many different sections of the satellite. Operation of most of the accelerator and RF transmission experiments may preclude simultaneous operation of many other instruments. The electromagnetic pulse from discharging capacitor banks may seriously disrupt instruments that are not turned off.

Optical instruments are severely affected by a small amount of dust in the FOV. These may cause thermal emission in the IR or scattering of radiation in the UV-vis-IR. A single 1 μm particle in FOV may preclude observations. Mechanical movements on the shuttle will raise dust. A particular dust particle may require a complete orbit to drift out of the FOV. Solenoids to change filters, scanning mechanisms, boom deployment, pallet movement, etc. will contribute to the dust. Some data processing techniques may eliminate the effects of dust in the near field. Baffling to eliminate thermal or scattered radiation from the parts of the shuttle will be difficult.

Outgassing, gas releases, contaminants from alkali metal releases and rocket motors can lead to a number of degradations. The scattering of radiation from various gases can lower sensitivities. These contaminants will affect mass spectrometer observations. Gases and dust may freeze out on cyrogenically cooled surfaces and may contaminate optical windows by sticking to their surface.

There is some threshold value of each of these environments above which observations become degraded. In general, the AMPS vehicle can meet none of the criteria now used for no significant effect. Very careful design can reduce effects. Many instruments may need to be placed in subsatellites. For example, can cold or hot plasma measurements be made from shuttle?

Data Processing

High time, space, and observable resolution on many instruments lead to extremely large amounts of data to be managed. Two-dimensional resolution may consist of 10^6 discrete elements. Temporal resolution may require measurements every 10^0 to 10^{-3} seconds. Each physical measurement may require 10^1 to 10^3 numbers of 10 bit accuracy. Thus, a single instrument may require 10^8 to 10^{12} numbers of 10 bit accuracy. Thus, a single instrument may require 10^8 to 10^{12} bits per second. A shuttle payload may require data management of 10^9 to 10^{13} bits per second.

Let us look at the data handling problems if one could send 10^{11} bits per second back to the ground. This adds up to 10^3 standard 2400 ft. magnetic tapes per second. Thus, a one week shuttle mission would produce:

6×10^8 tapes
3×10^4 semitrailer trucks
1.4×10^2 acres of storage space.

How would this data ever get processed and distributed to the investigators?

Even if we assume that the maximum shuttle telemetry is 50 MHz (equivalent to about 8 standard TV systems), we will still fill several magnetic tapes every second adding up to more than 10^6 tapes per mission (1 week).

It is obvious that real time data processing must reduce the data to be transmitted by many orders of magnitude. Experiment and instrument design must take the requirement for real time data processing into account.

Orbit

An orbit of about 350-400 km altitude is not very effective for investigating space physics problems, unless wide ranging subsatellites are employed. The orbit simply does not sample the best regions of space.

BIBLIOGRAPHY

Much of the material in this manuscript is taken from the following three books:

MAGNETOSPHERIC PARTICLES AND FIELDS, 1976, (Ed. by Billy M. McCormac) D. Reidel Publishing Company, Dordrecht, Holland, in press.

MAGNETOSPHERIC PROCESSES, 1974, (Ed. by Billy M. McCormac) D. Reidel Publishing Company, Dordrecht, Holland.

EARTH'S MAGNETOSPHERIC PROCESSES, 1972, (Ed. by Billy M. McCormac) D. Reidel Publishing Company, Dordrecht, Holland.

AAS 75-133

CHALLENGES IN THE ATMOSPHERIC SCIENCES[*]

Robert E. Dickinson[+]

This paper discusses two major challenges to atmospheric sciences in the next decade. These are stratospheric chemistry and climate change. Both subjects are ripe for further study. The ability to predict consequences of human activities in these areas is needed to avoid either unnecessary economic restrictions or environmental disasters.

INTRODUCTION

I shall discuss here two areas of atmospheric sciences that will be both scientifically very challenging and also of a considerable societal importance in the next decade. These are: 1) stratospheric chemistry and its effect on climate and the biosphere, 2) long-term climatic change. I shall be discussing these topics from a broad scientific perspective and not referring to particular observing platforms. However, observations from satellites in general, and AMPS in particular will certainly play a significant role in meeting the challenges provided by these problem areas. Indeed, if this were not so, I would expect that NASA would have severe difficulties in gaining public support for many of the other scientific problems it will be considering in the future. These areas that I shall discuss have already evoked strong public interest, and indeed much of the current political justification for NASA's budget appears to hinge on promises of advances in these subjects.

---

[*] Prepared for technical papers that will later be published in the Proceedings of the American Astronautical Society.

[+] Dr. R. E. Dickinson is affiliated with the National Center for Atmospheric Research which is sponsored by the National Science Foundation.

If this be true, I believe that two criteria must be met for the space program to meet the public expectations of significant progress in these areas: 1) That scientific research actually focusing on these problems should not be but a small fraction of the research justified by their importance. In other words credibility of the research programs devoted to these subjects will be greatly weakened if they consist largely of studies only superficially related to the more crucial questions that must be answered to secure practical results; 2) The fraction of the available support devoted to analysis and interpretation of data and related theoretical studies should be increased. It will no longer be possible for instruments to be built and flown without extensive and rapid exploitation of the data collected. If it is not obvious that the data to be obtained meets an important need, it will not be possible to build the instruments. In the past, satellite data of urgent scientific importance such as the global distribution of ozone have been unavailable for many years because of inadequate programs for their reduction and analysis. It seems unlikely that such an approach will in the future be tolerated by the scientific public.

The scientific areas that I shall be discussing here are characterized both by feedback couplings between many physical processes and by long chains of causal relationships. Some of these connections are now reasonably well understood, others still but poorly understood. The filling in of the poorly understood connections may require in some cases as much as a decade of research or more. The establishing of these connections, I believe, will be a major challenge to atmospheric science in the 80s. Let us now turn to the particular subject areas that I am discussing.

STRATOSPHERIC CHEMISTRY

The major focus of stratospheric chemistry today is on the global distribution of stratospheric ozone. It is generally believed that the formation of an ozone layer several billion years ago was essential for the evolution of life on earth. Conversely, total removal of the ozone layer would surely lead to a rapid destruction of the biosphere. The reason for ozone's great importance to the biosphere, is its absorption of most of the ultraviolet radiation between 220 and 310 nm, radiation which is very damaging to various kinds of living tissue in the amounts put out by the sun. (A popular review of current issues in stratospheric ozone chemistry is given in[1].)

Many of the basic chemical processes responsible for maintaining ozone at its natural levels have been identified only in the last few years. Useful surface observations of total ozone in a column, on the other hand, have been obtained for a long time, and these have been available in a global observing system for nearly the last two decades (with, however, quite poor coverage over oceans and the total Southern Hemisphere). From this data, we have learned that total ozone, like other meteorological quantities, has large variations with latitude and season, and fluctuates significantly from day to day.

The current improvements in understanding of ozone chemistry have pointed to the importance of odd nitrogen and odd chlorine in catalytic reactions which reduce the amount of ozone below that maintained by reactions involving oxygen species alone, the classical Chapman ozone chemistry. Chapman chemistry alone would largely control the distribution of ozone if the stratosphere were not coupled to trace gases produced in the troposphere, including not only nitrogen and chlorine compounds but water vapor and methane as well. I shall simplify the discussion here by referring only to the roles of the nitrogen and chlorine compounds.

There are both natural and human sources of these chemical species that perturb ozone after being transported in one form or another from the troposphere to the stratosphere. The most important of these at present is nitrous oxide, N_2O, that is produced by dentrifying bacteria at the earth's surface from nitrates in the soil and oceans and is transported to the stratosphere, where a small fraction of it is oxidized to nitric oxide, one of the species that actively destroys ozone. Without this upward flux of nitrous oxide and subsequent conversion of NO, ozone concentrations in the stratosphere would be order of 50% larger than is observed. However, such basic questions as the relative contributions of land and ocean surfaces to the flux of N_2O into the atmosphere are still poorly understood[1]. Equality between the rates of dentrification and nitrogen fixation in the soil is necessary for a long-term global balance. Projections of artificial fertilizer production growing by the year 2000 to rates larger than the natural nitrogen fixation rates over land have led some scientists recently to give warnings of this possible threat to ozone.

It was the possibility raised a few years ago of essentially an equaling of the natural production of NO in the stratosphere by a large fleet of SSTs that has motivated much of the Department of Transportation's recent Climatic Impact Assessment Program[2]. Concern over the possible occurrence of similar effects from chlorine in the fuel of space shuttle boosters has shown these to be relatively negligible, but has also led to the recent realization that there are a number of anthropogenic chlorine compounds which reach the stratosphere in sufficient abundance to significantly diminish ozone concentrations. There have also been recent discoveries of natural sources of chlorine to the stratosphere, though these have a much smaller effect than do the natural oxides of nitrogen.

Looking to the 80s, I think it will probably not be possible to prevent entirely the release into the stratosphere of gases produced by human activities and leading to reduction of ozone concentrations. In some cases these activities will be regarded as of great societal value and the releases as unavoidable. However, they will only be possible if all consequences can be satisfactorily determined and weighed against the benefits so derived. If there is some degradation of the global environment by a human activity, the costs will generally be born by different people, indeed different nations than those who derive the primary benefits. Some form of compensation is thus ultimately required which can only be determined if the costs are known. Where scientific understanding cannot provide satisfactory estimates of costs, there will be strong disagreements as to the wisest action to take, and the permitted level of the activity in question can only be decided on the basis of power, political or otherwise, rather than on the basis of equity to all parties involved.

A translation of these conclusions into scientific requirements suggests that one challenge of the 80s will be an understanding of all aspects of the ozone cycle and its climatic and biospheric interactions in highly quantitative terms. Progress in the direction is already evident as far as the ozone chemistry itself goes. Presumably by the next decade global observations of the more important chemical species affecting stratospheric ozone will become available, and theoretical models predicting changes of ozone consequent to various levels of species emissions will be reasonably accurate. I can't be as optimistic as regards the capabilities for predicting climatic and biological effects resulting from changes of ozone or in biological effects influencing the ozone layer. In these areas there appears to be many fundamental but yet unsolved problems and a much lower level of activity in terms of people involved and scientific support. Yet, it is already these areas which limit the certainty with which the effects of perturbations of the natural ozone level may be evaluated.

An understanding of the biological effects, including crop production, of changes in solar UV radiation resulting from changes in ozone, will have to be achieved if an understanding of the dependence of the distribution of ozone on trace chemical species is to be fully utilized. Thus this area will certainly be a major challenge in the 80s. The whole question of the biospheric sources and sinks of crucial atmospheric trace gases will also receive much attention. Such questions are not, strictly speaking, atmospheric science and so will not here be discussed further.

The other field of inquiry, that of climatic change, is obviously much broader than just an evaluation of the effects of changes in stratospheric species on climate and is the other challenge to atmospheric sciences that I am discussing today.

CLIMATE CHANGE

The occurrence of climate change in the past is too well known to need detailed discussion here. Likewise, the effects of these changes on human civilization have been extensively discussed elsewhere. I shall merely point out that many human activities, especially agriculture, have been optimized to the present climate, and that departures in either direction from mean conditions are likely to be economically harmful. I shall also here avoid attempting to precisely define climate except to point out that surface temperature averaged over seasons or longer is an important component of both popular conceptions of climate and any technical definition.

There is a great practical need for developing the capability to predict any climate change that would result from human perturbations. We do know that various human activities are rapidly approaching the level where they can have significant impact on climate. However, our present capability for quantitative evaluation of these effects is still quite inadequate. I have previously mentioned the possibility that changes in the distribution of ozone resulting from human activities may lead to climate change. Other prominently mentioned possible causes of climate change are the projected increases of CO_2 due to fossil fuel combustion, the modification of global cloud and aerosol cover by industrial processes, and global thermal pollution[3,4,5]. Again, many of the activities that threaten climate change are highly desirable components of our economy. Without a quantitatively accurate theory to predict climatic effects of human activity, we may have to face prospects of on the one hand unnecessarily severe economic restrictions, and on the other hand of possible climatic disasters. I thus see as one of the major scientific challenges of the next decade, the development of practical theories of climate change. I can't be as confident of the success of this endeavor as of the development of predictive capabilities for stratospheric chemistry. The present attitude among workers in the field is that we must first study the various physical processes involved in determining climate so that their connections are reasonably well understood. Only then, will it be possible to say whether an effective predictive capability is achievable.

Models are the tools both for studying the coupling between various climatic parameters and for attempting prediction of responses to perturbations of the system. Let us consider the ingredients of the present crude models to see which physical processes most sensitively influence climate and which are most poorly known. There is, indeed, a large hierarchy of models of different intent and application. I refer the reader elsewhere[6] for detail. Here I shall merely comment on some of the most important features of current modeling efforts.

Any thoughtful attempt to develop a model of climate change must first confront the questions of the time scale and the spatial scales to be considered. A farmer would find most useful a prediction of seasonal anomalies, that is, the departures from mean conditions over some forthcoming period of, say, three months. He needs such a prediction at his locality. Climatological data indicate that variations of a given month from long-term monthly means are significantly larger than departures from mean conditions averaged over several years or more. Likewise, there are large latitudinal and longitudinal variations in climatic anomalies, so that global mean climatic variations are considerably smaller than local variations. Yet it appears that for time scales of decades or more variations of global mean climate may be most simply modeled.

There is some hope of providing seasonal forecasts with detailed spatial resolution using three-dimensional General Circulation Models (GCMs). Only simpler 2-D and 1-D models are capable of examining climate change over many years. These models indicate that variations in cloudiness, sea surface temperatures, and polar ice cover may be crucial factors determining global mean climate change. These factors may also be important for determining more local changes but their couplings to GCMs are yet poorly understood.

As I mentioned earlier, surface temperature is perhaps the most important climatic parameter. Many of the 1-D and 2-D models attempt mainly to predict changes in global mean surface temperature resulting from changes in atmospheric composition or surface conditions. These models respond primarily to changes in the surface and atmospheric heat budgets. These variations in heat budget are generally relatively easy to determine. The difficult question is the determination of the changes of various other climatic variables which in turn couple to surface temperature. For example, ice and snow cover reflect much more solar radiation than do other surface types. As mean temperatures increase, ice and snow cover would be expected to decrease, hence leading to an amplification of the temperature increase. However, reliable inclusion of this positive feedback in models is difficult.

A yet more difficult question is that of the changes in cloud radiative properties as mean conditions change. Predictions of either positive or negative feedback on surface temperature are found in current scientific discussions, and it is not possible to choose between the two. As far as we know, variations of cloud properties may either reduce or increase by factors of 2 the response of surface temperature from that which would occur without variations. Satellite observations give some information about cloud radiative properties but necessarily emphasize the outgoing radiation from the top of the atmosphere. Certainly, good observations of surface radiative fluxes and perhaps also fluxes within the atmosphere will be needed to assist in determining the relationship of clouds to climate.

CONCLUSIONS

In summary, the areas of stratospheric chemistry and climate change are two areas that are very challenging now, and I believe will be increasingly so in the 80s. If this country is not able to provide the necessary resources to meet these challenges, I think it will be politically difficult to justify a continuing space program.

REFERENCES

1. P. J. Crutzen, "Estimates of Possible Variations in Total Ozone Due to Natural Causes and Human Activities," Ambio, Vol. 3, No. 6, 1974 pp. 201-210.

2. CIAP Report of Findings, The Effects of Stratospheric Pollution by Aircraft, A. J. Grobecker, S. C. Coroniti, R. H. Cannon, Jr., Eds., DOT-TST-75-50, Dec. 1974.

3. R. E. Dickinson, "Climatic Effects of Stratospheric Chemistry," Canad. J. Chem., Vol. 52, No. 8, 1974 pp. 1616-1624.

4. W. W. Kellogg and S. H. Schneider, "Climatic Stabilization: For Better or for Worse?" Science, Vol. 186, Dec. 1974 pp. 1163-1172.

5. S. H. Schneider and R. D. Dennett, "Climatic Barriers to Long-Term Energy Growth," Ambio, Vol. 4, No. 2, 1975 pp. 65-74.

6. S. H. Schneider and R. E. Dickinson, "Climatic Modeling," Rev. Geophys. Space Phys., Vol. 12, No. 3, Aug. 1974 pp. 447-493.

AAS 75-134

AMPS SCIENCE OBJECTIVES AND PHILOSOPHY

E. R. Schmerling
NASA Headquarters
Washington, DC

The Space Shuttle will open a new era in the exploration of earth's near-space environment, where the weight and power capabilities of Spacelab and the ability to use man in real time add important new features. The AMPS is conceived of as a facility, where flexible core instruments can be flown repeatedly to perform different observations and experiments. The twin thrusts of remote sensing of the atmosphere below 120 km and active experiments on the space plasma are the major themes, and have broader implications in increasing our understanding of plasma physics and of energy conversion processes elsewhere in the Universe.

INTRODUCTION

When the space shuttle becomes operational, we will have explored and measured our near-space environment for over two decades. It is expected that by the early 1980's, we will have available a great deal of data on the major atmospheric constituents, the charged particle population, and the electric and magnetic fields, as well as their first-order variation with position, time and solar activity. It is, consequently, anticipated that the main thrust of research in Atmospheric and Space Physics will shift away from large-scale data gathering and towards quite specific cause-and-effect studies aimed at the detailed understanding of the physical mechanisms which control our environment. It is here that the high weight-carrying capability of the shuttle and the availability of man become important new features, and the one to four-week operational period is sufficient for most of the specific studies considered. This paper outlines a concept which is presently under detailed study called the Atmospheric, Magnetospheric and Plasmas-in-Space (AMPS) project. The AMPS is conceived with two principal objectives: the remote sensing of the atmosphere below 120 km, with emphasis on dynamics and minor constituents, and the performance of active laboratory-style experiments on the space plasma in the magnetosphere. The combination of atmospheric and plasma physics on one vehicle is not fortuitous, but has arisen because the atmospheric instrumentation will perform the diagnostics for some of the plasma experiments, and the plasma instruments can excite atmospheric species of interest to the aeronomers, even gases carried up and released which are relevant to the aeronomy of other planets and comets. The AMPS project is intended to provide a laboratory in space, equipped with flexible core instruments which can operate together, and perform many different experiments with only minor modifications from flight to flight. The AMPS is thus a facility, which will be available to many users.

SCIENCE OBJECTIVES

The detailed physical understanding of many of the features found in the terrestrial environment still elude us, so that we may still pose significant questions for study such as: How do solar wind particles enter the magnetosphere, and how are they precipitated out?

One of the important functions of AMPS is to provide the means for testing specific hypotheses by carrying out controlled experiments. The first part of the above question can be approached by controlled releases of tracers outside the magnetosphere, near selected possible entry zones, with the main AMPS containing the mass spectrometers and detectors to locate the inward streaming of the tracer particles. The second part of the question is more complex, and closely related to the broader problem of how instabilities are triggered in plasmas and stored energy released not only in our own magnetosphere, but in the flare-producing regions of the sun, in pulsars, and elsewhere in the Universe. To investigate a broad range of plasma interactions, it is proposed to incorporate means for stimulating the magnetospheric plasma by injecting electromagnetic waves, by injecting streams of charged particles, and by injecting cold plasma. These interactions are to be observed by carefully measuring the effects: detecting the perturbations in the ambient particle spectra, observing the airglow emissions, and receiving the radio waves. Some of these observations can be made from the Spacelab, some from a boom, and others will require sub-satellites a suitable distance away.

Electric fields, boundaries between open and closed magnetic field lines, and the mechanism of auroral production can well be investigated by emitting electron beams from an accelerator. Here again, it is clear that the atmospheric excitation can be observed with the photometers and spectrometers also used for observing the atmosphere, and the presence of man can be very helpful in adjusting the instruments in real time to maximize the effects and optimize the data collected. A detailed investigation of the energization of charged particles in the earth's magnetosphere is of great interest since the discovery that both Jupiter and Earth emit low energy cosmic rays, and it may help us understand the origin of cosmic rays in other parts of the Universe.

The lower atmosphere provides an instructive example of how apparently minor details can only be neglected at our peril. A few short years ago, many scientists believed that the atmosphere was well understood in principle, apart from a few unimportant minor details. We now know that constituents present in concentrations of a few parts per billion can exert a major controlling influence on the ozone in the stratosphere via a series of catalytic reactions, and understanding these processes has become a matter of urgency.

The measurement of minor constituents, their fluctuations and the dynamic motions by means of remote sensing with lidar, microwaves and other optical techniques is another important objective of AMPS. Which constituents? It is quite possible that some of them, such as the oxides of nitrogen, may be well understood by the 1980's. The instruments, however, should be designed so that they may be retuned to other radicals and re-flown at short notice. It now takes at least four to five years to design, build and fly a spacecraft to make the appropriate measurements to test a new hypothesis. With AMPS, once the facility is available, such lead times should be reduced by an order of magnitude. In addition, the incident solar flux can be measured accurately with sensors which can be recalibrated after each flight, and auxiliary equipment can be readily installed -- such as, for example, a radio atmospheric noise receiver to test hypotheses which relate concentrations of minor constituents to thunderstorm activity.

GENERAL CONCLUSIONS

AMPS represents an important new tool for investigating a broad range of problems in our near-space environment. Some of these will help us understand the processes which control our atmosphere and magnetosphere, but many of them also have broader implications in increasing our understanding of processes which occur elsewhere in the Universe.

The facility concept represents a challenge in devising new ways of doing business in both the design and the operational stages. It is clear that the core instruments must be designed with a maximum of flexibility but a minimum of cost, and cannot be left to the whims of a single Principal Investigator. On the other hand, the close involvement of the potential user is essential at every step. It is hoped to set up Instrument Design teams by an Announcement of Opportunity which will select scientists who wish to participate in the operation of AMPS. During the operational phase, additional proposals will be solicited for use of the facilities. Proposers may then elect to use only the core instruments on the facility, or may wish to add some of their own instrumentation, for which they will be encouraged to take full responsibility, subject only to the requirements of the safety of the men and machines on Spacelab.

REFERENCES

1. <u>Scientific Uses of the Space Shuttle</u>, National Academy of Sciences, 1974

2. <u>Report of the AMPS Working Group</u>, NASA/Marshall Space Flight Center, 1975

AAS 75-135

AMPS PROGRAM STATUS

Rein Ise*

The Atmospheric, Magnetospheric and Plasmas in Space (AMPS) Payload on Spacelab/Shuttle missions provides a significant new capability for study of the near-earth space environment. The large payload capability of the Shuttle, in combination with the interactive involvement of scientists in the conduct of experiments, permits a wide variety of investigations employing active perturbations of the processes which control this region.

Preliminary definition studies of the AMPS scientific instrumentation and payload configuration are being completed and the program is proceeding into the final definition phase. The AMPS Scientific Definition Working Group has developed the preliminary scientific objectives, instruments interface functional requirements and experiment operational requirements. Requests for proposals have been released for the final payload definition effort and an Announcement of Opportunity for scientific participation is in preparation. Initial hardware effort is expected to begin by FY 1977 in preparation for launch on one of the early Spacelab flights in the 1980's.

---

* Manager, AMPS Task Team, NASA Marshall Space Flight Center

ORGANIZATION: PROGRAM DEVELOPMENT	MARSHALL SPACE FLIGHT CENTER ATMOSPHERIC, MAGNETOSPHERIC & PLASMAS IN SPACE	NAME: Rein Ise
		DATE: August 1975

AMPS SCIENTIFIC OBJECTIVES

- PERFORM SPECIFIC CAUSE-AND-EFFECT STUDIES AIMED AT THE DETAILED UNDERSTANDING OF THE PHYSICAL MECHANISMS WHICH CONTROL THE NEAR-EARTH ENVIRONMENT

- ANSWER FUNDAMENTAL QUESTIONS ABOUT THE EFFECTS OF THE SUN AND MAN ON THE EARTH'S NATURAL ENVIRONMENT

- CONDUCT BASIC PLASMA PHYSICS EXPERIMENTS IN SPACE TO INCREASE OUR UNDERSTANDING OF PLASMA PHYSICS AND OF ENERGY CONVERSION PROCESSES ELSEWHERE IN THE UNIVERSE

ORGANIZATION: PROGRAM DEVELOPMENT	MARSHALL SPACE FLIGHT CENTER ATMOSPHERIC, MAGNETOSPHERIC & PLASMAS IN SPACE	NAME: Rein Ise
		DATE: August 1975

AMPS PAYLOAD CONCEPT

- VERSATILE REUSABLE LABORATORY
- EVOLUTIONARY THROUGH A SERIES OF MISSIONS
- INTERNATIONAL PARTICIPATION
- USE OF SHUTTLE/SPACELAB CAPABILITIES
- USE OF EXISTING CONCEPT/HARDWARE
- INTERACTIVE INVOLVEMENT OF SCIENTISTS

ORGANIZATION: PROGRAM DEVELOPMENT	MARSHALL SPACE FLIGHT CENTER ATMOSPHERIC, MAGNETOSPHERIC & PLASMAS IN SPACE	NAME: Rein Ise
		DATE: August 1975

AMPS PROJECT SCOPE

AMPS (DEDICATED)

DEFINITION, DESIGN, DEVELOPMENT, INTEGRATION AND INITIAL LAUNCH OF A REUSABLE SCIENTIFIC FACILITY TO BE FLOWN WITH THE SPACELAB AS A DEDICATED MISSION ON OPERATIONAL SHUTTLE FLIGHTS FOR DETAILED STUDY OF THE NEAR-EARTH ENVIRONMENT AND CONDUCT OF PLASMA PHYSICS EXPERIMENTS IN SPACE.

FIRST SPACELAB

DEFINITION, DESIGN, DEVELOPMENT, INTEGRATION AND LAUNCH OF SELECTED AMPS INSTRUMENTS AND ASSOCIATED SUPPORTING SUBSYSTEMS ON THE FIRST SPACELAB MISSION.

AMPS FOLLOW-ON MISSIONS

REFURBISHMENT AND LAUNCH OF THE AMPS FACILITY WITH DIFFERENT SCIENTIFIC INVESTIGATIONS AT FREQUENCIES OF UP TO FOUR PER YEAR.

ORGANIZATION: PROGRAM DEVELOPMENT	MARSHALL SPACE FLIGHT CENTER ATMOSPHERIC, MAGNETOSPHERIC & PLASMAS IN SPACE	NAME: Rein Ise
		DATE: August 1975

AMPS PAYLOAD EVOLUTION

SHUTTLE ORBITAL FLIGHT TEST PROGRAM
- FLIGHTS 3 & 5
- ENVIRONMENT DETECTOR PACKAGES
- ORBITER WAKE & ENVIRONMENT STUDIES

FIRST SPACELAB
- LIDAR
- ELECTRON ACCELERATOR
- DIAGNOSTIC PACKAGE

DEDICATED AMPS
- FULLY OPERATIONAL FACILITY
- REUSABLE INSTRUMENTS
- DIFFERENT EXPERIMENTS ON EACH FLIGHT

ORGANIZATION: PROGRAM DEVELOPMENT	MARSHALL SPACE FLIGHT CENTER ATMOSPHERIC, MAGNETOSPHERIC & PLASMAS IN SPACE	NAME: Rein Ise
		DATE: August 1975

AMPS PROGRAM BACKGROUND

- SEVERAL STUDIES OF MAGNETOSPHERIC, ATMOSPHERIC AND AURORAL SORTIE MISSIONS PERFORMED IN 1972-1973
- 1973 WOODS HOLE SUMMER STUDY RECOMMENDED A SINGLE FACILITY TO ACCOMPLISH MAGNETOSPHERIC/IONOSPHERIC, ATMOSPHERIC AND PLASMA PHYSICS INVESTIGATIONS WITH THE SPACE SHUTTLE
- PHASE A FEASIBILITY STUDY OF SUCH A SCIENTIFIC FACILITY WAS PERFORMED BY MSFC IN 1974
- THE AMPS SCIENTIFIC DEFINITION WORKING GROUP WAS FORMED IN MARCH 1974
- RESPONSIBILITY FOR ADDITIONAL PAYLOAD DEFINITION EFFORT WAS ASSIGNED TO MSFC IN JANUARY 1975
- AMPS SCIENTIFIC OBJECTIVES WERE INCLUDED IN FIRST SPACELAB MISSION OBJECTIVES (SPRING 1975)
- RFP FOR PARALLEL PHASE B PAYLOAD DEFINITION STUDIES WAS RELEASED TO INDUSTRY IN JULY 1975

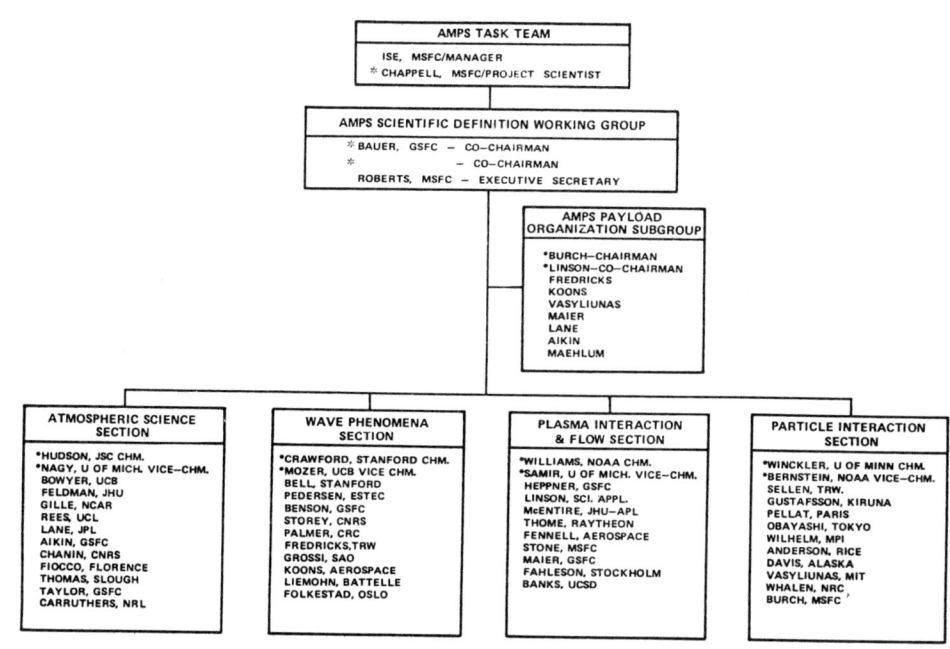

| ORGANIZATION: 2223-75 | MARSHALL SPACE FLIGHT CENTER
AMPS
DEFINITION ACTIVITIES | NAME: REIN ISE
DATE: AUGUST 13, 1975 |

MILESTONES	CY 1974				CY 1975				CY 1976				CY 1977			
	1	2	3	4	1	2	3	4	1	2	3	4	1	2	3	4
SCIENTIFIC DEFINITION																
WORKING GROUP MEETINGS		▲	▲		▲	▲		△	△							
SCIENCE SUPPORT STUDIES						▬▬▬▬▬▬▬▬▬▬			⊢ - - - - - - - - ⊣							
SCIENCE OBJECTIVES DOC						P▲		△F								
IFRD'S					P▲			△U								
EOR'S							▲	△U								
USER GUIDELINES									△							
PHASE B STUDY SUPPORT									⊢▬▬▬▬▬▬▬▬▬⊣							
AO RELEASE								△								
PAYLOAD DEFINITION																
PHASE A STUDY (MSFC)	▬▬▬▬▬▬▬▬															
SPECIAL STUDIES (MSFC)						▬▬▬▬▬▬▬										
PHASE B STUDIES (CONTR)																

NOTES: P — PRELIMINARY   IFRD — INSTRUMENT FUNCTIONAL REQUIREMENTS DOCUMENT
        U — UPDATE   EOR — EXPERIMENT OPERATIONAL REQUIREMENTS
        F — FINAL   AO   — ANNOUNCEMENT OF OPPORTUNITY

| ORGANIZATION: PROGRAM DEVELOPMENT | MARSHALL SPACE FLIGHT CENTER
ATMOSPHERIC, MAGNETOSPHERIC
&
PLASMAS IN SPACE | NAME: Rein Ise
DATE: August 1975 |

SUMMARY OF 1975/76 AMPS SCIENCE SUPPORT STUDIES

STUDY	RESPONSIBILITY	ORGANIZATION
LIDAR MEASUREMENTS	NAGY	U OF MICH
PLASMA WAKE EXPERIMENTS	SAMIR	U OF MICH
COHERENT/INCOHERENT RADAR FEASIBILITY	MOZER	UCB
MULTIPLE DETECTOR UV SPECTROMETER	FELDMAN	JHU
CURRENT NEUTRALIZATION DEFINITION	LIEMOHN	BATTELLE
LABORATORY STUDY OF ANTENNAS	FREDRICKS	TRW
VLF ANTENNA SURVEY	KOONS/FENNELL	AEROSPACE
REMOTE IMAGING SYSTEMS	LIND	U OF ALASKA
LIDAR ENGINEERING	HUDSON	JSC
CHEMICAL RELEASE (ENGINEERING)	HEPPNER	GSFC
CHEMICAL RELEASE (SCIENTIFIC)	LINSON	SCIENCE APPL
ATMOSPHERIC SCATTERING	AIKIN	GSFC
ELECTRON BEAM CHARACTERISTICS	BERNSTEIN	NOAA
ACOUSTIC GRAVITY WAVES	THOME	RAYTHEON
ENERGETIC ION DETECTOR	McENTIRE	JHU
TETHER EMF GENERATION	BANKS	UCSD
ACCELERATOR DEFINITION	SELLEN	TRW
ELECTRON ECHO EXPERIMENTS	WINCKLER	U OF MINN
TETHER FEASIBILITY (ELECTRODYNAMICS)	GROSSI	SAO
SCIENTIST/AMPS INSTRUMENT INTERFACES	ANDERSON	RICE

ORGANIZATION: PROGRAM DEVELOPMENT	MARSHALL SPACE FLIGHT CENTER ATMOSPHERIC, MAGNETOSPHERIC & PLASMAS IN SPACE	NAME: Rein Ise DATE: August 1975

DEFINITION OF ANTENNAS FOR VLF EXPERIMENTS IN SPACE

- OBJECTIVES
  - DEFINE THEORETICAL VLF ANTENNA PERFORMANCE IN SPACE
  - CONDUCT ANTENNA PERFORMANCE MEASUREMENTS
  - DEFINE ANTENNA CONCEPTS
- PRELIMINARY RESULTS
  - TRIANGULAR LOOP ANTENNA CONFIGURATIONS EXPLORED FOR LOW FREQUENCY WAVE EXPERIMENTS
  - LOW FREQUENCY WAVE TO PLASMA COUPLING ACHIEVED IN SCALED LABORATORY TESTS

ORGANIZATION: PROGRAM DEVELOPMENT	MARSHALL SPACE FLIGHT CENTER ATMOSPHERIC, MAGNETOSPHERIC & PLASMAS IN SPACE	NAME: REIN ISE DATE: AUGUST 1975

POSSIBLE LOOP ANTENNA CONFIGURATION FOR LOW FREQUENCY WAVE EXPERIMENTS

| ORGANIZATION: PROGRAM DEVELOPMENT | MARSHALL SPACE FLIGHT CENTER
ATMOSPHERIC, MAGNETOSPHERIC
&
PLASMAS IN SPACE | NAME: Rein Ise
DATE: August 1975 |

VACUUM CHAMBER TESTS OF ELECTRON ACCELERATOR BEAMS

- OBJECTIVES

  TO STUDY IN A LARGE VACUUM CHAMBER:
  - ELECTRON BEAM COUPLING TO THE AMBIENT GEOMAGNETIC FIELD
  - ELECTRON BEAM CONTROL (SPREADING AND FOCUSING)
  - INDUCED ELECTRIC FIELDS
  - COUNTER STREAMING IN THE ELECTRON BEAM

- PRELIMINARY RESULTS
  - DEMONSTRATED BEAM CONTROLLABILITY IN GEOMAGNETIC FIELD
  - OBSERVED CORRELATION OF COUNTERSTREAMING WITH PLASMA RESONANCE
  - ACHIEVED SUCCESSFUL LOW LIGHT LEVEL OBSERVATIONS OF BEAM CHARACTERISTICS

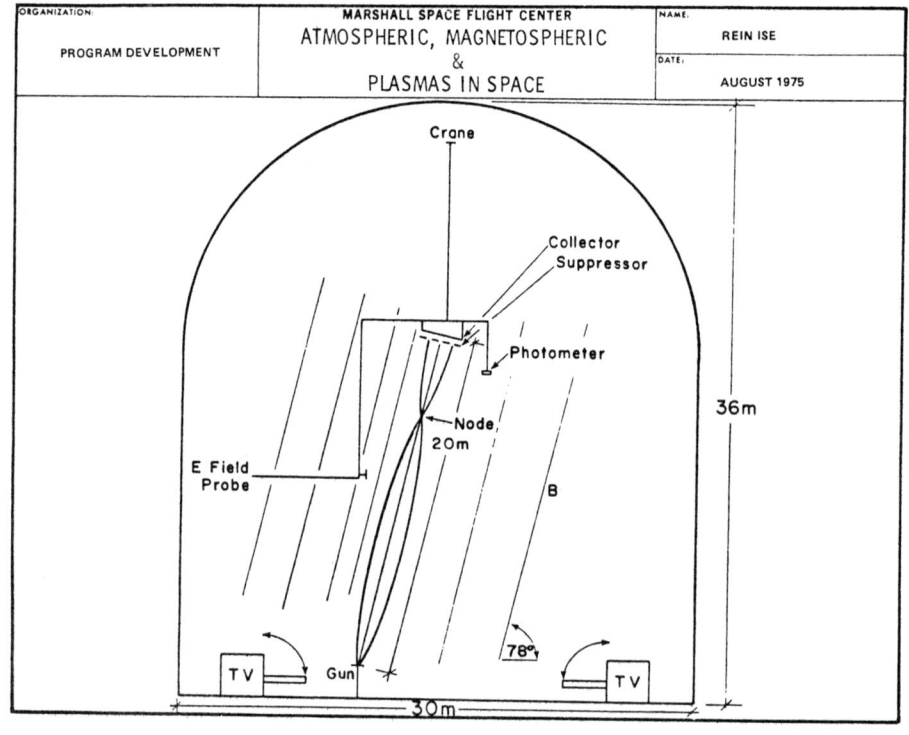

| ORGANIZATION: PROGRAM DEVELOPMENT | MARSHALL SPACE FLIGHT CENTER
ATMOSPHERIC, MAGNETOSPHERIC
&
PLASMAS IN SPACE | NAME: Rein Ise
DATE: August 1975 |

TYPICAL DEDICATED AMPS PAYLOAD ELEMENTS

SCIENTIFIC INSTRUMENTS
- PARTICLE ACCELERATORS
- HIGH POWER TRANSMITTERS
- LIDAR (LASER RADAR)
- CHEMICAL RELEASE DEVICES
- OPTICAL REMOTE SENSING INSTRUMENTS
- SUBSATELLITES

SUPPORTING SUBSYSTEMS
- GIMBAL AND POINTING SYSTEMS
- SPECIAL SUPPORT STRUCTURES
- BOOMS
- ENERGY STORAGE SYSTEMS
- EJECTION MECHANISMS
- CONTROLS AND DISPLAYS

SHUTTLE ORBITER AND SPACELAB RESOURCES
- ELECTRICAL POWER
- ENVIRONMENTAL CONTROL
- ATTITUDE STABILIZATION
- COMMUNICATIONS
- COMPUTERS
- LABORATORY AREA

| ORGANIZATION: PROGRAM DEVELOPMENT | MARSHALL SPACE FLIGHT CENTER
ATMOSPHERIC, MAGNETOSPHERIC
&
PLASMAS IN SPACE | NAME: REIN ISE
DATE: AUGUST 1975 |

AMPS INSTRUMENT ACCOMMODATIONS SUMMARY
(INITIAL SHUTTLE MISSIONS)

MISSION DURATION	7 DAYS NOMINAL
ORBIT	28.5° – 57° INCLINATION
	UP TO 450 Km CIRCULAR ALTITUDE
SPACELAB CONFIGURATION	PRESSURIZED MODULE PLUS UNPRESSURIZED 3 METER PALLETS
EXPERIMENTS OPERATION	2 SHIFTS, 24 HOURS
AMPS PAYLOAD WEIGHT	5500 Kg MAXIMUM
PALLET SURFACE AREA	17 m^2 PER PALLET
POINTING	ORBITER (COARSE) ± 0.5 DEGREES
	PLATFORM (FINE) 2 SEC P&Y, 40 SEC ROLL
AVAILABLE ELECTRICAL POWER	391 KWH AT 3.8 KW AV. AND 8.8 KW PEAK
HEAT REJECTION	APPROXIMATELY 4.85 KW
DATA MANAGEMENT	
COMPUTER	16–32 BITS, 64 K MEMORY
RECORDERS	26 HI RATE DIGITAL TRACKS AT 1.15 MBPS/TRACK
	6 MHz ANALOGUE
DATA TRANSMISSION	50 Mbps WIDE BAND
	2 Mbps DIGITAL, 4.2 MHz ANALOGUE
COMMUNICATIONS	DUPLEX VOICE AND DOWNLINK TV
VIEW PORTS	TOP AND AFT OF PRESSURIZED MODULE

NOTE: DIAGNOSTIC INSTRUMENT PLATFORMS SUCH AS BOOMS, SUBSATELLITES AND TETHERED SATELLITES ARE UNDER STUDY WITH CHARACTERISTICS TBD

ORGANIZATION: PROGRAM DEVELOPMENT	MARSHALL SPACE FLIGHT CENTER ATMOSPHERIC, MAGNETOSPHERIC & PLASMAS IN SPACE	NAME: Rein Ise
		DATE: August 1975

CANDIDATE INSTRUMENTS FOR FIRST SPACELAB MISSION

- ELECTRON ACCELERATOR
- LIDAR (EUROPEAN DEVELOPMENT)
- DIAGNOSTIC INSTRUMENTS
    - MAGNETOMETER (FLUXGATE)
    - MAGNETOMETER (SEARCH COIL)
    - ELECTRIC FIELD METER
    - ELECTRO-STATIC ANALYZER
    - LANGMUIR PROBE
    - ION MASS SPECTROMETER
    - SOLID STATE PARTICLE DETECTOR
    - ION TRAP
    - NEUTRAL MASS SPECTROMETER
    - LOW LIGHT LEVEL TV
    - UV IMAGING SYSTEM

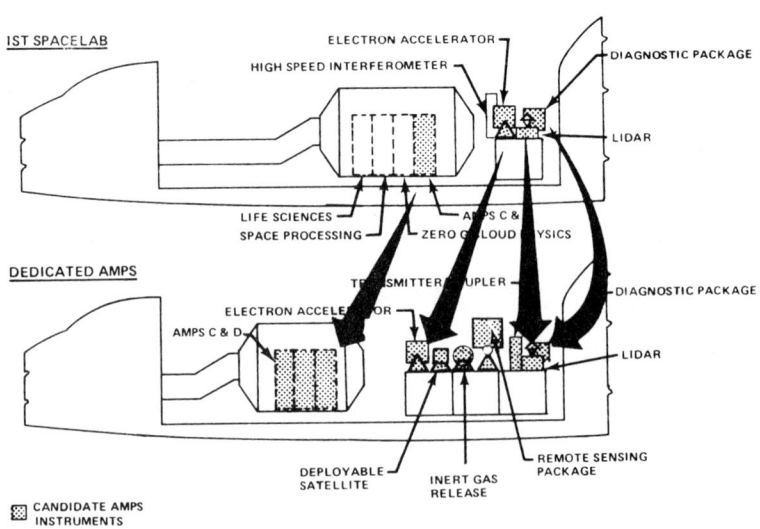

CANDIDATE SPACELAB MISSION CONFIGURATIONS

ORGANIZATION: PROGRAM DEVELOPMENT	MARSHALL SPACE FLIGHT CENTER ATMOSPHERIC, MAGNETOSPHERIC & PLASMAS IN SPACE	NAME: Rein Ise
		DATE: August 1975

CANDIDATE INSTRUMENTS FOR DEDICATED MISSION

- UV - VIS - NIR SPECTROMETER
- FAR IR SPECTROMETER
- CRYO LIMB SCANNER
- FABRY - PEROT INTERFEROMETER
- PLASMA ACCELERATOR
- ANTENNA/TRANSMITTER SYSTEM
- CHEMICAL RELEASES
- DIAGNOSTIC INSTRUMENTS
    - SPHERICAL ION PROBE
    - DRIFT METER
    - RUBIDIUM MAGNETOMETER
    - PLANAR ELECTRON PROBE
    - RF PROBE
    - ENERGETIC ION ANALYZER
    - HIGH SENSITIVITY ION SPECTROMETER

ATMOSPHERIC, MAGNETOSPHERIC & PLASMAS IN SPACE (AMPS)
OVERALL PROGRAM SCHEDULE
(PRELIMINARY)

DATA: KEN TURNER
STATUS AS OF AUG. 18, 1975

	FY-75	FY-76	FY-77	FY-78	FY-79	FY-80	FY-81
	1974	1975	1976	1977	1978	1979	1980	1981

KEY MILESTONES — 1st SPACELAB MSN. 8; 1st AMPS MSN. 19

- SCIENCE REQUIREMENTS & DEF.
    - SCIENCE WORKING GROUP — INITIAL IFRD, UPDATE, FINAL IFRD
- SCIENTIFIC INSTRUMENTS
    - FIRST SPACELAB — A.O., AWARD, C/D PROPOSAL, PREL DES, DESIGN/DEVELOPMENT, DEL
    - FIRST AMPS — AWARD, C/D PROPOSAL, PREL DES, DESIGN DEVELOPMENT, DEL
- AMPS PAYLOAD — FINAL REPORT
    - PHASE A (IN-HOUSE, MSFC)
    - DEFINITION (PHASE B) — RFP, AWARD, DEFINITION
    - DEVELOPMENT AND INTEGRATION — RFP, AWARD, DESIGN/DEVELOPMENT, MSN. 8 1st SPACELAB SUPPORT, MSN. 19 1st AMPS DEL, INTEGR & TEST, LAUNCH

485

ORGANIZATION: PROGRAM DEVELOPMENT	MARSHALL SPACE FLIGHT CENTER ATMOSPHERIC, MAGNETOSPHERIC & PLASMAS IN SPACE	NAME: Rein Ise
		DATE: August 1975

AMPS PHASE B STUDIES
SCOPE OF WORK

DEFINITION OF:

- OVERALL PAYLOAD DESIGN
- SYSTEMS AND SUBSYSTEMS INTERFACE REQUIREMENTS
- SYSTEM AND SUBSYSTEM PERFORMANCE REQUIREMENTS
- INTEGRATION AND OPERATIONS REQUIREMENTS
- PROGRAMMATIC REQUIREMENTS/PLANS AND COST ESTIMATES FOR THE DEVELOPMENT/OPERATIONS PHASE

ORGANIZATION: PROGRAM DEVELOPMENT	MARSHALL SPACE FLIGHT CENTER ATMOSPHERIC, MAGNETOSPHERIC & PLASMAS IN SPACE	NAME: Rein Ise
		DATE: August 1975

PLANNED NEAR-TERM MILESTONES

- START PHASE B PAYLOAD DEFINITION STUDIES

- RELEASE AO FOR SCIENTIFIC PARTICIPATION

- DEFINE AMPS PARTICIPATION IN SHUTTLE ORBITAL FLIGHT TESTS AND INITIAL SPACELAB FLIGHTS

AAS 75-215

SOME EXPERIMENTS WITH ENERGETIC PARTICLE INJECTORS

William Bernstein[*]
and
John R. Winckler[†]

The AMPS experiment payload will contain a comprehensive facility for the injection of energetic particles into the ambient medium utilizing "on board" electron and perhaps ion plasma accelerators. By "energetic" we include the energy range 1 keV - 40 keV. The anticipated power levels range from 1 kw, determined primarily by limitations in the detection thresholds for the injected particles, to conceivably as high a 1 MW determined primarily by technological problems in the flight of such high power instrumentation. The implementation of such a facility into the AMPS payload will be discussed in detail by J. M. Sellen; the very wide variety of possible experiments will have been considered in several "theme" talks. I will therefore limit this talk to a discussion of a few selected experiments which have been studied in somewhat greater detail by members of the Particle Interaction Section of the AMPS Working Group. I hope these descriptions will provide an understanding of both the experimental objectives and the necessary experimental configuration utilizing the AMPS vehicle.

<u>Measurements of the Characteristics of Magentic Field Aligned Electric Fields, Their Global Distribution, and Temporal Characteristics.</u>

Recent experimental studies of the energy and pitch angle distribution of precipitated electrons in one class of auroral events are consistent with particle acceleration by an electric field (total potential drop 1 - 10 kV) aligned parallel to the magnetic field at unknown altitudes but not far above the ionosphere (Evans, 1974). Direct measurements of such field configurations have not been obtained to date. Two theoretical

---

[*]Space Environment Laboratory, NOAA/ERL, Boulder, Colorado

[†]School of Physics & Astronomy, University of Minnesota, Minneapolis, Minnesota.

models have been proposed to explain how such an electric field could occur in a collisionless medium with classically infinite conductivity parallel to the magnetic field. In one model (Kindel and Kennel, 1971), current driven plasma turbulence is proposed to provide the required ohmic dissipation. This model leads to a low field (10 - 100 mV m^{-1}) and a long (100 - 1000 km) region of potential drop. In the alternate configuration, localized (< 1 km) regions of large electric field, called double layers (Block, 1972) are proposed to exist. In addition to particle acceleration and Joule dissipation effects, these field-aligned potential drops effectively decouple the convective motions of the outer magnetosphere and the ionosphere and would imply modifications in frozen-in field line concepts (infinite conductivity) of the magnetosphere-ionosphere system.

The basic features of an experiment to measure such an electric field configuration is shown in Fig. 1. An electron beam at selected pitch angles and energy is injected upwards from the AMPS toward the potential drop region. If the parallel energy of the injected beam in this region is greater than the potential drop, it will pass through into the outer magnetosphere. On the other hand, if it is less than potential drop, it will be reflected down the field line where it could be detected. Measurement of the parallel energy at which reflection disappears provides a measure of the total potential drop; measurement of the transit time provides the altitude of the drop; and estimates of the field strength can be obtained from the transit time variation with injected particle parallel energy.

Two methods of detection have been considered. The simplest is the detection of the reflected beam by observation of the optical emissions produced when the beam strikes the atmosphere. The use of a wide field of view, low light level imaging system on the AMPS would be useful for this purpose. On the other hand, large beam fluxes (0.1 - 1A) may be required to yield a detectable optical signal, particularly in the presence of natural precipitation; it is not clear whether the injection of such high density beams (orders of magnitude larger than the natural energetic particle density) will modify the characteristics of the region

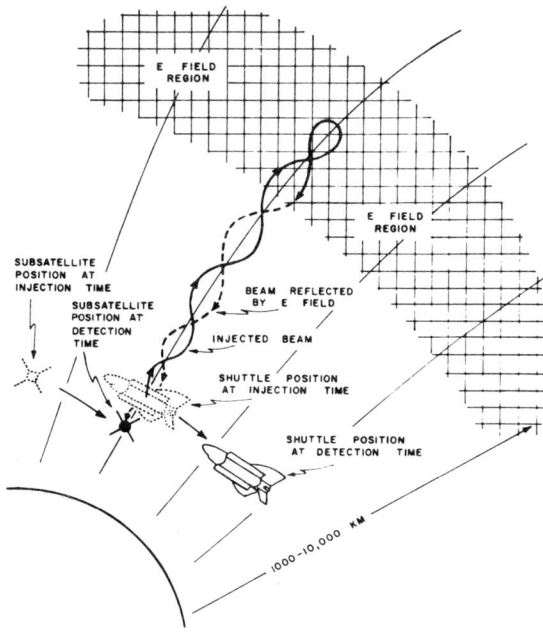

Fig. 1 Schematic Representation of an Experiment to Measure the Characteristics of Localized Regions with an Electric Field Aligned Parallel to the Geomagnetic Field from the AMPS

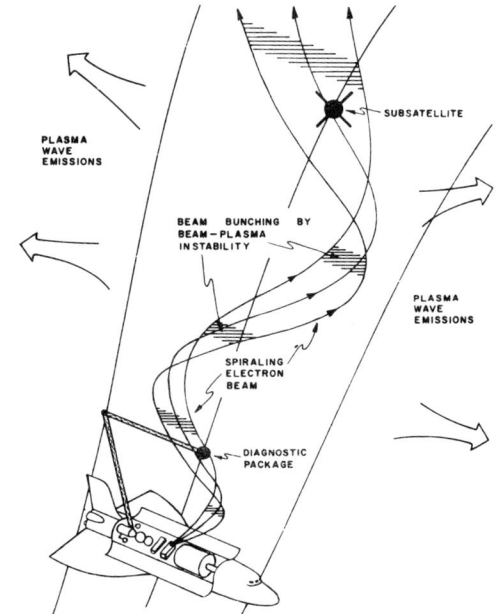

Fig. 2 Schematic Representation of an Experiment to Study Collective Beam-Plasma Interactions from AMPS

under study. Much lower injection densities are feasible if direct methods for detection of reflected beam particles are employed. However, as can be seen in Figure 1, the AMPS vehicle will have travelled several kms during the flight time of the injected beam. Thus direct detection cannot be performed with the AMPS itself and the use of a positioned subsatellite is required. In addition, the variable (magnitude and direction) $E_\perp B$ ambient electric field will give a variable drift velocity across the B field to the beam particles; this drift velocity must be considered in the placement of the subsatellite.

Clearly many of the problems associated with the direct detection technique can be reduced if the beam is launched from a rocket with a horizontal velocity < 0.1 of the AMPS orbiting velocity. Such an experimental configuration is now under construction. A positive result from this payload can confirm the existence of these electric field configurations, but any study of the general global distribution and temporal characteristics requires an orbiting vehicle. In these respects, the rocket experiments represent the necessary preliminary work prior to the AMPS experiments.

Study of the Interaction of Particle Beams with the Ambient Plasma

The configuration of a monoenergetic but cold particle beam immersed in a cold ambient plasma is a general example of an unstable plasma configuration. The consequence of the instability is the removal of the two peaked velocity distribution through wave particle interactions and the eventual merging of the beam and ambient plasma distribution. Beam-plasma interactions of this type are believed to be the origin of one class of observed radio noise such as Type III solar radio bursts, the Jovian decametric noise, and possibly the terrestrial kilometric noise. On the other hand, a very large class of injected beam experiments suggested for the AMPS program require the transmission of monoenergetic beams over very large distances (10^5 km) through the ionospheric and magnetospheric plasmas. These experiments are all based on the assumption that the introduction of the beam into the plasma does not lead to the occurrence of new collective interactions which modify the properties of the probing beam and ambient plasma. Thus there are two general reasons for the study of beam-plasma interactions: (a) simulation of an important astrophysical process, and (b) identification of limiting constraints on some beam injection experiments.

Figure 2 shows one version of how an experimental study of beam-plasma stability could be carried out from AMPS. Near field measurements (≤ 30 m) would be carried out using a diagnostics package mounted on the manipulator arm; the diagnostics would include sensors for beam energy, current distribution and a plasma wave monitoring system. Although increased sensitivity could be achieved by modulation of the injected beam current near the electron plasma frequency, f_{pe}, it is still questionable whether sufficient instability growth would have occurred in this short path length to permit clear-cut measurements. Therefore a second diagnostic package mounted on a subsatellite which could penetrate the beam at distances of several kms is included. A supplemental method of diagnosis is not shown in this figure and is based upon the ejection of a neutral gas jet (N_2 for example) from AMPS along the beam path to render the beam optically visible to a low light level energizing system.

A great deal of preliminary experimental information has already been obtained relative to the questions obvious in the prior discussion. The rocket-borne accelerator experiments of Hess et al (1971) and McEntire et al (1974) have indicated that at least a large fraction of the injected beam particles can propagate over large distances in the ionosphere and magnetospheric plasmas without major modification in their characteristics through beam-plasma instabilities. On the other hand, Figure 3 shows the wave spectra reported by Cartwright and Kellogg (1974) to have been stimulated by the electron beam injection of McEntire et al. Clearly waves at the ambient plasma frequency and the second harmonia of the electron cyclotron frequency are excited by the injected beam. Thus the expected interactions do occur, but the wave amplitudes appear insufficient to produce major perturbations in the beam velocity distribution.

Lastly, beam-plasma interactions at low particle densities and long (20 m) pathlengths have been studied in the very large, but now deactivated, NASA vacuum facility at Plum Brook, Ohio (Bernstein et al, 1975).

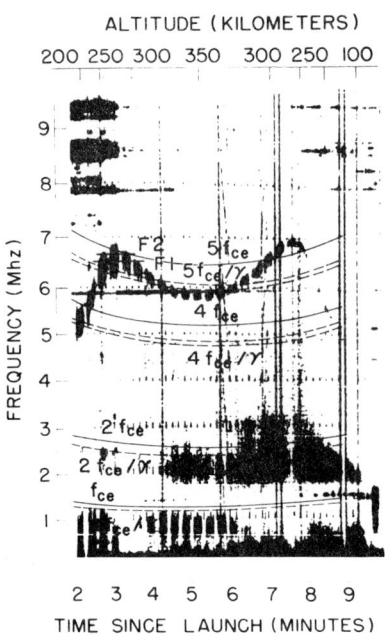

Fig. 3 Time Dependence of the Frequency Spectra of the Emissions Stimulated by Energetic Electron Beam Injection During the Flight of ECHO I

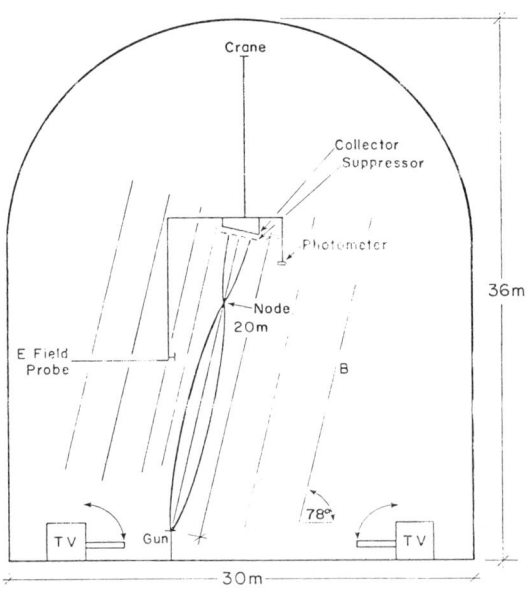

Fig. 4 Schematic Representation of the Electron Beam Experiments Performed in the Plum Brook Vacuum Facility

Figure 4 shows a schematic representation of the experiment in which an electron beam is injected parallel to the magnetic field; the diagnostic instrumentation included a low light level TV system to observe emissions from the ambient neutral gas excited by the beam, and an E field probe to detect plasma waves. Two results particularly revelent to the AMPS program have been obtained in this experiment: (a) low level wave amplitudes believed near f_{pe} are observed, but in the 20 m path length, the modification in beam velocity distribution is too small to be measured with the techniques employed; and (b) the beam was easily visible in the TV display even at the lowest pressures employed in the experiment, 5×10^{-7} mm Hg. Clearly the gas fluorescence technique could be a useful remote diagnostic for electron beam properties in the ambient plasma medium. Figure 5 is a photograph of the TV display obtained at a pressure of $\sim 10^{-5}$ mm Hg. The very sharp node shown here is also shown schematically in Figure 4. This refocusing of the beam at integral cyclotron wavelengths is characteristic of monoenergetic beams with a small spread in injection pitch angles injected $\sim$ parallel to a weak magnetic field so that the cyclotron radius of particles with maximum $v_\perp (R_{c\ max}) = \gg R_{source}$. The depth of the node is a measure of the spread in beam $v_{\|}$; the deep node in Figure 5 is consistent with the absence of significant modification in beam $v_{\|}$ by instability effects.

Studies of the Interaction of Energetic Particle Beams with the Atmosphere

In the past, extensive efforts have been directed toward the identification of the properties of the precipitating energetic particles (flux, energy, pitch angle distribution, species, spatial extent) from observations of the characteristics of the optical emissions (spectra, altitude profile, intensity, doppler shifts and broadening of selected lines, spatial and temporal distribution) together with x-ray fluxes produced in the interaction of the energetic particles with the atmosphere. The availability of energetic particle beams with known properties on board AMPS allows the study of these complex interactions with accurately known input parameters. Figure 6 represents one possible experimental configuration. In order to obtain altitude profiles, a triangulation technique is required. These supplemental observation sites could be the ground

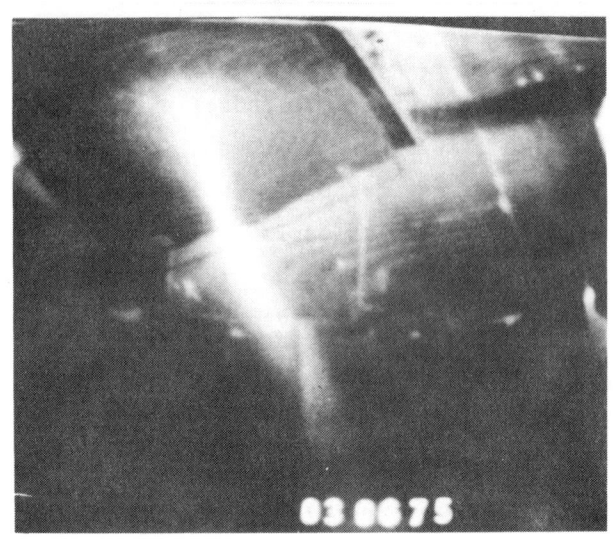

Fig. 5 Photograph of the Electron Beam Trajectory from One Frame of the Low Light Level TV System. Beam energy 600V, Beam Current 4ma and Ambient Pressure 1×10^{-5} torr.

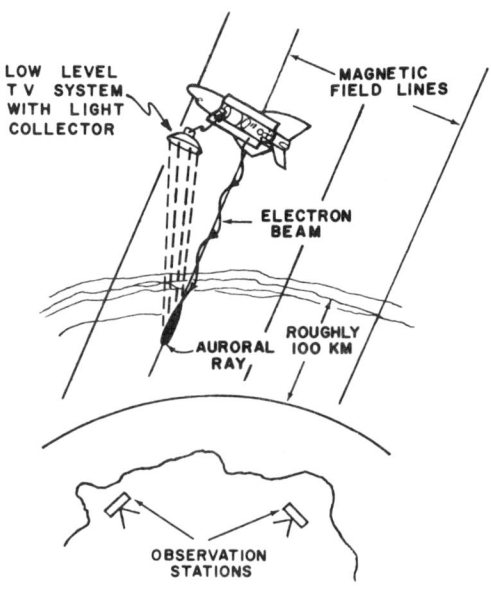

Fig. 6 An Experiment Configuration to Study the Interactions of Energetic Particle Beams Incident Upon the Atmosphere.

stations shown in Figure 6, or a properly placed subsatellite (not shown) having the required optical instrumentation.

Measurements of the Gross Magnetic and Electric Field Topology of the Magnetosphere

If it is assumed that the injection of energetic particle beams does not lead to the generation of new collective interaction modes, then these beams provide excellent traces for the determination of the gross magnetic and electric field topology. Provided the energy density of the injected particles remains less than the energy density of the magnetic field lines, the injected particles will follow the field lines subject to drifts associated with the curvature and gradients in the magnetic field and cross field drifts associated with E_1B. Therefore measurement of the spatial location of the beam particles after completion of a long flight path together with the required transit time for a given energy particle can provide necessary data into theoretical models of the magnetic field topology. Because the electric field produced drift is proportional to the transit time, the use of particle beams of different energy allows separation of the electric and magnetic drifts. Figure 7 represents the conjugate point configuration described by Davis et al (1973) in which an electron beam injected in one hemisphere was detected by the optical emissions produced when it impacted the atmosphere in the conjugate hemisphere. In the absence of ground-based observation sites in the conjugate hemisphere, either aircraft- or satellite-borne observation instrumentation would be required. In an alternative configuration (Fig. 8) described by McEntire et al (1974), the return flux (echo configuration) from the conjugate point is detected in the near vicinity of the injection point (for energetic electrons the combined drifts are not large). The return flux originates from either magnetic mirroring or atmospheric back-scattering in the conjugate hemisphere depending upon the relative magnetic field strengths in the two hemispheres and the injection pitch angle. Both techniques suffer from a major limitation because only end point detection techniques are employed. A wide variety of possible physical phenomena may prevent the injected beam from reaching the detection point. Consequently, the inability to detect either the conjugate

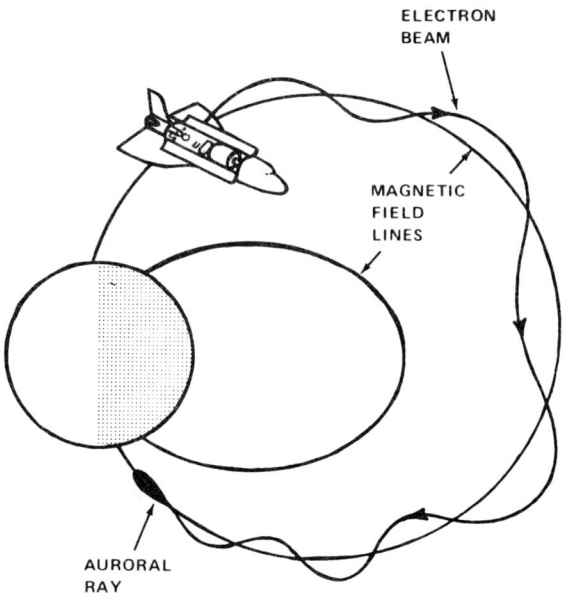

Fig. 7 Conjugate Point Configuration for Study of the Gross E+B Field Topology of the Magnetosphere from AMPS

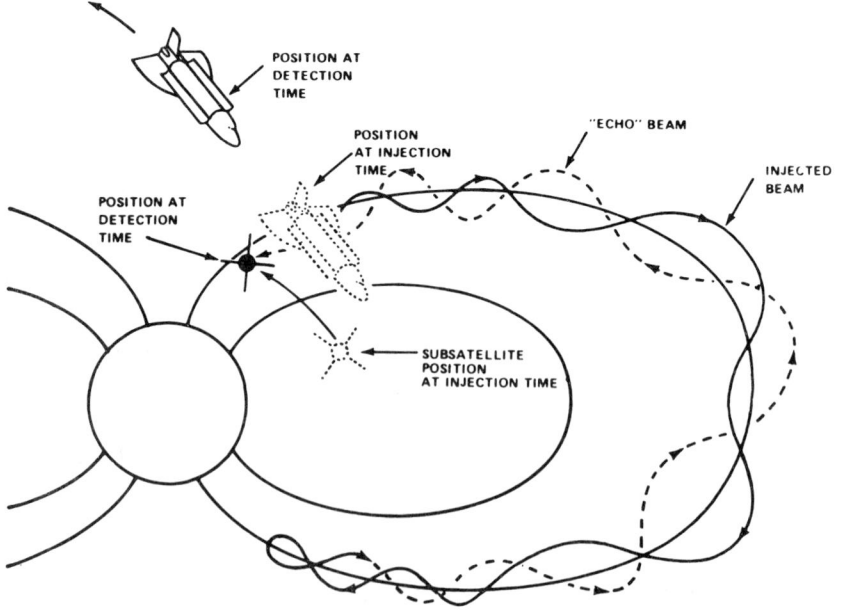

Fig. 8 Echo configuration for Study of the E+B Field Topology of the Magnetosphere from AMPS.

point or echo signal does not identify the specific process which causes
the loss of propagation. A more complete characterization of the field
topology would be obtained if the entire beam trajectory rather than just
the end points could be observed. One possible method involves the use
of ion beams which fluoresce under solar photon irradiation and which
could then be observed remotely over their entire trajectory. The use
of energetic Ba^+ injections, produced either in shaped charge releases
or a Ba-fueled ion engine configuration, appears attractive for this purpose. However, the low velocity characteristic of high mass ions tends
to over emphasize electric field drifts; thus low mass ions would be more
attractive. Observations of the beams from AMPS rather than from the
ground removes the general restriction that the fluorescence lines lie
in the visible spectrum.

This field line racing category of experiments is very active at the
present time using rocket-borne electron accelerators and Ba^+ shaped
charge releases. I will attempt to describe the echo experiments performed with rockets in somewhat more detail and to show how they may be
adapted to a possible AMPS configuration. Figure 9 shows a schematic
representation of the injection of an electron beam from a rocket launched
with selected trajectory, its reflection at the conjugate hemisphere,
drift in longitude during the flight and the subsequent interception
and detection of the return beam by the same rocket. Figure 10 illustrates
an echo pattern recorded by a scintillation detector showing the prompt
injection pulse caused by particle scattering in the vicinity of the
rocket and the delayed, lower intensity echo pulse, in this case, atmospheric backscatter rather than mirroring provided the reflection process with a consequent reduction in flux. At times, multiple echoes
corresponding to several conjugate reflections have been observed.

The very high AMPS orbital velocity in general precludes the direct
detection of the echo beam with AMPS-borne instrumentation. However, one
possible experimental configuration, applicable to those cases where the
particle mirror height occurs at a greater altitude in the conjugate
hemisphere so that the full beam is reflected, is to utilize the near
hemisphere atmosphere as a scintillation screen with detection of the

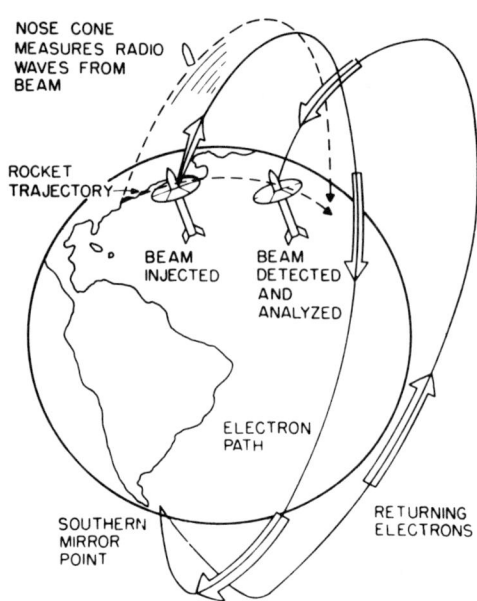

Fig. 9 Schematic Representation of the Echo Experiment Where Both the Electron Accelerator and Particle Detection System are Carried on the Same Rocket

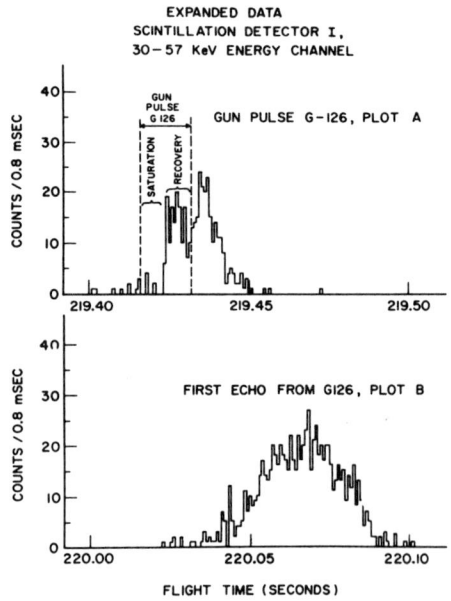

Fig. 10 Echo Pattern Observed During Echo I Flight Indicating Successful Interception of the Return Beam from the Conjugate Hemisphere

atmospheric emission by a low light level imaging system. Direct particle detection requires the use of an appropriately positioned subsatellite. Because the width of the return beam is small and $\sim$ to the particle cyclotron radius, the $E_1 B$ drifts can significantly modify the relative orientation of the return beam and subsatellite for detection. The proper placement therefore requires some knowledge of the existing $E_1 B$ configuration.

I will conclude with a very brief movie which illustrates several features of this talk. The first section shows the behavior of the injected electron beam in the Plum Brook experiment, the second section shows the atmospheric emissions produced by an energetic electron beam ejected from a rocket-borne accelerator and the final section shows the output from a computer program designed to model the interception of echo beams in both the rocket and AMPS configuration.

These are but a few of the many experiments which could be performed with an accelerator facility aboard AMPS. They are not simple experiments to implement and perform, but at the present time we have not been able to identify either physical or technological reasons which could severely limit their applicability. The next few years should see the development of new and useful experiments using this facility.

REFERENCES

1. Bernstein, W. et al (1975), "Laboratory Observations of RF Emissions at ω_{pe} and $(n + 1/2)\omega_{ce}$ in Electron Beam-Plasma and Beam-Beam Interactions," *J. Geophys. Res.* (in press).
2. Block, L. P. (1972), "Potential Double Layers in the Ionosphere," *Cosmo Electrodynamics 3*, 349.
3. Cartwright, D. G. and P. J. Kellogg (1974), "Observations of Radiation from an Electron Beam Artificially Injected into the Ionosphere," *J. Geosphys. Res.*, 79, 1439.
4. Davis, T. N., W. N. Hess, and E. M. Westcott (1973), "Initial Results of a Recent Electron Accelerator Experiment," *EOS*, 54, 436.
5. Evans, D. S. (1974), "Precipitating Electron Fluxes Formed by a Magnetic Field Aligned Potential Difference," *J. Geophys. Res.*, 79, 2853.
6. Hess, W. N., M. C. Trichel, T. N. Davis, W. B. Beggs, G. E. Kraft, E. Stassinopoulis, and E. J. R. Maier (1971), "Artificial Aurora Experiment and Principal Results," *J. Geophys. Res.*, 76, 6067.
7. Kindel, J. M., and C. F. Kennel (1971), "Topside Current Instabilities," *J. Geophys. Res.*, 76, 3055.
8. McEntire, R. W. (1974), "Electron Echo Experiment 1: Comparison of Observed and Theoretical Motion of Artificially Injected Electrons in the Magnetosphere," *J. Geophys. Res.*, 79, 2343.

AAS 75-217

AMPS EXPERIMENTS INVOLVING GAS RELEASES

Lewis M. Linson[*]

INTRODUCTION

The primary purpose of this paper is to acquaint the reader with some experiments involving the release of gases that have been suggested for the Spacelab. Most of these experiments fall into one of the two following categories: 1) active or modification experiments; or 2) tracer or seeding experiments. We shall concentrate in this report on experiments in the first category which include: ionospheric conductivity modification by either barium thermite releases or dual high-velocity releases; neutral wind generation of electric fields; acoustic-gravity-wave generation; whistler-duct formation in the plasmasphere; a test of Alfven's critical velocity hypothesis; and the creation of an artificial comet. Most of these experiments are in the conceptual stage and further scientific definition of them is required. For completeness we mention several experiments in the second category: tracer experiments to study transport and energization processes by releasing lithium in the solar wind upstream of the earth's bow shock or in the magnetotail or by releasing various gases (H_2, He, Li, N_2) in the polar wind; a series of barium releases in order to measure the neutral wind and ionospheric drifts over a global scale during the same time frame; and a study of magnetospheric instabilities that can be generated by seeding the nightside equatorial magnetosphere with cold plasma.

Science Applications, Inc., La Jolla, California

The principal features of the shuttle that these experiments exploit is its tremendous weight-carrying capacity (which allows consideration of releasing 1000's of kg of gas) and the large kinetic energy associated with its orbital velocity of 7.8 km/s (which amounts to 3×10^{10} J for 1000 kg of gas or the energy equivalent of 7 tons of TNT). A second feature of the large orbital velocity is that material may be rapidly dispersed over a large path length; it takes less than 13 sec for the shuttle to traverse 100 km.

MODIFICATION EXPERIMENTS

Ionospheric Conductivity Modification

The atmosphere, ionosphere, and magnetosphere are electrodynamically coupled due to ion-neutral collisions in the ionosphere and the high conductivity parallel to the earth's magnetic field lines which maps electric fields and velocities between the ionosphere and the magnetosphere. Small thermite releases of barium vapor, which produce both visible clouds of neutrals and ions, have allowed the mapping of ionospheric electric field patterns around the world. In this sense the small barium release acts like a potentiometer in measuring the local potential gradient of the complex electrodynamic circuit. Large barium releases can be used as an active (variable) circuit element to determine the circuit and generator characteristics of the ionosphere, magnetosphere system. One could imagine producing a long (~ 100 km) trail of enhanced ionozation which would contribute ~ 50 mho to the height-integrated conductivity. If this trail were oriented in a north-south direction, it would tend to block the magnetospheric convective flow and short out the electric field. If it were oriented in an east-west direction, it provides a path for the Hall current and shunts the magnetospheric tail current and has the same configuration that auroral arcs have. Two principal methods have been suggested for creating the enhanced ionization: a) deploy a trail of closely-spaced barium thermite releases; b) release two high-velocity clouds of neutral gas, the former providing the seed for the ionizing release immediately following. The energy for ionization is derived from sunlight in the first case while it comes from the orbital velocity of the gas at the time of release in the second case. It is estimated that either method will require a mass of order 10^3 kg and the deployment should be below 200 km altitude. Possible observable effects of these experiments include perturbations of the electric and magnetic fields, the initiation of electron precipitation, and ultimately, the triggering of an aurora or substorm.

"Open-circuiting", or Ionospheric Conductivity Depletion

The Hall current in auroral arcs is primarily an electron current. This current could be "open-circuited" by the release of gas such as SF_6 which has a high electron affinity. Although the concept and technique have not been explored in detail, the analogy to the conductivity enhancements discussed above is obvious.

Creating a Neutral-Wind-Driven Dynamo

Another method for changing a circuit parameter is to modify the electric field (in the earth-fixed frame) applied to a significant portion of the ionospheric-magnetospheric circuit by creating a man-made neutral-wind-driven dynamo. Hence the release of 10^4 kg of a gas should be able to induce the ambient air to flow at 200 m/s in an areal extent of 100 x 300 km^2 if only 2.5% of the original energy in the gas can remain in directed flow.

Generation of Acoustic-Gravity-Waves

Gas releases from Spacelab will perturb the atmosphere and launch acoustic gravity waves. Appropriate releases should make possible controlled experiments in which sources of known spatial extent, known energy input, known position, and known time of generation are created by a large gas release. Small "tracer" releases would be used for optical diagnosis of the perturbations in the neutral and ionized gasses and a network of phase sounders would be used to monitor the motion of the ionosphere. A possible first step would be a point source release at an altitude of around 200 km over a major ionospheric observatory, such as Arecibo, during twilight conditions most suitable for optical diagnostics. The diagnostics would concentrate on obtaining a clear picture of how much of the energy of the release is coupled to the radiated wave field, to establish the "radiation pattern" of the source, and to make possible one-to-one comparisons between observation and numerical mesh calculations.

Create Whistler Duct by Releasing Reactive Gases

From recent theoretical work, it appears that artificial ducts can be created through the injection of reactive gases which substantially deplete the ionosphere and the overlying plasmasphere. The possibility of creating such a plasma duct leads to the possibility of studying VLF propagation along the duct, the dynamics of the duct, and the processes by which the duct refills.

Test Alfven's Critical Velocity Hypothesis

Laboratory experiments have revealed that plasma and neutral gas moving relative to each other and transverse to a magnetic field can interact in an unexpectedly strong way when the relative velocity exceeds a certain value, v_{crit}, given by

$$v_{crit} = 2eV_{ion}/M$$

where M and V_{ion} are the mass and ionization potential of the neutral gas particles. When this condition is satisfied, the plasma motion is rapidly braked and the neutral gas is rapidly ionized even when the mean free path is so large that binary encounters would cause negligible braking. The existence of the critical velocity phenomenon was originally introduced as a hypothesis in Alfven's theory of the origin of the solar system. It will be highly desirable to conduct experiments in the ionosphere to test whether the phenomenon also takes place in the very different range of plasma parameters that occur in space. If the mechanism will operate in a tenuous plasma, it may have important consequences in many natural situations, e.g., the interaction of the solar wind with comets and with planetary atmospheres.

Creation of an Artificial Comet

With the advent of the shuttle it will become feasible to consider creating an artificial comet, i.e., injecting such a mass of plasma into the solar wind that the flow will be strongly disturbed for a few hours.

The optical phenomenon resulting from the deployment of 1000 kg of Ba - CuO thermite to 25 R_e within the supersonic solar wind along the flanks of the magnetosphere would consist of a "coma", the diamagnetic cavity, of angular diameter of 1° after 10^3 sec, and rays originating from the "coma" and bending back to form a tail. The surface brightness of the "coma" would be 10 kR and the visual magnitude would be $m_v = +1.8$. An important aspect of this experiment is the significant extension of spectral coverage and dynamic range that will be available with instrumentation aboard the Spacelab. There are very few cosmic phenomena which man may actually simulate in space. Cometary tails are within technical reach.

ACKNOWLEDGEMENT AND CONCLUSION

Many of the experiments summarized here have been suggested and discussed by my colleagues, principally Drs. P. Banks. U. Fahleson, J. Heppner, R. Mc Entire, and G. Thome, on the AMPS Science Definition Working Group. Work in several of these areas is in progress to bring these experiments beyond the conceptual stage to the planning stage. It is hoped that this presentation will stimulate many additional ideas from other members of the scientific community.

AAS 75-218

LIDAR SYSTEMS FOR AMPS
Richard D. Hake, Jr.*

A lidar consists essentially of a laser light source, generally pulsed but possibly continuous-wave, and a telescopic receiver. The strengths of the lidar technique are derived both from its ability to do certain things that cannot be done passively, and from its ability to make measurements without relying on sources of natural light. A single-ended lidar (laser and receiver collocated) is capable of acquiring, by traditional radar methods, information on the range dependence of the species that interacts with the laser beam, while a double-ended lidar (receiver or signal reflector at a remote location) can only acquire column-content information. The information content of a single-ended lidar is carried both by the strength and the wavelength dependence of the return signal, and can be recovered only when the electromagnetic scattering process is known. The most important scattering processes and their uses are given in Table 1, and some adaptative uses of these processes are given in Table 2.

Experiments proposed for early AMPS missions utilize mainly Rayleigh, Mie, and resonant scatter, along with velocity determinations based on these scattering mechanisms. Later AMPS missions call for increasing implementation of limb-scanning, column-content measurements using a retroreflecting subsatellite. Raman, including "resonant" Raman, has largely been dismissed from the stable of alternatives because of its low sensitivity, and fluorescence (except for resonance fluorescence) does not appear promising at the present time. DISC is such a new technique that its utility has yet to be evaluated, but it offers some promise for sulfate analysis in the stratospheric aerosol. Table 3 contains a list of the species that exhibit the most promise for single-ended lidar

---
* Stanford Research Institute, Menlo Park, California

Table 1
SCATTERING PROCESSES AND THEIR USES

Rayleigh	--	Bulk Atmospheric Density
Mie	--	Cloud, Aerosol Density
Hard Surface	--	Albedo
Resonant	--	Specific Constituent Densities
Raman	--	General Constituent Analysis
Fluorescence	--	Specific Constituent Densities

Table 2
SPECIALIZED TECHNIQUES

DIAL (Differential absorption lidar)--Density of material determined by preferential absorption of lidar energy at characteristic wavelengths

DISC (Differential scatter)--Aerosol composition determined from wavelength sensitivity of Mie scattering

Velocity determination--Line-of-sight component of velocity of scatterer determined from Doppler shift of return signal

applications at the current time, divided roughly into the categories of feasibility ranked as straightforward implementation (1), more complicated implementation (2), and extensive hardware development (3). The range of suggestions for a limb-scanning long-path lidar covers almost every molecular gas of interest in the atmosphere. System calculations are sorely needed in order to assess feasibility in this area.

One major focus of the first AMPS missions will be a study of the dynamics and chemistry of the atmospheric region around the turbopause. Fortuitously located in this altitude region is a layer of free atomic sodium, which happens to be a remarkably good target for lidar probing. Lidar studies of temporal and spatial density variations in the atomic sodium layer will provide information on atmospheric waves, and some input to the chemistry of the region.

Table 3

REPRESENTATIVE SINGLE-ENDED LIDAR EXPERIMENTS

Category	Objective	Species	Wavelength	System (all $1 m^2 R_x$)
1	Mesopause, turbopause, dynamics	Na	5890 Å	1 J, 0.02 Å nadir, 450 km
1	Stratospheric dynamics and albedo	Aerosols, Cirrus	5890 Å (6943)	1 J, ~ 1 Å nadir, 450 km
2	D-region dynamics	Na temperature	5890 Å	0.1 J, 0.001 Å nadir, 200 km
2	D-region composition, dynamics	Ca^+ Li K	3965 Å 6707 7665	0.5 J, 0.02 Å nadir, 200 km
		Noctilucent clouds	5880 Å	1 J, ~ 1 Å nadir, 200 km
2	Stratospheric dynamics and albedo	Aerosols (high accuracy)	5890 Å (6943)	1 J, 0.001 Å nadir, 200 km
3	D, E, F-region dynamics and chemistry	Fe Fe^+ FeO Mg Mg^+ MgO	3859 Å 2599 ~ 5790 2852 2796 ~ 5007	1 J, 0.02 Å nadir, 200-400 km
3	D-region (auroral) stratosphere chemistry	NO OH	2262 Å 3125	0.5 J, 0.1 Å nadir, 200 km
3	Stratospheric chemistry and albedo	O_3	2500-3000 Å	0.5 J, 1 Å nadir, 200 km
3	Auroral energetics	N_2^+	3914 Å	1 J, 1 Å nadir, 200 km
3	Atmospheric winds	Na Aerosols	5890 Å (6943)	1 J, 0.001 Å steerable, 200 km
3	Atmospheric and magnetospheric dynamics by release tracking	AlO Na Li Ba Ba^+	4844 Å 5890 6707 5535 4554	1 J, 0.02 Å steerable, 200 km

Lidar studies of the altitude profiles of temperature and bulk velocity in the layer will provide information on the energy structure and wind field of the region. Figure 1 is an example of the detailed system calculation that needs to be done for each species in Table 3 and shows, in the altitude region from 70 to 120 km, the signal level expected from a realistic lidar, and the uncertainty in that level under two limiting background conditions. The signal is large enough in a single pulse to give 1% definition at the peak of the layer density in a single pulse, offering hope of mapping out the wave structure in the layer on a fairly short horizontal scale. Also shown in Figure 1 are the molecular scattering (without uncertainty bounds) and atmospheric-aerosol scattering signal levels that come along with the sodium signal for free.

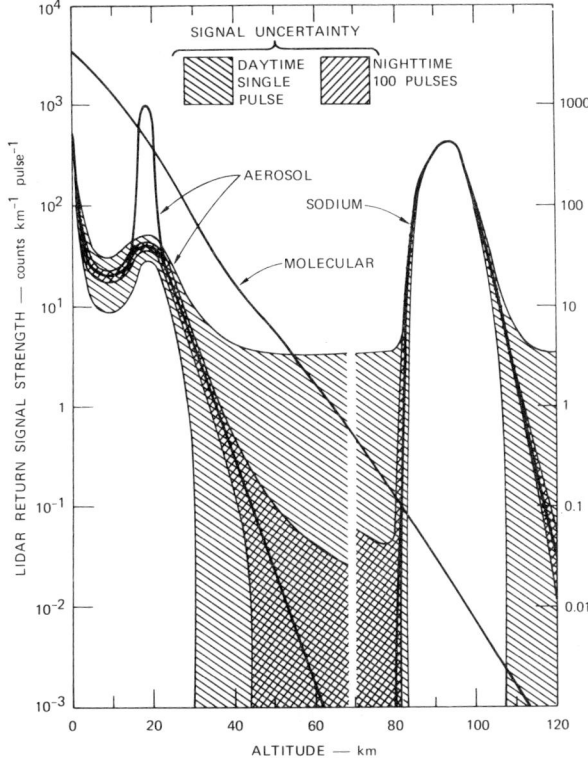

Fig. 1. Return Signal for 5890Å, 0.1-J, 1.0-m^2-Aperture Shuttle Lidar at 185 km

The aerosol, molecular density, and sodium density topics have all been addressed from ground-based lidars for some time. The technology is well in hand and extension to AMPS requires only flight qualification of equipment, along with the solution of certain thorny tradeoff questions such as how much real-time laser adjustment should be provided by the flight crew. When the topics of wind fields and sodium temperatures are addressed, however, the technological outlook is not so bright. Laser technology for producing high-energy pulses of the required linewidth (~0.001Å for optimum data acquisition) has not been fully demonstrated yet, and implementation aboard the shuttle is not so straightforward.

The cautionary remarks in this paper should be taken as such, however, and not as a pessimistic outlook. A great deal of work remains to be done in assessing the performance characteristics of AMPS lidar, and the potential rewards continue to appear substantial.

OBSERVATION OF SPACECRAFT GENERATED ELECTROSTATIC
FIELDS IN THE VICINITY OF THE ATS 6 SATELLITE

Elden C. Whipple, Jr.[*]

The occurrence of anomalies in the behavior of spacecraft instrumentation on geostationary satellites correlates well with the times of high spacecraft potentials. Anomalies on ATS 5 have been shown to be associated with differential charging of the spacecraft. Data from ATS 6 indicate the presence of a potential barrier in the vicinity of the spacecraft which turns back spacecraft-emitted electrons to the surface and keeps out low energy plasma electrons. The effect is most likely due to differential charging since the magnitude of the barrier is too large to be explained in terms of a secondary electron sheath about a uniformly charged spacecraft. Differential charging can occur whenever there are non-conducting surfaces and anisotropic charging mechanisms.

INTRODUCTION

There is recent evidence that anomalies in the behavior of spacecraft instrumentation on geosynchronous satellites correlate with large spacecraft potentials. Fredricks and Scarf[1] have studied the distribution of spacecraft anomalies as a function of local time. Figure 1 is taken from their paper, and it shows that anomalies tend to occur between midnight and 0600 local time. In addition, anomalies usually occur during or after periods of magnetic activity such as those associated with magnetic substorms. DeForest[2] has shown that the charging of the ATS 5 satellite to large potentials is almost always associated with magnetic substorms and the associated injection of hot plasma into the synchronous altitude region. At a recent symposium on spacecraft charging, Pike[3] and Shaw et al[4] presented evidence for the correlation of spacecraft anomalies and electrical discharges on spacecraft with magnetic substorms.

---

[*]Aeronomy Laboratory, National Oceanic Atmospheric Administration, Boulder, Colorado.

Lennartson, Reasoner and Chappel[5] at the same symposium showed that spacecraft charging events tended to occur between 0100 and 0700 local time. Figure 2 is taken from their paper and shows approximately the same distribution for charging events as does Fig. 1 for spacecraft anomalies.

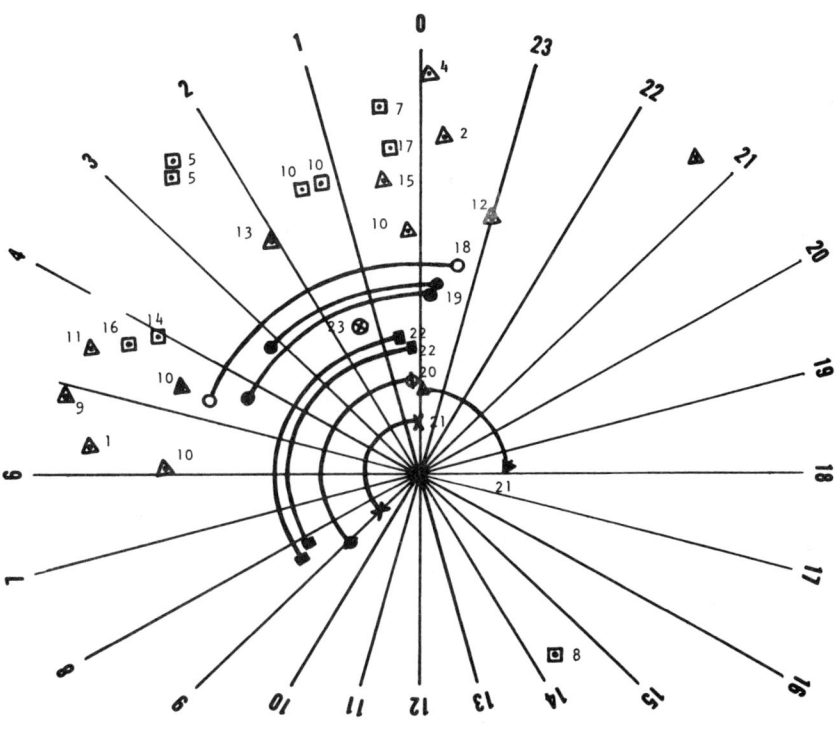

Fig. 1. Distribution of anomalies observed by several geostationary satellites. Rays are local time at spacecraft in hours, numbers identify events.

Some of the electrical discharges discussed by Shaw et al[4] were attributed to differential charging of the satellite. DeForest[6] has also attributed some of the anamolies observed on ATS 5 to differential charging. For example, cavities which were shielded from the sun would probably charge to different potentials than surfaces exposed to sunlight. The variation of surface potentials with time would depend in a complicated way on the variation of particle and solar fluxes as the satellite spins, coupled with the various time constants for charging which can vary from a fraction of a second to several minutes.

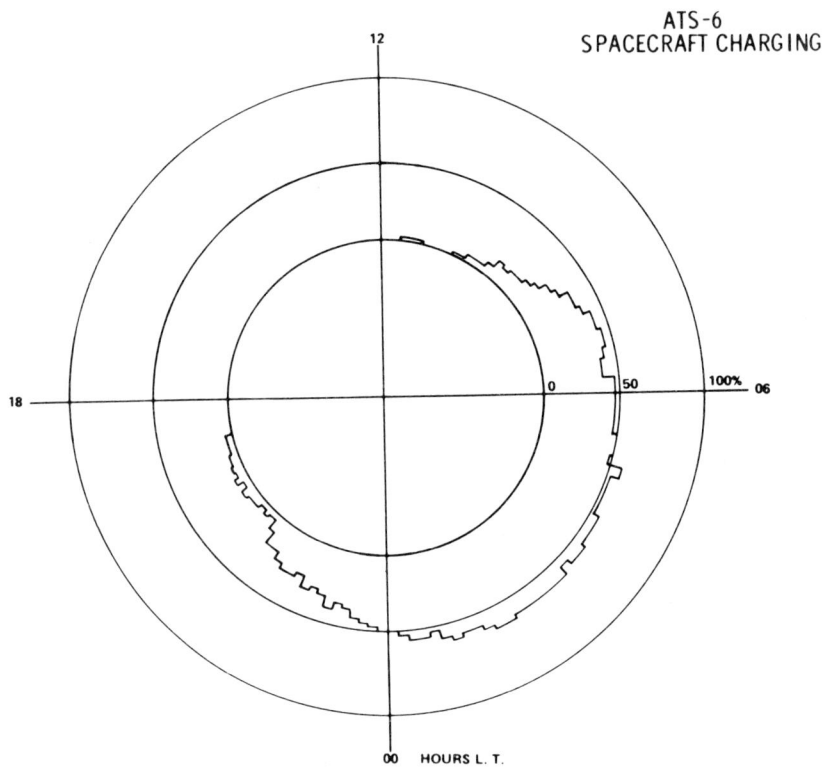

Fig. 2. Distribution of charging events observed on the ATS 6 satellite. (Taken from Reference 5.)

In this paper we present evidence that differential charging is also occurring on the ATS 6 spacecraft.

THE ATS 6 DATA

The ATS 6 satellite is shown in Figure 3. It was launched on May 29, 1974, into a synchronous orbit where it was stationed at approximately 94° West longitude. The data to be described was obtained by the

University of California at San Diego Auroral Particle Experiment which is located in the module behind the large parabolic reflector. The experiment is described in more detail elsewhere.[7]

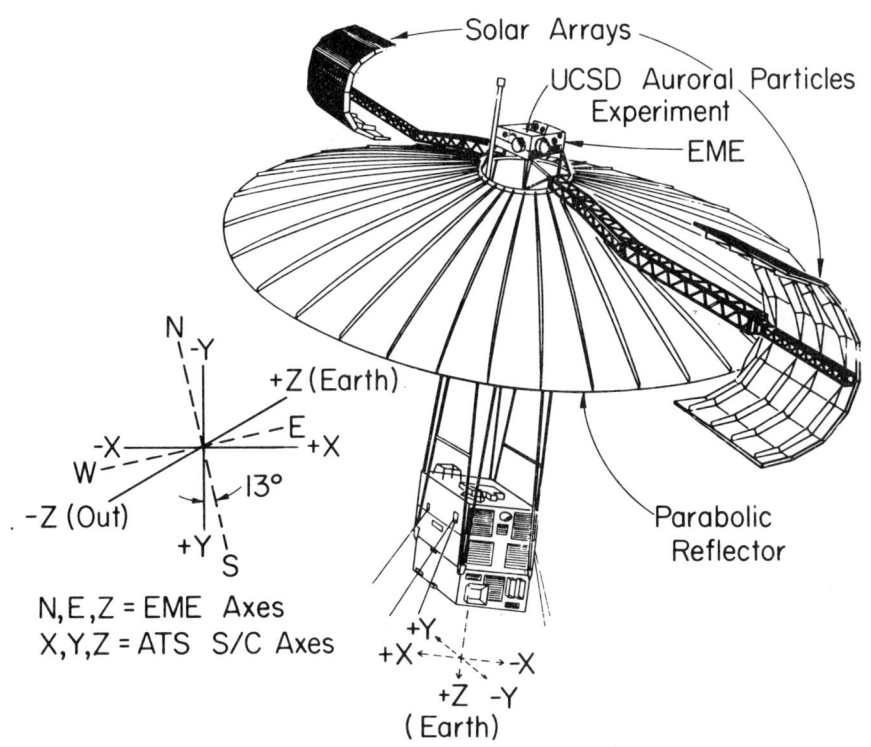

Fig. 3. The ATS 6 Satellite

Some data from the electron detectors are shown in Figure 4. The counting rates have been converted into electron distribution functions as a function of the electron kinetic energy. A Maxwellian distribution would appear as a straight line on this plot. It is apparent that there are at least two distinct distributions present; one with a mean kinetic energy of about 6 eV, and the other with a higher mean energy of about 320 eV. There is, in addition, a much higher component with a mean energy near 8 keV. This data was taken under eclipse conditions when it was possible to infer from the ion data (not shown) that the spacecraft potential was about −2000 volts.

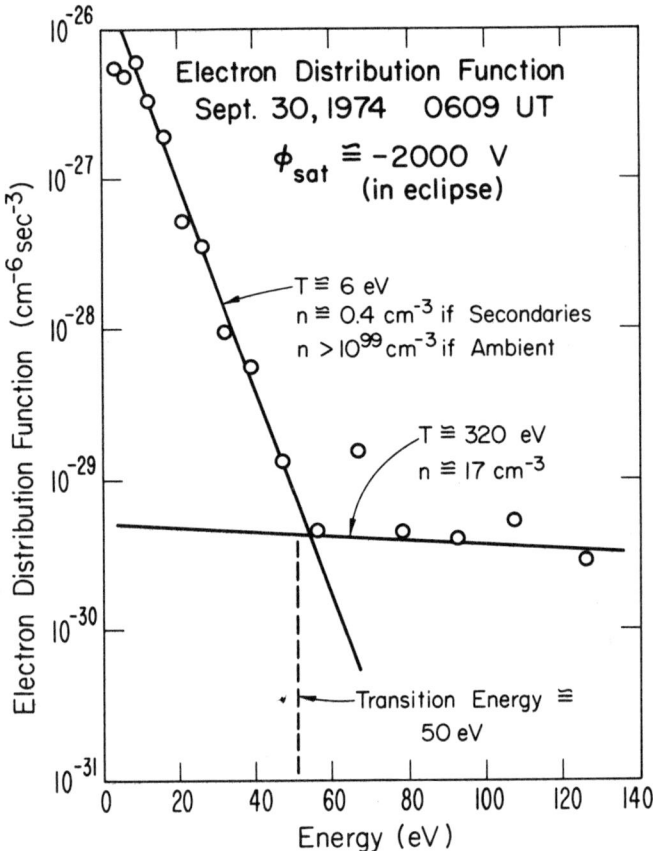

Fig. 4. Electron distribution functions obtained from ATS 6. The mean energy at 320 eV was obtained by considering other data points in addition to those shown in the figure.

The sharp transition between the 6 eV and the 320 eV electron distributions leads to the inference that the two distributions have different sources; if both distributions were present in the plasma, the transition would be more gradual. This inference is confirmed by the calculated electron densities. The potential difference between the particle source and the detector must be taken into account in inferring the electron density, with the result that the density of the 6 eV electrons would be greater than 10^{99} cm^{-3} if their source is the ambient plasma. On the other hand, if the source is the spacecraft itself, their density is calculated to be about 0.4 cm^{-3}. This latter density and the energy of these electrons are consistent with spectra and yields for secondary electrons emitted as a result of the impact of the higher energy primary electrons from the ambient plasma[8]. We conclude that the electrons corresponding to this lower energy distribution are secondary electrons emitted from the spacecraft which are returning to the spacecraft surface.

The particle detectors of the UCSD experiment rotate through 180°; one pair of electron detectors rotating in the north-south plane, and another pair in the east-west plane. The low energy secondary electrons are detected at all rotational angles. Since the spacecraft is charged to a negative potential, this behavior of the secondaries implies that there is a potential barrier in the plasma sheath external to the spacecraft which reflects the spacecraft-emitted electrons. The barrier returns the low energy secondaries to the spacecraft and prevents plasma electrons from reaching the spacecraft unless they have sufficient energy to overcome the barrier. The behavior of the electron counting rate at a constant energy as the detectors rotate can be shown to be consistent with the interpretation of the low energy electrons as secondaries.

INTERPRETATION OF THE ATS 6 DATA

Additional examples of this phenomenon from the ATS 6 spacecraft have been presented elsewhere[9]. The effect occurs in sunlight as well as in darkness; in sunlight the low energy electron distribution is similar to typical photoelectron spectra[10]. The effect is seen consistently during a large fraction of the observing time.

It is apparent that at times there can be a potential barrier in the vicinity of the spacecraft with a magnitude of at least a few tens of volts. This barrier blocks the passage of both outgoing spacecraft-emitted electrons and incoming ambient electrons. At the position of the barrier the electrostatic potential must be a minimum, and hence it follows from Poisson's equation that the space charge there must be negative.

Several suggestions have been advanced to explain the existence of such a barrier. Guernsey and Fu[11] and others have suggested that the photoelectrons or secondary electrons themselves could contribute enough negative space charge to create the existence of such a potential barrier. However, it can be shown that the magnitude of the observed barrier is much too large to be explained in this way[12]. If the potential barrier were due to the spacecraft-emitted electrons, its magnitude would be of the order of the mean electron energy. In the case of the ATS 6 data, the mean electron energy is only a few eV, whereas the magnitude of the barrier is some tens of eV in terms of potential energy.

Another suggestion has been that the motion of the ambient plasma with respect to the spacecraft creates a wake behind the satellite. This would have the effect of preferentially removing positive ions from the wake region, so that a negative space charge region could be created with an associated potential barrier. However, the plasma motions at synchronous altitude are in general not sufficiently persistent in magnitude and direction to account for the observations.

The remaining suggestion seems to be the most likely explanation; namely, that the surfaces of the satellite are differentially charged to different potentials. For example, if the solar arrays (see Fig. 3) were charged to a more negative potential than the rest of the satellite, then a "ridge" of negative potential would be formed (i.e., more negative than the surroundings) in the space between the two solar arrays above the spacecraft. Secondary electrons or photoelectrons emitted from the arrays could more easily reach this region than electrons from the rest of the spacecraft, because of the smaller difference in potential. Hence it would be possible for a potential minimum to form.

The conditions for differential charging of different regions of the spacecraft are that the surfaces be non-conducting and that the different regions be exposed to different charging currents. In this region of space, the important charging mechanisms are the incident flux of ambient electrons and ions, and the emission at the spacecraft surfaces of photoelectrons and secondary electrons produced by primary particle impacts. The ambient particle distributions are usually strongly dependent upon their pitch angle with respect to the magnetic field. Hence it is easy for regions of the spacecraft with different orientations to experience different particle fluxes. In addition, the asymmetries of the spacecraft itself will introduce shadowing effects. DeForest[6,11] has emphasized the fact that cavities in the spacecraft surface might easily charge to very different potentials from the rest of the spacecraft when they are aligned, for example, with the magnetic field.

In conclusion, it is important that electrostatic charging be taken into account in the future design and construction of spacecraft. The resulting electrostatic fields not only perturb scientific experiments but also cause anomalous behavior in other spacecraft instrumentation.

ACKNOWLEDGMENTS

Part of this work was done at the University of California at San Diego at the invitation of Dr. Carl McIlwain and Dr. Sherman DeForest whom I thank for many helpful discussions. I also thank Dr. David Reasoner for a copy of his figure on the distribution of charging events.

REFERENCES

1. R. W. Fredricks and F. L. Scarf, "Observations of Spacecraft Charging Effects in Energetic Plasma Regions," in Photon and Particle Interactions with Surfaces in Space, edited by R. J. L. Grard, pp. 277-308, Reidel Pub. Co., Dordrecht-Holland, 1973.

2. S. E. DeForest, "Spacecraft Charging at Synchronous Orbit," J. Geophys. Res., vol. 77, pp. 651-659, 1972.

3. C. P. Pike, "A Correlation Study Relating Spacecraft Anomalies to Environmental Data," EOS, vol. 56, p. 408, 2975.

4. R. R. Shaw, J. E. Nanevicz, and R. C. Adamo, "Observations of Electrical Discharges Caused by Differential Satellite Charging," EOS, vol. 56, p. 408, 1975.

5. W. Lennartsson, D. L. Reasoner, and C. R. Chappell, "Relationship Between the Local Time Dependencies of ATS-6 Spacecraft Charging Events and Warm Plasma Occurrences," presented at AGU 1975 Spring Annual Meeting, Washington, D. C.

6. S. E. DeForest, "Electrostatic Potentials Developed by ATS-5," in Photon and Particle Interactions with Surfaces in Space," edited by R. J. L. Grard, pp. 263-276, Reidel Publ. Co., Dordrecht-Holland, 1973.

7. C. E. McIlwain, R. LaQuey, B. Mauk, E. Strein, W. Atha, and P. Gifford, "Handbook for the UCSD/ATS-F and G Auroral Particles Experiment," University of California at San Diego, La Jolla, California, 1974.

8. E. C. Whipple, Jr., and L. W. Parker, "Effects of Secondary Electron Emission on Electron Trap Measurements in the Magnetosphere and Solar Wind," J. Geophys. Res., vol. 74, pp. 5763-5774, 1969.

9. E. C. Whipple, Jr., "Observation of Photoelectrons and Secondary Electrons Reflected from a Potential Barrier in the Vicinity of ATS-6," to be published in J. Geophys. Res., 1975.

10. R. J. L. Grard, "Properties of the Satellite Photoelectron Sheath Derived from Photoemission Laboratory Measurements," J. Geophys. Res., vol. 78, pp. 2885-2906, 1973.

11. R. L. Guernsey and J. H. M. Fu, "Potential Distribution Surrounding a Photoemitting Plate in a Dilute Plasma," J. Geophys. Res., vol. 75, pp. 3193-3199, 1970.

12. E. C. Whipple, Jr., "Theory of the Spherically Symmetric Photosheath: A Thick Sheath Approximation and Comparison with the ATS-6 Observation of a Potential Barrier," submitted to J. Geophys. Res., 1975.

AAS 75-222

MEASUREMENT OF STATIC ELECTRIC FIELDS
AT THE SURFACE OF SATELLITES[*]

J. E. Nanevicz[†]

Surface charges on spacecraft are of increasing concern. A field mill capable of providing accurate indications of surface field in a space environment was developed, tested in laboratory simulation facilities, and flown on rockets and a synchronous-orbit satellite. The design and testing of the field mill are presented together with flight test results.

INTRODUCTION

Experimenters concerned with the study of the stratosphere and ionosphere historically have used balloons, rockets, and satellites to carry electric field measuring instrumentation to successively higher altitudes. Since the objective of these measurements was the gathering of information about ambient conditions, great care was exercised in the design of the instruments to render them as free as possible from vehicle-induced perturbations. Highly specialized scientific vehicles were used, and certain of the sensors were mounted on long, thin booms to remove them from the vicinity of the main body of the vehicle.

More recently, satellite users are finding that electrostatic processes occurring as the result of the interaction of the vehicle with its environment can affect the functioning of payloads and even of the

---

[*] Prepared for technical papers that will later be published in the Proceedings of the American Astronautical Society.

[†] Dr. J. E. Nanevicz is Program Manager affiliated with Stanford Research Institute, Menlo Park, CA. (Bldg. 406A) (415) 326-6200.

basic satellite systems themselves.[1,2] Accordingly, there is an awakening of interest in obtaining information regarding the electrostatic environment in the immediate vicinity of the vehicle itself. Information of this sort can be of particular concern in the case of large complex systems such as AMPS which carry a variety of sensors and systems whose operations can be influenced by the electrostatic environment.

The dominant charging processes active on a spacecraft are illustrated schematically in Figure 1. Negative currents are produced by incident

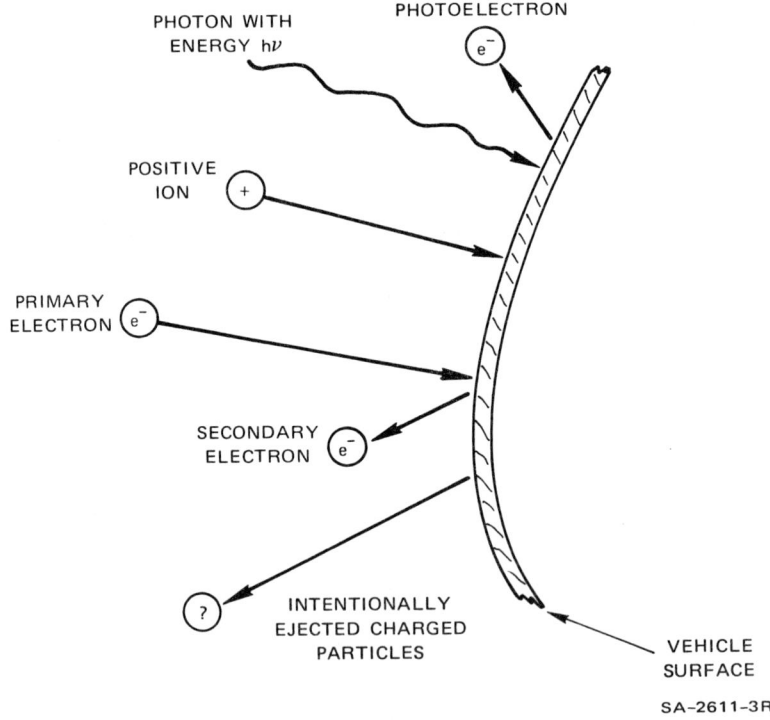

Fig. 1 Major Spacecraft Charging Processes

primary electrons and by reattracted secondary electrons and photoelectrons. Positive currents are produced by incident positive ions and by secondary electrons and photoelectrons that totally escape the surface. At any moment the magnitudes of these currents are determined by (1) the angle of incidence of or by lack of solar illumination, (2) the number densities and energy spectra of the local electrons and positive ions, and (3) the secondary emission and photoemission properties and surface potential of the surface involved.

In addition, as in the case of several of the experiments proposed for the AMPS payload, currents may be produced by the intentional ejection of charged particles from such devices as electron or ion guns or plasma thrusters.

The equilibrium potential of any point on a vehicle's surface is reached when the sum of the positive and negative currents flowing to that point is zero.

In general, the structure of a practical spacecraft makes it difficult to specify the spacecraft's surface electrostatic characteristics. Of necessity, vehicle surfaces are made of a wide diversity of materials largely dictated by mechanical and thermal considerations. For example, for some satellites, the requirement that the satellite surface function as a black body at infrared wavelengths requires that second-surface mirrors be used on exterior thermal-control panels. This means that the outer surfaces of the thermal-control panels are electrical insulators. Similarly, in the case of photocells, the active photocell material is covered with a transparent insulating material, which results in an insulating outer surface on the photocells. Thermal-blanket materials are made of many layers of plastic film, such as mylar, coated on one side with aluminum. Unless great care is exercised, successive layers of aluminum may not be bonded together or the system may not be connected to the vehicle frame.

It is evident from the above discussion that a satellite can be considered to consist of an electrically-bonded metal frame covered with various materials, all of which can acquire electrostatic charge as the result of processes in the ionosphere or the magnetosphere. The polarity and magnitude of charging on one surface may be quite different from charging on other surfaces exposed to the same environment, as the result of differences in the photoelectron and secondary-electron-emission characteristics of the surface materials. Thus, potential differences can exist between parts of the satellite. If one of the parts in question is an unbonded metal substructure, all of the charge it acquires is available to flow in any arc stemming from a voltage breakdown to the frame. Such breakdowns can thus be intense sources of noise. They are possible in the case of thermal-blanket materials when the conducting layers are deliberately or inadvertently not bonded to the frame.

Similarly, the insulating outer surfaces of thermal-control panels can become electrically charged with respect to the frame. Static charging of these surfaces, which must remain clean for thermal-control purposes, can have deleterious effects as the result of two different processes. First, electrostatic charge residing on the thermal-control surfaces will attract ionized or electrically charged particles back to the surface where they will be deposited. Second, if the electrostatic surface-charge density is sufficiently high, electrical breakdowns can occur from the insulating surface to the vehicle frame. These electrical breakdowns cause the liberation of various materials that are then free to settle out on nearby optical surfaces, thereby accelerating the rate of surface contamination considerably beyond what one would expect on the basis of simple outgassing.

The magnitudes of the effects discussed above, of course, depend upon such parameters as ambient ion and electron density and energy. These in turn depend upon altitude, local time, and whether or not an ionospheric disturbance is in progress.

It is important to observe that, in general, the environment of the spacecraft includes charged particles and energetic photons as well as the electric fields at the surface of the vehicle. Accordingly, any instrument to be used for electric field measurements must be capable of functioning in this complex environment.

METHODS FOR MEASURING STATIC ELECTRIC FIELDS

Various schemes have been proposed and implemented for the measurement of static electric fields. Perhaps the oldest is the generating voltmeter or "field mill". With this system, motion of an electrical conductor is used to modulate the ac field lines terminating on a sensor electrode, thereby generating an ac signal proportional to the dc field. Various methods of achieving the desired parametric modulation have been employed ranging from schemes whereby the sensor electrode is completely covered and uncovered by a rotating or vibrating vane to schemes for achieving modulation using a moving piston[3,4] and even to a scheme using a vibrating diaphragm.[5]

Since field mills require moving electrical conductors exposed to the static electric field, there has been considerable incentive to develop devices capable of performing the desired measurement without the need for moving parts. Techniques considered included the Stark effect in which ammonia gas was used in an absorption cell.[6] It was found that changes accumulating on the walls of the glass cell containing the ammonia produced a marked hysterisis in the system.[6] Other alternatives to the field mill have included the use of beams of moving charged particles that are deflected by the electric field to be measured.[7,8] Field magnitude is determined either by measuring the changes in current magnitude arriving at elements of an array of collecting electrodes[7] or by measuring the deflection voltage required to keep collection current constant in a feed back system.[8]

DESIGN OF FLIGHT TEST FIELD METER

Since the first flight test program required that surface field measurements be made on a rocket vehicle continuously from engine ignition to payload separation at synchronous orbit, the use of electron beam devices for field measurement was immediately ruled out. Experience with field mills on aircraft and sounding rockets[9] indicated that field mills could be made sufficiently rugged to function in the severe environment associated with rocket engine operation, and that they could be designed to give meaningful outputs in a plasma environment. Accordingly, an especially rugged field mill system was designed, built, qualified and used on a pair of Titan III C launches.[10] Later the same system was repackaged and flown on a non-NASA synchronous-orbit satellite.[11]

The field meter developed for the flight tests was of the rotating vane design. The detector head is mounted in a hole in the skin in such a manner that the meter vanes are exposed to the exterior of the vehicle as shown in Figure 2. Movement of the grounded rotor shown in the figure causes the stator to be alternately exposed to and shielded from the exterior environment. The alternating component of the stator short-circuit current in response to a true field is

$$I_f = \frac{1}{2} j\omega\epsilon_o EA \tag{1}$$

where ω is the radian frequency with which the stator is covered and uncovered, ϵ_o is the permittivity of free space, E is the field strength of the electric field terminating on the stator (when the stator is uncovered), and j is the imaginary unit $\sqrt{-1}$. The alternating component of the stator short-circuit current in response to a convection current is nominally

$$I_c = \frac{1}{2} JA \tag{2}$$

where J is the current density flowing toward the stator (when the

(a) INTERIOR OF ROCKET

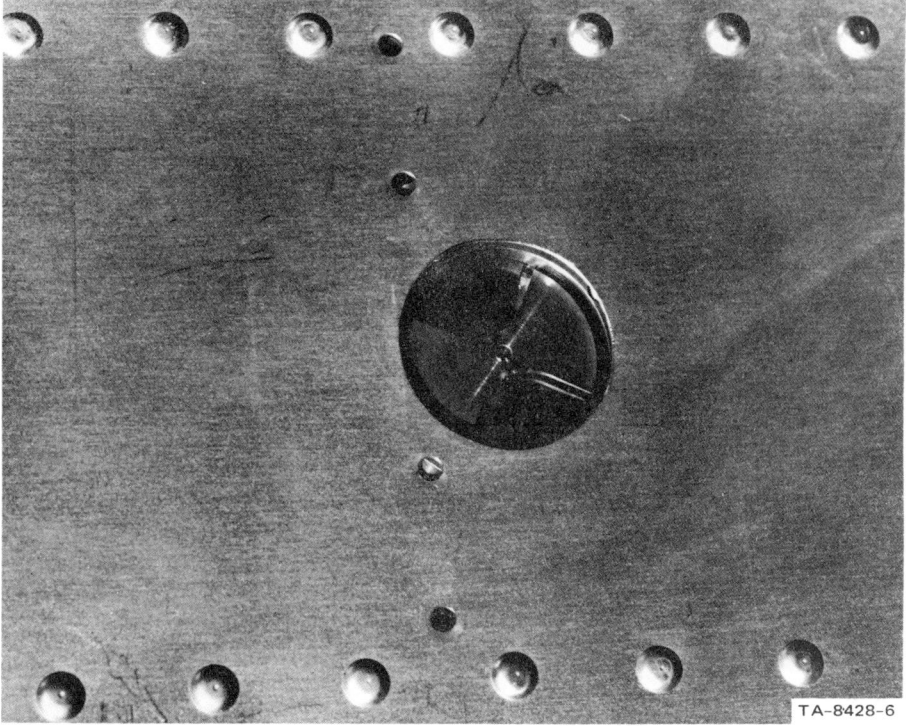

(b) EXTERIOR OF ROCKET

Fig. 2   Field-Meter Installation on Titan-III Rocket

stator is uncovered) and A is the surface area of the stator. Since the response to convection currents is 90 degrees out of phase with the response to an electric field, a coherent detector using a stable reference signal synchronized with the rate at which the rotor covers and uncovers the stator can be used to discriminate between the field response and the current response of the stator. (In the present field-meter system the reference signal is generated by an auxiliary set of vanes within the detector head).

Similar field-meter systems using coherent detectors have often been used in airborne and ground field measurements because the coherent detector permits one to determine polarity of the field as well as field strength. In these systems, however, only the "in-phase" component of the stator signal is normally detected. That is, the response to the electric field is detected and the response to convection currents, which is usually negligible at altitudes below 60 kilometers, is rejected. By using a second coherent detector adjusted to respond to stator signals in phase-quadrature with the field-produced signals, the response of the field-meter to convection currents as well as electric fields can be obtained. Although this quadrature response of the field-meter is not necessarily of interest, it does provide a basis for evaluating the behavior of the field-meter in the ionosphere. In the present field-meter system, therefore, both the "in-phase" and "quadrature" components of the stator signal were detected.

As is indicated in the block diagram of Figure 3, the signal generated in the stator is amplified by two sets of amplifiers in series to provide two sensitivity levels for the system. The stator is a circular disc located beneath two sets of similarly shaped, grounded, sectored vanes. The upper (rotor) vanes are driven by a 28-V, 400-Hz motor and serve as a shutter in conjunction with the lower (iris) vanes to cover and uncover the stator. When the two sets of vanes are

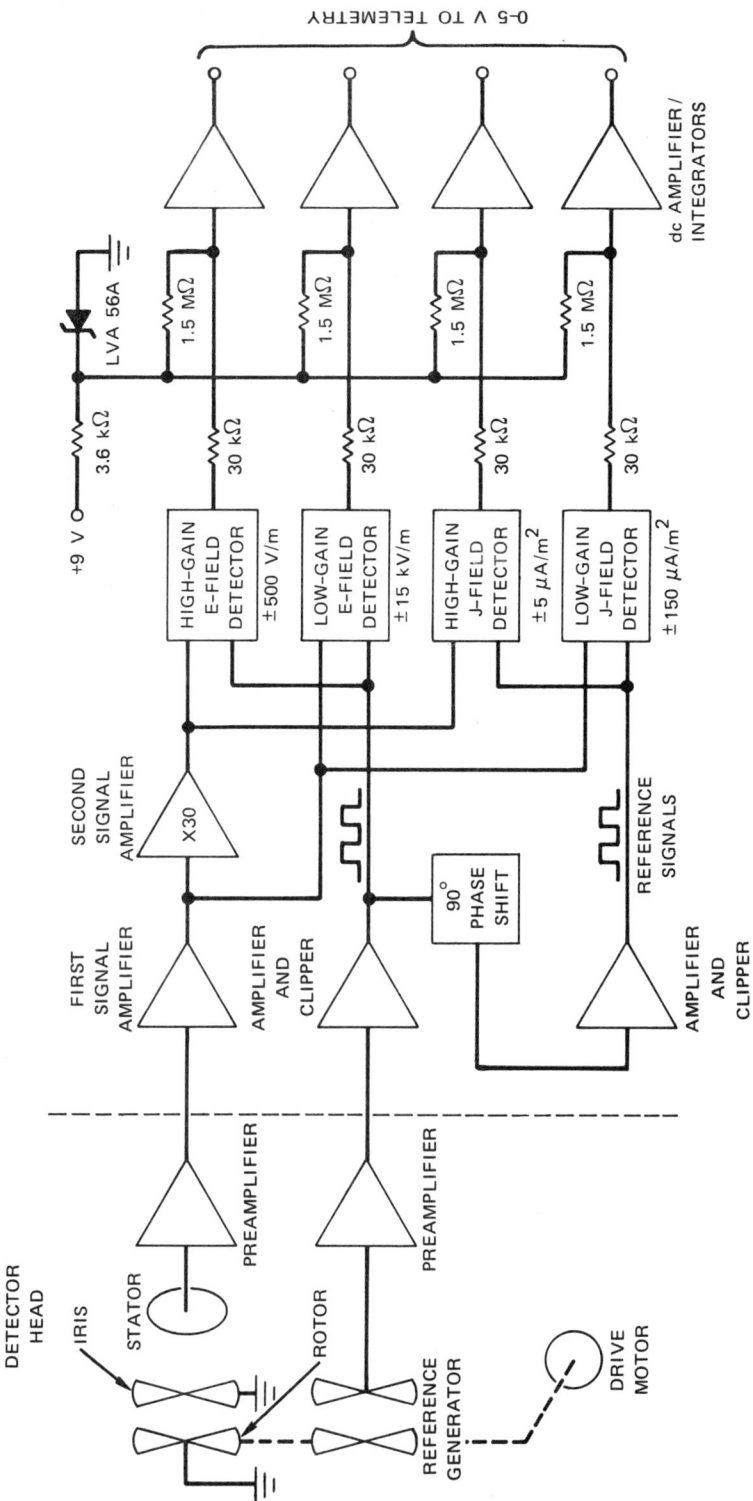

Fig. 3 Block Diagram of Electrostatic-Field Detector System

aligned, the stator is exposed; when the two sets of vanes are positioned at 90° with respect to one another, they form a continuous shield over the stator. The phase of the signal channel output with respect to the reference channel is adjusted simply by changing the position of the iris vanes. In this way, phase can be adjusted and undesired signals nulled out while the field-meter is operating.

The output from each signal amplifier is fed to two synchronous detectors, resulting in a high-gain and low-gain output for both E-field and J-field. As was indicated earlier, the reference signal is generated electrostatically by a vane structure mounted on the inboard end of the motor shaft. The E-field reference signal is fed directly from the reference stator to an amplifier-clipper which produces a square-wave output to the E-field synchronous detectors. To generate the J-field reference signals, the output from the reference stator is fed to a 90°-phase-shift circuit that drives the amplifier-clipper.

A typical calibration record is shown in Figure 4. The output is

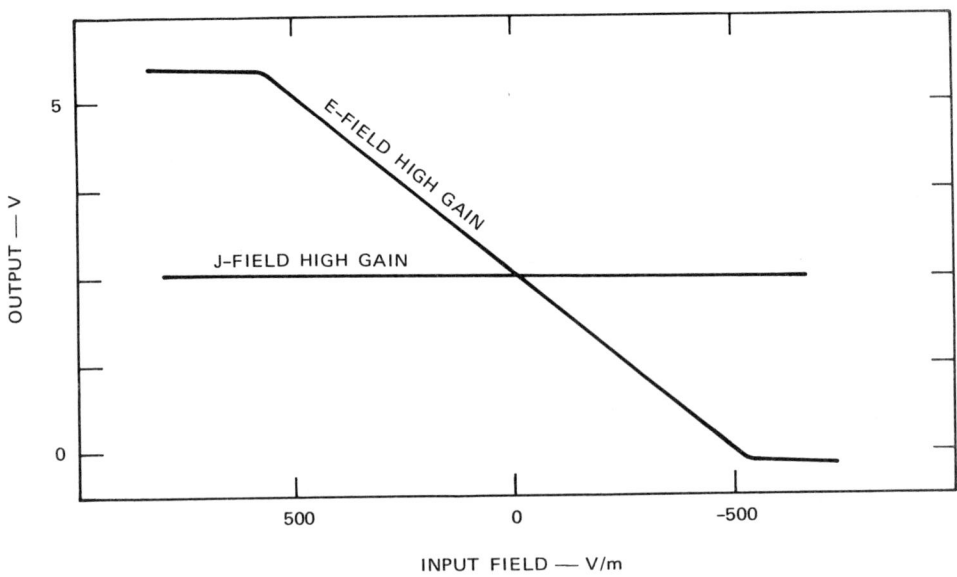

Fig. 4   Calibration Curves of Flight-Test Field-Meter System in Response to a True Electric Field

linear with input over the 0-to-5-V range and saturates rapidly outside this range. The degree to which the J-field channel is decoupled from true electric field is evident in the figure. A barely perceptible change in output occurs when the E-field is varied over the channel's dynamic range. The system sensitivities shown in Table I were chosen to permit accurate measurement of fields approaching breakdown potentials of mirror surfaces on the satellite ($\approx$ 2 kV).

Table 1

NOMINAL SPECIFICATIONS FOR SRI SATELLITE
ELECTROSTATIC FIELD SENSORS

Parameters Measured	Nominal Range	Nominal Sensitivity*
Electric field (high-gain)	±500 V/m ($\approx$ ±30 V)†	±4 V/m ($\approx$ 0.24 V)†
Electric field (low-gain)	±15 kV/m ($\approx$ ±900 V)†	±120 V/m ($\approx$ 7.2 V)†
Convection current (high-gain)	±5 $\mu A/m^2$	±40 nA/m^2
Convection current (low-gain)	±150 $\mu A/m^2$	±1.2 $\mu A/m^2$

*Equivalent of one telemetry count variation.
†Potential of surrounding mirror surface to produce equivalent output.

LABORATORY SIMULATIONS OF SPACE CONDITIONS

To verify the functioning of the fieldmeter system under space environmental conditions, laboratory tests were made to simulate various features of the space environment. In the first test illustrated in Figure 5, the field meter detector was mounted on an aluminum ground plane in a vacuum chamber equipped with a Krypton lamp ultra violet source to simulate solar UV illumination of the system. Provisions were made to apply bias voltage between the fieldmeter - mounting plate assembly and the chamber walls to simulate a net charge on the vehicle.

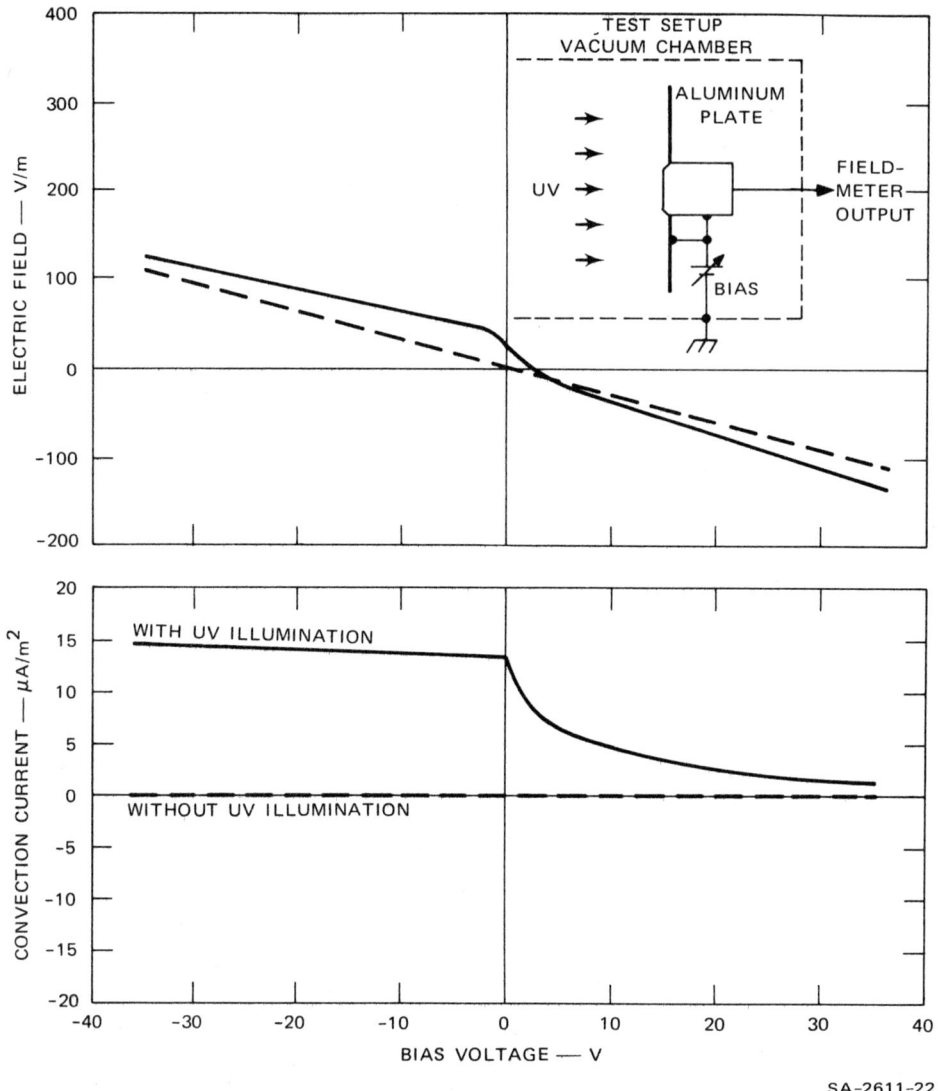

Fig. 5 Test Setup and Response of Electric-Field Detector Mounted in an Aluminum Panel Under UV Illumination

As can be seen in Figure 5, without UV illumination, the electric field is linearly proportional to the difference in potential between the detector head and its surroundings, and the convection current is zero. With the UV source turned on, there is little change in the electric-field reading. The convection current channel, however, indicates a uniform photoemission current for negative detector-head bias voltages. For positive bias voltages photoelectrons are re-attracted to the detector head and the convection current decreases. Essentially, this test demonstrated that the system functioned as expected under the simulated conditions. Furthermore, the data in Figure 5 allow a reasonably simple physical interpretation.

For the next series of tests, shown in Figure 6, the field meter was installed in a panel of quartz thermal-control mirrors to simulate the situation that exists when a field meter is installed on an insulating panel on a satellite. In this case, in addition to the photoemission currents measured with negative bias, large negative currents are measured when positive bias voltages in excess of 4 volts are applied to the test panel. It is assumed that the large negative currents are due to collection, by the positively biased field meter, of not only the photoelectrons emitted by its surface but also the photoelectrons emitted by the surrounding quartz surfaces. (Since the quartz mirrors are good insulators, the potential of the front surface is independent of the substrate potential).

The readings of the electric field channel are obviously very markedly influenced by the presence of the insulating surfaces in the vicinity of the field meter detector head. For a given applied bias voltage, the field meter output was increased by a factor of roughly 3 by the presence of the insulators. A rationale for accounting for this behavior of the E-field channel is as follows: In general as soon as the bias is applied, any free charges in the chamber will tend to be accelerated by the field and moved in such a direction as to minimize the potential energy of the system. With an insulating outer surface on the mounting panel, these charges will accumulate on the mirror surfaces and will tend to depress the potential of the surface to that of the chamber wall. (Photoelectrons emitted elsewhere are collected with positive bias, and photoelectrons emitted from the mirror surfaces are driven away from the mirrors when negative bias is applied.)

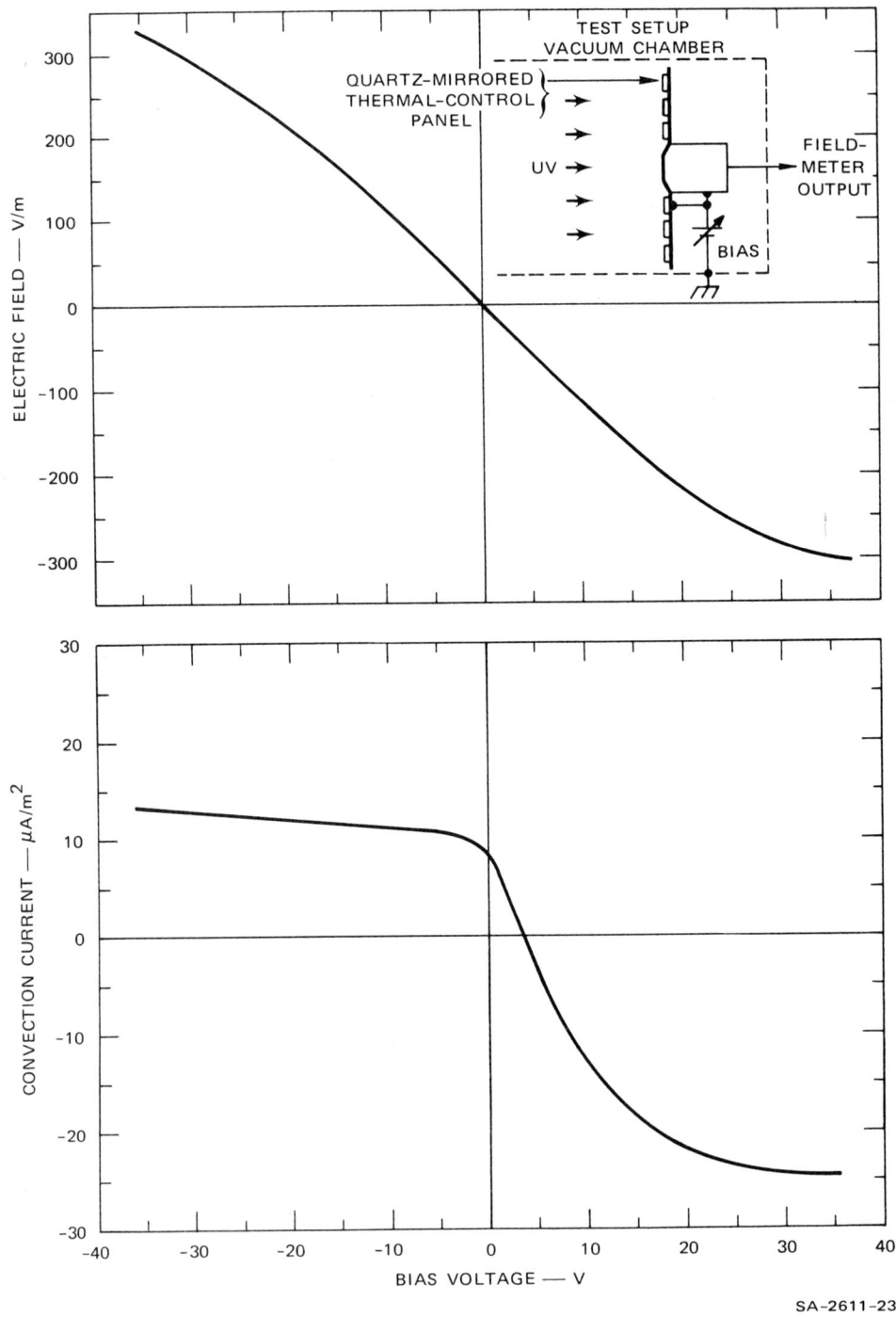

Fig. 6 Test Setup and Response of Electric-Field Detector Mounted in a Thermal-Control Panel Under UV Illumination

With the walls effectively moved closer to the mill, a given bias voltage will produce a higher electric field.

The results of the experiments illustrated in Figure 5 and 6 indicate that the field meter gives an accurate indication of processes occuring in an environment typical of that encountered in space. The same results also demonstrate that given a set of readings, one must be careful to consider all the possible active processes capable of producing the reading in devising a physical exploration for the observed results.

EXAMPLES OF FLIGHT TEST DATA

Data obtained from the field meter installed on the Titan III C-20 rocket when it was in the general altitude regime planned for the AMPS program are shown in Figure 7. During the telemetry window shown in the figure, the vehicle is being rotated by the firing of altitude control rockets (ACS rockets). The orientation of the vehicle with respect to the sun is shown by the insolation vector cone and clock angles. (The field meter is in the illuminated hemisphere for clock angles between 44° and 224°.)

The ion-current-density "J-field" channel field-meter data near the beginning of the record can be interpreted as being indicative of photoelectric current emission from the field-meter vanes. Prior to the beginning of the record, the clock angle of the insolation vector is greater than 224° so that the field meter is in the hemisphere shaded from the sun. As the clock angle decreases below 224°, the field meter becomes progressively more exposed to the sun, and the field-meter current-density channel indicates progressively increasing positive charge arriving on the stator vanes (or negative charge leaving). The current density increases to a maximum value of 3×10^{-5} A/m^2 = 3×10^{-9} A/cm^2 at time 0837. From Ref. 12 we find that for gold (the field-meter vanes were gold-plated) the integrated photoelectron flux for solar irradiation is 2.9×10^{-9} A/cm^2. Since the measured current density

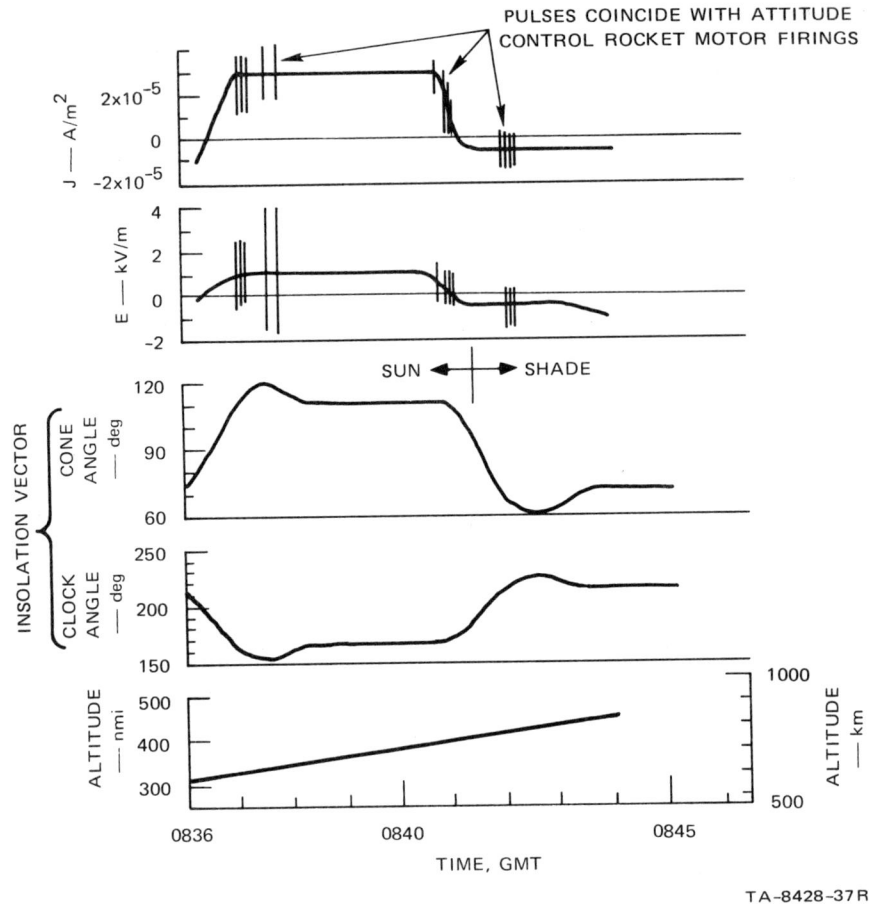

Fig. 7   Titan IIIC-20 Data From 310 to 450 nmi Altitude

is in agreement with the predicted photoelectron flux, it appears that the J-field reading from 0837 to 0841 GMT can be attributed to photoelectric emission from the field-meter stator vanes. At 0841 the vehicle is maneuvered to shield the field meter from the sun and the photoelectric current decreases. Beyond 0841, the field meter J channel indicates negative charge arriving. This is undoubtedly photoelectron current emitted on the sunlit side of the vehicle.

The electric field during the telemetry window of Figure 7 varies both in magnitude and polarity. At the beginning of the record at 0836:20 the field is zero. By 0837 the field has increased to 1 kV/m, with the vehicle negative with respect to the plasma at the field meter location. At 0841 the field changes sign following vehicle reorientation. Since the E-field variation is similar to the J-field variation, it is tempting to attribute the E-field readings to imperfect isolation between the E- and J-field channels. Later on in the flight, however, the two channels of the field meter do not change in unison. Accordingly, it must be assumed that the indications of local electric field at the field meter position on the surface of the vehicle must be accepted as real.

It is evident in the field meter records of Figure 7 that there are numerous transient changes in the field meter output signals at the general times when vehicle maneuvers are taking place. These transients are associated with firing of the ACS rocket motors to achieve desired vehicle orientation during the telemetry window. This correlation with ACS motor firings is significant since it indicates that even the small rocket motors used in the ACS system produce appreciable perturbation of the plasma in the vicinity of the vehicle.

On the non-NASA synchronous-orbit satellite discussed in Refs. 11 and 13, the field meter was deliberately installed in a thermal control mirror panel in the general manner shown in Figure 6. In this way the field meter recordings were sensitive to mirror surface potential. An important result of the satellite program was the demonstration that RF noise pulses (presumed to be caused by breakdowns of insulating surfaces) occur at times when surface-to-substrate potentials are 50 V or less. As a consequence, processes by which insulation breakdown can be produced with low surface voltages are presently being postulated for investigation.

CONCLUSIONS

In connection with the operation of a complex system such as AMPS it is often necessary to obtain information about the structure of the electrostatic fields in the vicinity of the vehicle. This requirement generally stems from compatibility considerations. For example static electric fields modify the way in which particles deposit on critical surfaces. Static charging is also of concern in connection with instruments and processes sensitive to the particle environment. In addition, it has been shown that static charging is capable of affecting the electromagnetic noise environment on the vehicle.

In general, with the development of new plastic materials with good optical and mechanical properties, their insulating qualities have simultaneously (inadvertently) been improved so that many of the materials used in building space vehicles are insulators of an excellence that was difficult to achieve in the past. The presence of large quantities of insulating material in many regions of a spacecraft can only aggravate the problems of electrostatic charge storage. Accordingly it is important to have available tools to study electrostatic field structures in critical locations on a vehicle.

The field mill described here is suitable for this purpose. It should be recognized that the physical embodiment of the detector head might profitably be changed to make it lighter and to eliminate the requirement for a rotating motor. (It should be recalled that the present unit was designed with the requirement that it function at liftoff.) Otherwise, the field mill concept using carefully-designed synchronous detectors is capable of providing information about surface electric fields even in a complex plasma environment.

Such detector heads could be mounted at places of interest on the vehicle. Alternatively one could use a mill mounted on a moveable assembly to scan surfaces of interest.

REFERENCES

1. A. Rosen, "Large Discharges and Arcs on Spacecraft," Aeronautics and Astronautics, vol. 13, No. 8, June 1975 pp. 36-44

2. S. E. De Forest, "Spacecraft Charging at Synchronous Orbit," J. Geophys. Res., vol. 77, No. 4, February 1972

3. Von W. Gohlke and U. Neubert, "Bemerkungen zur Hoch-und Hochstspannunsmessung," Zeitschrift fur technische Physik, No. 10, 1940 pp. 217-222

4. Von W. Gohlke and U. Neubert, "Messungen an einem Schwingvoltmeter," Zeitschrift fur technische Physik, vol. No. 23, 1942 pp. 70-76

5. R. H. Hertel, "Vibrating-Piston Electrostatic Field Sensor," Proc. Instr. Soc. Amer. Twelfth National ISA Aerospace Instrumentation Symposium, Philadelphia, Pennsylvania, 2-4 May 1966 pp. 179-186

6. P. K. Shumaker and B. W. Sherman, "The Use of the Stark Effect in Electric Field Measurement," EASCON '68 Record, pp. 345-360

7. H. S. Ogawa, J. M. Sellen, and R. K. Cole, "Electron Emissive Surface Electric Field Meter," Rev. Sci. Instr. vol. 38, No. 8, August 1967 pp. 1117-1123

8. B. W. Sherman, R. H. Manka, H. R. Anderson M. C. Terry and J. D. Droppleman, "A Lunar-Surface Electric-Field Detector" EASCON '68 Record, pp. 339-344

9. J. E. Nanevicz and E. F. Vance, "A Rocket-Borne Electrostatic Field Strength Meter," Proc. Instr. Soc. Amer. Twelfth National ISA Aerospace Instrumentation Symposium, Philadelphia, Pennsylvania, 2-4 May 1966 pp. 169-178

10. J. E. Nanevicz, "Results of Titan III Flight Electrostatic Experiments," <u>Lightning and Static Electricity Conference Papers</u>, Las Vegas, Nevada, AFAL-TR-72-325, Air Force Avionics Laboratory, AFSC Wright-Patterson AFB, Ohio, 12-15 December 1972 pp. 106-131

11. J. E. Nanevicz, R. C. Adamo, and R. R. Shaw, "Electrical Discharges Caused by Satellite Charging at Synchronous Orbit Altitudes" <u>Proceedings 1975 Conference on Lightning and Static Electricity</u> at Culham Laboratory, England, The Royal Aeronautical Society, 4 Hamilton Place, London WIVOBQ, England, 14-17 April 1975

12. B. Feuerbacher and B. Fitton, "Experimental Investigation of Photoemission from Satellite Surface Materials," <u>J. Appl. Phys.</u>, vol. 43, No. 4, April 1972 pp. 1563-1572

13. J. E. Nanevicz, R. C. Adamo, and W. E. Scharfman, "Satellite Lifetime monitoring" Final Report Contract F04701-71-C-0130 P. O. 126192, SRI Project 2611 for Aerojet Electrosystems Co. by Stanford Research Institute, Menlo Park, California, March 1974

AAS 75-228

PRELIMINARY SHUTTLE PAYLOAD CONTAMINATION ASSESSMENT[*]

E. B. Ress[+]
R. O. Rantanen
L. E. Bareiss

The on orbit contaminant induced environment about the Shuttle Orbiter and various Spacelab carrier configurations is currently being computer modeled for NASA by Martin Marietta Aerospace, Denver Division. A key element in establishing contamination control for any spacecraft system is the detailed understanding of the on orbit contamination induced atmosphere. This paper addresses the extent of the molecular induced atmosphere for major contamination sources about the Shuttle Orbiter and a payload carrier (Spacelab configuration - long module and a 3 meter pallet) in the Shuttle Orbiter payload bay. The major sources considered are the external surface materials mass loss, pressurized habitation area leakage, attitude control engines (25 lb thrust Vernier Control System - VCS engines), and supplemental flash evaporator venting.

The material presented herein describes the molecular induced atmosphere of the Shuttle Orbiter and a Spacelab carrier configuration and compares it to the ambient atmosphere at various orbital altitudes. The predictions presented represent near maximum values for the sources considered and are subject to change as the source characteristics are better defined or are changed from existing baseline definitions. One assumption that carries throughout the predictions is that the molecules

---

[*] Concurrent activities being conducted for the Shuttle Orbiter and Spacelab carrier under NASA contracts NAS9-14212 (JSC) and NAS8-31574 (MSFC), respectively.

[+] The authors are employed in the Contamination Analysis, Systems Engineering Department; Martin Marietta Aerospace, P.O. Box 179, Denver, Colorado 80201.

leaving a source travel to the point where the fluxes and densities are calculated without experiencing a collision with other induced molecules or the ambient atmosphere molecules. This assumption is not overly restrictive since the mean free paths of the majority of induced species are very large over the majority of the range of orbital altitudes considered. The larger molecules (outgassed species) will be influenced more at lower altitudes (200-300 km) with the net effect being dictated by the direction of the spacecraft velocity vector and the position or viewing direction of a specific payload. However, the values presented can occur at all altitudes for specific situations and do indicate the magnitude of the contaminant species.

The results are based on near maximum flow conditions of the contaminant sources. The mass loss of external surfaces is treated as two different source types. These are outgassing (long term bulk effect) and offgassing (loss of adsorbed and absorbed volatile species during early vacuum exposure). The outgassing and offgassing rates modeled relate to periods when the external surfaces of the Shuttle Orbiter and Spacelab carrier are near a maximum temperature profile. The outgassing rate at a fixed temperature is essentially constant as a function of time. The offgassing results are indicative of a rate corresponding to a 10-hour point of on orbit vacuum exposure. The 10-hour point was chosen for presentation as being representative of that point during a mission when initial payload operation might be expected to commence. The cabin leakage source is considered constant and is diffusely emitted from the forward end of the payload bay for the Shuttle Orbiter and uniformly from the surfaces of the Spacelab carrier tunnel and crew modules (the pallets do not have a leakage characteristic). The supplemental flash evaporator and VCS engine source rates correspond to periods of actual on time and are highly mission dependent.

Figure 1 presents the density of each of the sources considered as a function of distance out of the payload bay for a line-of-sight perpendicular to the payload bay. Also indicated in Fig. 1 for comparison is the medium ambient atmosphere density at various orbital altitudes.

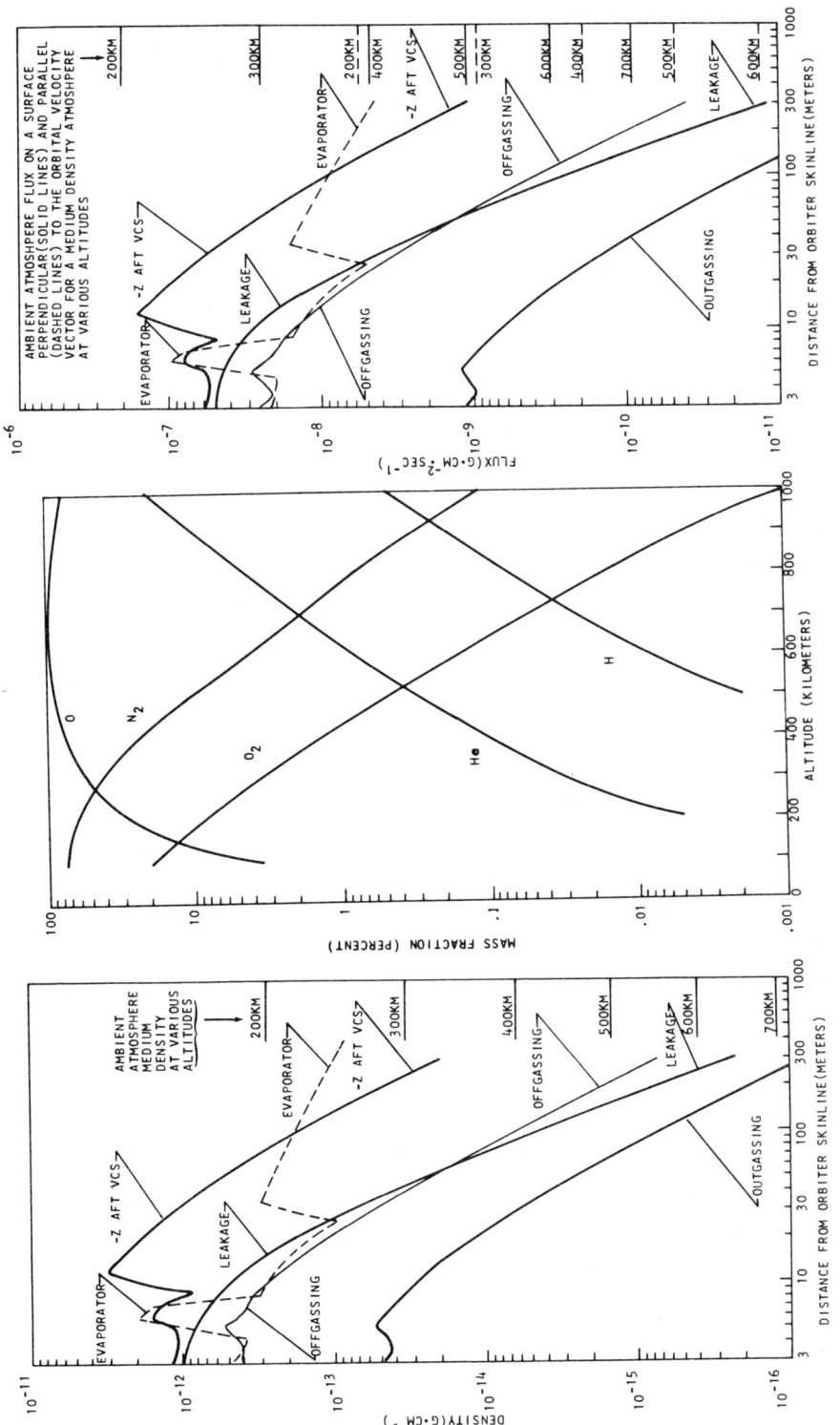

Figure 1. Contaminant Density as a Function of Distance from Orbiter Skinline

Figure 2. Percent by Mass of Ambient Atmosphere Constituents

Figure 3. Contaminant Flux as a Function of Distance from the Orbiter Skinline

Figure 2 shows the percent by mass of the major constituents of a medium density ambient atmosphere as a function of altitude. From these figures, the constituents of the ambient atmosphere can be compared to the constituent fractions of the contaminant sources. The ambient atmosphere density itself can vary one to two orders of magnitude from the medium density values presented as a result of sunspot activity and the day/night portions of an orbit. By integrating under the curves in Fig. 1, the mass column or molecular number column density can be obtained for each source.

Knowledge of the flux levels of the induced atmosphere as a function of distance from the Shuttle Orbiter is important for deploying payloads, subsatellite positioning, or in the use of boom mounted sensors. Figure 3 presents the average flux levels of the induced atmosphere constituents shown in Fig. 1 as a function of distance from the modeled configurations. The flux values presented correspond to the flux arriving at a surface viewing back toward the payload bay perpendicular to the line-of-sight along which the fluxes were calculated. For comparative purposes, the flux of a medium density ambient atmosphere on a surface perpendicular and parallel to the velocity vector at various orbital altitudes is included in Figure 3. The perpendicular flux component is dependent on the orbital velocity and ambient atmospheric density of each altitude while the parallel flux component is dependent on the ambient density and the mean thermal velocity of the ambient atmosphere at each altitude.

Depending upon the orbital altitude a payload will be flown at, the induced atmosphere generated from the Shuttle Orbiter and a Spacelab carrier can significantly alter the ambient atmosphere resulting in the potential of significant contamination levels. However, these can be minimized by recognizing the extent and duration of the induced environment and by proper on orbit constraints and positioning of payloads in and around the Shuttle Orbiter and Spacelab carrier. This requires a basic understanding of the contamination sources and their potential influence on payload surfaces and payload sensitivities to the induced environment so that mission success can be optimized. Such is the goal of the contamination computer modeling currently in progress.

AAS 75-229

AMPS DATA MANAGEMENT CONCEPTS

P.N. Metzelaar*

Several typical AMPS (Atmospheric, Magnetospheric, and Plasma in Space) experiments were formulated to permit exploration of AMPS data management concepts. Design studies were conducted detailing these experiments in terms of the applicable procedures, data processing and displaying functions. The design study began with the physical phenomena to be observed in the environment and the instruments required to observe them. Setup and control functions for these instruments were studied, as well as the displays needed to obtain experimenter control over the operation. A crucial factor in the design was the assumptions that were made regarding the degree of flexibility that will best serve the experiment's success.

The design concepts arrived at make extensive use of three elements—the computer, the experimenters' alphanumeric keyboard, and a cathode ray tube display. They permit both automatic repetitive measurement sequences and experimenter controlled step-by-step procedures believed most appropriate for AMPS. In addition to the presentation of these concepts, some examples are shown of: (1) the choices that these design concepts would give to the experimenter in the displays and controls that he uses, and (2) the types of computer software needed to make the system work.

---

* P.N. Metzelaar is with TRW Systems Group, Redondo Beach, California 90278. These studies were supported in part by NASA by contract to TRW.

INTRODUCTION

For several years, TRW has been concerned with AMPS experiments and with the establishment of an integrated Spacelab AMPS payload that can efficiently serve to make these experiments successful. This paper reports some results of recent TRW studies on the data management aspects of the AMPS payload.

SYSTEM DESCRIPTION

The AMPS payload will carry perhaps 40 to 80 scientific instruments and mounting booms, platforms and other support equipment of various types, as well as up to four spaceborne experimenters to run the experiments. Essentially, the data management system is the link between the experimenters in the Spacelab and the instrumentation on the Spacelab pallet that is exposed to the AMPS environment. Some characteristics of this data management link are now becoming defined. The European design of the experiment part of the command and data management system presents us basically with the capabilities of a modern minicomputer. This computer is provided with 65,536 16-bit words of core memory, with a read only tape unit mass memory of 131 million bits, plus one to three ASCII character keyboards for experiment control and one to three cathode ray tubes for general purpose displays. Anything that the spaceborne minicomputer system cannot do will have to be done by general or special purpose equipment. The general purpose equipment will probably be common physics laboratory equipment such as oscilloscopes and wave spectrum analyzers. Of the available six racks, general purpose equipment will probably occupy four or five racks. Any special purpose equipment will probably have to fit within one or two racks. It is therefore very likely that special purpose equipment should be installed only where the general purpose equipment is marginal or cannot do the job.

STATEMENT OF THE PROBLEM

By examining the needs of the spaceborne experimenter, we can discover the criteria that the design must meet. His job is to set up the instruments properly, verify that the instruments are working correctly, and from time to time make those adjustments that he believes will enhance the value of the data. The experimenter must also coordinate his actions with the ground processing system and he must prevent bad, useless, or excessively redundant data from entering the major processing facilities on the ground, where major expense may be incurred in removing this data from the system.

Many of the sensors can produce hundreds of times more information output than the data management system can process. The data management system, in turn, can produce hundreds of times more information than the spaceborne experimenter can usefully respond to. A means of reducing the amount of data displayed to only that essential for controlling the experiment is necessary.

The problem of the data management system designer, therefore, is to design a system which will: (1) provide the experimenter with the power to select the proper set of data on which to base his actions from the mass of data available, (2) display this data to him in suitable form, (3) give him the means to control the data acquisition and flow to downstream recording and processing facilities, and (4) stay within other Spacelab constraints such as rack space, weight and power.

APPROACH

The data management concepts problem is so complex that a simulation or modeling approach seems most suitable. The first step in this approach is to formulate some representative candidate experiments and some promising data management concepts to handle their requirements. We have taken this first step for five experiments and are reporting the results in this paper.

CONCLUSION

As a result of this work, five AMPS experiments were defined sufficiently well so that implementation and verification of the data management concepts by simulation can be undertaken.

This definition effort took into account the following:

1) <u>Environmental Computation</u>. Environment equations were defined (orbits, magnetic fields, electron densities, etc.) to simulate the experiment input data to the instruments.

2) <u>Instrument Control</u>. Instrument control parameters were identified.

3) <u>Processing Requirements</u>. Processing steps were defined for control parameters and for the data processed for display.

4) <u>Display Formats</u>. Detailed formats were defined for use in each experiment simulation.

5) <u>Procedures</u>. The steps were defined for setting up and running each experiment.

6) <u>Data Management Concepts</u>. Concepts were selected for managing the variety of experiments and instruments to be used.

The candidate data management concepts that were selected in this study are:

1) <u>General Purpose Equipment</u> (computers, cathode ray displays, keyboards) is used where its flexibility recommends it. Special purpose equipment is used where general purpose equipment has marginal performance or cannot do the job.

2) <u>Output-Driven System</u>. The experimenter can have processed and displayed only what he wants to see and can use.

3) <u>Hierarchical Organization</u>. Via computer programming, the experiments, instruments, and their input, output and control parameters can be organized in their natural hierarchical order.

4) <u>Ease of Operation</u>. Prompted display systems, backup values, and automatic sequences will make for more accurate, faster experiment operations.

5) <u>Flexibility</u>. Experimenter has the ability to start, stop, change parameters, or instruments. Flexible software-driven displays help achieve this.

6) <u>Data Monitoring Capability</u>. Experimenter has experiment data displays for review and adjustment of parameters and the ability to inhibit transmission of bad data for ground processing.

It is hoped that the methods, examples and data management concepts presented in this paper will be of assistance to those who are now in the process of formulating AMPS Experiments.

ACKNOWLEDGMENT

Many colleagues at TRW have contributed to the development of the experiment models and data management concepts presented. In particular, the author would like to acknowledge the contributions of Mr. W. V. Neisius, Dr. N. L. Sanders, Dr. J. L. Vogl and Dr. R. L. Wax.

AAS 75-230

EXPERIMENT INTEGRATION OF AMPS

Robert Witholder*

Analytical integration of AMPS experiments will be concerned with the technical communications link or interchange between three program areas that proceed in parallel to arrive at a compatible experiment facility. The three areas are: 1) scientific requirements; 2) instrument development; and 3) facility development. All three areas have their multi-faceted characteristics and yet have a common interest. Integration must insure that these areas play-together, but remain a silent partner.

The facility integration will deal with sizing problems of weight, C.G., power, etc. of some 60-80 instruments which, as a simple sum total, exceed the Spacelab capability. In addition consideration of operational variables of various experiment modes increases the integration problem.

In satisfying the instrument interface requirements with existing spacecraft systems, it is found that some significant requirements are of concern. These are: 1) high energy sources; 2) compatible development; 3) crew/instrument/experiment interaction; 4) evolution of instrument systems; and 5) instrument data handling.

Other integration problems at the subsystem level are of special significance to the AMPS Facility. They are the: center-of-gravity, data management, free flyer, safety contamination, cryogenics, and control schemes.

INTRODUCTION

AMPS is currently in the analytical integration phase. Analytical integration of AMPS experiments is concerned with the technical communications

---
* Martin Marietta Corporation

link or interchange between three program elements and the solution of problems as the result of this technical interchange. The three program elements are: 1) scientific community; 2) instrument development; and 3) facility development.

The AMPS integration challenge is to converge these three elements upon a low cost facility, but maintain its scientific flexibility to conduct experiments of the future. The problem is simple to state but complex in terms of the number of variables and interfaces involved. We are talking (presently) about 50 international scientists (growing to hundreds as the program enters the flight phase); a minimum of 100 major instruments (not including subsystems), which will be developed in every possible mode (new development, off-the-shelf, modified; NASA in-house, other NASA centers, Foreign Space Agencies, corporations, universities); and a major facility that can fly up to 4 flights per year over a ten year period.

In addition there are a multiplicity of general goals that must find their way into the designs, integration, and operation. Some of the key goals are:
- a) low cost
- b) facility type instruments
- c) utilize the building block concept
- d) minimum change between flights
- e) international participation at all levels
- f) maximum use of Spacelab systems (when cost effective)
- g) provide real time crew/scientist interaction with data
- h) flexibility to adjust to changing scientific interpretation
- i) address complex coupling between the atmosphere and magnetosphere.

To meet the integration challenge, integration is involved at three levels: 1) organization of the program elements to converge upon a facility; 2) instrument development; and 3) subsystem development and operation.

FACILITY DEVELOPMENT AND INTEGRATION

A simple organization of the program elements is shown in figure 1. Up to this point in time the program has undergone a "brainstorming" session by the science community and feasibility analysis (AMPS ØA Study). The next phase is the preliminary design definition, ØB. There are three key inputs to this phase: 1) the science requirements (objectives, instrument concepts, working group minutes); 2) Spacelab/Shuttle accommodations; 3) continual inputs from the science community.

The key integration tool of ØB for preliminary design definition is the compatibility analysis. This tool is used to make a finer analysis (relative to the feasibility analysis) of the scientific requirements to determine if they fit the program constraints (accommodations, cost, and goals). The problem would be academic from an analytical integration point-of-view if we could fly in a single facility all instruments that have been defined by the scientist. However, this is not the case for AMPS (not enough payload weight, volume, power available). Thus at the outset of ØB we know that all the instruments will not "fit" on the first facility.

When this occurs the integration role is to communicate the problems with the science community and provide facility options that do perform meaningful science and show planning that satisfies the objectives on multiple flights by the addition of other instruments, subsystems, and more observations. With this visibility, NASA and the science community can exercise their review function and decide which facility concept should proceed into the next program phase, C/D.

To develop these facility options there are various approaches that can be used to re-scope the facility. Figure 2 shows two possible approaches. From the instrument and subsystem point of view you go to a minimum number of instruments and subsystems that can do meaningful experimentation. The instruments would in general be hardmounted (use existing orbiter attitudes) dedicated C&Ds for each instrument and minimum interfaces with the orbiter/spacelab system. This would be characterized as the "volkswagen" concept as opposed to the "cadillac" which flys the maximum systems without regard to cost.

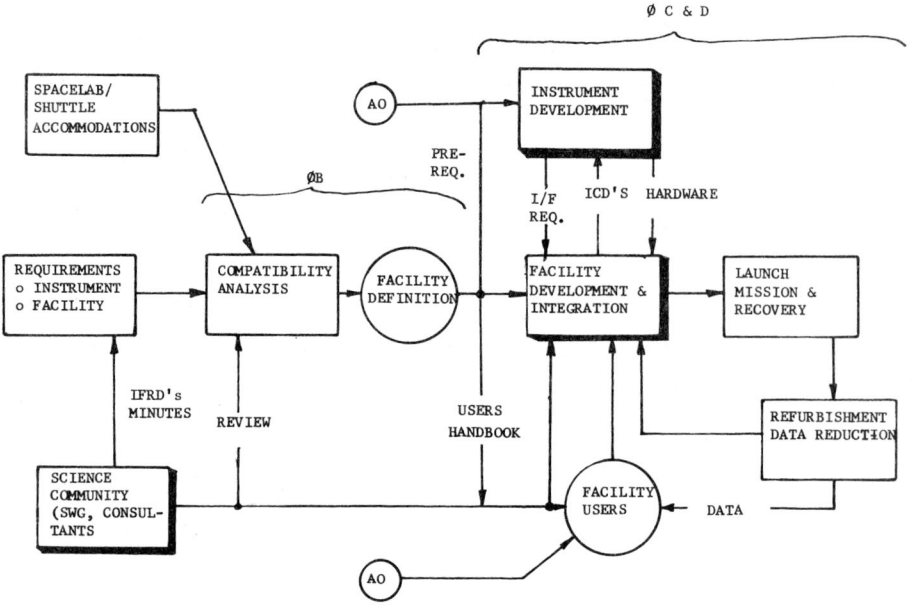

Fig. 1 Integration of Program Elements

(1) VOLKSWAGEN VS CADILLAC

(2) FACILITY COST

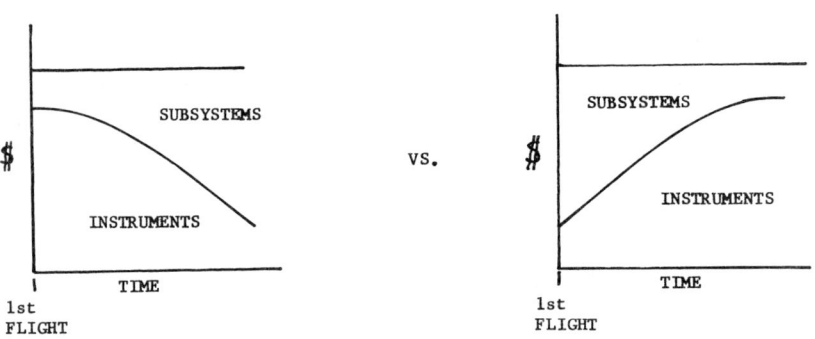

Fig. 2 Facility Integration Challenge

Two additional options would be to develop major subsystems on the early flights and add instruments as the facility grows. This is accomplished by analysis of all instrument interface requirements and determining the maximum demands of the instruments. Subsystems for these peak requirements are designed and are developed early in the program. This demands that a large percentage of the program funds be applied to subsystems early in the program. The opposite approach would be to spend a greater percentage of the funds on the major instrument systems and add subsystems as the program progresses. The subsystems would be added to improve instrument performance and expand experimentation in a meaningful method.

These options or "ideas" (and others) will undergo refinement using compatibility analysis much like that shown in figure 3. This will develop during ØB compatible payload concepts that perform meaningful science, meet cost limitations, and have evolution possibilities. It will then be the task of NASA and the science community to decide which facility should proceed into ØC&D (design, development, and flight). For the chosen facility, instrument developers will be provided preliminary characteristics for performance of instruments that are compatible with the facility concept. This link is necessary for follow-up in ØC&D of the ØB preliminary definition and for controlling cost of the facility.

This begins with a requirements review at the instrument level followed by a similar review at the facility level. This series of reviews baselines the requirements for the instruments and facility as the basis for preliminary design. With these requirements in hand, the preliminary design of hardware and integration of the facility proceeds in parallel. The key integration communication media at this point are the instruments interface requirements. The facility integration task which develops preliminary ICDs which define in detail the mechanical, electrical data, control interfaces to satisfy the interface requirements. This data must be available in time to support the detail design of the instrument and normally occurs at Preliminary Design Review. Once the ICDs are agreed to, then final design proceeds in parallel. The final design review called "Critical Design Review" reviews the detail design drawings prior to manufacture or assembly of the facility. At this stage precise performance characteristics of the facility are available to hold an Announcement-of-Opportunity for the facility users. The proposals are submitted and those accepted are used to make up the detailed flight plan. The hardware is delivered and physical integration begins. The hardware is tested to be compatible with its interfaces, with other instruments, and with the Spacelab/Orbiter systems.

The flight is flown and data collected. The data is then distributed to the users.

INSTRUMENT INTEGRATION

In working the AMPS program over the past year a number of instrument integration concerns or challenges have been surfaced which will be resolved during the ØB effort.

High Energy Sources

A number of the major instrument systems (Lidar, Accelerators, Plasma Guns, Transmitters) have high peak power demands that exceed the power available at a given time from the Spacelab/Orbiter systems. To satisfy these demands a number of subsystems are being considered (batteries, capacitor banks, flywheel motor-generators) to resolve this problem. Cost and Safety will be the key drivers in evaluating these alternatives.

Compatible Development

A major challenge to a complex program is to develop instruments that are compatible with the facility from the beginning of design, to delivery and flight. It should not be an independent function but definitely related. I have discussed ways of accomplishing this earlier in the paper. It essentially requires that the key performance and characteristics of the instrument that the ØB facility definition and cost were based upon must be implemented in the specification for the instrument prior to development. If a change in these basic performance and characteristics is proposed by the instrument developer, this change must be viewed or impacted by all parties prior to implementation.

Crew/Instrument/Experiment Interaction

A key problem here is to define the degree of the interaction with the data. The two extremes for the situation are easily defined: 1) viewing and operation on all data taken by the instruments in conducting experiments; or 2) record/dump the data for analysis on the ground after the mission. To converge on a meaningful compromise will require more meaningful criteria from the science community.

Evolution of Instrument Systems

A major challenge for integration will be to develop for each major instrument system the specific areas in the instrument that should be modularlized or made sub-assemblies in order to facilitate changes to the use of the instrument in the future which incurr major cost impacts. For the LIDAR System the separation of the LIDAR receiver assembly from the LASER transmitter has obvious advantages when changing the LASER for survey of different constituents in the atmosphere.

Instrument Data Handling

It is desirable that facility type instruments for AMPS conform to the program goals of being of general use on multiple flights to the scientific community and still provide crew interaction with the data. Figure 4 illustrates the problem with these objectives relative to instrument design. The raw signal from the detector must be reproducible to the scientist as a minimum. If operations on the raw signal are allowed before a record of the signal is made, then the basic data is lost forever. For AMPS, a reasonable criteria would be to record to the best sampling ability of the data handling system the raw signal generated by the instrument. This would allow multiple operations or manipulation of the data by other scientists that use the instrument in the future. Real-time interaction with this data during a given mission can still be provided by a parallel link and a record of this data made also. This approach would maximize the general nature of the instrument for future flights and minimizes basic changes from one flight to another. The parallel operations on the data could be an add-on module to the basic system.

SUBSYSTEM INTEGRATION CONCERNS

Center-of-Gravity (C.G.)

This is an integration concern for all manned shuttle payloads that have been considered. Various AMPS payload concepts have had varying

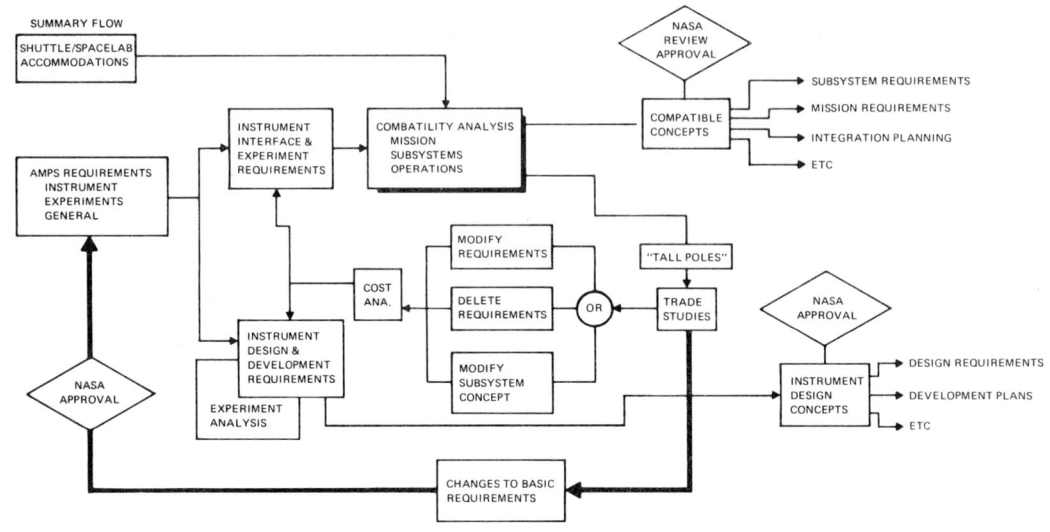

Fig. 3 AMPS Compatibility Analysis

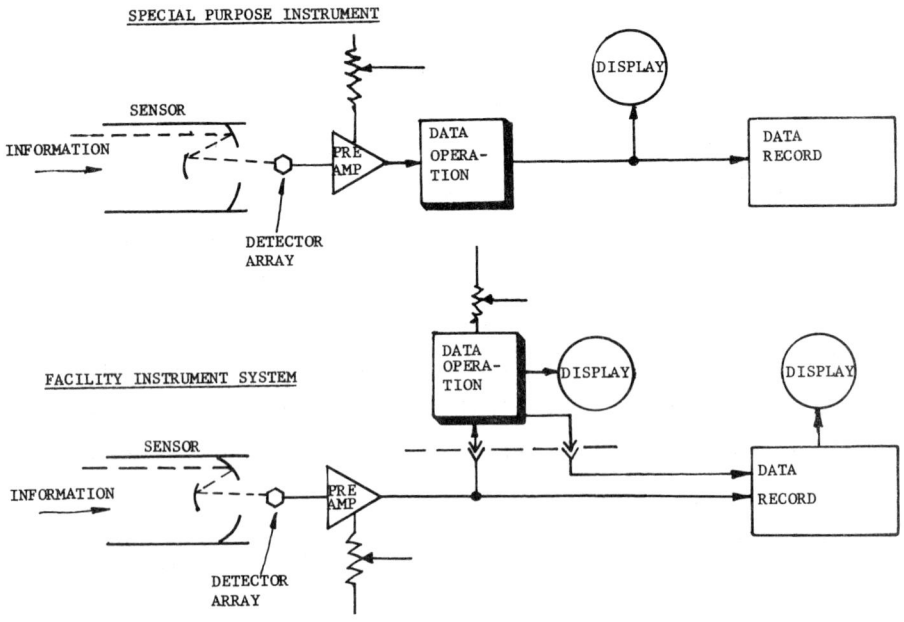

Fig. 4 Instrument Data Handling

success with meeting this constraint. The approach generally involves placing the heavier instruments to the rear portion of the payload bay without dropping the total weight of instruments flown or volume of instruments that can meaningfully be flown in the bay. As weights generally change in a program, the C.G. problem will probably get worse rather than better with the payloads suffering the most.

Data Management

The orbiter/TDRSS data dump coverage is quoted at being in the 80-90% mission time bracket with a maximum blackout period of 20-45 minutes. This approach assumes use of the orbiter Ku-band communications link (in the nose of the orbiter) and control of the attitude of the orbiter for dumping data to the TDRSS. On the surface, it appears (with a few exceptions) that AMPS payloads would be compatible with this situation. Preliminary analysis of AMPS attitude constraints on the orbiter for experiments will drop the data coverage to as low as 45% (based upon orbiter being in Z-LV attitude during the experimentation period) with maximum blackout periods of up to 3 hours. This would indicate a severe loss of AMPS data or experiment time (the Spacelab CDMS Tape Recorders can only record for 20-45 minutes and the magnetic tapes cannot be changed). The need for a dedicated AMPS recording system or changes to the CDMS tape recorders will be analyzed.

Free Flyers

The general approach for AMPS has been to assume usage of existing satellites (such as the Atmospheric Explorer) to satisfy this mission for AMPS. Presuming that the satellite control problems can be satisfied, there still exists basic problems of volume, power, and surface area available for AMPS instrumentation. The volume available in the AE for instruments is $\approx .2m^3$ (using 50% packing efficiency thus drops to $.1m^3$) while the complement of instruments on AMPS for the satellite have a total volume of $\approx .4m^3$ (or $\approx .8m^3$ if 50% packing efficiency is considered). Power provided by the AE for instruments is of the order of 4000 watt-min over an orbit while the requirement of AMPS is approximately twice this.

Recovery of the subsatellites into the bay (as well as their deployment) may require the control of the subsatellites in the orbiter. This would involve a complex control interface with the orbiter in addition to a less-than-optimum approach to scientist/experiment interaction (data from the instruments on-board the subsatellite would be displayed in the Spacelab while control of the position of the subsatellite is in the orbiter).

Safety

There are four areas of concern relative to the degree of implementation. AMPS payloads will have releases (shaped charges), the release boosters, separation/ejection devices, and the recontact problems (the release booster, subsatellites, etc.). There are solutions readily available for handling these problems but the concern is the degree to which measures are to be taken. At present, the critical categories of hardware needs to be refined for Spacelab to classify possible failures relative to crew safety, mission safety, or experiment failure. Then the instrument systems can be dealt with according to these failure modes. A system failure mode that does involve crew safety require more precautionary measures than a mode that impacts only the experiment data collection.

Contamination

Contamination of the instruments and the production of a detrimental environment about the orbiter will impact the experiments performed. This will be a concern for integration until more data is available: 1) to indicate the amount of contamination during a given mission (must know how often attitude control propulsion systems are fired); and 2) it is determined that the amount of contaminant available does impact the performance of the instruments.

Control Schemes

The Spacelab CDMS system can certainly be used to control the instruments, manage data, etc., provided the cost of implementing this approach on the first facility is within the funding limits of AMPS. A major trade study to develop the cost is shown in figure 5. This trade study will be performed during the ØB. The back-off position from maximum use of the CDMS for instruments is to provide dedicated C&D panels (upgraded development units) on the first flights of the instrument and as operational modes for instruments become more firm

	DEDICATED C & D for 1st Useage of Instrument System	vs		Use of CDMS for control of instrument system on initial flight of facility

Pro	Con		Pro	Con
o USES DEVELOPMENT C&D, OFF-SHELF EQUIPMENT o PROVIDES FLEXI-BILITY DURING INSTRUMENT DEVELOPMENT o MINIMUM SOFTWARE FOR 1st FACILITY o LOW IMPACT TO SYSTEMS TEST o INDEPENDENT ACCEPTANCE TEST	o MAXIMUM WEIGHT AND VOLUME o TAKES MORE POWER	COST?	o MAXIMUM USE OF SPACELAB o CENTRALIZED CONTROL	o MAXIMUM SOFTWARE IMPACT o ADDITIONAL CONTRACTOR INTERFACE o LIMITED FLEXIBILITY FOR CONTROL o PUTS TIME ON FLIGHT CDMS o SIMULATOR NEEDED AT DEVELOPMENT SITE

Fig. 5 C&D Integration Trade Study

or understood, the control of that operation be delegated to the CDMS.

CONCLUSION

AMPS itself will provide a new integration challenge that is not insurmountable, but has sufficient new problems to make it an interesting and worthwhile adventure to integration engineers. We will be able to apply many of the lessons learned from past programs as well as invent new approaches to solve problems.

AAS 75-231

ULTRAVIOLET REMOTE SENSING OF
ATMOSPHERIC OZONE
FROM PAYLOADS USING SHUTTLE CAPABILITIES

A. J. Krueger, D. F. Heath[*]

H. A. Roeder [+]

Previous atmospheric ozone observations using an ultraviolet double monochromator to measure the absorption of backscattered solar radiation from 255 nm to 340 nm have provided information on the global total ozone distribution as well as its vertical profile. This information can be used to structure the comprehensive ozone observations that are needed to provide a better understanding of the ozone formation and destruction mechanisms. Some potential ozone threats, especially that from photodissociation of fluorocarbons, make Shuttle-based observations especially timely. Because of the nature of the global ozone transport process, measurements must be performed from space on a global scale with sufficient resolution and accuracy to determine the atmospheric total ozone. A better understanding of ozone processes in global regions inaccessible from previous platforms is needed to complete this ozone data base. This paper describes the potential for Shuttle payloads to provide the required information.

INTRODUCTION

The importance of trace atmospheric constituents in the chemistry of stratospheric ozone has recently been recognized. Observed ozone densities can be explained if catalytic destruction processes, involving such constituents as nitric oxide, hydroxyl and free chlorine, are presumed to take place in the upper atmosphere. The natural abundance of these constituents is less than one part per billion, a level that can be significantly altered by technological products. For example, supersonic transports of present design, could almost triple the stratospheric nitrogen oxide content by the year 2005 if used in commercial fleets. Estimation of the effects on ozone is very difficult because of poorly understood interactions

---
[*] NASA Goddard Space Flight Center, Greenbelt, Maryland 20771
[*] Beckman Instruments, Inc., Anaheim, California 92804

between atmospheric circulation, chemistry, deposition locations and removal mechanisms. However, an ozone decrease would almost certainly occur with attendant increases in skin cancer due to additional ultraviolet sunlight reaching the surface.

A possible new threat to the ozone layer is catalysis by chlorine that results from photodissociation of fluorocarbons in the stratosphere. These compounds, which are commercially produced for refrigerants and aerosol spray can propellants, are inert in the lower atmosphere. They accumulate and slowly mix into the stratosphere where the catalytic cycle is initiated by the ultraviolet photodissociation. An assessment of the effects on ozone includes not only studies of the chemical reaction rates but measurement of the composition and transport processes in the stratosphere on a global scale. The ozone field itself carries a great deal of information about the transport so that global surveys of ozone are of primary importance. Natural sources of chlorine compounds, such as volcanic eruptions, must also be evaluated before the significance of pollutants can be determined. The space shuttle can serve as a valuable platform for observing ozone and related meteorological phenomena and, in conjunction with other space platforms, can play an essential role in an ozone monitoring system.

OZONE SENSING WITH BACKSCATTERED SUNLIGHT

The Backscatter Ultraviolet (BUV) technique has been shown to be an effective method for measurement of atmospheric ozone (Heath, et. al. 1973). In the first flight of a dedicated sensor for ozone on Nimbus 4, more than five year's data have been accumulated with accuracies suitable for geophysical investigations and a climatological survey (Krueger, et. al. 1973, Mateer, et. al. 1971). This technique as shown in Fig. 1, makes use of ultraviolet sunlight which is scattered by air molecules, reflected at the surface, and absorbed by ozone. At wavelengths shorter than 295 nm, the scattering occurs at wavelength dependent levels in the atmosphere above 25 km such that the vertical ozone distribution can be inferred. At wavelengths longer than 310 nm, the scattering occurs in the troposphere and at the surface so that the total amount of ozone can be estimated.

The instrument required for measurement of the earth radiance is a double Ebert-Fastie monochromator shown in Fig. 2. Twelve preselected wavelengths between 255 and 340 nm are serially sampled by positioning the diffraction gratings with a cam. A double monochromator is required for rejection of stray

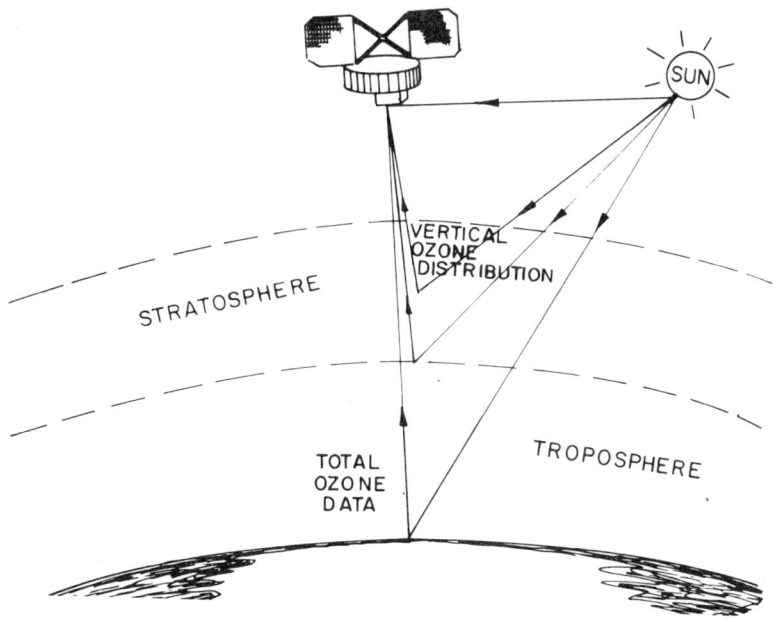

Fig. 1. Ozone is measured from space by sensing ultraviolet sunlight scattered in the stratosphere (vertical distribution) or scattered from the troposphere and surface (total amount).

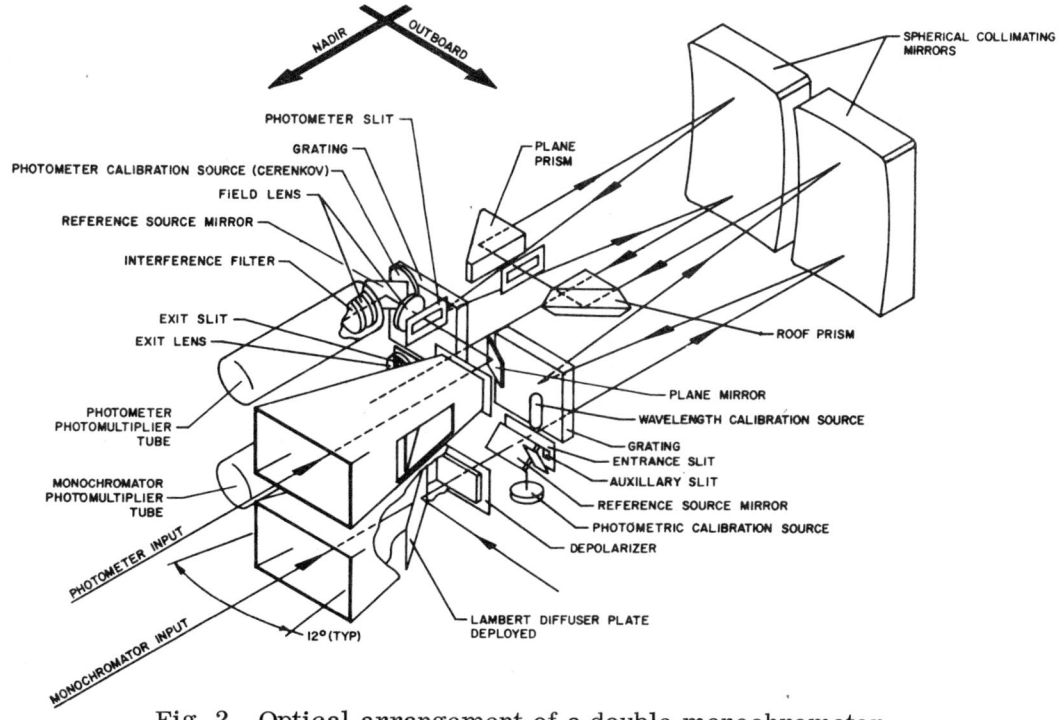

Fig. 2. Optical arrangement of a double monochromator for ozone measurements.

light which would contaminate the low radiance signals at the shortest wavelengths. A parallel photometer channel outside the ozone absorption band monitors surface reflectivity for the total ozone measurement.

The Nimbus 4 instrument takes data along the ground track of the spacecraft. With 13 orbits per day, the effective spatial resolution is capable of resolving only the large-scale global patterns. From modulations along the track, the total ozone particularly appears to be influenced by much smaller scale dynamic processes. A second-generation instrument, the Solar-Backscatter Ultraviolet/Total Ozone Mapping Spectrometer (SBUV/TOMS) will be flown on Nimbus G. That instrument will have a single monochromator with a spatial scanner for 50 km resolution mapping of the total ozone field in addition to the nadir-viewing vertical profile sounder (BUV).

The high spatial frequencies in the ozone field appear to diminish with altitude indicating that the total ozone modulations are due to tropospheric influences on the stratosphere. Other apparently unrelated modulations, however, appear in the upper stratosphere above 30 km. These unexplained modulations exist in the tropics and in the winter hemisphere at high latitudes. Such effects, due perhaps to variability of catalysts, to transport, or to additional ozone sources can be explored with a spatial scanning vertical sounder.

SHUTTLE APPLICATIONS

Space shuttle missions offer new capabilities for sensor evaluation, survey of unique geophysical phenomena and for calibration of monitoring sensors. Understanding of the dynamics controlling mixing of trace contaminants into the stratosphere and the transport of ozone requires new tools. A design study in progress will recommend the optimum scanning technique for a high resolution 3-D ozone mapper. The design is intended for a shuttle-based BUV imaging spectrometer to meet survey mission requirements. Comparisons are being made between point scanners and spatial multiplex scanners based on sensitivity, instantaneous field of view, simplicity, cost and data processing complexity. The study includes Hadamard spatial and spectral scanning, image- and object-plane scanning, image tube systems, photomultiplier and mosaic arrays, conventional monochromators and combinations thereof. Preliminary findings favor a linear photomultiplier array for spatial scanning with either a Hadamard mask or a conventional monochromator for spectral scanning. Figure 3 depicts this system as

Fig. 3. Scan pattern of a 3-D High Resolution Ozone Mapper for Space Shuttle applications.

proposed for evaluation on shuttle flights. As shown, a 15 element pushbroom scanner perpendicular to the flight path is translated by vehicle motion. The spectral scanner senses ozone at 3 to 5 stratospheric levels with 25 km spatial resolution.

A number of survey missions with this High Resolution Ozone Mapper (HROM) are of great interest. Initial candidates are investigations of 1) tropical cellular features at high levels, 2) high-latitude winter hemisphere ozone enhancements, 3) effects of local contaminant sources (eg, volcanoes, nuclear tests) on ozone and 4) detailed structure near tropopause breaks.

The third category of shuttle-based ozone missions, calibration of monitoring sensors, is essential for assessment of secular trends in the global ozone burden and the influence of large scale pollution sources. Detection of ozone changes less than 1% per year places severe constraints on system calibration stability. UV calibration standards are not practical for space flight because of high power requirements. Shuttle missions, however, can carry a baseline standard UV monochromator. Simultaneous, collinear measurements of the earth and sun with the standard sensor and the monitor instrument are adequate for transfer of the calibration. Periodic shuttle flights at 4 to 6 month intervals could then maintain the monitor calibration such that trends outside the natural variations of ozone could be rapidly detected.

In summary, the inadvertent modification of atmospheric ozone by pollutants is now recognized as a real possibility. Determination of these changes and understanding of the associated physical processes requires improved observing capabilities. Space shuttle missions are necessary for ozone sensor evaluation, for geophysical investigations, and for maintenance of calibration of monitoring sensors.

References

Heath, D. F., C. L. Mateer, and A. J. Krueger, "The Nimbus 4 Backscatter Ultraviolet (BUV) Atmospheric Ozone Experiment — Two Years Operation," PAGEOPH, 106-108, 1238-1253 (1973).

Krueger, A. J., D. F. Heath and C. L. Mateer, "Variations in the Stratospheric Ozone Field Inferred from Nimbus Satellite Observations," PAGEOPH, 106-108, 1254-1263 (1973).

Mateer, C. L., D. F. Heath and A. J. Krueger, "Estimation of Total Ozone from Satellite Measurements of Backscattered Ultraviolet Earth Radiance," J. Atmos. Sci., 28, 1307-1311 (1971).

AAS 75-232

DESIGN OF HIGH-VOLTAGE INSULATION SYSTEMS
FOR AEROSPACE EQUIPMENT

W. G. Dunbar[*]

A program plan is needed before a specification can be adequately prepared for the system, equipment, and circuits. This program plan should include pre-flight testing, storage, boost environment, the spacecraft constraints, and constraints or unusual characteristics of equipment aboard the spacecraft. When two or more vehicles are launched together or subsequently connected in tandem, the environmental and electrical characteristics of each vehicle should be included.

A good high-voltage program plan includes a requirements plan and a design-and-test plan. The requirements plan includes evaluation of historical data applicable to the experiments and the spacecraft, operational and mission constraints, and the test and test equipment requirements. Historical data for spacecraft equipment operating at voltages up to 10 kilovolts is abundant. Likewise, materials, designs, and manufacturing techniques for this voltage region are readily available. For voltages over 10 kilovolts information is scarce, and often needed is research and development, tailored to the constraints and requirements unique to the spacecraft and equipment aboard the spacecraft. An example is the Shuttle, a duo spacecraft, and AMPS, one of its assigned payloads. The pressure environment surrounding AMPS and Shuttle must be known whenever AMPS is near Shuttle. Contributing to the atmosphere around Shuttle will be gasses escaping from the crew compartment, orbit-keeping propulsion products, and outgassing from some of the materials exposed to space.

---

[*] Boeing Aerospace Company

High-voltage testing becomes hard to define for several reasons. First, the supplier of electronic components may lack some test equipment or test experience within his design organization, necessitating compromises in the hierarchy of testing; second, there are several levels of testing to be performed with difficult-to-evaluate options on when to perform what tests; third, test equipment sensitivity is affected by the equipment being tested and the connection thereto. Some equipment and experiments actually can be designed so that they test themselves. All these elements must be defined in the requirements plan by the spacecraft designer and his customer before preliminary design is undertaken.

A test plan should be developed for each high-voltage item aboard the spacecraft. It should contain the vehicle constraints and requirements that constrain the item; for example, pressure, temperature and outgassing products other than air. Testing should be sequenced with materials selection and application, and packaging to avoid delays and costly overruns from improper application of a specific material. Such sequencing requires that the insulating and conductor materials be selected and tested early in the program to establish their adequacy and life-stress capability. Particularly important to watch is dense parts packaging where mechanically stressed insulation must withstand wide temperature variations. Some insulations crack when subjected to temperatures lower than $-20°C$. With high electric fields between parts, cracked insulation is a precursor to partial discharges and ultimate failure.

DESIGN GUIDES

Each supplier of high-voltage equipment uses his own design guides for the electronics design. However, he usually depends upon technical reports and packaging design guides for parts selection, electrical manufacturing processes and packaging. This may lead to difficulty. For example, technical papers, supplier information, and design guides may not provide such subtleties as the effects of frequency, cure time and temperature, bond-line dielectric characteristics, outgassing, and radiation. These elements must be understood by the packaging designer or determined by testing. Even then, measurements with an insulating compound between flexible shallow aluminum disks may give results far different from those obtained in an actual design. For instance, encapsulants between long thin plates or filaments should have a length-to-thickness ratio less than 6 to 1; otherwise, cracking can be expected.

HIGH-VOLTAGE DESIGNS

The three basic high-voltage designs for spacecraft are: (1) conformally coated open construction, (2) solid encapsulation, and (3) pressurization with liquid or gas. Solid encapsulants, conformal coatings, and pressurized containers have been used successfully in spacecraft for circuits in the 2 to 15 kilovolt range. Their technology can be extended to higher voltages. However, solid encapsulation is heavy.

Conformal coatings are very thin, with 90 to 98 percent of the voltage drop between electrodes occurring across the vacuum gap. These gaps will range from a fraction of a millimeter to several centimeters long. The field lines between a high-voltage conductor and ground plan can be derived. In the high pressure region the conformal insulation on the conductor appears thick compared to the gap, and thus sustains high voltage. At low pressures the insulation value of the conformal coating is insignificant, its effect being merely to reduce the supply of free electrons from the conductor.

Pertinent to high-voltage design is solder balling. A ball of solder or rounded cap should be attached to each sharp point to reduce dielectric field stress. Not only do joints and terminals require solder balls, but also sharp bends in a wire and any thin-film wire passing or crossing another wire should be solder balled. Solder balls should be specified for all high-voltage designs, whether pressurized, open, or solid. In open construction at least 0.1 millimeter conformal coating should cover all solder balled joints and filaments. This conformal coating must stay free of cracking or crazing throughout the mission, even after temperature cycling and exposure to radiation and plasma.

CIRCUIT PROTECTION

For open-construction designs, one of the most effective protection devices is a pressure sensor, such as an ion gauge, located in line with the outgassing ports of the high voltage circuits. The ion gauge may be interconnected to lock off the high-voltage circuits when this pressure exceeds a level predetermined by test and analysis. One pressure sensor may be sufficient for several high-voltage circuits when the outgassing ports have a common path. More pressure sensors are required when high-voltage experiments are located on opposite sides of the spacecraft.